MARINE FISH FARMING

MARINE FISH FARMING

Edited by:

Bruno Augusto Amato Borges

Delve
PUBLISHING

www.delvepublishing.com

Marine Fish Farming

Bruno Augusto Amato Borges

Delve Publishing

2010 Winston Park Drive,

2nd Floor

Oakville, ON L6H 5R7

Canada

www.delvepublishing.com

Tel: 001-289-291-7705
001-905-616-2116

Fax: 001-289-291-7601

Email: orders@arclereducation.com

ISBN: 978-1-77361-038-2 (Hardcover)

DECLARATION

Some content or chapters in this book are open access copyright free published research work, which is published under Creative Commons License and are indicated with the citation. We are thankful to the publishers and authors of the content and chapters as without them this book wouldn't have been possible.

ABOUT THE EDITOR

Bruno Augusto Amato Borges

Bruno Augusto is Aquaculture Engineer from Federal University of Santa Catarina. His focus is on writing projects and development of new technologies in aquaculture. He is currently project service provider, consultant and develops research in the field of Biofloc Technology (BFT) for freshwater fish farming.

TABLE OF CONTENT

SECTION 3 – DISEASES

SECTION 4 – FEED AND NUTRITION

SECTION 5 – GENETIC IMPROVEMENT

SECTION 6 – FUTURE SUSTAINABILITY AND ENVIRONMENTAL IMPACTS

LIST OF CONTRIBUTORS

Bruno Augusto Amato Borges
Aquaculture Engineer
Florianopolis, Brazil

Christophe Béné
Institute of Development Studies, University of Sussex, Sussex, UK
International Center for Tropical Agriculture, Cali, Colombia

Manuel Barange
Plymouth Marine Laboratory, Plymouth, UK

Rohana Subasinghe
Food and Agriculture Organization, Rome, Italy

Per Pinstrup-Andersen
Cornell University Ithaca, Ithaca, USA

Gorka Merino
AZTI, Pasaia, Gipuzkoa, Spain

Gro-Ingunn Hemre
Institute of Nutrition and Seafood Research, Bergen, Norway

Meryl Williams
17 Agnew Street, Aspely, Queensland, Australia

T. Håstein
National Veterinary Institute, Norway

A. Hellstrøm
National Veterinary Institute, Sweden

G. Jonsson
Veterinary Officer for Fish Diseases, Keldur, Vesturlandsveg, IS-112, Reykjavik, Iceland

N. J. Olesen
Danish Veterinary Laboratory, Aarhus, Denmark

E. R.- Pärnänen
National Veterinary and Food Research Institute (EELA), Finland.

Marek S. Ziętara
Department of Molecular EvolutionUniversity of GdańskGdańskPoland

Dar'ya Lebedeva
Institute of BiologyKarelian Research Centre, RASPetrozavodskRepublic of Karelia, Russia

Gabriela Muñoz
Facultad de Ciencias del Mar y de Recursos NaturalesUniversidad de ValparaísoViña del MarChile

Jaakko Lumme
Department of BiologyUniversity of OuluOuluFinland

Sadok Khouadja
Département de Microbiologie, Faculté de Pharmacie, Laboratoire d'AnalyseTraitement et Valorisation des Polluants de l'Environnement et des ProduitsMonastirTunisia

Faouzi Lamari
Département de Microbiologie, Faculté de Pharmacie, Laboratoire d'AnalyseTraitement et Valorisation des Polluants de l'Environnement et des ProduitsMonastirTunisia

Amina Bakhrouf
Département de Microbiologie, Faculté de Pharmacie, Laboratoire d'AnalyseTraitement et Valorisation des Polluants de l'Environnement et des ProduitsMonastirTunisia

MA Freeman
Institute of Ocean and Earth Sciences & Institute of Biological Sciences,

University of Malaya, Kuala Lumpur, Malaysia
Institute of Aquaculture, University of Stirling, Scotland, UK

M Eydal
Institute for Experimental Pathology, University of Iceland, Keldur, Iceland

M Yoshimizu
Laboratory of Microbiology, Hokkaido University, Hakodate, Japan

K Watanabe
Department of Aquatic Bioscience and Industry Faculty of Bio-Industry, Tokyo University of Agriculture, Abashiri, Japan

AP Shinn
Institute of Aquaculture, University of Stirling, Scotland, UK

K Miura
Hokkaido Central Fisheries Experiment Station, Yoichi, Japan

K Ogawa
Laboratory of Fish Diseases, The University of Tokyo, Tokyo, Japan

Haakon Hansen
Norwegian Veterinary Institute, Oslo, Norway.

Trygve T Poppe
Norwegian Veterinary Institute, Oslo, Norway.
Norwegian University of Life Sciences, School of Veterinary Medicine, Ås, Norway.

Turhan Markussen
Norwegian Veterinary Institute, Oslo, Norway.

Egil Karlsbakk
Institute of Marine Research, 5817 Nordnes Bergen, Norway

Saraya Tavornpanich
Section for Epidemiology, Department of Health Surveillance, Norwegian Veterinary Institute, Sentrum, Norway

Mathilde Paul
UR AGIRs, Centre de coopération internationale en recherche agronomique pour le développement (CIRAD), Montpellier, France

Hildegunn Viljugrein
Section for Epidemiology, Department of Health Surveillance, Norwegian Veterinary Institute, Sentrum, Norway
Centre for Ecological and Evolutionary Synthesis (CEES), Department of Biology, University of Oslo, Blindern, Norway

David Abrial
UR 346Institut National de la Recherche Agronomique (INRA), Saint-Genès-Champanelle, France

Daniel Jimenez
Section for Epidemiology, Department of Health Surveillance, Norwegian Veterinary Institute, Sentrum, Norway

Edgar Brun
Section for Epidemiology, Department of Health Surveillance, Norwegian Veterinary Institute, Sentrum, Norway

Toshio Takeuchi
Tokyo University of Marine Science and Technology, Tokyo, Japan

Torfinn Moldal
Norwegian Veterinary Institute, Sentrum, Norway

Guro Løkka
Norwegian School of Veterinary Science, Norway

Jannicke Wiik-Nielsen
Norwegian Veterinary Institute, Sentrum, Norway

Lars Austbø
Norwegian School of Veterinary Science, Norway

Bente E Torstensen
National Institute of Nutrition and Seafood Research, Nordnes, Norway

Grethe Rosenlund
Skretting ARC, Stavanger, Norway

Ole Bendik Dale
Norwegian Veterinary Institute, Sentrum, Norway

Magne Kaldhusdal
Norwegian Veterinary Institute, Sentrum, Norway

Erling Olaf Koppang
Norwegian School of Veterinary Science, Norway

Luca Tacchi
Institute of Biological and Environmental Sciences, University of Aberdeen, Tillydrone Avenue, Aberdeen AB24 2TZ, UK.

Christopher J Secombes
Institute of Biological and Environmental Sciences, University of Aberdeen, Tillydrone Avenue, Aberdeen AB24 2TZ, UK.

Ralph Bickerdike
BioMar Ltd, Grangemouth Docks, Grangemouth FK3 8UL, UK.

Michael A Adler
BioMar Ltd, Grangemouth Docks, Grangemouth FK3 8UL, UK.

Claudia Venegas
AVS Chile S.A., Casilla 300, Puerto Varas, Chile.

Harald Takle
AVS Chile S.A., Casilla 300, Puerto Varas, Chile.
Nofima, Aas, Norway.

Samuel AM Martin[1]
Institute of Biological and Environmental Sciences, University of Aberdeen, Tillydrone Avenue, Aberdeen AB24 2TZ, UK.

Sharen Bowman
The Atlantic Genome Centre, Halifax, Canada

Sophie Hubert
The Atlantic Genome Centre, Halifax, Canada
Department des Sciences, Universite Sainte Anne, Pointe-de-l'Eglise, Canada

Brent Higgins
The Atlantic Genome Centre, Halifax, Canada

Cynthia Stone
The Atlantic Genome Centre, Halifax, Canada

Jennifer Kimball
The Atlantic Genome Centre, Halifax, Canada

Tudor Borza
The Atlantic Genome Centre, Halifax, Canada

Jillian Tarrant Bussey
The Atlantic Genome Centre, Halifax, Canada
The Atlantic Genome Centre, NRC Institute for Marine Biosciences, National Research Council, Halifax, Canada

Gary Simpson
The Atlantic Genome Centre, Halifax, Canada

Catherine Kozera
The Atlantic Genome Centre, Halifax, Canada
The Atlantic Genome Centre, NRC Institute for Marine Biosciences, National Research Council, Halifax, Canada

Bruce A. Curtis
The Atlantic Genome Centre, Halifax, Canada
Department of Biochemistry and Molecular Biology, Dalhousie University, Halifax, Canada

Jennifer R. Hall
Ocean Sciences Centre, Memorial University of Newfoundland, St. John's, Canada

Tiago S. Hori
Ocean Sciences Centre, Memorial University of Newfoundland, St. John's, Canada

Charles Y. Feng
Ocean Sciences Centre, Memorial University of Newfoundland, St. John's, Canada

Marlies Rise
Ocean Sciences Centre, Memorial University of Newfoundland, St. John's, Canada

Marije Booman
Ocean Sciences Centre, Memorial University of Newfoundland, St. John's, Canada

A. Kurt Gamperl
Ocean Sciences Centre, Memorial University of Newfoundland, St. John's, Canada

Edward Trippel
Fisheries and Oceans Canada, St. Andrews, Canada

Jane Symonds
National Institute of Water and Atmospheric Research, Ruakaka, New Zealand

Stewart C. Johnson
Pacific Biological Station, Fisheries and Oceans Canada, Nanaimo, Canada

Matthew L. Rise
Ocean Sciences Centre, Memorial University of Newfoundland, St. John's, Canada

Sofia Morais
Institute of Aquaculture, University of Stirling, Stirling FK9 4LA, Scotland, UK.

Tomé Silva
Aquaculture Research Group, Centre of Marine Sciences (CCMAR), University of Algarve, 8005-139 Faro, Portugal.

Odete Cordeiro
Aquaculture Research Group, Centre of Marine Sciences (CCMAR), University of Algarve, 8005-139 Faro, Portugal.

Pedro Rodrigues
Aquaculture Research Group, Centre of Marine Sciences (CCMAR), University of Algarve, 8005-139 Faro, Portugal.

Derrick R Guy
Landcatch Natural Selection Ltd., Cooperage Way Business Centre, Alloa FK10 3LP, Scotland, UK.

James E Bron
Institute of Aquaculture, University of Stirling, Stirling FK9 4LA, Scotland, UK.

John B Taggart
Institute of Aquaculture, University of Stirling, Stirling FK9 4LA, Scotland, UK.

J Gordon Bell
Institute of Aquaculture, University of Stirling, Stirling FK9 4LA, Scotland, UK.

Douglas R Tocher

Luke D Gardner
Biology Department, Hopkins Marine Station, Pacific Grove, Stanford University

Nishad Jayasundara
Biology Department, Hopkins Marine Station, Pacific Grove, Stanford University

Pedro C Castilho
Biology Department, Hopkins Marine Station, Pacific Grove, Stanford University

Universidade Federal Rural de Pernambuco (UFRPE/UAST)

Barbara Block
Biology Department, Hopkins Marine Station, Pacific Grove, Stanford University

Leon Cantas
Department of Food Safety and Infection Biology, Norwegian School of Veterinary Science
Norwegian Food Safety Authority, District Office for Romerike

Thomas WK Fraser
Department of Production Animal Clinical Sciences, Norwegian School of Veterinary Science

Per Gunnar Fjelldal
Institute of Marine Research (IMR), Matre Research Station

Ian Mayer
Department of Production Animal Clinical Sciences, Norwegian School of Veterinary Science

Henning Sørum
Department of Food Safety and Infection Biology, Norwegian School of Veterinary Science

Kevin A Glover
Institute of Marine Research, Bergen, Norway

Abdullah S Madhun
Institute of Marine Research, Bergen, Norway

Geir Dahle
Institute of Marine Research, Bergen, Norway

Anne G E Sørvik
Institute of Marine Research, Bergen, Norway

Vidar Wennevik
Institute of Marine Research, Bergen, Norway

Øystein Skaala
Institute of Marine Research, Bergen, Norway

H Craig Morton
Institute of Marine Research, Bergen, Norway

Tom J Hansen
Institute of Marine Research, Bergen, Norway

Per G Fjelldal
Institute of Marine Research, Bergen, Norway

John A. Theodorou
Department of Fisheries and Aquaculture Technology, Technological Educational Institute (T.E.I.) of Western Greece, Mesolonghi, Greece

Costas Perdikaris
Department of Fisheries, Regional Unit of Thesprotia, Igoumenitsa, Greece

Barbara Venou
Department of Fisheries and Aquaculture Technology, Technological Educational Institute (T.E.I.) of Western Greece, Mesolonghi, Greece

John P. Volpe
Seafood Ecology Research Group, School of Environmental Studies, University of Victoria, Victoria, BC V8W 2Y2, Canada

Jennifer L.M. Gee
Seafood Ecology Research Group, School of Environmental Studies, University of Victoria, Victoria, BC V8W 2Y2, Canada

Valerie A. Ethier
Seafood Ecology Research Group, School of Environmental Studies, University of Victoria, Victoria, BC V8W 2Y2, Canada

Martina Beck
Seafood Ecology Research Group, School of Environmental Studies, University of Victoria, Victoria, BC V8W 2Y2, Canada

Amanda J. Wilson

Seafood Ecology Research Group, School of Environmental Studies, University of Victoria, Victoria, BC V8W 2Y2, Canada

Jenna M.S. Stoner
Seafood Ecology Research Group, School of Environmental Studies, University of Victoria, Victoria, BC V8W 2Y2, Canada

Kevin Alan Glover
Section of Population Genetics and Ecology, Institute of Marine Research

Cino Pertoldi
Department 18/Section of Environmental Engineering, Aalborg University
Department of Biosciences, Aarhus University

Francois Besnier
Section of Population Genetics and Ecology, Institute of Marine Research

Vidar Wennevik
Section of Population Genetics and Ecology, Institute of Marine Research

Matthew Kent
Center for Integrative Genetics, Department of Animal and Aquacultural Sciences, Norwegian University of Life Sciences

Øystein Skaala
Section of Population Genetics and Ecology, Institute of Marine Research

PREFACE

The cultivation of marine fish comes over the years gaining greater importance, based on the decline of fishing production and increased demand for this protein source. According to Ward and Myers (2005) in the past 50 years the fishing activity in high-scale in the oceans has interfered in the marine ecosystems, generating major ecological changes and dramatically reducing various fish stocks.

Currently, due to growing demand for this type of food worldwide, there is a pressure on the scientific community and producers in search of improvements in the system and management. It is known that the main challenges today are fighting diseases, studies about the best place for the cultivation of a particular species, the search for replacement of protein source in the feed and breeding with the goal of finding the best features expressed by any species. Therefore, there is a huge field to evolve and develop the activity, which is relatively new when compared to other animal farming systems. In addition, an important factor to be considered with respect to the cultivation of marine fish is the legislation involved, which in most countries is an obstacle to the development of the activity due to high bureaucracy involved to acquire areas for production, especially in off-shore or near-shore systems.

Some aspects are fundamental when it seeks to establish a marine fish farming, including the selection of species, area and cultivation system, the feed being used and your commercial availability, the availability of seed in the region, transportation and logistics, technologies employed in rearing, fattening and breeding, the knowledge of diseases and prophylactic and therapeutic treatments and funding and qualified personnel.

This book is intended for professionals, students and researchers who are interested in meeting some aspects of the current situation of production of marine fishes worldwide.

Editor

SECTION I

CHALLENGES, TRENDS AND CURRENT SITUATION IN MARINE FISH FARMING

CHAPTER

1

CHALLENGES, TRENDS AND CURRENT SITUATION IN MARINE FISH FARMING

Bruno Augusto Amato Borges

Aquaculture Engineer Florianopolis, Brazil

WORLDWIDE PRODUCTION AND MAIN SPECIES

The cultivation of marine fish comes over the years gaining greater importance, based on the decline of fishing production and increased demand for this protein source. Shepherd and Bromage (1988) reported that the first records of the cultivation of marine species date back to the year 1400 in Indonesia. However, only from 1960 marine fish farming began to gain prominence through the advances obtained in Japan. According to Ward and Myers (2005) in the past 50 years the fishing activity in high-scale in the oceans has interfered in the marine ecosystems, generating major ecological changes and dramatically reducing various fish stocks. Marine fish farming has been one of the largest annual growth rates between the various sectors of aquaculture (FAO, 2011). Currently, the production of marine fish by aquaculture is over 2 million tonnes. Already diadromous fish production, which covers much of the salmonids, reaches almost 5 million tonnes (FAO, 2016). This group of individuals has high commercial value because of the quality and enjoyment of the flesh by the market, in addition to being considered a stem food importance to human health. Because the commercial importance of salmon aquaculture, we decided to put it in the book.

But as any productive activity, marine fish farming has many challenges to overcome, as improvements in management, feed formulation, environmental impacts and waste.

The management and use of technologies is extremely important to be assessed, due to the high technological level involved in most enterprises, mainly related to the coast making own logistics of production and harvest. One of the main discussions is the protein source used in the diets, since most farmed species and high market value are carnivorous, positive results in weight gain and growth can only be achieved with feed made from animal protein sources. This causes a problem that currently there are difficult to overcome. In addition, the cultivation of marine fish raises a number of environmental impacts, such as modifying the installation location of cultivation with the increase of organic matter from the feed remains and fecal material deposited in farm sediments (Kondo, 2012).

The main species currently cultured around the world are: Atlantic salmon *(Salmo salar),* milkfish (*Chanos chanos),* atlantic cod (*Gadus morhua)*, blue fin tuna (*Thunnus thynnus),* cobia (*Rachycentron canadum),* European sea bass (*Dicentrarchus labrax*), gilt-head sea bream (*Sparus aurata*), yellowtail (*Seriola quinqueradiata*), barramundi (*Lates calcarifer*), Japanese flounder (*Paralichythys olivaceus*), turbot (*Scophthalmus maximus*), red sea bream (*Pagrus major*), among others.

The salmon is one of the main fish species produced in the world. Is an anadromous fish, but because of your importance we decided to add it to the book. According to Torrissen (2011) salmon culture has become a "super-commodity" during the last decade, with Norway being the largest producer followed by UK, Chile and Canada. Began in Norway in the late 70 and soon popularized by the countries of the North Atlantic, pacific Canada, Chile and Australia (Torrissen, 2011). According to Sandersen

(2015) development in marine aquaculture is dependent on several factors, including availability of feed, healthy marine environments and suitable locations. A major and enduring debate in Norwegian aquaculture involves the question of access to suitable new production sites (Hersoug 2013). When it comes to technology, the Atlantic salmon is the species that is more developed in studies and new trends of production. Genetic improvement, nutrition, feeding and handling are examples of areas where this species is reference.

Milkfish is one of the oldest and most important aquaculture species in Asia, particularly in Indonesia, the Philippines and Taiwan. Production has been increasing since the 1950s, and milkfish continues to be an important aquaculture species, with large investments made since the 1970s by some private investors, aid donors and governments (Nelson and Marygrace 2010).

Atlantic cod is a species that has been overexploited by the capture fishery (Bowman et al., 2011). Therefore, it is extremely important that the cultivation of this species grow to meet the global demand. It is a new industry that still has significant challenges including developing production methods and protocols that ensure a stable and profitable production. Around 2000 the cod aquaculture developed, also this time mainly in Norway but also in United States of America, Canada, United Kingdom and Iceland. At this time juvenile production technology was further developed and large hatcheries, based on intensive production protocols similar to those used for sea bass/sea bream, were used (FAO, 2004-2017).

Atlantic bluefin tuna is one of the high value aquaculture species, inhabits the North Atlantic Ocean and the Mediterranean Sea and is considered to be an endangered species, largely a consequence of overfishing (Trumbic, 2015). Unlike other cropping systems, the bluefin tuna is based only on the capture and fattening of individuals in captivity, making the production dependent on natural stocks. Only a few study centers around the world are developing knowledge related to the complete cycle of cultivation of this species, with breeding, spawning and hatchery. The three main producer countries of fattened blue fin tuna are Spain, Croatia and Malta. Greece, Italy, Morocco, Portugal, Tunisia and Turkey also produce, but at lower amounts (FAO, 2015-2017).

Cobia aquaculture was first reported in 1975 with the collection of wild caught eggs off the coast of North Carolina. This species is distributed worldwide in warm marine waters, except for the central and eastern Pacific, resulting in a very large potential area suitable for the production of native species. In 2014, the world production of this species for aquaculture was 40,329 t (FAO, 2007-2017). China, Taiwan and Viet Nam are major producers, while there are research and development projects in several other countries.

The European sea bass Dicentrarchus labrax is one of the most important (Palaiokostas et al., 2015) and extensively farmed fish species in the Mediterranean, resulting in steadily increasing pressure on producers (Sarropoulou et al., 2009). In 2014, the world production reached 156.449 tonnes (FAO, 2016). This species can be produced in various cropping systems. Currently are used extensive tanks with water renewal by the tide or cages near-shore or off-shore, depending on the investment and technologies used, in addition to the productivity that if you want to achieve.

The gilthead sea bream (Sparus aurata) and red sea bream (Pagrus major) are marine fish of the family Sparidae. Sparids are of great importance for aquaculture, being excellent food fish, with high commercial value S. aurata is one of the most prominent, with an average cultured production of 100 million metric tonnes per year. The great importance of the gilthead sea bream for marine aquaculture has increasing number of studies in many different areas such as immunology, endocrinology, bone morphology, and muscle physiology (Farraresso, 2008).

Yellowtail (Seriola quinqueradiata), a member of the order Perciformes, is a popular and important species in Japanese aquaculture (Aoki et al., 2015). In Japan this species is the most cultured fish, its meat being relished as sashimi. The aquaculture production of Seriola quinqueradiata constituted about 57 percent of the total farmed marine finfish production in Japan in 2003 (FAO, 2006-2017). In 2014, the global production of this species reaches 135,998 tonnes (FAO,2016).

Barramundi (Lates calcarifer) is a high-value aquaculture fish endemic and widely farmed in south-east Asia and Australia and with increasing aquaculture interest in Europe and North America (Tucker et al., 2002). It is also called Asian seabass or the giant sea perch, belonging to the family Latidae is widely distributed in the coast and freshwater of the tropical Indo-west Pacific. Because of good meat quality and relatively high market value, it has become an attractive commodity of both large and small-scale aquaculture enterprises. It is cultivated in Thailand, Malaysia, Singapore, Indonesia, Hong Kong, China and Australia in several types of culture in fresh or marine water (cages or ponds). It is a diadromous fish, inhabiting rivers before returning to the estuaries to spawn.

Japanese flounder, Paralichthys olivaceus, is widely distributed along the coast of Northeast Asia and is one of the most economically important marine species in the region (Castaño-Sánchez et al., 2010). The production of this species began in 1985 and is gaining importance in the world market, reached in 2014 annual productivity of almost 45000 tonnes (FAO, 2016).

The turbot Scophthalmus maximus is one of the most promising European aquaculture species, is naturally distributed in European waters, from Northeast Atlantic to the Arctic Circle. The total production of this species in the year 2014 was around 70 tonnes (FAO, 2016). It is a fish with a high value and market take-up, besides having a great quality of meat. The main methods of production of this species are in tanks with open-circuit pumped seawater or in cages. Spain, with its highly suitable oceanographic conditions, is today the major producer worldwide but turbot is also currently farmed in Denmark, Germany, Iceland, Ireland, Italy, Norway, Wales (UK), and Portugal (FAO, 2005-2017).

Currently, due to growing demand for this type of food worldwide, there is a pressure on the scientific community and producers in search of improvements in the system and management. It is known that the main challenges today are fighting diseases, studies about the best place for the cultivation of a particular species, the search for replacement of protein source in the feed and breeding with the goal of finding the best features expressed by any species. Therefore, there is a huge field to evolve and develop the activity, which is relatively new when compared to other animal farming systems. In addition, an important factor to be considered with respect to the cultivation of marine fish is the legislation involved, which in most countries is an obstacle to the development of the activity due to high bureaucracy involved to acquire areas for production, especially in off-shore or near-shore systems.

It is up to the concerned in the area get such improvements, in such a way that the activity grows and prospers in a sustainable manner and perennial. I'm a big believer in the potential of cultivation of marine fishes worldwide, and I know the importance and responsibility that the activity has in the current scenario of food production.

CULTURE AND MANAGEMENT

In general, compared to other aquatic animals cultivated, the cultivation of marine fish is relatively complex, especially when it comes to nutritional requirement, management and breeding. In addition, as there is little or no control in the estuarine and marine environment becomes extremely complicated adopting production techniques and standardized management to different regions and, consequently, water quality and zootechnical parameters of target species. Therefore, in addition to the search for places favorable for the cultivation of a particular species, there is the great challenge of developing technological packages to cultivate individuals adapted to the site, avoiding using management techniques already used for other species. The development of marine fish farming is based on mainly intensive cultivations and carnivorous species, with high market value.

Some aspects are fundamental when it seeks to establish a marine fish farming, including the selection of species, area and cultivation system, the feed being used and your commercial availability, the availability of seed in the region, transportation and logistics, technologies employed in rearing, fattening and breeding, the knowledge of diseases and prophylactic and therapeutic treatments and funding and qualified personnel.

There are currently several types of cultivation used throughout the world. Recirculating systems, marine water collection in laboratories, in ponds excavated using the tidal regime for water renewal, in cages near and away from the coast (near and off-shore systems), among others. What will define the system to be used are the available resources and access to and quality of the environment.

Compared to traditional systems of cultivation in ponds, recirculating systems the benefit possession related to reduction of water consumption, reducing also the generation of effluents. This system enables high productivity and still meets the concepts of a responsible and environmentally correct aquaculture (Blancheton, 2000). In addition, this system delivers high control over the environment and production, allowing produce year-round, can still be installed and located close to consumer markets. According to Kubitza (2006), the main factor of failure related to this system of cultivation is the difficulty of maintenance and scalability, in addition to ignorance of basic participles that governing your operation.

Offshore aquaculture installations have been drawing increasing attention from researchers, industry and policy makers as a promising opportunity for large-scale expansion of the aquaculture industry (Troell et al., 2009). This type of cultivation, as well as near-shore farming, enables less control to the producer in relation to environmental aspects and quality of water, but due to the flow of currents and tides are in constant water exchange, keeping your quality almost always constant. In addition, typically requires a higher degree of technology for the achievement of most of the processes, such as transportation, food and harvest management.

To start a production of marine fish, is extremely important to define what kind of culture will be deployed, to analyze what is the appropriate management also in the various processes that involve the cultivation. Food management, reproduction, population and harvest are some examples of important processes that can define the success or failure of an enterprise if not studied previously or run correctly.

Furthermore, as in any undertaking, must be given a lot of importance in drawing up the project properly and scale the system responsibly, taking into account all factors that include cultivation (location, access, availability of resources, the local population, production, consumer market, among others).

DISEASES

Among the main challenges in the production of marine fish in the world, one of the most important affecting productivity in such a way as to compromise the majority of enterprises is the existence and combats the disease. It is the main factor that makes aquaculture an activity considered high-risk. The consequences of contracting a disease in its cultivation vary from total mortality until the presence of differentiated features in, losing totally the market acceptance.

Normally, the manifestation of disease in marine fish cultivation occurs due to an imbalance of water quality, nutrition, management and as a result, the welfare of farmed organisms. The disease-causing organisms in marine fish are already present in the environment, but in good condition of nutrition and water quality they don't manifest, therefore, there must be some concern in control and maintain the proper environment for the cultivation. The physical and chemical characteristics of water that must be balanced and the breakdown of this balance could exert negative influences on animals.

In General, there are many diseases that can affect the cultivation of marine fish in the world. In this chapter, will not be addressed in detail each disease, will only be discussed your importance and how to avoid your appearance in cultivation. Currently there is a strong discussion about acting on prevention of the disease and no cure. It is common knowledge the factors that can trigger an illness, such as lack of well-being for individuals caused mainly by high stocking density, nutritional quality of the diet and the environment.

Currently, in some countries are already adopted measures that seek to monitoring and surveillance of diseases that affect crops of marine fish. Along with the creation of the notifiable diseases, is a great tool for controlling the development and spread of disease outbreaks. In Nordic countries, the list of diseases under monitoring and surveillance varies. All the countries runs a surveillance program for Viral Haemorrhagic Septicaemia (VHS) and Infectious Haematopoietic Necrosis (IHN) based on EU regulations (Ariel et al., 1999) and a monitoring program for Infectious Pancreatic Necrosis (IPN) on a national level (Håstein et al., 2001).]

The clinical signs of diseases include observation of white dots or necrosis of the epidermis, high production of mucus, anorexia, cutaneous hemorrhage and enlarged in gill filaments, branchial, lack of appetite, erratic or disoriented swimming, among others.

I believe that the best way to avoid such problem is to change the mentality of the producer related to good management practices and production, acting on prevention. An individual who feeds nutritionally well and lives in a suitable environment will hardly contract some disease. The main trigger of disease outbreaks is the high density followed the poor quality in nutrition, both factors to minimize production costs and make more profit.

FEED AND NUTRITION

As previously stated, the proper nutrition in all phases of cultivation is one of the main pillars that prevent the emergence and outbreaks of disease in marine fish species cultivated, along with management and water quality.

Most of the cultivated species are carnivorous, so the main ingredients in the formulation of your diet are fish meal and fish oil, from the fishing industry. It is currently debatable and several studies have been carried out in order to replace animal protein for vegetable protein in the formulation of commercial feed (Regost et al., 1999) (Bell et al., 2001) (Sitjà-Bobadilla et al., 2005) (Millamena, 2002) (Kaushik et al., 2004).

I believe it is extremely important to realize this kind of research to activity become less and less dependent on animal protein source without harming final productivity.

GENETIC IMPROVEMENT

Since the establishment and technical domain about reproduction and hatchery of marine fish, has begun a process of genetic improvement, as the best individuals were selected for mating and reproduction in search of the best possible offspring. Currently, this method is known as selective breeding, and is widely used by several companies. The basic principle of this method of selection is to use individuals that express the desired trait and formed couples who will keep such trait to future generations. However, it's not as simple as it seems. Should be avoided crossing of individuals with similar genetics, which favors the inbreeding. It is important to have individuals from different populations in search of greater genetic variability, which makes the method more complex because they must be formed several populations.

With advances in studies in this area, there are currently several selection methods, including individual selection, pedigree selection, family selection, within family selection, combined selection and indirect selection. All methods with the same objective: increase the genetic value of the population for a determinate trait.

In Europe, the main marine fish (including anadromous salmon) that have selective breeding programs are the Atlantic salmon, European seabass, gilthead seabream and turbot (Janssen et al., 2016).

The genetic improvement is fundamental to the development of the production of marine fish. A large number of enterprises use genetically enhanced individuals in your production. The trend now is to improve the techniques and developing programs for species that do not have. The first generations are the ones obtained genetic gain, greatly improving production rates. For Atlantic salmon, in ten generations was achieve a total of 196% genetic gain (Solberg et al., 2013). For turbot (Danacher, & Garcia-Vazquez, 2007) and seabass (Thorland et al., 2015) the genetic gain achieved was 45% and 65% in three and two generations, respectively.To the seabream was achieved a 30% genetic gain in three generations (Knibb, 2000). There are encouraging results that in large volume bring a significant increase in production.

FISH FARMING AND ENVIRONMENT

Any fish farm is a site of concentrated food production. The environmental impact of marine fish farming depends very much on species, culture method, stocking density, feed type, hydrography of the site and husbandry practices (Wu, 1995). In general, 85% of phosphorus, 80% of carbon and 70% of nitrogen input into a marine fish culture system as feed may be waste into the environment through feed waste, fish excretion and respiration.

The main discussion currently, beyond source of protein used in the formulation of commercial feed, is the waste of nutrients to the environment, causing also a local modification in the ecosystem. There are already multitrophic systems that reuse these excessive nutrients from leftover feed and excretion of fish for the production of seaweeds, showing very positive results. Phytoplankton is not the only source of organic food, seaweeds are also important primary producers, and there is a further input of dead organic matter (Tyler 1984). During the last decades, the evolution of multi-use planning of ocean space and resources has become a crucial step in achieving an ecosystem-based sea use management (Tammi and Kalliola, 2014).

The potential impacts of marine aquaculture are since aesthetic aspects to direct pollution problems (O'Sullivan, 1992). These environmental impacts include dispersion of organic matter (waste feed and faeces), and pharmaceutical agents (antibiotics) to the surrounding water body and local sediments, as well as genetic interactions between escaped farmed and wild fish. The main and most common environmental effects in cultivation of marine fish in cages result only local impacts. In this way, is of local interest to monitor and evaluate pollution parameters to seek ways to minimize them, getting to know the environmental support capacity and productivity that can be achieved without compromising the local ecosystems.

The term sustainability is gaining enough space over the last decade and has been inserted into some new ventures in a variety of ways. In aquaculture, one should opt for the cycling and reuse of effluent nutrients of cultivation, the lowest possible use of water (through different systems of cultivation), deployment in that area do not undermine the local ecosystem, among other decisions aimed to avoid losses, waste land unnecessary spending.

REFERENCES

1. Aoki, J., Kai, W., Kawabata, Y. et al.(2015). Second generation physical and linkage maps of yellowtail (*Seriola quinqueradiata*) and comparison of synteny with four model fish. BMC Genomics, 16: 406. DOI: 10.1186/s12864-015-1600-7.
2. Ariel E, Helgason S, Mortensen HF, Olesen NJ (1999). Report of the Third annual Meeting of EU National Reference Laboratories for Fish Diseases, Aarhus, Denmark 9.th-11th June 1999. 31 pp.
3. Blancheton, J.P. (2000). Developments in recirculation systems for Mediterranean fish species. Aquacultural Engineering, 22(1):17-31.
4. Bowman, S., Hubert, S., Higgins, B. et al. (2011). An Integrated Approach to Gene Discovery and Marker Development in Atlantic Cod (*Gadus morhua*). Marine Biotechnology, 13: 242. DOI: 10.1007/s10126-010-9285-z.
5. Castaño-Sánchez, C., Fuji, K., Ozaki, A. et al. (2010) A second generation genetic linkage map of Japanese flounder (*Paralichthys*

olivaceus). BMC Genomics, 11: 554. DOI: 10.1186/1471-2164-11-554.

6. Danacher, D. & Garcia-Vazquez, E. (2007). Turbot - Scophthalmus maximus. In: Svåsand, T., Crosetti, D., García-Vázquez, E., Verspoor, E. (Eds.), Genetic Impact of Aquaculture Activities on Native Populations. Genimpact Final Scientific Report. 55–61.
7. FAO. FISHSTAT PLUS: Universal software for fishery statistical time series. Version 2.3.2000. Fisheries Department Fishery Information, Data and Statistics Unit, FAO, Rome, Italy. 2011.
8. FAO. 2016. FAO yearbook. Fishery and Aquaculture Statistics. 2014. Rome, Italy.
9. FAO 2004-2017. Cultured Aquatic Species Information Programme. Gadus morhua. Cultured Aquatic Species Information Programme. Text by Håkon Otterå. In: *FAO Fisheries and Aquaculture Department* [online]. Rome. Updated 1 January 2004. [Cited 15 February 2017].
10. FAO 2005-2017. Cultured Aquatic Species Information Programme. Psetta maxima. Cultured Aquatic Species Information Programme . Text by Rodríguez Villanueva, J. L. & Fernández Souto, B. In: *FAO Fisheries and Aquaculture Department*[online]. Rome. Updated 4 May 2005. [Cited 15 February 2017].
11. FAO 2006-2017. Cultured Aquatic Species Information Programme. Lates calcarifer. Cultured Aquatic Species Information Programme. Text by Rimmer, M.A. In: *FAO Fisheries and Aquaculture Department* [online]. Rome. Updated 3 June 2006. [Cited 15 February 2017].
12. FAO 2006-2017. Cultured Aquatic Species Information Programme. Lates calcarifer. Cultured Aquatic Species Information Programme. Text by Rimmer, M.A. In: *FAO Fisheries and Aquaculture Department* [online]. Rome. Updated 3 June 2006. [Cited 15 February 2017].
13. FAO 2007-2017. Cultured Aquatic Species Information Programme. Rachycentron canadum. Cultured Aquatic Species Information Programme. Text by J.B. Kaiser, J.G. Holt. In: *FAO Fisheries and Aquaculture Department* [online]. Rome. Updated 23 May 2007. [Cited 15 February 2017]. http://www.fao.org/fishery/

culturedspecies/Rachycentron_canadum/en

14. FAO 2015-2017. Cultured Aquatic Species Information Programme. Thunnus thynnus. Cultured Aquatic Species Information Programme. Text by De la Gándara, F. In: *FAO Fisheries and Aquaculture Department* [online]. Rome. Updated 2 October 2015. [Cited 15 February 2017].
15. Ferraresso, S., Vitulo, N., Mininni, A.N. et al. (2008). Development and validation of a gene expression oligo microarray for the gilthead sea bream (*Sparus aurata*). BMC Genomics, 9: 580. DOI: 10.1186/1471-2164-9-580.
16. Bell, G. J. et al. (2001). Replacement of Fish Oil with Rapeseed Oil in Diets of Atlantic Salmon (Salmo salar) Affects Tissue Lipid Compositions and Hepatocyte Fatty Acid Metabolism. The Journal of Nutrition, 131: 1535-1543.
17. Håstein, T., Hellstrøm, A., Jonsson, G. et al. (2001). Surveillance of Fish Diseases in the Nordic Countries. Acta Veterinaria Scandinavica. 42 (Suppl 1): S43
18. Hersoug, B. 2013. The battle for space: The position of Norwegian aquaculture in integrated coastal zone planning. In Global challenges in integrated coastal zone management, ed. E Moksness, E Dahl, and J Støttrup, 159–168. West Sussex, UK: Wiley.
19. Janssen, K., et al., Impact of selective breeding on European aquaculture, Aquaculture (2016), http://dx.doi.org/10.1016/j.aquaculture.2016.03.012.
20. Kaushik, S.J., Covès, D., Dutto, G., Blanc, D. (2004). Almost total replacement of fish meal by plant protein sources in the diet of a marine teleost, the European seabass, *Dicentrarchus labrax*. Aquaculture, v. 230, Issues 1-4, 391-404.
21. Knibb, W. (2000). Genetic improvement of marine fish — which method for industry? Aquaculture Research. 31, 11–23.
22. Kondo, R., Shigematsu, K., Kawahara, N., Okamura, T., Yoon, Y.H., Sakami, T., Yokoyama, H., Koizumi, Y. (2012). Abundance of sulphate-reducing bacteria in fish farm sediments along the coast of Japan and South Korea. Fisheries Science, 78: 123-131. DOI: 10.1007/s12562-011-0439-3.

23. Kubitza, F. (2006). Sistemas de recirculação: sistema fechado com tratamento e reuso da água. Panorama da Aquicultura, 16 (95): 15-22.

24. Millamena, O.M. (2002). Replacement of fish meal by animal by-product meals in a practical diet for grow-out culture of grouper *Epinephelus coioides*. Aquaculture, v. 204, Issues 1-2, 75-84.

25. Nelson AL and Marygrace C. 2010. Chanos chanos. Cultured Aquatic Species Information Programme. Retrieved from http://www.fao.org/fishery/culturedspecies/Chanos_chanos/en.

26. O'Sullivan, A.J. (1992). Aquaculture and user conflicts. Aquaculture and the environment. Spec. Publ. Eur. Aquacult. Soc., 16 (1992) 405–412.

27. Ole Torrissen , Rolf Erik Olsen , Reidar Toresen , Gro Ingunn Hemre , Albert G.J. Tacon , Frank Asche , Ronald W. Hardy & Santosh Lall (2011) Atlantic Salmon (*Salmo salar*): The "Super-Chicken" of the Sea?, Reviews in Fisheries Science, 19:3, 257-278, DOI: 10.1080/10641262.2011.597890.

28. Palaiokostas, C., Bekaert, M., Taggart, J.B. et al. (2015). A new SNP-based vision of the genetics of sex determination in European sea bass (*Dicentrarchus labrax*). Genetic Selection Evolution (2015) 47: 68. DOI: 10.1186/s12711-015-0148-y.

29. Regost, C., Arzel, J., Kaushik, S.J. (1999). Partial or total replacement of fishmeal by corn gluten meal in diet for turbot (*Psetta maxima*). Aquaculture, v. 180, Issues 1-2, 99-117.

30. Sandersen, H.T. & Kvalvik, I. (2015). Access to aquaculture sites: A wicked problem in Norwegian aquaculture development. Maritime Studies, 14: 10. DOI: 10.1186/s40152-015-0027-8.

31. Sarropoulou, E., Sepulcre, P., Poisa-Beiro, L. et al. (2009). Profiling of infection specific mRNA transcripts of the European sea bass *Dicentrarchus labrax*. BMC Genomics, 10: 157. DOI: 10.1186/1471-2164-10-157.

32. Shepherd, J., Bromage, N. (1988). Intensive fish farming. Oxford: BSP Professional Books. 404 p.

33. Sitjà-Bobadilla, A. et al. (2005). Effect of fish meal replacement by plant protein sources on non-specific defence mechanisms and

oxidative stress in gilthead sea bream (*Sparus aurata*). Aquaculture, v. 249, Issues 1-4, 387-400.

34. Solberg, M.F., Skaala, O., Nilsen, F., Glover, K.A. (2013). Does domestication cause changes in growth reaction norms? A study of farmed, wild and hybrid Atlantic salmon families exposed to environmental stress. PLoS ONE 8, 1–11.
35. Tammi, I., Kalliola, R. (2014). Spatial MCDA in marine planning: experiences from the Mediterranean and Baltic Seas. Mar. Policy, 48, 73-83.
36. Thorland, I., Kottaras, L., Refstie, S., Papanna, K., Papaharisis, L., Rye, M. (2015). Response to Selection for Harvest Weight in European Sea Bass (Dicentrarchus labrax) in Greece. ISGA XII, Santiago de Compostella, Spain.
37. Trumbić, **Ž.,** Bekaert, M., Taggart, J.B. et al. (2015). Development and validation of a mixed-tissue oligonucleotide DNA microarray for Atlantic bluefin tuna, *Thunnus thynnus* (Linnaeus, 1758). BMC Genomics, 16: 1007. DOI: 10.1186/s12864-015-2208-7.
38. Troell, M., Joyce, A., Chopin, T. (2009). Ecological engineering in aquaculture — Potential for integrated multi-trophic aquaculture (IMTA) in marine offshore systems. Aquaculture, 297, 1-4, 1-9
39. Tucker, J.W., Russell, D.J., Rimmer M. (2002). Barramundi culture: A success story for aquaculture in Asia and Australia. World Aquaculture, 33 (3): 67-72.
40. Tyler, I.D. (1984). A carbon budget for Creran, a Scottish sea loch. Ph.D. thesis. University of Strathclyde.
41. WARD, P.; MYERS, R. A. Shifts in open-ocean fish communities coinciding with the commencement of commercial fishing. Ecology, New York, v. 4, n. 86, p. 835-847, apr. 2005.
42. Wu, R.S.S. (1995). The environmental impact of marine fish culture: Towards a sustainable future. Marine Pollution Bulletin, 31, Issues 4-12, 159-166.

SECTION II

WORLDWIDE TRENDS

CHAPTER

2

FEEDING 9 BILLION BY 2050 – PUTTING FISH BACK ON THE MENU

Christophe Béné[1,2], Manuel Barange[3], Rohana Subasinghe[4], Per Pinstrup-Andersen[5], Gorka Merino[6], Gro-Ingunn Hemre[7], Meryl Williams[8]

[1]Institute of Development Studies, University of Sussex, Sussex, UK

[2]International Center for Tropical Agriculture, Cali, Colombia

[3]Plymouth Marine Laboratory, Plymouth, UK

[4]Food and Agriculture Organization, Rome, Italy

[5]Cornell University Ithaca, Ithaca, USA

[6]AZTI, Pasaia, Gipuzkoa, Spain

[7]Institute of Nutrition and Seafood Research, Bergen, Norway

[8]17 Agnew Street, Aspely, Queensland, Australia

ABSTRACT

Fish provides more than 4.5 billion people with at least 15 % of their average per capita intake of animal protein. Fish's unique nutritional properties make it also essential to the health of billions of consumers in both developed and developing countries. Fish is one of the most efficient converters of feed into high quality food and its carbon footprint is lower compared to other animal production systems. Through fish-related activities (fisheries and aquaculture but also processing and trading), fish contribute substantially to the income and therefore to the indirect food security of more than 10 % of the world population, essentially in developing and emergent countries. Yet, limited attention has been given so far to fish as a key element in food security and nutrition strategies at national level and in wider development discussions and interventions. As a result, the tremendous potential for improving food security and nutrition embodied in the strengthening of the fishery and aquaculture sectors is missed. The purpose of this paper is to make a case for a closer integration of fish into the overall debate and future policy about food security and nutrition. For this, we review the evidence from the contemporary and emerging debates and controversies around fisheries and aquaculture and we discuss them in the light of the issues debated in the wider agriculture/farming literature. The overarching question that underlies this paper is: how and to what extent will fish be able to contribute to feeding 9 billion people in 2050 and beyond?

Keywords: Fish, Food security and nutrition, Micro-nutrient deficiency , Fisheries, Aquaculture

INTRODUCTION

Fish is critically important to food security and good nutrition (Allison 2011; Thilsted 2012; Beveridge et al. 2013). Fish and other aquatic foods are high in protein and contain many essential micronutrients. The fishery and aquaculture sectors are the source of income for millions of women and men in low-income families (Béné 2006), thus contributing directly and indirectly to their food security (Béné et al. 2007; Allison 2011; World Bank/FAO/WorldFish 2012). Yet, the potential contributions of fish to food security and nutrition (FSN) are all but ignored in the international debate, as if a firewall stood between the discussions about the role of fish and the broader debates about FSN issues. A recent review of international development and research agencies working on FSN revealed for instance that "fish is strikingly missing from strategies for reduction of micronutrient deficiency, precisely where it could potentially have the largest impact" (Allison et al. 2013: 45).

The purpose of this paper is to make a case for a closer integration of fish into the overall FSN debate. As such the discussion is directed not to those in the fisheries and aquaculture communities who have been relentlessly advocating for fish as an entry point in this debate (see e.g., Béné et al. 2007; Hall et al. 2011; Thilsted 2012), but to the vast majority of the international experts who are influencing the wider debate and policies on FSN.

The World Committee on Food Security (CFS) took an important step in the right direction in 2012 by requesting that the High Level Panel on Food Security (HLPE) undertook an in-depth study of the role of sustainable fisheries and aquaculture for FSN. The report (HLPE 2014) presents the strongest case yet for incorporating fish into the debates on how to achieve FSN for all and suggests a set of activities and policies to be pursued.

Building on this report and a large body of literature, our ambition is to bring fish onto the table. A number of papers have succeeded in presenting a very comprehensive evidence-based argument for supporting fish as a central element in the FSN debate (see e.g., Prein and Ahmed 2000; Roos et al. 2003; 2007; Kawarazuka and Béné 2011; Beveridge et al. 2013). Here, we aim to review the evidence from the contemporary and emerging debates and controversies around fisheries and aquaculture. In doing this we do not go into the micro-level issues concerning social, gendered and micro-economic aspects of fish, FSN at the local levels – for those we refer readers to the HLPE report (HLPE 2014).

Some of the underlying questions in the present study are closely linked to the issues discussed in the wider agriculture/farming literature. The latest estimate suggests for instance that, in 2009, fish accounted for 17 % of the global population's intake of animal protein and 6.5 % of all protein consumed (FAO 2014). In the context of this Special Issue on "Feeding 9 Billion", the key questions are: a) Can we maintain these consumption rates, given the projected growth in human population and the growing environmental challenges facing the earth? and b) Could fish - one of the most efficient converters of feed into high quality food – be a substitute for other sources of animal protein? Finally, in the context of the current discourse about future resource scarcity, should fish be considered a more environmentally friendly source of protein than the other livestock production systems?

Fish is more than just a source of animal protein. Fish contains several essential amino acids, especially lysine and methionine. The lipid composition of fish, with the presence of long-chain, poly-unsaturated fatty acids (LC-PUFAs), is unique. Fish is also an important source of essential micronutrients – vitamins D, A and B, and minerals (calcium, phosphorus, iodine, zinc, iron and selenium), which makes it particularly attractive in the current fight against malnutrition in low income and food deficient countries (LIFDCs). Some countries (e.g., Zambia, Brazil, or Chile) have already recognized this potential and have included fish in their national school-feeding programmes. Should this be more systematically considered in countries with high levels of malnutrition to complement or substitute the technology-dependant (and expensive) bio-fortification programmes?

In addressing these issues we must first acknowledge that fish production sectors and associated value chains have been remarkably dynamic in the last three decades – what some refer to as the 'blue revolution'. Every second fish we consume is now produced in aquaculture. This new situation is a result of the stagnation of capture fisheries following decades of expansion, some of it carried out in an unsustainable manner (FAO 2012), combined with impressive growth rates in the aquaculture industry, subject to some major adjustments to overcome the early challenges such as fish disease and negative environmental impacts. Critics of aquaculture have also pointed to the use of fishmeal and fish oil produced from wild fish, to feed farmed fish. These concerns are valid and need to be taken (along with their solutions) into account in this discussion.

Understanding fish production in relation to food security and nutrition

The fastest growing food-supply industry in the world

In 2011, 173 Mt of fish were extracted from the global marine and inland water ecosystems, of which 7 to 10 Mt were discarded prior to landing and 12 Mt were lost at the post-harvest stage. The production from capture fisheries and aquaculture available amounted therefore to 154

Mt, of which about 131 Mt were utilized directly for human consumption (Fig. 1) (HLPE 2014). Thanks to aquaculture and fisheries, the global supply of fish has grown by a factor of 8 since 1950. By comparison, even after the Green Revolution the world rice production increased only by a factor of 3. In effect, fish production has been the fastest growing food industry in the world for the last 40 years and is expected to remain so in the near future. The fish supply per capita has more than tripled in the last half century, from 6 kg/year in 1950 to 18.8 kg/year in 2011 – Fig. 2. In fact, the world fish supply has effectively been growing faster than the world's population (FAO 2012). These global figures mask, however, some important regional variations: fish consumption is the lowest in Africa (9.1 kg per capita in 2009), while Asia accounts for almost two-thirds of total consumption (20.7 kg per capita).

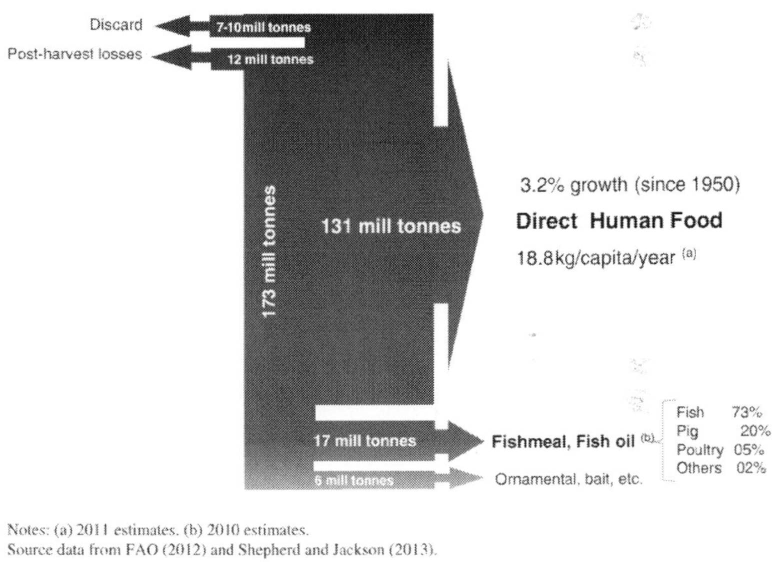

Figure. 1: World fish utilization. Notes: **a** 2011 estimates. **b** 2010 estimates. Source data from FAO (2012) and Shepherd and Jackson (2013).

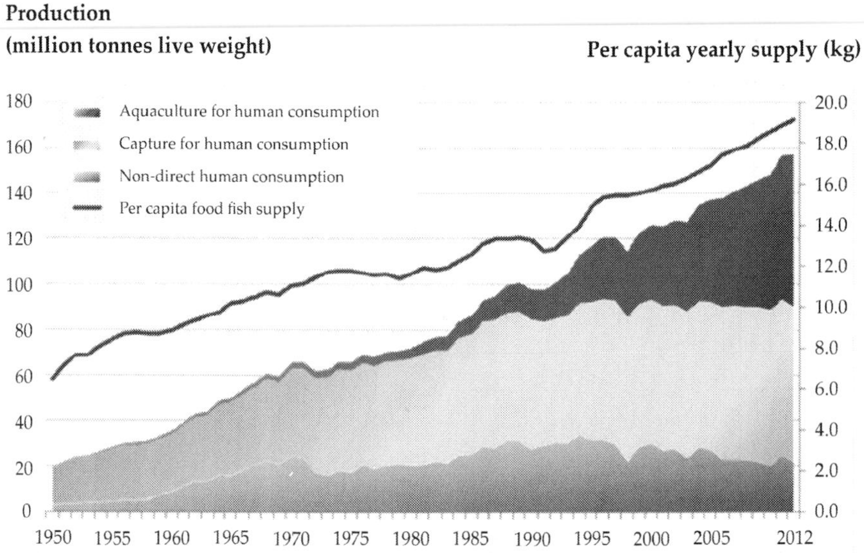

Figure. 2: Relative contribution of aquaculture and capture fisheries to production and food fish supply. Source: FAO Statistics and Information Branch of the Fisheries and Aquaculture Department.

Fish as the Largest Source of Animal Protein

Fish is a major source of animal protein, overshadowing most other sources. In 2010 it represented a source twice as important as poultry, and three times larger than cattle (Fig. 3). Today capture fisheries and aquaculture provide 3 billion people with almost 20 % of their average per capita intake of animal protein, and a further 1.3 billion people with about 15 % of their per capita intake (HLPE 2014). This share can exceed 50 % in some countries. In West Africa, Asian coastal countries, and many small island states, the proportion of total dietary protein from fish can reach 60 % or more (e.g., Gambia, Sierra Leone, Ghana, Cambodia, Bangladesh, Indonesia, Sri Lanka, or the Maldives) (FAO 2014).

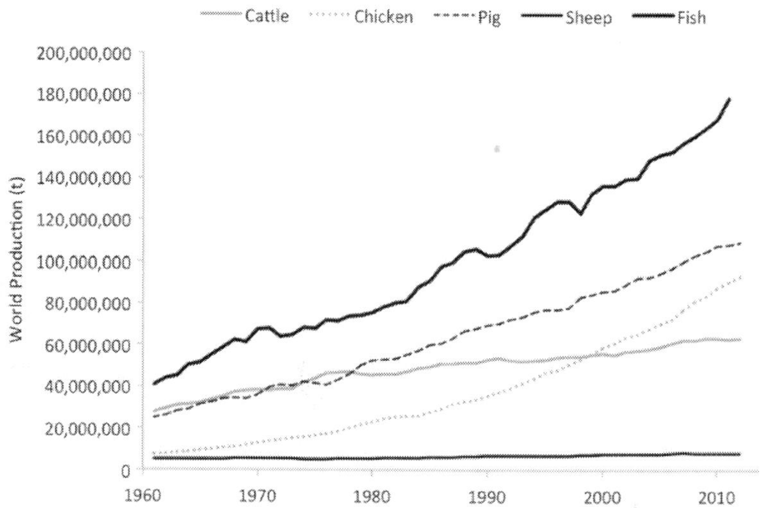

Figure. 3: World Production of the main sources of animal protein over the period 1960–2010. Source: FAO Stat.

The geography of fish as a source of protein is also significant in the FSN discussion. Twenty-two of the 30 countries where fish contribute more than one-third of the total animal protein supply were officially referred to as LIFDCs in 2010 (Kawarazuka and Béné 2011). In other words, almost three-quarters of the countries where fish is an important source of animal protein are poor (income-wise) and food-deficient. Yet in these LIFDCs, even if fish is a substantial proportion of the food intake, undernourishment can still occur as total food intake is often insufficient.

Fish - beyond Protein...

Some point out, however, that the main contribution of fish to FSN may not be in relation to its protein content, but its lipid and micro-nutrient content.

The lipid composition of fish is unique, having LC-PUFAs, with many beneficial effects for child development and adult health (Thilsted et al. 1997; Larsen et al. 2011; Richardson and Montgomery 2005). Among fish species that are cheaper in developing countries, small pelagic fish such as anchovy and sardine are some of the richest in LC-PUFAs (USDA 2011), especially compared to large freshwater fish such as carp and tilapia. When its rich nutrient content is preserved, fish can provide protective effects against a wide range of health issues. LC-PUFAs for instance provide protection against diseases such as stroke, high blood pressure or coronary heart disease (Miles and Calder 2012; Rangel-Huerta et al. 2012).

Complementing its fatty acid content, fish is also known to be an important source of essential micronutrients – vitamins D and B, and minerals (Roos et al. 2003; 2007; Bonham et al. 2009). Lipid-rich fish also contain vitamin A. Recent research showed that fish species consumed whole with bones, heads, and viscera play a critical role in micronutrient intakes of people as these parts are where most micronutrients are concentrated. Some of these small fish (such as mola (*Amblypharyngodon mola*), darkina (*Esomus danricus*), sardines and pilchards, anchovy, seabass, tilapia) contain high levels of minerals such as calcium, phosphorus, iodine, zinc, iron and selenium, which are low in other foods. The potential contribution that fish (even in small quantity) can therefore offer to address multiple micronutrient deficiencies is now being recognized (e.g., Roos et al. 2007; Kawarazuka and Béné 2011; Thilsted 2012). For instance, the high level of iodine found in fish can help prevent iodine deficiency which is known to cause cretinism (stunted growth and mental retardation).

CURRENT ISSUES IN RELATION TO FISH CONTRIBUTION TO FOOD SECURITY

Fishmeal and Fish Oil in Aquaculture

In addition to being used directly as human food, fish also contributes indirectly to human nutrition when it is used as fishmeal for aquaculture and poultry/livestock feeds (Tacon and Metian 2009). In 2011, 23 Mt of fish – essentially small pelagic fish species such as anchovy, herring, mackerel and sardine – have been destined to non-direct human consumption, of which 75 % (17 Mt) was reduced to fishmeal and fish oil for aquaculture, poultry and other livestock feeding (cf. Fig. 1). In 2010, 73 % of the total world fishmeal was used to feed farmed fish, followed by pigs (20 %), poultry (5 %) and others (2 %) (Shepherd and Jackson 2013).

From a FSN perspective, the use of fishmeal for farmed fish (and livestock) raises important issues. Leaving aside the debate on the role of small pelagic fish in supporting larger fish, birds and marine mammals in the ecosystem (Smith et al. 2010), is fishmeal the most efficient way to use fish (especially low-cost small pelagic fish rich in LC-PUFA) or would these fish contribute more to food security if a larger share of them was eaten directly by people? Indeed, despite some substantial improvement in the last decade, the rate of conversion of fishmeal to fish is still a source of concern (Troell et al. 2014). On average, for every kg of farmed fish produced, 0.7 kg of wild fish is needed (Tacon and Metian 2008). This average figure, however, masks important differences: while for omnivorous farmed fish, the rate is down to an acceptable 0.2 to 1.41 kg of wild fish per 1.0 kg of farmed fish, for carnivorous farmed fish, the figure is higher: 1.35 to 5.16 kg to produce one kg of farmed fish (Boyd et al. 2007).

Fish Losses and Implications on Food Security and Nutrition

The global discards of fish (fish caught but dumped overboard due to low quality, damage or spoilage, non-targeted species or below regulation size) were estimated to be around 7.3 Mt in 2005, 80 % of which coming from industrial fleets. In contrast, small-scale fisheries generate less wastage in the form of discards (about 2 Mt a year, that is, 4 % of their estimated landing[1]) (Kelleher 2005).

With increasing fish scarcity and increasing fish prices, species previously deemed commercially inferior are progressively integrated into consumer eating habits and markets. Most shrimp trawlers, which used to discard up to 95 % of their catch, are now landing more bycatch for human consumption (Béné et al. 2007). In small-scale fisheries where discarding fish is rare, substantial quantities and quality are lost due to post-harvest mishandling during transport, storage, processing, on the way to markets and waiting to be sold. Especially in developing countries, where access to electricity and cold chain can still be an issue, fish post-harvest losses can be significant. Estimated at 10–12 Mt in 2005, the total accounts for 10 % of global capture and culture fisheries' production (Béné et al. 2007). While physical losses in small-scale fisheries are less than 5 %, economic losses can be substantial. In Africa, some estimates

(FAO Focus nd) put post-harvest losses in some cases to levels as high as 20–25 %. In aquaculture, waste streams in value chains have been large in the past but are tending to decline rapidly as competitive pressures force innovation (Arthur et al. 2013).

Sustainability of Fisheries and Implications For Food Security

Because total fish production (availability) is an important dimension of food security, a key issue in this debate is the environmental sustainability of capture fisheries. The extent to which capture fisheries have exceeded sustainable levels has generated strong expert and public opinion debates (Worm et al. 2006; Pauly et al. 2005; Hilborn 2013) and many media headlines, scientific papers and environmental campaigns have been in the last two decades framed around the idea that world fisheries resources are in crisis due to overfishing (see e.g., Pauly et al. 1998; Myers and Worm 2003). FAO has expressed a more nuanced but nonetheless concerned view about the state of world marine fisheries (e.g., FAO 2012), acknowledging the granularity in the state of resources worldwide. The current consensus is that global fisheries would be more productive if the levels of overfishing were reduced (Srinivasan et al. 2010), and the environmental sustainability of fisheries were recognized to be a *sine qua none* condition for FSN (HLPE 2014).

An important point in this debate, however, is that improving FSN through fisheries would depend on stock recovery and also on access to and distribution of the harvest, as well as gender consideration (de Schutter 2012; Williams 2010). Indeed overfishing per se is only one aspect of the problem. Other economic activities, such as oil drilling, coastal development, pollution, or dams and water flow management have significant negative impact on aquatic habitats (Halpern et al. 2008).

Indirect Contribution to Food Security and Nutrition through Livelihood Support

A critical pathway to enhance FSN is through the income that people generate from engaging in remunerated activities. In this respect aquaculture and fisheries (especially through the number of small-scale operators engaged in fishing, aquaculture, processing, and trading businesses – see e.g., Béné et al. 2010) play a critical role in low income and emergent countries. Altogether it is estimated that between 660 and 820 million people (fishers, fish-farmers, fish traders, workers in fish processing factories, and their families) depend on fish-related activities as a source of income (Allison et al. 2013; HLPE 2014). This represents more than 10 % of the world population. For most of these households, the revenues generated may not necessarily be very high (Neiland and Béné 2004; Allison et al. 2011; Béné and Friend 2011), but it is often the main component of their livelihood, which allows them to secure accessibility to food (Heck et al. 2007; Béné et al. 2009; Eide et al. 2011).

FEEDING 9 BILLION BY 2050: WHERE DOES FISH STAND?

The global human population is expected to exceed 9 billion by 2050 (UN estimates), increasing the pressure on the food sectors to maximize production and reduce waste. Production increase must occur in a sustainable way and in a context where key resources, such as land and water, are likely to be scarcer and where climatic change impact will intensify. The fish-production sector is no exception.

In this context two key questions emerge. First, will fisheries and aquaculture be able to maintain the current global fish consumption rate of 18 kg per capita per year, and the equivalent regional values, and if not, how will society address the needs of expected winners and losers (Barange et al. 2014)? So far technological and institutional innovations have ensured that the combined production of fish through fisheries and aquaculture has been faster than the world population's demand of fish. The question is now whether we can keep up this pace with another 2 billion people added in the next 35 years, and how the four dimensions of food security (availability, accessibility/affordability, utilization and stability) will be balanced to ensure that fish go to those who need them most (Merino et al. 2012).

The second key question is whether sustainable fisheries and aquaculture will be able to help address the bigger food security issue that will affect the world in the coming decades? In particular, could aquaculture become a substitute form of protein for some of the less efficient food production systems, or even be used to compensate for the decline in farming systems' productivity that is predicted as a consequence of the impact of climate change?

How Much Fish do we Need?

Global Drivers of Fish Demand

World population is often presented as a key driver for the growth in seafood demand and for fisheries development. In reality, a more important driver for fish (and other animal-source food) consumption is income (Speedy 2003). Demand for fish as food is particularly high in the wealthier strata of societies, including in the low-income countries, and as income will continue to increase in highly populated countries such as China and India, demand levels are likely to rise more strongly (Garcia and Rosenberg 2010). Overall, a large increase in the number of people moving into the middle class, particularly but not exclusively in Asia, is likely to result in a very large expansion in the demand for fish.

Income is, however, not the only driver of fish demand. It is recognized that urbanization is also an important factor increasing animal-source food consumption in general and fish consumption in particular. Delgado et al. (1997) suggest that changes in food preference driven by urbanization alone have, in the past, accounted for an extra 5.7–9.3 kg per capita consumption of meat and fish per year. Similarly, Betru and Kawashima (2009) and Toufique and Belton (2014) present data from Ethiopia and Bangladesh, indicating that urbanization affects animal food consumption rates independently of income.

These different factors explain the rapid increase in demand for meat, milk, and fish in the emerging economies of Asia. In China, for example, the demand for fish is likely to increase from 24.4 kg per person per year in 2000 to 41 kg per person per year by 2030 (World Bank 2014).

Demand Projection

Modelling exercises have been conducted recently with the objective of estimating the projections of fish demand and supply. These modelling exercises include the World Bank-FAO-IFPRI Fish 2030 analysis elaborated on the IMPACT model developed by the International Food Policy Research Institute (IFPRI) (World Bank 2014); the OECD-

FAO Agricultural outlook model built on the combined multimarket, partial equilibrium AgLink-CoSiMo model (OECD-FAO 2013; Lem et al. 2014); and a series of peer-reviewed articles relying on various types of modelling and projection tools (Rice and Garcia 2011; Merino et al. 2010, 2012; Barange et al. 2014). The time horizon of these different analyses is not always the same. For instance, the OECD-FAO outlook model works over a 10 year-projection period (i.e., up to 2023 as per the last iteration of the model), while the World Bank-FAO-IFPRI model runs until 2030; Merino et al. 2010 used a 20 year simulation, calibrated on a 1997–2004 data set (meaning technically that their projection runs until 2024), while Merino et al. (2012) and Barange et al. (2014) proposed a projection up to 2050. A strict comparison of the different projections is therefore difficult.

A bigger issue is that very few of these studies integrated information on the drivers of changes (the combination of urbanization and increase in income), to estimate with accuracy the future demand for fish across the world. Many studies assume constant consumption rates in the future (e.g., Barange et al. 2014) or fixed nutritional targets (Rice and Garcia 2011). Others worked directly with projected fish consumption (e.g., OECD-FAO 2013; Merino et al. 2012), that is, by dividing the projected supply by the projected population. None of these approaches therefore offers an appropriate basis for estimating the actual demand for fish. The World Bank-FAO-IFPRI study uses regional fish consumption rates to estimate the global demand but the report does not clarify how these figures are constructed.[2] Aggregating the regional figures at the global level, the report estimates that the world demand for fish will be around 152 Mt in 2030 (World Bank 2014).

Merino et al. (2012) estimated the expected production capacity of marine ecosystems exploited under maximum sustainable yield principles, and projected aquaculture production requirements to achieve a range of food consumption targets. They concluded that between 125 and 210 Mt of fish by 2050 will be necessary to maintain fish consumption at around 15–20 kg per capita per year. Starting from a different angle, Rice and Garcia (2011) sought to estimate the need for additional fish necessary to supply 20 % of the dietary protein requirement to feed a 9 billion population by 2050. On this basis, an additional production of 75 Mt of fish from fisheries and aquaculture would be needed above the 2006 production level (144 Mt), that is, approximately 215 Mt. This represents an almost 50 % increase in production with respect to the 2006 level. While this figure is above other projections, it has the advantage that it starts from an actual estimate of the future needs (as opposed to the estimate of apparent consumption rates), but without factoring in future production potentials or market responses to the fishmeal/fish oil demand (see below).

In summary, the current understanding of the global drivers of fish demand (urbanization and increase in living standards in developing and emerged countries) is relatively well established but not all of the current models have integrated these drivers comprehensively. In comparison, the efforts to better understand the ability of the world to produce fish in the future (i.e., the supply side of the equation) have been more elaborate.

The Future of Fisheries Production and the Impact of Climate Change

A consensus has emerged in the literature that the doom-and-gloom rhetoric that had driven the discussion surrounding the state of marine fisheries in the late 2000s (Garcia and Grainger 2005; Caddy and Seijo 2005) was exaggerated (Grafton et al. 2010; Hilborn 2010) and that although the situation remains concerning in respect to many stocks, we are not likely to face the global collapse that had been announced by some biologists (e.g., Myers and Worm 2003; Worm et al. 2006; Pauly 2009). Instead, the downward trend of overfished stocks may have been reined in (Fig. 4).[3] Reflecting this, most of the projections proposed in the recent literature estimate that the global fisheries' landings are likely to be stable in the short to medium term. The OECD-FAO model for instance estimates that capture fisheries will be 5 % higher by 2024 than it is was in 2013, that is, around 96 Mt (OECD-FAO 2013) while the World Bank-FAO-IFPRI model estimates that this will be around 93 Mt in 2030. These figures are at a global scale however, and some regional outlooks are for good stock rebuilding, while others are for a worsening in overfishing.

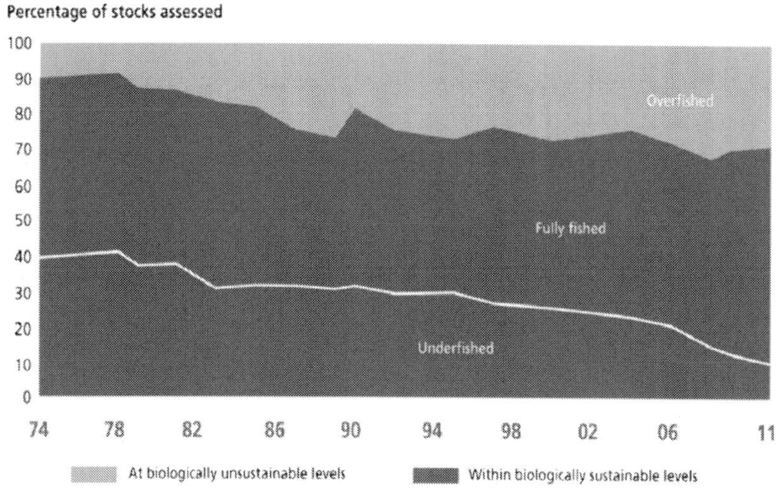

Figure. 4: Global trends in the state of world marine fish stocks (1974–2011). Source: FAO (2012).

Yet another key factor for which much uncertainty remains is the impact of climate change. Unlike most terrestrial animals, aquatic animal species are poikilothermic (cold-blooded) and changes in aquatic habitat temperatures will more rapidly and significantly influence distribution, prey availability, metabolism, growth and reproduction, with stronger impact on fishing and aquaculture distribution and productivity (Cheung et al. 2009). At the same time however, the interconnectedness of aquatic systems allows many species to change spatial distribution more easily as ecosystems shift, to remain in their zones of preference. Clearly, therefore, the impact of global climate change on ocean capture fisheries will be important. Biological predictions based on ocean-atmosphere general circulation models (OA-GCMs) have already demonstrated that the physical and chemical properties of the oceans will be modified, affecting the productivity, distribution, seasonality and efficiency of food webs, from primary producers to fish (Steinacher et al. 2010; Cheung et al. 2009, 2011). Some of these earlier GCM models however, were limited by their coarse resolution, too low to capture the processes that dominate the dynamics of the world's coastal and shelf regions, where most fisheries operate. More advanced models are now available (e.g., Merino et al. 2012; Blanchard et al. 2012; Barange et al. 2014).

Overall, and with few exceptions, the conclusion of all these models is that although climate change will alter the present geographical distribution of shelf-sea ecosystems productivity, in most of the regions and EEZs, the overall potential impact is projected to be low to moderate. Barange et al. (2014), for instance, used a high resolution shelf-sea physical-biological model that allowed them to scale down the analysis and gave greater confidence in predicting the consequences at national scales. They conclude that by 2050, estimates of national fish production should remain on average within ±10 % of the present yields.

The Contribution of Aquaculture to Future Fish Supply

The second element on the supply side is aquaculture. Discussion of the rise of aquaculture has so far largely focused on its contribution to global aquatic animal food supplies, ignoring the resultant changes in species composition of the fishes consumed, how it is farmed, and the implications for food and nutrition security (Kawarazuka and Béné 2011; UNHRC 2012). As a consequence, our understanding mainly concerns the question of the ability of the aquaculture industry to maintain its rate of growth. In this regard, most of the recent analyses agree that the era of exponential growth is over and while the sector will still continue to grow, the projected rate of growth is expected to decelerate. The main causes of this slower growth are likely to be freshwater scarcity, lower availability of locations for optimal production, and high costs of fishmeal, fish oil and other feeds (FAO 2012). Nonetheless, the World Bank-FAO-IFPRI model suggests this rate will still be sufficient to maintain a steady rise, reaching the point where it will equal global fisheries production by 2030 – around 93 Mt (World Bank 2014) (Fig. 5). Technical innovation, improved farm and animal health management, and improved and more efficient germplasm will be responsible for this increased growth, in combination with the continued expansion of fish-farming. Combined with a projected capture fish production that will remain fairly stable over the 2000–2030 period (see above), the global fish supply is projected to rise to 187 Mt by 2030 (World Bank 2014). These figures are consistent with the projections proposed by OECD-FAO in which the global fish production will reach 181 Mt in 2022, of which 161 Mt would be destined for direct human consumption (OECD-FAO 2013).

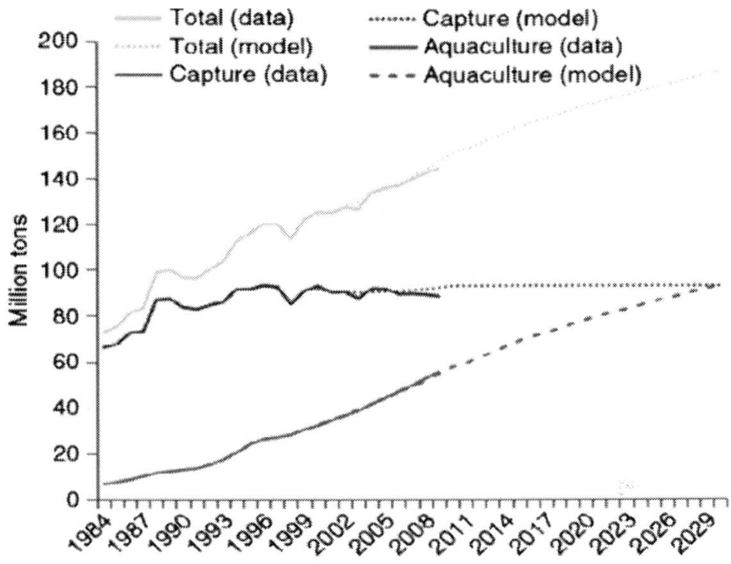

Data source: FAO FishStat and IMPACT projection model.

Figure. 5: Global fish production: data and projection 1984–2030 from the IMPACT model (World Bank 2014). Data source: FAO FishStat and IMPACT projection model.

A key element in this discussion is the importance of fishmeal and fish oil and how markets and technological innovations will respond to price signals (Merino et al. 2010, 2012; FAO 2012; World Bank 2014; Troell et al. 2014).[4] The consensus is that the use of fishmeal in aquaculture feeds is expected to decrease in percentage with time, thanks to increasingly effective replacements, including plant proteins, waste products from fish and terrestrial animals and use of better/improved breeds of aquatic animals with better feed conversion (Tacon et al. 2011). Formulated feeds are a significant factor in production costs, and this is a strong incentive to develop technology that will make feeds more affordable and sustainable. Overall, the proportion of fish used for non-direct human consumption has decreased from 30 % in the early 1990s to 15 % in 2010 and the World Bank analysis concluded that the projected expansion of aquaculture will be achieved with only an 8 % increase in the global fishmeal supply during the 2010–2030 period (World Bank 2014).

Putting All the Pieces Together

Given the very rough projections for fish demand and the more elaborate projections for fish supply just reviewed above, will fisheries and aquaculture be able to maintain their current contribution to food security in the future? The answer that emerges from the literature is that as far as food availability and demands are concerned, it is a conditional yes. All the projection models currently available seem to agree that the overall fish consumption rate could be maintained, in other words, that the fisheries and aquaculture aggregated growth will keep up with population growth rates. In fact, the OECD-FAO even estimates that the world fish consumption will increase by another 10 % and reach 20.9 kg per capita per year by 2023. The World Bank-FAO-IFPRI report is slightly more conservative and estimates that the per capita fish consumption will remain at around 18 kg per year in 2030. Merino et al. (2012) and Barange et al. (2014) reached the same conclusion for 2050, but this is essentially due to the underlying assumptions of their models.

All these analyses, however, stress the fact that for this outcome to occur, several specific conditions must be satisfied: capture fisheries will need to be exploited according to sustainable principles; very significant technological development will need to take place in aquaculture feeds to reduce fishmeal dependency and in farm management and germplasm to improve the overall efficiency of aquaculture; and discards, waste and losses will need to be reduced. Some of these conditions could be particularly challenging and a 'business-as-usual' approach is expected to fail.[5]

In addition to whether or not the world (and the markets) will be responsible enough to ensure that these conditions are satisfied, the question of who will be the winners and losers is vital to the other key FSN issues, which are: access to livelihoods in fish value chains and affordability of fish. Unfortunately, the consensus is that in the coming decades, the current situation regarding the imbalance between consumers in developed and developing countries is unlikely to change. While the present figures indicate the lowest fish consumption per capita in Africa (9.1 kg in 2009), the different models project that this imbalance will deteriorate further. The OECD expects that per capita consumption in Africa will decrease by a further 10 % by 2024, while that of Asia will show the highest growth rate (+14 %, OECD-FAO 2013). The World Bank-FAO-IFPRI projection (World Bank 2014) is even more alarming with per capita fish consumption expected to decline in sub-Saharan Africa by 1 % annually to 5.6 kg in 2030.[6] Yet at the same time developing countries will account for more than 91 % of the total increase in fish consumption. Even so, their annual per capita fish consumption will still remain below that of more developed regions (19.8 vs. 24.2 kg) (World Bank 2014).

A last important point needs to be mentioned. In view of the importance of income growth and urbanization as drivers of fish demand, the objective of merely maintaining fish production growth rate at the same level as the growth in world population would not be enough to prevent fish price from increasing. The urban population with the highest income growth will increase fish consumption while low-income people will experience reductions in their fish consumption. This is an important distributional and food security consequence. To avoid that, fish production needs to expand faster than population growth.

CAN FISH CONTRIBUTE FURTHER TO THE 2050 GLOBAL FOOD SECURITY CHALLENGE?

The arguments presented above are mainly constructed around a sectorial perspective. At least three other trans-sectorial (or systemic) arguments should be considered in relation to the wider debate of 'feeding 9 billion by 2050'.

First, in terms of animal protein availability, with 18.2 kg per capita per year, fish is providing 115, 133, and 189 % more protein per capita than pig, poultry and beef respectively. In fact, fish (combining capture fisheries and aquaculture) has been the main contributor to the 61 % increase in the world per capita consumption of animal protein for the period 1969–2009 (Table 1). As economic development is expected to continue driving an increasing trend in animal protein demand, (OECD-FAO 2013) and aquaculture is projected to remain the fastest growing food commodity sector, this sector will soon become even more central in the future food security of the world population.

Table 1: World per capita meat and fish food supply (kg per capita per year) 1969–2009

	1969	1979	1989	1999	2009
Pig meat	9.4	11.5	13.1	15.1	15.8
	25.2 %	27.7 %	28.2 %	28.1 %	26.3 %
Poultry meat	3.8	5.6	7.3	10.7	13.6
	10.2 %	13.5 %	15.7 %	19.9 %	22.7 %
Bovine meat	10.8	10.7	10.3	9.7	9.6
	29.0 %	25.8 %	22.2 %	18.1 %	16.0 %
Mutton and goat meat	1.8	1.5	1.7	1.8	1.9
	4.8 %	3.6 %	3.7 %	3.4 %	3.2 %
Meat, other	0.8	0.8	0.7	0.8	0.9
	2.1 %	1.9 %	1.5 %	1.5 %	1.5 %
Capture	10	10.4	11	10.5	10
	26.8 %	25.1 %	23.7 %	19.6 %	16.7 %
Aquaculture	0.7	1	2.4	5.1	8.2
	1.9 %	2.4 %	5.2 %	9.5 %	13.7 %
Capture and Fisheries	10.7	11.4	13.4	15.6	18.2
	28.7 %	27.5 %	28.8 %	29.1 %	30.3 %
Total	37.3	41.5	46.5	53.7	60

Source: FAOStat

Figures in percent are the respective contribution of each sector to the total figure

Second, in term of efficiency, fish in aquaculture systems are very efficient converters of feed into protein – in fact far more efficient than most terrestrial livestock system (Fig. 6a). For instance, poultry converts about 18 % of their consumed food and pigs about 13 %, as compared with 30 % in the case of fish (Hasan and Halwart 2009). Production of 1 kg of beef protein requires 61.1 kg of grain, production of 1 kg of pork protein requires 38 kg of grain, while fish only requires 13.5 kg (Hall et al. 2011). Most of this difference comes from two biological characteristics of fish which give them great advantages over land-based livestock in growth performance: (i) the fact that fish are poikilotherms and therefore do not expend energy maintaining a constant body temperature; and (ii) the fact that, because finfish are physically supported by the aquatic medium, fewer resources are used on bony skeletal tissues, and a larger part of the food they eat is effectively allocated to body growth.

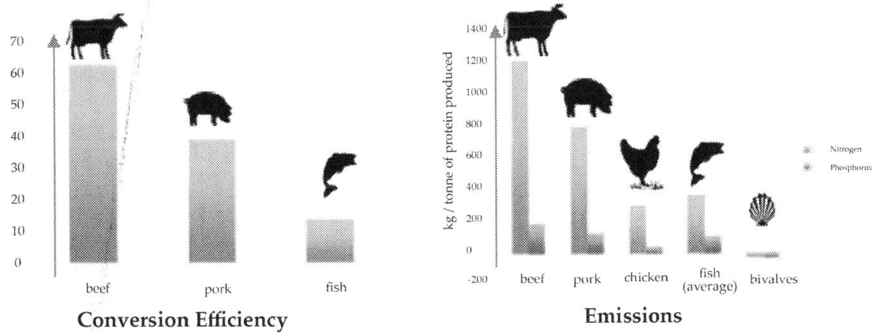

Source: Data for fish are derived from Hall et al. (2011). Data for beef, pork and chicken are derived from Flachowsky (2002) in Poštrk (2003).

Fig. 6: **a** Feed and protein conversion efficiency of beef, pork and fish; **b** Nitrogen and phosphorous emissions for animal production systems. Source: Data for fish are derived from Hall et al. (2011). Data for beef, pork and chicken are derived from Flachowsky (2002) in Poštrk (2003).

Third, in terms of carbon footprint, aquatic animal production systems have a lower carbon footprint per kilogram of output compared with other terrestrial animal production systems (Hall et al. 2011). As a consequence, nitrogen and phosphorous emissions (kg of nitrogen and phosphorus produced per tonne of protein produced) from aquaculture systems are much lower than those in beef and pork production systems and slightly higher than that of poultry (Fig. 6b). In fact, some aquaculture production systems such as farming of bivalves absorb nitrogen and phosphorous emissions from other systems.

All these reasons are important arguments for giving fish far greater attention in the food security debate and in the current discussion about how to feed 9 billion by 2050. The CFS has started to recognize this and the report that was commissioned in 2012 indeed recommends that fish (a) becomes an integral element in inter-sectoral national FSN policies and programmes and (b) should be more systematically included in countries' nutritional programmes and interventions aiming at tackling micronutrient deficiencies especially among children and women (HLPE 2014: 18 Recommendation 1a and 1b).

CONCLUSION

Fish is already making a major contribution to human food supply and to the support of FSN for more than 660 million fish-workers and their families. It also provides more than 4.5 billion consumers with at least 15 % of their average per capita intake of animal protein. In addition, because fish is more nutritious than staple foods such as cereals, providing in particular essential fatty acids and micronutrients, it can play an extremely important role in improving the nutritional status of individuals, in particular those at risk such as children and women.

Yet limited attention has been given so far to fish as a key element in FSN strategies at national level and in wider development discussions and interventions. Part of the problem might have been that specialists in fisheries debates have concentrated predominantly on questions of biological sustainability and on the economic efficiency of fisheries, neglecting issues linked to its contribution to reducing hunger and malnutrition and to supporting livelihoods (Kawarazuka and Béné 2010, 2011). On the other end of the spectrum, and with too few exceptions, most (non-fishery) food security experts and decision-makers seem unfamiliar with these facts and, therefore, unaware of the critical role that aquaculture is likely to play in the future. The problem is particularly pronounced in the current debate about how to make the food system more nutrition sensitive, i.e., how to change and improve the food systems in order to advance nutrition. As a consequence, fish has so far been only marginally included in the international FSN debate. Many nutritional programmes are still not aware of, or not recognizing and building on the potential of fish for the reduction of micronutrient deficiency.

In this paper, we make the case that fish deserves more attention in food policies due to its importance in the food basket, its unique nutritional properties, its higher efficiency of production and carbon footprint compared to other forms of animal production systems. We acknowledge some challenges, however, especially in making fish more affordable for the poor or in maintaining – or restoring – the environmental sustainability of the sectors. We also stress that the ability to meet the potential average fish demands of 9 billion people masks inequalities and inequities in who eats the fish and who benefits from the value chains. At present, people in Africa, the poor in many societies, and women and minority groups, including small-scale fishing and aquaculture communities are in tension with large corporations and production units over access to fish and fish-related employment (HLPE 2014). We have shown that the best available projections for fish supply and demand are relatively positive in terms of the capacity to meet future demands, although more solid modelling is still needed to better incorporate demand projections. Climate change impacts on fish production also create uncertainty in the projections.

In conclusion, fish should certainly be on the menu.

FOOTNOTES

1. Overall, small scale fisheries land approximately 40 Mt annually, compared to 50 Mt by large scale fisheries (HLPE 2014).

2. The report explains that "For the subsequent years in the simulation [after 2000 which was used as the base year for the calibration], these intercept values [between supply and demand] are changed according to the exogenous growth rates specified for each of the supply and demand functions" (World Bank 2014: 23, our emphasis). It is not clear from the report how the exogenous growth rates in regional demand have been computed in the model.

3. According to the FAO, the number of stocks fished at unsustainable levels decreased from 32.5 % in 2008 to 28.8 % in 2011 (FAO 2014).

4. For instance, one of the 6 selected scenarios of the World Bank-FAO-IFPRI's analysis is specifically focusing on these issues (Scenario 2 "Expanded Use of Fish Processing Waste in Fishmeal and Fish Oil Production") (World Bank 2014) and one of the selected issues of the 2012 FAO SOFiA report was on "Demand and supply of aquafeed and feed ingredients for farmed fish and crustaceans: trends and future prospects" (FAO 2012: 171–182).

5. Merino et al. (2012) estimate that the fish used in aquaculture feed to produce one unit of output would have to be reduced by at least 50 % from current levels to secure sustainability. If not, demand will push the price of fishmeal products up to levels where the short term economic incentive to exploit small pelagic (the main source of this fishmeal) beyond their maximum sustainable yield would be high, potentially leading to increases in fishing capacity and rapid depletion of resources.

6. Some other regions of the world are also expected to face lower fish

consumption rate per capita: Japan, Latin America, Europe, and Central Asia.

NOTES

Acknowledgments

This paper was part of a workshop sponsored by the OECD Co-operative Research Programme on Biological Resource Management for Sustainable Agricultural Systems.

REFERENCES

1. Allison, E. H. (2011). *Aquaculture, fisheries, poverty and food security. Working Paper 2011–65*. Penang: WorldFish Center.
2. Allison, E. H., Béné, C., & Andrew, N. L. (2011). Poverty reduction as a means to enhance resilience in small-scale fisheries. In R. S. Pomeroy & N. L. Andrew (Eds.), *Small-scale fisheries management – frameworks and approaches for the developing world* (pp. 216–238). Wallingford: CABI.
3. Allison, E. H., Delaporte, A., & Hellebrandt de Silva, D. (2013). *Integrating fisheries management and aquaculture development with food security and livelihoods for the poor. Report submitted to the Rockefeller Foundation.* Norwich: School of International Development, University of East Anglia.
4. Arthur, R., Béné, C., Leschen, W., & Little, D. (2013). Fisheries and aquaculture and their potential roles in development: an assessment of the current evidence. London, UK: Marine Resources Assessment Group Limited (MRAG). (http://r4d.dfid.gov.uk/pdf/outputs/fisheries/61091-Fisheries_and_Aqua_Evidence_Review.pdf).
5. Barange, M., Merino, G., Blanchard, J. L., Scholtens, J., Harle, J., Allison, E. H., et al. (2014). Impacts of climate change on marine ecosystem production in fisheries-dependent societies. *Nature Climate Change, 4*, 211–216.
6. Béné, C. (2006). *Small-scale fisheries: assessing their contribution to rural livelihoods in developing countries. FAO Fisheries Circular, No. 1008*. Rome: Food and Agriculture Organization (FAO).

7. Béné, C., & Friend, R. (2011). Poverty in small-scale inland fisheries: old issues, new analysis. *Progress in Development Studies, 11*(2), 119–144.

8. Béné, C., Macfadyen, G., & Allison, E. H. (2007). *Increasing the contribution of small-scale fisheries to poverty alleviation and food security. FAO Fisheries Technical Paper, No. 481*. Rome: FAO.

9. Béné, C., Steel, E., Kambala Luadia, B., & Gordon, A. (2009). Fish as the "bank in the water" - evidence from chronic-poor communities in Congo. *Food Policy, 34*, 104–118.

10. Béné, C., Hersoug, B., & Allison, E. H. (2010). "Not by rent alone": analysing the pro-poor functions of small-scale fisheries in developing countries. *Development Policy Review, 28*(3), 325–358.

11. Betru, S., & Kawashima, H. (2009). Patterns and determinants of meat consumption in urban and rural Ethiopia. *Livestock Research for Rural Development*, 21(9/143).

12. Beveridge, M. C. M., Thilsted, S. H., Phillips, M. J., Metian, M., Troell, M., & Hall, S. J. (2013). Meeting the food and nutrition needs of the poor: the role of fish and the opportunities and challenges emerging from the rise of aquaculture. *Journal of Fish Biology, 83*, 10671084.

13. Blanchard, J., Jennings, S., Holmes, R., Harle, J., Merino, G., Allen, I., et al. (2012). Potential consequences of climate change on primary production and fish production in 28 large marine ecosystems. *Philosophical Transactions of the Royal Society, B: Biological Sciences, 367*(1605), 2979–2989.

14. Bonham, M. P., Duffy, E. M., Robson, P. J., Wallace, J. M., Myers, G. J., Davidson, P. W., et al. (2009). Contribution of fish to intakes of micronutrients important for foetal development: a dietary survey of pregnant women in the Republic of Seychelles. *Public Health Nutrition, 12*(09), 1312–1320.

15. Boyd, C. E., Tucker, C., McNevin, A., Bostock, K., & Clay, J. (2007). Indicators of resource use efficiency and environmental performance in fish and crustacean aquaculture. *Reviews in Fisheries Science, 15*, 327–360.

16. Caddy, J. F., & Seijo, J. C. (2005). This is more difficult than we thought! The responsibility of scientists, managers and stakeholders

to mitigate the unsustainability of marine fisheries. *Philosophical Transactions of the Royal Society London B: Biological Sciences, 360*(1453), 59–75.

17. Cheung, W. W. L., Lam, V., Sarmiento, J., Kearney, K., Watson, R., Zeller, D., & Pauly, D. (2009). Large-scale redistribution of maximum fisheries catch potential in the global ocean under climate change. *Global Change Biology, 16*, 24–35.

18. Cheung, W. W. L., Dunne, J., & Sarmiento, J. L. P. D. (2011). Integrating ecophysiology and plankton dynamics into projected maximum fisheries catch potential under climate change in the Northeast Atlantic. *ICES Journal of Marine Science, 68*, 1008–1018.

19. de Schutter, O. (2012). *The right to food - Note to the General-Secretary from the Special Rapporteur on the right to food.* New York: United Nation, Sixty-seventh session General Assembly.

20. Delgado, C. L., Crosson, P., & Courbois, C. (1997). *The impact of livestock and fisheries on food availability and demand in 2020. MSSD Discussion Paper, No.19.* Washington, DC: IFPRI.

21. Eide, A., Bavinck, M., & Raakjęr, J. (2011). Avoiding poverty: Distributing wealth in fisheries. In J. Svein & A. Eide (Eds.), *Poverty mosaics: Realities and prospects in small-scale fisheries*(pp. 13–25). Dordrecht: Springer.

22. FAO. (2012). *The state of world fisheries and aquaculture 2012.* Rome: FAO.

23. FAO. (2014). *The state of world fisheries and aquaculture 2014.* Rome: FAO.

24. FAO (nd). Post-harvest losses in artisanal fisheries. Focus fisheries and food security. Rome: FAO. http://www.fao.org/focus/e/fisheries/proc.htm. Accessed Sept 2014.

25. Flachowsky, G. (2002). Efficiency of energy and nutrient use in the production of edible protein of animal origin. *Journal of Applied Animal Research, 22*(1), 1–24.

26. Garcia, S. M., & Grainger, R. J. R. (2005). Gloom and doom? The future of marine capture fisheries. *Philosophical Transactions of the Royal Society, B: Biological Sciences, 360*, 21–46.

27. Garcia, S. M., & Rosenberg, A. A. (2010). Food security and

marine capture fisheries: characteristics, trends, drivers and future perspectives. *Philosophical Transactions of the Royal Society, B: Biological Sciences, 365*(1554), 2869–2880.

28. Grafton, R. Q., Hilborn, R., Squires, D., & Williams, M. J. (2010). Marine conservation and fisheries management: At the crossroads pp. 3–19. In R. Q. Grafton, R. Hilborn, D. Squires, & M. J. Williams (Eds.), *Handbook of marine fisheries conservation and management*. Oxford: Oxford University Press.
29. Hall, S. J., Delaporte, A., Phillips, M. J., Beveridge, M., & O'Keefe, M. (2011). *Blue frontiers: managing the environmental costs of aquaculture*. Penang: The WorldFish Center.
30. Halpern, B. S., Walbridge, S., Selkoe, K., Kappel, C. V., Micheli, F., D'Agrosa, C., et al. (2008). A global map of human impact on marine ecosystems. *Science, 319*(5865), 948–952.
31. Hasan, M. R., & Halwart, M. (Eds.). (2009). *Fish as feed inputs for aquaculture; practices sustainability and implications. FAO Fisheries and Aquaculture Technical Paper. No. 518*. Rome: FAO.
32. Heck, S., Béné, C., & Reyes-Gaskin, R. (2007). Investing in African fisheries: building links to the millennium development goals. *Fish and Fisheries, 8*(3), 211–226.
33. High Level Panel of Experts. (2014). *Sustainable fisheries and aquaculture for food security and nutrition. A report by the high level panel of experts on food security and nutrition of the committee on world food security*. Rome: FAO.
34. Hilborn, R. (2010). Apocalypse forestalled: why all the world's fisheries aren't collapsing. *Science Chronicles*, 5–9.
35. Hilborn, R. (2013). Environmental cost of conservation victories. *Proceedings of the National Academy of Sciences, 110*(23), 9187.
36. Kawarazuka, N., & Béné, C. (2010). Linking small-scale fisheries and aquaculture to household nutritional security: a review of the literature. *Food Security, 2*(4), 343–357.
37. Kawarazuka, N., & Béné, C. (2011). The potential role of small fish species in improving micronutrient deficiencies in developing countries: building evidence. *Public Health Nutrition, 14*(11), 1927–1938.

38. Kelleher, K. (2005). *Discards in the world's marine fisheries – an update. FAO Fisheries Technical Paper. No. 470*. Rome: FAO.

39. Larsen, R., Eilertsen, K., & Elvevoll, E. O. (2011). Health benefits of marine foods and ingredients. *Biotechnology Advances, 29*, 508–518.

40. Lem, A., Bjorndal, T., & Lappo, A. (2014). *Economic analysis of supply and demand for food up to 2030 – Special focus on fish and fishery products. FAO Fisheries and Aquaculture Circular, No. 1089*. FAO: Rome.

41. Merino, G., Barange, M., Mullon, C., & Rodwell, L. (2010). Impacts of global environmental change and aquaculture expansion on marine ecosystems. *Global Environmental Change, 20*, 586–596.

42. Merino, G., Barange, M., Blanchard, J. L., Harle, J., Holmes, R., Allen, I., et al. (2012). Can marine fisheries and aquaculture meet fish demand from a growing human population in a changing climate? *Global Environmental Change, 22*, 795–806.

43. Miles, E. A., & Calder, P. C. (2012). Influence of marine n-3 polyunsaturated fatty acids on immune function and a systematic review of their effects on clinical outcomes in rheumatoid arthritis. *British Journal of Nutrition, 107*(Supplement S2), S171–S184.

44. Myers, R. A., & Worm, B. (2003). Rapid worldwide depletion of predatory fish communities. *Nature, 423*, 280–283.

45. Neiland, A.E., & Béné, C. (2004). *Poverty and small-scale fisheries in West Africa* (eds.). Published by Kluwer Academic Publishers for the Food and Agriculture Organization, 254 p.

46. OECD-FAO (2013). Agricultural outlook 2013–2022. Paris: Organisation for Economic Co-operation and Development (OECD) and the Food and Agriculture Organization (FAO) of the United Nations.

47. Pauly, D. (2009). Aquacalypse now: the end of fish. *The New Republic, 240*, 24–27.

48. Pauly, D., Christensen, V., Dalsgaard, J., Froese, R., & Torres, F. (1998). Fishing down marine food webs. *Science, 279*(5352), 860–863.

49. Pauly, D., Watson, R., & Alder, J. (2005). Global trends in world fisheries: impacts on marine ecosystems and food security. *Philosophical Transactions of the Royal Society, B: Biological Sciences, 360*(1453), 5–12.

50. Poštrk, V. (2003). The livestock revolution: dietary transition: global rise in consumption of animal food products. *Environmental Science*. Lund. Master: 50 pp. Lund, Sweden.

51. Prein, M., & Ahmed, M. (2000). Integration of aquaculture into smallholder farming systems for improved food security and household nutrition. *Food and Nutrition Bulletin, 21*(4), 466–471.

52. Rangel-Huerta, O. D. R., Aguilera, C. M., Mesa, M. D., & Gil, A. (2012). Omega-3 long-chain polyunsaturated fatty acids supplementation on inflammatory biomakers: a systematic review of randomised clinical trials. *British Journal of Nutrition, 107*(Supplement S2), S159–S170.

53. Rice, J. C., & Garcia, S. M. (2011). Fisheries, food security, climate change, and biodiversity: characteristics of the sector and perspective on emerging issues. *ICES Journal of Marine Science, 68*(6), 1343–1353.

54. Richardson, A. J., & Montgomery, P. (2005). The Oxford-Durham study: a randomized, controlled trial of dietary supplementation with fatty acids in children with developmental coordination disorder. *Pediatrics, 115*(5), 1360–1366.

55. Roos, N., Islam, M. M., & Thilsted, S. H. (2003). Small indigenous fish species in Bangladesh: contribution to vitamin A, calcium and iron intakes. *Journal of Nutrition, 133*, 4021S–40126S.

56. Roos, N., Wahab, M. A., Chamnan, C., & Thilsted, S. H. (2007). The role of fish in food-based strategies to combat Vitamin A and mineral deficiencies in developing countries. *The Journal of Nutrition, 137*(4), 1106–1109.

57. Shepherd, C. J., & Jackson, A. J. (2013). Global fishmeal and fish-oil supply: inputs, outputs and markets. *Journal of Fish Biology, 83*(4), 1046–1066.

58. Smith, M. D., Roheim, C. A., Crowder, L. B., Halpern, B., Turnipseed, M., Anderson, J., et al. (2010). Sustainability and global seafood. *Science, 327*(5967), 784–786.

59. Speedy, A. W. (2003). Global production and consumption of animal source foods. *The Journal of Nutrition, 133*(11), 4048S–4053S.

60. Srinivasan, U. T., Cheung, W. W. L., Watson, R., & Sumaila, U. R. (2010). Food security implications of global marine catch losses due to overfishing. *Journal of Bioeconomics, 12*(3), 183–200.

61. Steinacher, M., Joos, F., Frölicher, T. L., Bopp, L., Cadule, P., Cocco, V., et al. (2010). Projected 21st century decrease in marine productivity: a multi-model analysis. *Biogeosciences, 7*, 979–1005.

62. Tacon, A. G. J., & Metian, M. (2008). Global overview on the use of fish meal and fish oil in industrially compounded aquafeeds: trends and future prospects. *Aquaculture, 285*, 146–158.

63. Tacon, A. G. J., & Metian, T. M. (2009). Fishing for feed or fishing for food: increasing global competition for small pelagic forage fish. *AMBIO: A Journal of the Human Environment, 38*(6), 294–30.

64. Tacon, A. G. J., Hasan, M. R., & Metian, M. (2011). *Demand and supply of feed ingredients for farmed fish and crustaceans: trends and prospects. FAO Fisheries and Aquaculture Technical Paper. No. 564*. Rome: FAO.

65. Thilsted, S. H. (2012). The potential of nutrient-rich small fish species in aquaculture to improve human nutrition and health. In R. P. Subasinghe, J. R. Arthur, D. M. Bartley, S. S. de Silva, M. Halwart, N. Hishamunda, et al. (Eds.) (2010), *Farming the waters for people and food*. Proceedings of the Global Conference on Aquaculture. Phuket, Thailand, (pp. 57–73). Rome: FAO and Bangkok: NACA.

66. Thilsted, S. H., Roos, N. & Hassan, N. (1997). The role of small indigenous fish species in food and nutrition security in Bangladesh. *WorldFish Centre Quarterly*, 82–84.

67. Toufique, K. A., & Belton, B. (2014). Is aquaculture pro-poor? Empirical evidence of impacts on fish consumption in Bangladesh. *World Development, 64*, 609–620.

68. Troell, M., Naylor, R. L., Metian, M., Beveridge, M., Tyedmers, P. H., Folke, C., et al. (2014). Does aquaculture add resilience to the global food system? *Proceedings of the National Academy of Sciences, 111*(37), 13257–13263.

69. UNHRC (2012). The Rights to Fish for Food. New York, NY: United Nations Human Rights Commission. http://www.srfood.org/images/stories/pdf/officialreports/20121030_fish_en.pdf/.

70. USDA (United State Department of Agriculture) (2011). *National nutrient data base*(http://ndb.nal.usda.gov/).

71. Williams, M. J. (2010). Gender dimensions in fisheries management. In R. Q. Grafton, R. Hilborn, D. Squires, M. Tait, & M. Williams (Eds.), *Handbook in marine fisheries conservation and management* (pp. 72–86). Oxford: Oxford University Press.

72. World Bank. (2014). *Fish to 2030 Prospects for Fisheries and Aquaculture. World Bank Report, No. 83177-GLB*. Washington, DC: World Bank.

73. World Bank/FAO/WorldFish. (2012). *Hidden harvest: The global contribution of capture fisheries. World Bank Report, No. 66469-GLB*. Washington: World Bank.

74. Worm, B., Barbier, E. B., Beaumont, N., Duffy, E., Folke, C., Halpern, B. S., et al. (2006). Impacts of biodiversity loss on ocean ecosystem services. *Science, 314*(5800), 787–790.

SECTION III

DISEASES

CHAPTER

3

SURVEILLANCE OF FISH DISEASES IN THE NORDIC COUNTRIES

T. Håstein[1], A. Hellstrøm[2], G. Jonsson[3], N. J. Olesen[4], E. R.- Pärnänen[5]

[1]National Veterinary Institute, Norway,

[2]National Veterinary Institute, Sweden,

[3]Veterinary Officer for Fish Diseases, Keldur, Vesturlandsveg, IS-112, Reykjavik, Iceland,

[4]Danish Veterinary Laboratory, Aarhus, Denmark,

[5]National Veterinary and Food Research Institute (EELA), Finland.

ABSTRACT

Due to the increasing importance of disease problems in the fish farming industry and the impact disease may have on both feral and farmed fish in the Nordic countries, monitoring and surveillance on diseases have for many years been considered to be of socioeconomic importance. All the Nordic countries have a national legislation as basis for their surveillance and disease control in aquatic animals and regulations listing notifiable diseases of concern to the countries. The list of diseases vary between the countries. In addition, Denmark, Finland and Sweden are ruled by [3] as regards placing on the market of aquaculture animals and products. The surveillance for viral diseases in all the Nordic countries has mainly been based on the testing procedures given in the EU Commission [2].

The list of diseases to be monitored or surveyed for varies between the Nordic countries. All the Nordic countries run a surveillance programme for Viral Haemorrhagic Septicaemia (VHS) and Infectious Haematopoietic Necrosis (IHN) based on EU regulations and a monitoring programme for Infectious Pancreatic Necrosis (IPN) on a national level. Iceland and Norway have also established a surveillance programme for Viral Nervous Necrosis (VNN) on a national level, while Finland and Sweden has established a surveillance programme for Spring Viraemia of carp (SVC) on a national level in order to obtain additional guarantees within the EU for that disease. Norway is the only country that has established a control programme for *Herpesvirus scophthalmi* in turbot, for *Anguillicola* spp in eel and for pasteurellosis in sea bass. These programmes are due to start year 2000. Monitoring of Bacterial Kidney Disease (BKD) takes place in all the Nordic countries; in some of the countries it is being compulsory and in others it is carried out on a voluntary basis. Due to the importance of the haptorworm *Gyrodactylus salaris* for wild stocks of Atlantic salmon, a surveillance programme for this parasite is under consideration in Norway, while in Finland regular monitoring for the parasite takes place in certain regions recognised to be free of the parasite in order to obtain additional guarantees within the EU. Similarly, Finland, Iceland and Sweden run a national control programme for furunculosis for the same reason and Finland has asked for additional guarantees for certain areas in which the disease do not occur.

The measures taken when disease occur vary from restrictions on movements to stamping out procedures as well as disinfection procedures and other means to establish control of a disease. The ultimate goal is to eradicate or keep the level of disease to a minimum.

INTRODUCTION

On a global scale, fish and fishery products are the main food supply for human beings and these products constitutes more than 80% of the total amount of seafood consumed. It is widely known that the supplies of fish from traditional fisheries are more or less constant [4] and that the shortage in fish and fish products has to be met by aquaculture. The potential for the aquaculture industry to meet the challenge as regards food security has been clearly demonstrated by the rapid expansion of the fish farming industry world wide [14]. Fig. 1 shows the development of aquaculture compared to global fisheries in the period 1987–1997 (FAO). The figures show that the total capture in fisheries is relatively constant at some 90 millions metric tonnes while there has been an increase in aquaculture production from 10 millions metric tonnes to 28 millions metric tonnes in the same period.

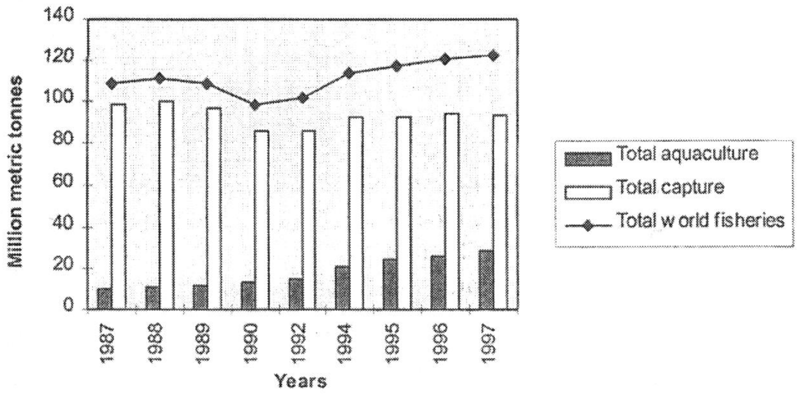

Figure 1: Aquaculture and global fisheries production (FAO information).

The way fish farming is conducted and the species farmed vary considerably in the world depending on geographical conditions, water resources, temperature etc.

FISH FARMING IN THE NORDIC COUNTRIES

In the Nordic countries, the fish farming industry is mainly based on salmonid species, Atlantic salmon (*Salmo salar*) and rainbow trout (*Oncorhynchus mykiss*), but in the later years several different marine species such as halibut (*Hippoglossus hippoglossus*), turbot (*Scophthalmus maximus*) and sea bass (*Dicentrarchus labrax*) have gained increased importance.

Table 1 gives an overview over the fish species farmed in the Nordic countries as well as the size and production of these species.

Table 1: Fish farming in the Nordic countries; No. of farms, species and production

Country	No. of farms	Species kept under farming conditions	Total production all species Metric tons
Denmark	478	Eel, Rainbow trout	46.000
Finland	333	Arctic char, Baltic salmon rainbow trout, whitefish	16.597
Iceland	40	Atlantic salmon, Arctic char, rainbow trout, halibut, Sea bass	3.669
Norway	1169 (salmonids)	Atlantic salmon, Arctic char, brown trout, Cod, eel, halibut, turbot, sea bass	390.764
	381 (others)		
Sweden	177	Arctic char, Baltic salmon, rainbow trout, grass carp	6.200

THE RATIONAL FOR DISEASE SURVEILLANCE

Over the last decades, several emerging or serious diseases in fish have been diagnosed in farmed and feral populations, creating large problems in the fish farming industry and thus being the subject of surveillance and monitoring programmes in many countries [6]. The differences between infectious diseases in fish and those of terrestrial animals means that the approach to the problems and the eradication efforts differ as the diseases may spread effectively through flowing water [7].

Following is the described basis for and the results of monitoring and surveillance programmes as well as established measures on some of the diseases considered to be of major concern in the Nordic countries.

BASIS FOR THE MONITORING AND SURVEILLANCE IN THE NORDIC COUNTRIES

All the Nordic countries have a national legislation as basis for their surveillance and disease control in aquatic animals, as well as regulations listing notifiable diseases of concern to the countries. The list of notifiable diseases vary from country to country. In addition to their national legislation, Denmark, Finland and Sweden are ruled by [3] as regards placing on the market of aquaculture animals and products, but even Iceland and Norway have accepted the principles laid down in the Directive.

The monitoring and surveillance for viral diseases in all the Nordic countries has mainly been based on the testing procedures given in the EU Commission [2] while for the bacterial and parasitic diseases, standard diagnostic procedures for such diseases has been used for screening purposes.

The list of diseases under monitoring and surveillance varies between the Nordic countries. All the countries runs a surveillance programme for Viral Haemorrhagic Septicaemia (VHS) and Infectious Haematopoietic Necrosis (IHN) based on EU regulations [1] and a monitoring programme for Infectious Pancreatic Necrosis (IPN) on a national level. The Nordic countries have also a clinical monitoring for infectious salmon anaemia (ISA) as no valididated surveillance methods so far have been established. Iceland and Norway has also established a surveillance programme for Viral Nervous Necrosis (VNN) on a national level, while Finland and Sweden have established a surveillance programme for Spring Viraemia in carp (SVC) on a national level in order to obtain additional guarantees within the EU for that disease. Norway is the only country that have established a control programme for *Herpesvirus scophthalmi* in turbot, for Anguillicola spp in eel and for pasteurellosis in sea bass which will start year 2000.

Monitoring for Bacterial Kidney Disease (BKD) takes place in all the Nordic countries; in some of the countries being compulsory and in others carried out on a voluntary level.

Due to the importance of the haptorworm *Gyrodactylus salaris* for wild stocks of Atlantic salmon, a surveillance programme for this parasite is under consideration in Norway, while in Finland regular monitoring for the parasite takes place in certain regions recognised to be free of the parasite to obtain additional guarantees within the EU. Similarly Finland, Iceland and Sweden runs a national control programme as regards furunculosis for the same reason and Finland has asked for additional guarantees for certain areas in which the disease due not occur.

AIM OF SURVEILLANCE

The aim of the monitoring and surveillance programmes for fish diseases in question in the Nordic countries is either to document and maintain freedom of disease, to eradicate a disease or to keep a disease under control within certain bonds. For some of the diseases that may be present on a low level such as IPN in Finland and Sweden, the ultimate aim will probably be eradication.

SURVEILLANCE PROGRAMMES AND TARGET POPULATIONS

The main target population for the monitoring and surveillance programmes is salmonids for most of the diseases under surveillance.

The size of the target populations is given in Table 1. All types of farms are included in the survey such as hatcheries, brood stock farms as well as grow out farms, table 2 shows the diseases under surveillance and the target species for the disease in question.

Table 2: Diseases and target species under surveillance in the Nordic countries

Disease	Target species	Country
Viral Diseases		
Infectious salmon anaemia (ISA)	Salmonids	Denmark, Finland, Iceland, Norway, Sweden
Infectious pancreatic necrosis (IPN)	Salmonids	Denmark, Finland, Iceland, Norway, Sweden
Infectious haematopoietic necrosis (IHN)	Salmonids	Denmark, Finland, Iceland, Norway, Sweden
Viral haemorrhagic septicaemia (VHS)	Salmonids	Denmark, Finland, Iceland, Norway, Sweden
Spring viremia in Carp (SVC)	Carp	Finland, Sweden
Viral nervous necrosis (VNN)	Halibut	Iceland, Norway,
Herpesvirus scophthalmi infection	Turbot	Norway
Bacterial diseases		
Bacterial kidney disease (BKD)	Salmonids	Denmark, Finland, Iceland, Norway, Sweden
Furunculosis	Salmonids	Finland, Iceland, Sweden

Pasteurellosis	Sea bass	Norway
Parasitic diseases		
Gyrodactylus salaris	Salmonids	Finland, Norway
Anguillicola spp	Eel	Norway

Denmark, Norway and Scotland run a surveillance programme with the aim of deriving information on VHS virus and other viral infections in wild marine populations is carried out in European coastal waters from the English channel in the south to the Spitsbergen area as well as in the Baltic sea [9, 12]. Similar examinations were carried out by Sweden in the Baltic sea and Kattegat in 1998 and the beginning of 1999.

ORGANIZATION

The basis for the different surveillance and monitoring programmes in formal terms is partly based on EU regulations, OIE criteria or criteria derived from national legislation. For some of the diseases such as IHN and VHS, the participation is compulsory in all the Nordic countries as regards approval and maintenance of disease free status for the two diseases either for the whole country, for zones or for farms in non approved zones. In Denmark a control program for VHS started as early as in 1965, while in Finland, Iceland, Norway and Sweden an official surveillance program started in 1995, 1993, 1994 and 1993 respectively. For other diseases such as furunculosis the participation is compulsory in some countries such as Sweden, while it is voluntary or even non existing on a regular basis in other countries such as Norway.

In Finland, the surveillance programme for *G. salaris* has been compulsory for the last three years.

SAMPLING

The sampling of fish for surveillance vary between the Nordic countries depending on the infrastructure of the health control in the different countries. In Sweden the National Fish Health Service (Fiskhälsan FH AB) are responsible for all sampling, while in Norway, the responsible Authorithy is the district veterinary officers (DVO) or local fish health services operating under instruction of the DVO. In Finland the municipal veterinarians do the necessary sampling when inspecting the farms which is done biannually.

In Iceland, the responsible authority as regards sampling is under the supervision of the Veterinary Officer for Fish Diseases.

In Denmark, the Danish Veterinary and Food Administration do the necessary sampling and control.

Analyses of the diseases under surveillance are carried out at official veterinary laboratories i.e. National Veterinary Institute in Norway and Sweden, the Veterinary Officer for Fish Diseases, Keldur, in Iceland, Danish Veterinary Laboratory in Denmark and National Veterinary and Food Research Institute (EELA) in Finland.

TEST AND TESTING PROCEDURES

The surveillance for viral diseases in all the Nordic countries has mainly been based on the testing procedures given in the EU Commission [2]. According to the decision, fish farms are inspected clinically biannually and samples for virological examinations are collected by rotation in 50% of the fish farms each year to document freedom for IHN and VHS. In Denmark, Finland, Iceland and Sweden, all or some of the same samples are used for IPN testing. In Denmark, the control of IPN is based on the monitoring of IPN free broodstock farms and ongrowing farms as IPN is considered to be endemic in non controlled farms. In Norway testing for IPN virus is done in a specific monitoring programme in brood stocks of Atlantic salmon and rainbow trout as IPN virus neutralisation is used in the examinations for IHN/VHS virus.

The examination procedures given in the [10] are the basis for examinations as regards BKD, *Gyrodactylus salaris* and VNN while for furunculosis, pasteurellosis and *Anguillicola* spp, standard bacteriological or parasitological procedures are used, respectively. The frequency in the testing of these diseases is varying between the countries.

MEASURES TAKEN WHEN POSITIVE DIAGNOSIS OCCUR

In all the Nordic countries, it is the Veterinary Authorities that is responsible for the implementation of measures that will be used in order to control a given notifiable disease. The implementation involves both central and regional veterinary officers.

In all the Nordic countries except non-approved zones in Denmark, stamping out procedures followed by cleaning, disinfection and fallowing will be carried out if VHS is diagnosed [11, 13]. In non-approved zones in Denmark, a stamping out program will be started if 1/3 of the farms in a river system agrees to an eradication program to get rid of the disease.

Since IHN and SVC is considered to be absent in all the Nordic countries, stamping out procedures will be applied to these diseases. As regards IPN, the detection of the disease or the virus as such, will be subject to restrictions followed by a sanitation programme laid down by the Board of Agriculture in Sweden, while in Norway the mere detection of IPN virus will not bring about any restrictions. Clinical IPN will, however, lead to restrictions on movement of live fish in Norway as long as mortality and signs of the disease occur. In Finland, isolation of IPN virus will lead to restrictions on movement of live fish in continental zones, while clinical IPN probably will result in more strong measures. These have so far not been determined. Prevention may be achieved by avoiding introduction of disease free eggs and/or fish into disease free farms as well as using protected water supply (e.g. spring-, borehole water).

In Iceland, detection of furunculosis is also followed by stamping out because the disease is recognised as a List A disease, while in Sweden active measures are taken only if the disease occur in freshwater, but not in sea water where the disease is considered to be enzootic. In Norway, restrictions are laid upon farms if the furunculosis is diagnosed. Hatcheries are not allowed to sell live fish while affected fish farms in sea water may keep the fish until reaching marketable size for slaughtering.

In all the Nordic countries an infected farm may restock after fallowing if no signs of infection appear after a sanitation programme has been carried out.

ECONOMIC COMPENSATION

In Sweden, all expenses will be paid by the Board of Agriculture for IHN, VHS, SVC and IPN as these diseases are included in the epizootic act. Similarly, in Finland, the government will pay for all expenses if IHN and VHS are detected, probably also for SVC. In Norway affected farm(s) will have to pay themselves for any measures imposed by the authorities for the time being, because no compensation is granted.

In Denmark expenses will probably be paid in case of IHN and ISA outbreaks and in case of a VHS outbreak in approved zones free of VHS, expenses might be covered as well.

CERTIFICATION

Health certificates and/or transportation documents is needed in connection with deliveries of live fish for stocking into grow out farms and restocking into rivers. Additionally, in Norway it is not allowed to move fish prior to slaughter, when they are stocked in a sea water site. This is to prevent spread of disease if disease occur.

RECORD KEEPING

In all the Nordic countries, the record of findings are kept by the responsible authorities, both regionally and centrally. The diagnostic laboratories also keep the necessary documentation on the examinations carried out. Furthermore all farms have to keep records on events in the farms that can be requested by Competent Responsible Authorithy.

CURRENT STATUS

While still a few cases of VHS is diagnosed in Denmark each year, Finland, Iceland, Norway and Sweden are considered to be free from VHS. However, VHS was diagnosed in one farm in Norway and Sweden in 1998. All Nordic countries have status as free from IHN and SVC although only Finland and Sweden is carrying out a specific surveillance programme for SVC.

IPN is considered to be the main viral disease problem in Norway. The disease is endemic in Denmark but not considered as a mayor problem. The disease level of IPN is low in Finland and Sweden and so far not reported in Iceland. Iceland is considered to be free from classical furunculosis, while the disease appears to be endemic in Denmark, Finland, Norway and Sweden. However, some areas in Finland are reported free of furunculosis and the authorities has established so-called protective zones to which live fish and fish eggs only can be transported if special permission is granted.

BKD has been reported from all the Nordic countries but the level appear to be relatively low [5, 8]. In Iceland. the disease has occasionally been detected in wild Atlantic salmon returning to spawning grounds while it has not been detected in farms for the last 4 years. BKD was for the first time detected in wild salmon in the Baltic sea in 1999, so it is quite apparent that a reservoir of the disease exists in the wild.

Gyrodactylus salaris has been described from Denmark, Finland, Norway and Sweden, but it is only in Norway that disease caused by this parasite has given high mortality in feral fish populations in Norway. The parasite has now totally been detected in 40 rivers and 37 farms/hatcheries since its first detection.

Anguillicola crassus has so far been reported from Denmark, Norway and Sweden, but the distribution in the wild is not known.

CONCLUSION

Although some differences occur, all the Nordic countries have established appropriate surveillance and monitoring for fish diseases of concern to the fish farming industry. Due to these systems and good management practises, the fish disease situation in the Nordic countries is generally good compared to other countries in the world.

REFERENCES

1. Ariel E, Helgason S, Mortensen HF, Olesen NJ: Report of the Third annual Meeting of EU National Reference Laboratories for Fish Diseases, Aarhus, Denmark 9th–11th June 1999. 1999, 31-
2. Decision 96/240/EC: Laying down the sampling plans and diagnostic methods for the detection and confirmation of certain fish diseases.
3. Directive 91/67/EEC: Placing on the market of aquaculture animals and products.
4. FAO Aquaculture production and statistics 1988–1997: Fishery Information. Data and Statistics Unit, FAO Fisheries Department, 203-
5. Heuer OE, Lorenzen E, Korsholm H, Hansen RG, Olesen NJ: BKD i Danmark: Status efter to sæsoner. Ferskvandsfiskeribladet. 1999, 97 (5): 116-118.
6. Håstein T: Disease problems, use of drugs, resistance problems and preventive measures in fish farming world wide. Sustainable Fish

Farming. Edited by: Reinertsen H, Haaland H. 1995, 183-194.

7. Håstein T, Hill BJ, Winton JR: Succesful aquatic animal disease emergency programmes. Rev sci tech Off int Epiz. 1999, 18 (1): 214-227.
8. Lorenzen E, Korsholm H, Olesen NJ, Heuer OE: Påvisning af bakteriel nyresyge (BKD) hos regnbueørred i Danmark. Ferskvandsfiskeribladet. 1997, 95 (7): 160-168.
9. Mortensen HF, Heuer OE, Lorenzen N, Otte L, Olesen NJ: Isolation of Viral Haemorrhagic Septicaemia Virus (VHSV) from Wild Marine Fish Species in the Baltic Sea, Kattegat, Skagerrak and the North Sea. Virus Research. 1999, 63: 95-106. 10.1016/S0168-1702(99)00062-3.
10. OIE Diagnostic Manual for Aquatic Animal Diseases: 1999, 251-
11. Olesen NJ, Korsholm H: Control measures for viral diseases in aquaculture: Eradication of VHS and IHN. Bull Eur Ass Fish Pathol. 1997, 17: 229-233.
12. Olesen NJ: Isolering og karakterisering af VHS virus fra saltvandsfisk. Ferskvandsfiskeribladet. 1997, 95 (6): 129-130.
13. Olesen NJ: Sanitation of Viral Haemorrhagic Septicaemia (VHS). J Appl Ichthyol. 1998, 14: 173-177.
14. Roberts RJ, Muir JF: 25 years of world aquaculture: Sustainability, a global problem. Sustainable Fish Farming. Edited by: Reinertsen H, Haaland H. 1995, 167-181.

CHAPTER

4

A MONOGENEAN FISH PARASITE, GYRODACTYLUS CHILEANI N. SP., BELONGING TO A NOVEL MARINE SPECIES LINEAGE FOUND IN THE SOUTH-EASTERN PACIFIC AND THE MEDITERRANEAN AND NORTH SEAS

Marek S. Ziętara[1], Dar'ya Lebedeva[2], Gabriela Muñoz[3], Jaakko Lumme[4]

[1]Department of Molecular EvolutionUniversity of GdańskGdańskPoland

[2]Institute of BiologyKarelian Research Centre, RASPetrozavodskRepublic of Karelia, Russia

[3]Facultad de Ciencias del Mar y de Recursos NaturalesUniversidad de ValparaísoViña del MarChile

[4]Department of BiologyUniversity of OuluOuluFinland

ABSTRACT

Gyrodactyluschileani n. sp. is the first *Gyrodactylus* species reported from Chile. It is an ectoparasite living on fins and skin of a small fish, the Chilean tidal pond dweller *Helcogrammoides chilensis* (Cancino) (Perciformes: Tripterygiidae). A phylogenetic analysis based on 5.8S+ITS2 of rDNA placed the new species close to marine *Gyrodactylus* species found in Europe: *G. orecchiae* Paladini, Cable, Fioravanti, Faria, Cave & Shinn, 2009 on gilthead seabream *Sparus aurata* L. from the

Adriatic and Tyrrhenian Sea fish farms (Perciformes: Sparidae), and an undescribed species on the black goby *Gobius niger* L. from the North Sea (Perciformes: Gobiidae). A morphological description of the latter species is unavailable. These geographically distant parasite samples on different host families form a new well supported *Gyrodactylus orecchiae* lineage. Using molecular phylogenetics, it is shown that the marine species groups of *Gyrodactylus* may have a worldwide distribution.

INTRODUCTION

The monogenean flatworm genus *Gyrodactylus* Nordmann, 1832 is one of the most species-rich genera. The number of described species was 409 in the latest checklist (Harris et al., 2004; see also GyroDB - http://www.gyrodb.net/), but the real number of species is estimated to be about 20,000 (Bakke et al., 2002), which is certainly an underestimate if the resolution of molecular methods is utilised and the classical criteria for delineating species are retained (Ziętara & Lumme, 2003).

Malmberg (1970) subdivided the genus into six subgenera. Two of these subgenera remain perfectly valid, i.e. those infecting Eurasian freshwater fish *G.* (*Limnonephrotus*) Malmberg, 1964 and *G.* (*Gyrodactylus*) Nordmann, 1832. The four other subgenera (*Metanephrotus*Malmberg, 1964, *Mesonephrotus* Malmberg, 1964, *Neonephrotus* Malmberg, 1964 and *Paranephrotus* Malmberg, 1964) are on the other hand paraphyletic or polyphyletic using molecular criteria (Matějusová et al., 2003; Vanhove et al., 2011).

The genus itself is not monophyletic and it is recognised to be a catch-all taxon within the Gyrodactylidae Beneden & Hesse, 1864 (Kritsky & Boeger, 2003). The genera: *Acanthoplacatus*Ernst, Jones & Whittington, 2001, *Diplogyrodactylus* Přikrylová, Matějusová, Musilová, Gelnar & Harris, 2009, *Fundulotrema* Kritsky & Thatcher, 1977, *Gyrdicotylus* Vercammen-Grandjean, 1960, *Gyrodactyloides* Bykhovskiy, 1947 and *Macrogyrodactylus* Malmberg, 1957, subgenera and species groups or single species, are all paraphyletic within the family Gyrodactylidae (see Vanhove et al., 2011).

The mechanism resulting in such a phylogeny is the basal radiation of *Gyrodactylus* (*sensu lato*) generating new species groups, some marine, some freshwater, and some inhabiting both environments. The latter are perhaps the most interesting, and include examples where the host lineage has moved from marine to freshwater, and the parasite has followed. The best known examples are two species, *G. lotae* Gusev, 1953 and *G. alexgusevi* Ziętara & Lumme, 2003, on the freshwater burbot *Lota lota* (L.), a gadid, and two species, *G. hrabei* Ergens, 1957 and *G. mariannae* Winger, Hansen, Bachmann & Bakke, 2008, on an inland cottid, *Cottus poecilopus* Heckel, in Europe. The nearest relatives of all four of these species are parasites of marine hosts (Ziętara & Lumme, 2003; Rokicka et al., 2009).

To enlarge the geographical range of the *Gyrodactylus* spp., we describe the first *Gyrodactylus*species from Chile, which was among 215 fish ectoparasitic species listed by Muñoz & Olmos (2007). Two other records of *Gyrodactylus* spp. on small intertidal fishes found in rocky areas, *Scartichthysviridis* (Valenciennes) (see Muñoz & Randhawa, 2011) and *Sicyasessanguineus*Müller & Troschel (see Muñoz & Zamora, 2011), await molecular analyses and further description. The species described below is also the first molecularly defined *Gyrodactylus*reported from the Southern Pacific. Together with two other species of *Gyrodactylus* from Europe, which are related by ITS rDNA, it forms a new marine species group, which has crossed the equator during its evolutionary history.

MATERIALS AND METHODS

Fish were collected from rocky pools during low tides in July, 2009. Nineteen of 32 specimens were infected (prevalence 59.4%). Five ethanol-preserved worms were picked from the fins and skin of different hosts. The haptor was cut off, softened and cleared in 120 μg/ml proteinase K. It was then prepared for a microscopy on a slide with a saturated concentration of ammonium picrate in glycerine (Malmberg 1970). The remainder of the body was used for molecular analysis.

The molecular methods used in the present study were as previously described (Ziętara & Lumme, 2002, 2003; Ziętara et al., 2008). The ITS rDNA fragment encompassing ITS1-5.8S-ITS2 and short fragments of flanking 18S (15 bp) and 28S rRNA (9 bp) genes was amplified with primers *ITS1F* (5′-GTTTC CGTAG GTGAA CCT-3′) and *ITS2R*

(5′-GGTAA TCACG CTTGA ATC-3′) and sequenced with two additional primers *ITS1R* (5′-ATTTG CGTTC GAGAG ACCG-3′) and *ITS2F* (TGGTG GATCA CTCGG CTCA-3′). The PCR program was 3 min at 95°C and 35 cycles of 40 sec at 94°C, 30 sec at 48°C and 1 min at 72°C, followed by a final elongation step of 7 min at 72°C and an indefinite hold at 4°C.

The phylogenetic placement of the new species was estimated by aligning the 5.8S+ITS2 segment of the ribosomal DNA with selected sequences published in GenBank. Only predominantly marine lineages of gyrodactylid species were included. The alignment was made using MUSCLE (Edgar, 2004) as implemented in MEGA5 (Tamura et al., 2011) and corrected manually. The phylogenetic tree was estimated from Maximum Composite Likelihood distances (Tamura et al., 2004) using the Neighbor Joining algorithm (Saitou & Nei, 1987) with the pairwise deletion option. The validity of the branches was evaluated by bootstrapping 500 replicates (Felsenstein, 1985). Bootstrap values lower than 70% were hidden.

For the species description, the preserved haptors were photographed and measured under a microscope using the measurements of Gusev et al. (1985). Measurements of *G. orecchiae*Paladini, Cable, Fioravanti, Faria, Cave & Shinn, 2009 by Paladini et al. (2009) were included for comparison. The North Sea species on *Gobius niger* remains undescribed (Huyse et al., 2003).

PHYLOGENETIC CHARACTERISATION USING THE 5.8S+ITS2 RDNA FRAGMENT

As this report focuses on the new species from Chile and its nearest relatives, we present a phylogeny based on 5.8S+ITS2 sequences of marine representatives of *Gyrodactylus* (Fig. 1), omitting all exclusively freshwater subgenera and species groups, which have already been well analysed by Vanhove et al. (2011). The new species from Chile (JQ045347) clustered with high (98%) bootstrap support with two geographically distant species. These were *G. orecchiae*(FJ013097), described from Mediterranean cultured gilthead seabream *Sparus aurata*(Perciformes: Sparidae) by Paladini et al. (2009), and an undescribed species sequenced and discussed by Huyse et al. (2003) from the black goby *Gobius niger*

(Perciformes: Gobiidae, AY338452) in the North Sea. The relatedness of these two European species was noted by Paladini et al. (2009), but they did not place the species into a wider phylogenetic context. The phylogeographic coverage of this lineage is now extended significantly to include the Southern Hemisphere and the Pacific Ocean.

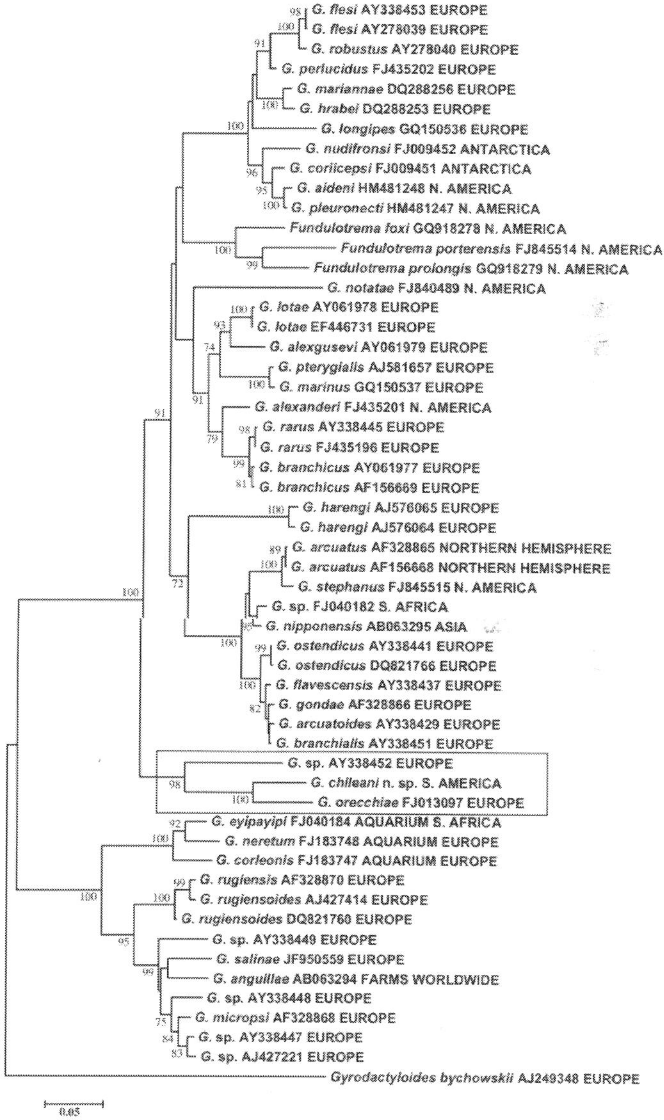

Figure.1: Hypothetical phylogenetic position of *Gyrodactylus chileani* n. sp. based on 5.8S+ITS2 rDNA sequences of marine (or of marine origin) gyrodac-

tylids. Bootstrap support less than 70% is omitted. *Gyrodactylus* sp. FJ040182 from *Lizarichardsonii* (Smith); *Gyrodactylus* sp. AY338452 from Gobius *niger* (see Huyse et al., 2003); *Gyrodactylus*sp. AY338449 from *Pomatoschistusnorvegicus* (Collett), a species provisionally called *G.* cf. *longicactylus* (see Huyse et al., 2003); *Gyrodactylus* sp. AJ427221 from *Pomatoschistuslozanoi* (de Buen), a species provisionally called *G.* cf. *micropsi* (see Huyse & Volckaert, 2002); and *Gyrodactylus* sp. AY338447 from *P. lozanoi*, a species provisionally called *G.* cf. *micropsi* (see Huyse et al. 2003)

The complete ITS rDNA (including ITS1+5.8S rDNA+ITS2) sequence is not available for *Gyrodactylus* sp. from *Gobiusniger* (see Huyse et al., 2003) and therefore the complete ITS segment of the rDNA was aligned only for *Gyrodactylusorecchiae* and the new species from Chile. The Maximum Composite Likelihood distance was estimated to be 15.4% between these two species, but it must be stressed that the ITS1+5.8S+ITS2 segments of the rDNA are not optimal for distance estimation, due to problems with the reliable alignment of ITS rDNA, from less related *Gyrodactylus* species with distances above 10%.

REMARKS ON THE SYSTEMATIC RELATIONSHIPS OF THE NEW TAXON

Subgeneric position. On the basis of the molecular sequence of the ITS region of the ribosomal DNA, the new species cannot be placed in any of the subgenera previously suggested by Malmberg (1970), which are already rendered poly- and paraphyletic by the inclusion of molecular data (Fig. 1). The Chilean *Gyrodactylus* species is placed within the mixed basal group of "short ITS" (Cable et al., 1999) species. The "short ITS" clade receives 100% bootstrap support in Fig. 1 as the sister group of the *G. rugiensis, G. micropsi* and *G. eyipayipi* species groups, which are the only marine groups (100% bootstrap support) among the "long ITS clade", and a sister clade of the subgenus *G.* (*Limnonephrotus*) (not demonstrated in Fig. 1, but see the phylogeny of Vanhove et al., 2011).

New species group. The new species clusters with *G. orecchiae* from the Mediterranean Sea (100% bootstrap support) and with an undescribed species on black goby from the North Sea. The three species cluster together with 98% bootstrap support and are relatively basal in the badly resolved clade (91% bootstrap support) of mostly marine, but also some derived freshwater, species. These species have previously been assigned to *G.* (*Mesonephrotus*), *G.*(*Metanephrotus*) and *G.* (*Paranephrotus*), or are not assigned at all. The considerable geographical spread of the three species assigned to this new *G. orecchiae* species lineage and the systematic diversity of their hosts suggest that a more comprehensive sampling may add numerous new species to this group.

Gyrodactylus chileani n. sp.

Type-host: *Helcogrammoides chilensis* (Cancino) (Perciformes: Tripterygiidae); local name 'Trombollito de tres aletas' (three-finned trombollito).

Type-locality: Tidal ponds at Las Cruces, Valparaiso (33°30′S, 71°37′W), El Tabo, Chile. Sampling date July 22nd, 2009.

Site: Fins and skin.

Type-material: Slides of 5 isolated opisthaptors and 1 complete specimen of *G. chileani* n. sp. were deposited in Finnish National History Museum in Helsinki University. Holotype: MZH 118095; paratypes: MZH 118096 (complete specimen); MZH 118097-MZH 118100 (isolated haptors).

Molecular data: Definitive identification is based on the nucleotide sequence of the internal transcribed spacers of the nuclear ribosomal DNA. The accession number for the ITS1-5.8S-ITS2 and short flanking segments of 18S and 28S rDNA from the holotype specimen is JQ045347. The holotype was cut in two, the haptor was mounted on a slide, and the remainder was used for DNA analysis. The ITS rDNA sequence was repeated from all of the five specimens that were measured.

Etymology: The name of the species is derived from the country name, as it is the first *Gyrodactylus* species described from Chile.

Description (Figs. 2, 3; Table 1)

[Five specimens were measured. Table 1 contains ranges and mean values of all measurements. Total length of anchor not measured as root is folded.] Shaft 23–27 μm; point 17–20 μm. Ventral bar length 16–20 μm, width 3–5 μm; membrane length 9–12 μm. Dorsal bar length 12–17 μm, width 2–3 μm. Marginal hook total length 21–23 μm; sickle length 4–5 μm. Shapes of hamuli resemble those of *G. orecchiae* and also those of *G. jirovecii* species goup of Eurasian freshwaters, which belong, according to DNA evidence, to subgenus *G.* (*Limnonephrotus*). Hamulus roots folded and points extend almost to half length of shafts (Figs. 2A, 3A). Ventral bar with much smaller processes than in *G. orecchiae* (Figs. 2A, 3B). Dorsal bar with narrow attachments (Fig. 2A). Marginal hooks clearly more delicate than in *G. orecchiae;* toe triangular in shape, not rhomboid as in *G. orecchiae*. Shaft of marginal hook points downwards and extends beyond toe. Heel pronounced, but smaller than in *G. orecchiae* (Figs. 2B, C, 3C).

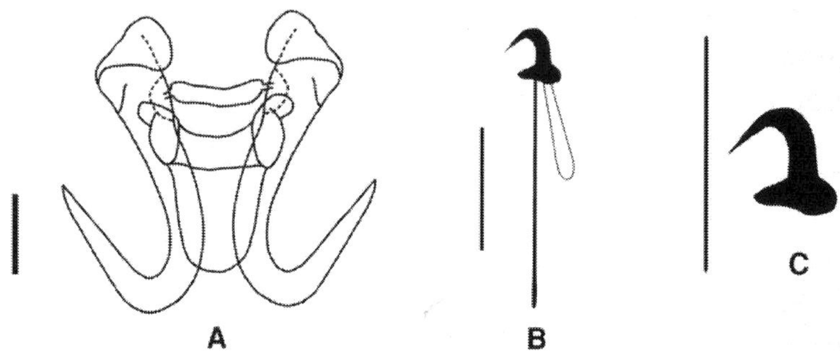

Figure.2: Haptoral morphology of *Gyrodactylus chileani* n. sp. A. hamuli, dorsal and ventral bars; B. marginal hook; C. marginal hook sickle. *Scale-bar*: 10 μm.

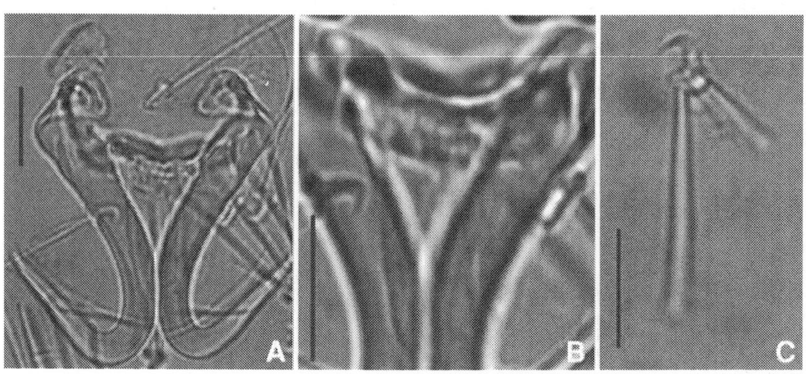

Figure.3: Photographs of *Gyrodactylus chileani* n. sp. Holotype. A. hamuli; B. ventral bar; C. marginal hook. *Scale-bar*: 10 μm.

Table 1: Measurements (in micrometres) of the haptoral hard parts of *Gyrodactylus chileani* n. sp.

Specimens	Hamuli		Dorsal bar		Ventral bar		Marginal hook		
	Shaft length	Point length	Length	Width	Length	Width	Membrane length	Total length	Sickle length
All (5)	23–27 (25)	17–20 (18)	12–17 (14.5)	2–3 (2.5)	16–20 (18)	3–5 (4)	9–12 (11)	21–23 (22)	4–5 (4.4)
Holotype	27	19	14	2	16	5	12	22	4.5
G. orecchiae	21	16	16	2	22	5	11	18	3.5

Mean values in parentheses. Measurements of *G. orecchiae* Paladini, Cable, Fioravanti, Faria, Cave & Shinn, 2009 from Paladini et al. (2009)

DISCUSSION

An overt phylogenetic revision of monogenean parasites of fishes was published by Perkins et al. (2009), with an informative title "Looks can deceive…". This seems to be a general rule among this class of parasites, certainly extending to the specific level among *Gyrodactylus* spp. With respect to the geographical coverage, the few molecular phylogenetic studies which have been attempted during the last decade are encouraging but far from satisfactory in terms of the strategic planning of a systematic revision (Cable et al., 1999, 2005; Harris & Cable, 2000; Ziętara et al., 2000, 2002, 2008; Huyse & Volckaert, 2002, 2005; Ziętara & Lumme, 2002, 2003, 2004; Boeger et al., 2003; Huyse et al., 2003, 2004, 2006; Matějusová et al., 2003; Huyse & Malmberg, 2004; LeBlanc et al., 2006; Malmberg et al., 2007; Kuusela et al., 2008; Rokicka et al., 2009; Vanhove et al., 2011). Due to the small fraction of the genus analysed in the above works, little can be said about its systematics. Additionally, a global systematic revision is certainly premature and unwarranted when less than 2% of the suspected species have been found and described; nevertheless, as the genus appears not to be monophyletic (Kritsky & Boeger, 2003, Vanhove et al., 2011), some kind of revision is needed. An alternative strategy towards the global systematics of the family Gyrodactylidae could be an extraction of well-supported monophyletic lineages based on their DNA, such as the nine lineages visible in present phylogeny, and their further supplementation by morphological characters, but leaving the genus *Gyrodactylus* as polyphyletic until all of its evolutionary lineages are resolved.

Such a new lineage is revealed in the present study. It consists of only three *Gyrodactylus*species at present. The hosts of the species belong to the order Perciformes, but the suborders and families differ. These are: *Gobius niger* (black goby) [suborder Gobioidei: family Gobiidae]; *Helcogrammoides chilensis* [Blennioidei: Tripterygiidae]; and *Sparus aurata* (gilthead seabream) [Percoidei: Sparidae]. The parasites on *S. aurata* were recorded in fish farms from Croatia and Albania (Paladini et al., 2009) or Italy (Paladini et al., 2011b), so it is not known whether the fish is the natural host of *Gyrodactylus orecchiae.* Considering the species richness of these host families (e.g. Hickey et al., 2009), we may expect that the three parasite species mentioned here are just a small subsample. They are all marine monogeneans and the geographical distance indicates that it is likely that at least the coastal waters of all seas contain related parasite species.

The phylogenetic hypothesis constructed in the present study demonstrates that the marine *Gyrodactylus* species groups, and in some cases even the species, are quite widely distributed. The *G. orecchiae* lineage extends from the Mediterranean and North Sea to the South-Eastern Pacific. *Gyrodactylus arcuatus* Bychowsky, 1933 has been confirmed on three-spined sticklebacks *Gasterosteus aculeatus* L. from Puget Sound (USA), Iceland, the Mediterranean and Black Seas, the Sea of Japan, Belgium (Cable et al., 1999; Ziętara et al., 2000), Finland (Ziętara et al., 2002, 2008), Norway (Ziętara et al., 2008; Hansen et al., 2012), Poland (Ziętara et al., 2000), Russian Karelia (Ziętara et al., 2008), Sweden (Ziętara et al., 2000; Huyse et al., 2003) and the UK (Cable et al., 1999; Ziętara et al., 2000). Its very close relatives include *Gyrodactylus nipponensis* Ogawa & Egusa, 1978 from Japan (Hayward et al., 2001) and *G. stephanus* Mueller, 1937 from the NW Atlantic (King & Cone, 2009) on other host fishes. On three-spined sticklebacks, *G. alexanderi* Mizelle & Kritsky, 1967 has also been confirmed using molecular data in both Atlantic (Hansen et al., 2012) and Pacific populations (Rokicka et al., 2009), but *G. avalonia* Hanek & Threlfall, 1969 and *G. canadensis* Hanek & Threlfall, 1969 await molecular confirmation.

The *G. eyipayipi*, *G. micropsi* and *G. rugiensis* species groups also have a very interesting connection across the equator: *G. eyipayipi* Vaughan,

Christison, Hansen & Shinn, 2010 has been recorded on greater pipefish *Sygnathus acus* L. in South African waters (Vaughan et al., 2010), and *G. micropsi* Gläser, 1974 and *G. rugiensis* Gläser, 1974 on the gills and fins of *Pomatoschistusmicrops* Krøyer, respectively, have been reported from European waters (Ziętara et al., 2002; Huyse et al., 2003, 2006). The related *G. corleonis* Paladini, Cable, Fioravanti, Fiara & Shinn, 2010 and *G. neretum* Paladini, Cable, Fioravanti, Fiara & Shinn, 2010 (Paladini et al. 2010) were found on pipefish *Syngnathustyphle* L. and *S. scovelli*Evermann & Kendall, respectively, from aquaria in Europe. *Gyrodactylus salinae* Paladini, Huyse &Shinn, 2011 on *Aphasiusfasciatus* Valenciennes has a Mediterranean distribution (Paladini et al., 2011a), and several species on gobies, provisionally referred to as *G.* cf. *micropsi* and *G.* cf. *longidactylus,* occur in the North Sea (Huyse & Volckaert, 2002; Huyse et al., 2003). Lastly, *G. anguillae* Ergens, 1960 is a cosmopolitan species on eels (see Hayward et al. 2001).

It has been already reported that the two Antarctic gyrodactylids, *G. coriicepsi* Rokicka, Lumme & Ziętara, 2009 and *G. nudifronsi* Rokicka, Lumme & Ziętara, 2009, are related to the European *Gyrodactylus* fauna, indicating the evolutionary continuum of the marine species in the Northern and Southern Hemispheres (Rokicka et al., 2009). This lineage also accommodates two recently described species, i.e. the European *G. longipes* Paladini, Hansen, Fioravanti & Shinn, 2011, reported as a mixed infection with *G. orecchiae* from *Sparusaurata*in the waters of Bosnia-Herzegovina and Italy (Paladini et al., 2011b), and the North American *G. aideni* Mullen, Cone, Easy & Burt, 2010 along with *G. pleuronecti* Cone, 1981, a species recently tagged by ITS rDNA (Mullen et al., 2010).

So far, the few *Gyrodactylus* species sampled from the coasts of the Southern Pacific and Southern Atlantic have been found to have quite close relatives in the species groups studied in the Northern Hemisphere. Connections between the Northern Pacific and Northern Atlantic are even tighter, including common species such as *G. arcuatus*. This finding supports the hypothesis of Boeger et al. (2003) suggesting the origin of viviparous gyrodactylids in South American waters and a further expansion from that continent via the marine environment. Such a scenario requires a mixing of the marine gyrodactylid fauna as demonstrated in the present paper.

REFERENCES

1. Bakke, T. A., Harris, P. D., & Cable, J. (2002). Host specificity dynamics: observation of gyrodactylid monogeneans. *International Journal for Parasitology, 32*, 281–308.

2. Boeger, W. A., Kritsky, D. C., & Pie, M. R. (2003). Context of diversification of the viviparous Gyrodactylidae (Platyhelminthes, Monogenoidea). *Zoologica Scripta, 32*, 437–448.

3. Cable, J., Harris, P. D., Tinsley, R. C., & Lazarus, C. M. (1999). Phylogenetic analysis of *Gyrodactylus* spp. (Platyhelminthes: Monogenea) using ribosomal DNA sequences. *Canadian Journal of Zoology, 77*, 1439–1449.

4. Cable, J., Oosterhout, C., Barson, N., & Harris, P. D. (2005). *Gyrodactylus pictae* n. sp. (Monogenea: Gyrodactylidae) from the Trinidadian swamp guppy *Poecilia picta* Regan, with a discussion on species of *Gyrodactylus* von Nordmann, 1832 and their poeciliid hosts. *Systematic Parasitology, 60*, 159–164.

5. Edgar, R. C. (2004). MUSCLE: Multiple sequence alignment with high accuracy and high throughput. *Nucleic Acid Research, 32*, 1792–1797.

6. Felsenstein, J. (1985). Confidence limits on phylogenies: an approach using the bootstrap. *Evolution, 39*, 783–791.

7. Gusev, A. V., Dubinina, M. N., Pugachev, O. N., Raykova, E. V., Khotenovskiy, I. A., & Ergens, R. (1985). *Opredelitel' parazitov presnovodnykh ryb fauny SSSR*, Vol. 2. *Paraziticheskie mnogokletochnye*. St Petersburg: Nauka, pp. 269–347. (In Russian).

8. Hansen, H., Jørgensen, A., & Mo, T. A. (2012). Spin-off from routine parasite diagnostics of Atlantic salmon; first report of *Gyrodactylus alexanderi* in Norway. *The Bulletin of the European Association of Fish Pathologists, 32*, 14–18.

9. Harris, P. D., & Cable, J. (2000). *Gyrodactylus poeciliae* n. sp. and *G. milleri* n. sp. (Monogenea: Gyrodactylidae) from *Poecilia caucana* (Steindachner) in Venezuela. *Systematic Parasitology, 47*, 79–85.

10. Harris, P. D., Shinn, A. P., Cable, J., & Bakke, T. A. (2004). Nominal species of the genus *Gyrodactylus* von Nordmann 1832

(Monogenea: Gyrodactylidae), with a list of principal host species. *Systematic Parasitology, 59*, 1–27.

11. Hayward, C. J., Iwashita, M., Ogawa, K., & Ernst, I. (2001). Global spread of the eel parasite *Gyrodactylus anguillae* (Monogenea). *Biological Invasions, 3*, 417–424.
12. Hickey, A. J. R., Lavery, S. D., Hannan, D. A., Baker, C. S., & Clements, K. D. (2009). New Zealand triplefin fishes (Family Tripterygiidae): contrasting population structure and mtDNA diversity within a marine species flock. *Molecular Ecology, 18*, 680–696.
13. Huyse, T., & Malmberg, G. (2004). Molecular and morphological comparisons between *Gyrodactylus ostendicus* n. sp. (Monogenea: Gyrodactylidae) on *Pomatoschistus microps*(Krøyer) and *G. harengi* Malmberg, 1957 on *Clupea harengus membras* L. *Systematic Parasitology, 58*, 105–113.
14. Huyse, T., & Volckaert, F. A. M. (2002). Identification of a host-associated species complex using molecular and morphometric analysis, with the description of *Gyrodactylus rugiensoides*n. sp. (Gyrodactylidae, Monogenea). *International Journal for Parasitology, 32*, 907–919.
15. Huyse, T., & Volckaert, F. A. M. (2005). Comparing host and parasite phylogenies: *Gyrodactylus* flatworms jumping from goby to goby. *Systematic Biology, 54*, 710–718.
16. Huyse, T., Audenaert, V., & Volckaert, F. A. M. (2003). Speciation and host-parasite relationships in the parasite genus *Gyrodactylus* (Monogenea, Platyhelminthes) infecting gobies of the genus *Pomatoschistus* (Gobiidae, Teleostei). *International Journal for Parasitology, 33*, 1679–1689.
17. Huyse, T., Malmberg, G., & Volckaert, F. A. M. (2004). Four new species of *Gyrodactylus* von Nordmann, 1832 (Monogenea, Gyrodactylidae) on gobiid fishes: combined DNA and morphological analyses. *Systematic Parasitology, 59*, 103–120.
18. Huyse, T., Pampoulie, C., Audenaert, V., & Volckaert, F. A. M. (2006). First report of *Gyrodactylus* spp. (Platyhelminthes: Monogenea) in the western Mediterranean sea: molecular and morphological descriptions. *Journal of Parasitology, 92*, 682–690.

19. King, S. D., & Cone, D. K. (2009). Morphological and molecular taxonomy of a new species of *Fundulotrema* and comments on *Gyrodactylus stephanus* (Monogenea: Gyrodactylidae) from*Fundulus heteroclitus* (Actinopterygii: Cyprinodontiformes) in Nova Scotia, Canada. *Journal of Parasitology, 95*, 846–849.

20. Kritsky, D. C., & Boeger, W. A. (2003). Phylogeny of the Gyrodactylidae and the phylogenetic status of *Gyrodactylus* Nordmann, 1832 (Platyhelminthes, Monogenoidea). In: Combes, C., & Jourdanne, J. (Eds) *Taxonomie, écologie et évolution desmétazoires parasites*. (*Livre Hommage à Louis Euzet*). Vol. II. Perpignan: Presses Universitaires de Perpignan, pp. 37–58.

21. Kuusela, J., Ziętara, M. S., & Lumme, J. (2008). Description of three new European cryptic species of *Gyrodactylus* Nordmann, 1832 supported by nuclear and mitochondrial phylogenetic characterization. *Acta Parasitologica, 53*, 120–126.

22. LeBlanc, J., Hansen, H., Burt, M., & Cone, D. K. (2006). *Gyrodactylus neili* n. sp. (Monogenea, Gyrodactylidae), a parasite of chain pickerel *Esox niger* Lesueur (Esocidae) from freshwaters of New Brunswick Canada. *Systematic Parasitology, 65*, 43–48.

23. Malmberg, G. (1970). The excretory systems and the marginal hooks as a basis for the systematics of *Gyrodactylus* (Trematoda, Monogenea). *Arkiv för Zoologi, 23*, 1–235.

24. Malmberg, G., Collins, C. M., Cunningham, C. O., & Jalali, B. J. (2007). *Gyrodactylusderjavinoides* sp. nov. (Monogenea, Platyhelminthes) on *Salmo trutta trutta* L. and *G. derjavini* Mikailov, 1975 *S. t. caspius* Kessler, two different species of *Gyrodactylus* – combined morphological and molecular investigations. *Acta Parasitologica, 52*, 89–103.

25. Matějusová, I., Gelnar, M., Verneau, A. O., Cunningham, C. O., & Littlewood, D. T. J. (2003). Molecular phylogenetic analysis of the genus *Gyrodactylus* (Platyhelminthes: Monogenea) inferred from rDNA ITS region: subgenera versus species groups. *Parasitology, 127*, 603–611.

26. Mullen, A. J., Cone, D. K., Easy, R., & Burt, M. D. B. (2010). Taxonomy and host-specificity of *Gyrodactylus aideni* n. sp. and *G. pleuronecti* (Monogenea: Gyrodactylidae) from *Pseudopleuronectes*

americanus (Walbaum) in Passamaquoddy Bay, New Brunswick. *Canada. Systematic Parasitology, 77*, 233–239.

27. Muñoz, G., & Olmos, V. (2007). Revisión bibliográfica de especies ectoparásitas y hosperados de sitemas acuáticos de Chile. *Revista de Biología Marina y Oceanografia, 42*, 89–148.
28. Muñoz, G., & Randhawa, H. S. (2011). Monthly variation in the parasite communities of the intertidal fish *Scartichthys viridis* (Blenniidae) from central Chile: are there seasonal patterns? *Parasitology Research, 109*, 53–62.
29. Muñoz, G., & Zamora, L. (2011). Ontogenetic variation in parasite infracommunities of the clingfish *Sicyases sanguineus* (Pisces: Gobiesocidae). *Journal of Parasitology, 97*, 14–19.
30. Paladini, G., Cable, J., Fioravanti, M. L., Faria, P. J., Di Cave, D., & Shinn, A. P. (2009). *Gyrodactylus orecchiae* sp. n. (Monogenea: Gyrodactylidae) from farmed populations of gilthead seabream (*Sparus aurata)* in the Adriatic Sea. *Folia Parasitologica, 56*, 21–28.
31. Paladini, G., Cable, J., Fioravanti, M. L., Faria, P. J., & Shinn, A. P. (2010). The description of *Gyrodactylus corleonis* sp. n. and *G. neretum* sp. n. (Platyhelminthes: Monogenea) with comments on other gyrodactylids parasitising pipefish (Pisces: Syngnathidae). *Folia Parasitologica, 57*, 17–30.
32. Paladini, G., Huyse, T., & Shinn, A. P. (2011a). *Gyrodactylus salinae* n. sp. (Platyhelminthes: Monogenea) infecting the south European toothcarp *Aphanius fasciatus* (Valenciennes) (Teleostei, Cyprinodontidae) from a hypersaline environment in Italy. *Parasites and Vectors, 4*, 100–111.
33. Paladini, G., Hansen, H., Fioravanti, M. L., & Shin, A. P. (2011b). *Gyrodactylus longipes* n. sp. (Monogenea: Gyrodactylidae) from farmed gilthead seabream (*Sparus aurata* L.) from the Mediterranean. *Parasitology International, 60*, 410–418.
34. Perkins, E. M., Donnellan, S. C., Bertozzi, T., Chisholm, L. A., & Whittington, I. D. (2009). Looks can deceive: molecular phylogeny of a family of flatworm ectoparasites (Monogenea: Capsalidae) does not reflect current morphological classification. *Molecular Phylogenetics and Evolution, 52*, 705–714.

35. Rokicka, M., Lumme, J., & Ziętara, M. S. (2009). Two Antarctic *Gyrodactylus* species (Monogenoidea): description and phylogenetic characterization. *Journal of Parasitology, 95*, 1112–1119.

36. Saitou, N., & Nei, M. (1987). The neighbor-joining method: a new method for reconstructing phylogenetic trees. *Molecular Biology and Evolution, 4*, 406–425.

37. Tamura, K., Nei, M., & Kumar, S. (2004). Prospects for inferring very large phylogenies by using the neighbor-joining method. *Proceedings of the National Academy of Sciences (USA),101*, 11030–11035.

38. Tamura, K., Peterson, D., Peterson, N., Stecher, G., Nei, M., & Kumar, S. (2011). MEGA5: Molecular evolutionary genetics analysis using maximum likelihood, evolutionary distance, and maximum parsimony methods. *Molecular Biology and Evolution, 28*, 2731–2739.

39. Vanhove, M. P. M., Snoeks, J., Volckaert, F. A. M., & Huyse, T. (2011). First description of monogenean parasites in Lake Tanganyika: the cichlid *Simochromis diagramma* (Teleostei, Cichlidae) harbours a high diversity of *Gyrodactylus* species (Platyhelminthes, Monogenea). *Parasitology, 138*, 364–380.

40. Vaughan, D. B., Christison, K. W., Hansen, H., & Shinn, A. P. (2010). *Gyrodactylus eyipayipi*sp. n. (Monogenea: Gyrodactylidae) from *Syngnathus acus* (Syngnathidae) from South Africa. *Folia Parasitologica, 57*, 11–15.

41. Ziętara, M. S., Arndt, A., Geets, A., Hellemans, B., & Volckaert, F. A. M. (2000). The nuclear rDNA region of *Gyrodactylus arcuatus* and *G. branchicus* (Monogenea: Gyrodactylidae). *Journal of Parasitology, 86*, 1368–1373.

42. Ziętara, M. S., Huyse, T., Lumme, J., & Volckaert, F. A. (2002). Deep divergence among subgenera of *Gyrodactylus* inferred from rDNA ITS region. *Parasitology, 124*, 39–52.

43. Ziętara, M. S., Kuusela, J., Veselov, A., & Lumme, J. (2008). Molecular faunistics of accidental infections of *Gyrodactylus* Nordmann, 1832 (Monogenea) parasitic on salmon *Salmo salar* L. and brown trout *Salmo trutta* L. in NW Russia. *Systematic Parasitology, 69*, 123–135.

44. Ziętara, M. S., & Lumme, J. (2002). Speciation by host switch and adaptive radiation in a fish parasite genus *Gyrodactylus* (Monogenea: Gyrodactylidae). *Evolution, 56*, 2445–2458.

45. Ziętara, M. S., & Lumme, J. (2003). The crossroads of molecular, typological and biological species concepts: two new species of *Gyrodactylus* Nordmann, 1832 (Monogenea, Gyrodactylidae). *Systematic Parasitology, 55*, 39–52.

46. Ziętara, M. S., & Lumme, J. (2004). Comparison of molecular phylogeny and morphological systematics in fish parasite genus *Gyrodactylus* Nordmann, 1832 (Monogenea, Gyrodactylidae). *Zoologica Poloniae, 49*, 5–28.

CHAPTER

5

CHARACTERIZATION OF VIBRIO PARAHAEMOLYTICUS ISOLATED FROM FARMED SEA BASS (DICENTRARCHUS LABRAX) DURING DISEASE OUTBREAKS

Sadok Khouadja, Faouzi Lamari, Amina Bakhrouf

Département de Microbiologie, Faculté de Pharmacie, Laboratoire d'AnalyseTraitement et Valorisation des Polluants de l'Environnement et des ProduitsMonastirTunisia

ABSTRACT

Vibrio parahaemolyticus was isolated from the internal organs of diseased sea bass (*Dicentrarchus labrax*) cultured in a marine cage farm and an inshore fish farm in the summer 2010. All strains isolated from diseased sea bass were tested for virulence by intraperitoneal injection. The isolates ($n=6$) were pathogenic for sea bass. Only bacterial strains showing haemolytic activity were virulent. LD50 values ranged from 3.52×10^4 to 2.29×10^6 cfu fish^{-1}. In this study, we report the characterization and the virulence of *V. parahaemolyticus* strains isolated from sea bass originating from two different fish farms (marine cage farm and inshore fish farm in Tunisia).

Keywords: Vibrio parahaemolyticus, Outbreaks, Virulence properties

BACKGROUND

Vibrio parahaemolyticus is part of the natural estuarian microflora and coastal marine waters and can be present in seafood, especially shellfish and bivalve mollusks (DePaola et al. 2003; Zorrilla et al. 2003). Starting from 2000, an increase of outbreaks associated with the 03:K6 serovar has been observed in Asia (Matsumoto et al. 2000), North America and Chile (Martinez-Urtaza et al. 2005). This serovar was also reported from outbreaks or sporadic cases in Europe, France and Spain (Martinez-Urtaza et al. 2005; Quilici et al. 2005), and Africa and Russia (Nair et al. 2007).

The *Vibrio* pathogenic species produce various virulence factors including enterotoxin, haemolysin, cytotoxin, protease, lipase, phospholipase, siderophore, adhesive factor and/or haemagglutinins (Zhang and Austin 2005). The virulence of *V. parahaemolyticus* strains is commonly associated with expression of thermostable direct hemolysin (TDH) and TDH-related hemolysin (TRH), which are encoded by the *tdh* and *trh* genes. Therefore, the *tdh* gene, marked by a β-type haemolysis on Wagatsuma agar (Nishibuchi and Kaper 1995), and the *trh*gene, correlated to a positive urease tests (Okuda et al. 1997), serve as markers for pathogenic strains. Many extracellular proteases are suggested to play important roles in virulence of *Vibrio spp*. In a recent study (Lee et al. 2002), a protease was identified as the major virulence factor in a clinical *tdh* and *trh* negative *V. parahaemolyticus* isolate, and this enzyme showed cytotoxic activity on CHO and Vero cells. In this study, we reported the characterization of six *V. parahaemolyticus* isolated during outbreaks affecting sea bass (*Dicentrarchus labrax*) isolated in two different fish farms (inshore and offshore fish farms), and we demonstrate its implication in the disease experimental challenges.

METHODS

Sampling and Bacterial Isolation

Juvenile (weight 5 to 7 g, length 8 to 10 cm) and adults (weight 100 to 120 g, length 14 to 16 cm) *D. labrax* were obtained from two different fish farm, marine cage farm and inshore fish farm in summer 2010. Ten fish were collected from each size and each site (20 juveniles and 20 adults). The two fish farms are located in the biggest centre in Tunisia (Sousse, Centre of Tunisia). Samples from liver, spleen, kidney and external lesions of several fish from marine cage farm and inshore fish farm were cultured on alkaline peptone water (1% NaCl, pH 8.6) and incubated at 37°C for 18 to 24 h. A loopful of the enrichment culture was streaked onto thiosulphate-citrate-bile salt-sucrose agar used for the selective isolation of *Vibrio* strains (TCBS agar, Scharlau Microbiology, Spain). After 18 to 24 h of incubation at 37°C, the cultures giving pure green colonies were randomly selected then subcultured on tryptic soy agar (TSA, Difco, Spain) supplemented with 1% NaCl according to the protocol described by Hara-Kudo et al. (2001). The isolated bacteria were frozen at −80°C with 20% (*v*/*v*) glycerol for further analysis.

Phenotypic Characterization of Bacterial Strains

Standard procedures described in Bergey's Manual of Systematic Bacteriology (Alsina and Blanch 1994) were used for the species identification. Gram coloration, cell morphology, the oxidase, catalase, indole production, O-F test, motility (mannitol-motility agar, Pronadisa, Madrid, Spain) and susceptibility to the vibriostatic compound O/129 (10 and 150 μg/disc) were the first tests employed to identify the organisms belonging to the *Vibrio* genus. The API 20NE (bioMerieux, Marcy l›Etoile, France) procedure was modified in order to incorporate a 2.5% NaCl concentration in all microtubes. An initial bacterial suspension was therefore prepared in 5 mL of a 2.5% NaCl solution instead of the recommended 0.85% NaCl medium. Incubation time and temperature were maintained within the limits prescribed (30°C for 24 h) (Croci et al. 2007). Identification was obtained through the APILAB PLUS software (bioMerieux) and was considered acceptable when giving a probability equal or greater than 80% (Croci et al. 2007).

Halophilism tests were performed using tryptone broth composed of casein enzymic hydrolysate and different concentrations of NaCl (0%, 3%, 6%, 8% and 10% *w*/*v*). Additional characterization tests for the identification of *V. parahaemolyticus* (Hormansdorfer et al. 2000), arginine dehydrolase tests and O/129 susceptibility tests were performed (FDA and Drug Administration 1992). The reference strains listed in

Table 1 were used as positive controls.

Table 1: Biochemical and morphological characterization of isolates from sea bass*D.*and*V. parahaemolyticus*(ATCC 17802 and ATCC43996)

Tests	**Strains (*n*= 6)**	***V. parahaemolyticu-s*ATCC17802**	***V. parahaemolyti-cus*ATCC43996**
Growth on TCBS	Green	Green	Green
Gram stain	–	–	–
Cytochrome oxidase	+	+	+
0/129 sensitivity	+	+	+
Swarming	–	–	–
Growth at% NaCl	+	+	+
0%	–	–	–
3%	+	+	+
6%	+	+	+
8%	+	+	+
10%	+	+	+
Growth at	+	+	+
4°C	–	–	–
25°C	+	+	+
37°C	+	+	+
40°C	+	+	+
β-galactosidase (ONPG test)	+	–	–
Arginine dihydrolase	+	–	–
Lysine decarboxylase	+	+	+
Ornithine decarboxylase	+	+	+
Citrate	–	–	–
Production of H_2S	–	–	–
Urease	–	+	–
Tryptophan deaminase	V	+	+
Indol	V	+	+
VP test	–	–	–
Gelatin liquefaction	+	+	+
Acid from	+	+	+
Glucose	+	+	+

Mannitol	+	+	+
Inositol	–	–	–
Sorbitol	–	–	–
Rhamnose	–	–	–
Sucrose	–	–	–
MeLibiose	–	–	–
Amygdalin	–	–	–
Arabinose	+	+	+
Lactose	–	–	–
Cellobiose	–	–	–

V, variable; ONPG, ortho-nitrophenyl-β-galactoside; VP, Voges-Proskauer test; +, positive; –, negative.

Confirmation of the Species Level by Detection of *toxR* gene

DNA minipreparations were made from each strain using the Wizard genomic DNA purification kit (Promega, Madison, WI, USA). The DNA was later resuspended in TE (10 mM Tris-HCl, 100 mM EDTA, pH 7.8) buffer and stored at −20°C. The quantity of genomic DNA in each sample was measured at 260 nm using a spectrophotometer (Ultraspec 2100 Pro; Amersham Biosciences Europe GmbH, France). The DNA concentration of each sample was adjusted to (40 ng/μL) for polymerase chain reaction (PCR) by diluting DNA in distilled deionized water.

V. parahaemolyticus identification was confirmed with the presence of the outer membrane protein regulation operon gene (*toxR*) (Kim et al. 1999). The reaction for *toxR* analysis was performed in a total volume of 25 μL containing 2.5 μL of DNA sample, 15.25 μL of deionized water, 2.5 μL of 10 × buffer, 2 μL of 25 mM concentration of $MgCl_2$, 0.5 μL of 10 mM concentration of dNTP, 1 μL (10 μM) of each primer (ToxR-F GTCTTCTGACGCAATCGTTG and ToxR-R ATACGAGTGGTTGCTGTCATG) and 0.25 μL of 5 U/μL of Taq DNA polymerase (Promega). The PCR assay was performed at 94°C for 5 min followed by cycles of 94°C for 1 min, 63°C for 1 min and 72°C for 2 min in a DNA thermal cycler (Kim et al. 1999). The PCR fragments were subjected to electrophoresis on a 1.5% agarose gel in Trisborate-EDTA buffer (0.89 mol/L Tris, 0.89 mol/L boric acid, 0.02 mol/L EDTA, pH

8.0), visualized by ethidium bromide staining and photographed using Gel Doc XR apparatus (Biorad, Hercules, CA, USA).

Serum Sensitivity and Exoenzyme Productions

Enzymatic activities of isolates were analysed by spot inoculations on TSA 2% NaCl supplemented with skimmed milk (2% *w*/*v*), Tween 80 (1% *w*/*v*), egg yolk emulsion (2% *w*/*v*), fibrinogen (0.28% *w*/*v*) and elastin (0.1% *w*/*v*) (Alcaide et al. 1999). DNase production was assayed on DNase agar (Oxoid, Hampshire, England). Gelatinase potency was evaluated using nutrient gelatin differential medium (Oxoid) that tests the ability of an organism to produce an exoenzyme that hydrolyzes gelatin. Haemolysin potency was evaluated using a modified plate assay technique described previously (Quindos et al. 1994). Briefly, 10 µL of suspensions (10^8ufc mL^{-1}) were spotted onto human blood and fish blood agar made by mixing 70 mL of each blood with 1,000 mL (TSA) supplemented with 3% g NaCl (*w*/*v*). Fish blood was collected by caudal venous puncture from four or five healthy sea bass with weight over 100 g. The plates were incubated at 37°C for 24 h. Positive haemolytic potency was recorded by the presence of a distinct translucent halo around the inoculum area. The diameters of the zones of lysis and the colony were measured, and the ratio (equal to or larger than 1) was used as a haemolytic index to represent the intensity of haemolysin production by the tested strains. All the experiments were repeated three times. As phenotypic marker of the *trh* gene from *V. parahaemolyticus* isolates, the urease activity (Ure) was assayed through a conventional method, using Christensen's urea agar base (with 3% NaCl) which is used for rapid detection of urease activity (Hammack et al. 2001; Christensen 1946).

Infectivity Tests

The 50% lethal dose (LD_{50}) test (with batches of 20 fish per dose) was conducted by intraperitoneal (i.p.) injection as previously described (Alcaide et al. 1999). Briefly, juveniles of *D. labrax* (weight 7 g and length 80 mm), kindly provided by the hatchery of higher Institute of Sea Monastir, were injected with 50 µL of a bacterial suspension containing 10^8 to 10^2 cfu mL^{-1}. The six strains were used in virulence test to quantify their virulence; the strains were grown overnight in TSA 3% NaCl at 30°C, from which one colony was subcultured in 40-mL fresh medium

(TSB 3% NaCl) at the same condition for 16 h. The cells were harvested by centrifugation (5,000×*g*, 10 min), washed and resuspended in PBS (0.01 M) to OD_{600} of 0.2 to 0.9, so that the bacterial concentrations were 10^2 to 10^8 cfu mL^{-1} determined by dilution plate method. Sterile PBS was injected i.p. into fish as control. The water temperature of the experiment was adjusted to 25°C. *D. labrax* juveniles were acclimated in sea water of 25°C for 4 to 5 days. Before infection, the juveniles of sea bream were fed with commercial food (ALLER FUTURA EX., ALLER AQUA, Aller, Christiansfeld, Denmark).

The juvenile fish were maintained at 25°C, 37‰ salinity and under continuous aeration. The juveniles were kept without feeding during the virulence test. Mortalities were recorded daily for 7 days and were only considered infected if *V. parahaemolyticus* was recovered from the assayed fish. The LD_{50} was calculated by simple method for estimating 50% endpoints (Reed and Muench 1938).

Survival in Fish Serum

The strains of *V. parahaemolyticus* were tested for their ability to resist the bactericidal activity of sea bass natural serum in a plate plaque assay. Sea bass serum was collected by centrifugation of total blood at 3,000 × *g* for 20 min at 4°C and stored at −20°C. The tests were conducted in 96-well microtitre dishes by mixing 100 μL of fresh serum or heat inactivated serum (44°C, 20 min) with 100 μL of bacteria suspension (10^6 cfu mL^{-1}) in PBS (pH 7). The assays were made by taking samples (10 μL) of the mixture (serum and bacteria) every 30 min for 4 h of incubation at room temperature (30°C). Viable counts were determined by drop plate on TSA 1% NaCl. Survival curve of all *V. parahaemolyticus* strains was determined using an Excel program.

RESULTS

Biochemical and Genetic Identification

Table 1 describes the biochemical characterization of the strains isolated. The most outstanding character in the environmental strains is tryptophan deaminase and indole negative. According to API 20NE profiles (Table 2), four strains are identified as *V. parahaemolyticus*, and two are *Burkholderia gladioli*. The isolates which were initially identified as *V. parahaemolyticus* by conventional biochemical tests were further confirmed by PCR targeting species specific *toxR*(368 bp) gene (Figure 1). The two strains identified like *B. gladioli* were also positive for the *toxR* gene.

Table 2: Origin, hydrolytic enzymes and virulence factors of different strains isolated from internal organs diseased fish

Code	Biotype on API 20NE	Identification API 20NE (%)	Fish farm	Isolated from	Serum (D.L)	Hydrolytic enzymes								
						Lipase	DNase	Amylase	Gélatinase	Caseinase	Lecithin-ase	Haemolysis (D.L)	Haemolysis (H)	DL50
S1	7073345	99.9	I.F	Kidney	A	+	–	+	+	+	+	+	+	2.31×10^5
S2	5053345	99.9	O.F	Kidney	A	+	+	+	+	+	+	+	+	4.1×10^5
S3	7053304	99.9	O.F	Kidney	A	+	–	+	+	+	+	+	+	3.52×10^4
S4	7053345	99.9	O.F	Spleen	A	+	–	+	+	+	–	–	–	1.31×10^5
S5	7053734	Unacceptable	I.F	Spleen	B	+	–	+	+	–	+	–	–	2.29×10^6
S6	5053365	Unacceptable	I.F	Heart	B	+	+	+	+	+	+	+	+	Avirulent
ATCC 17802	5053345	99.9	–	Human	A	+	+	+	+	+	+	+	+	Avirulent
ATCC 43969	7053375	99.9	–	Human	B	+	+	+	+	+	+	+	+	ND

I.F, offshore farm; I.N, inshore farm; +, positive; –, negative; virulence, cfu $fish^{-1}$; ND, not determined; H, human erythrocytes; D.L, *Dicentrarchus labrax* erythrocytes; A, resisted to the serum; B, sensitive to the serum.

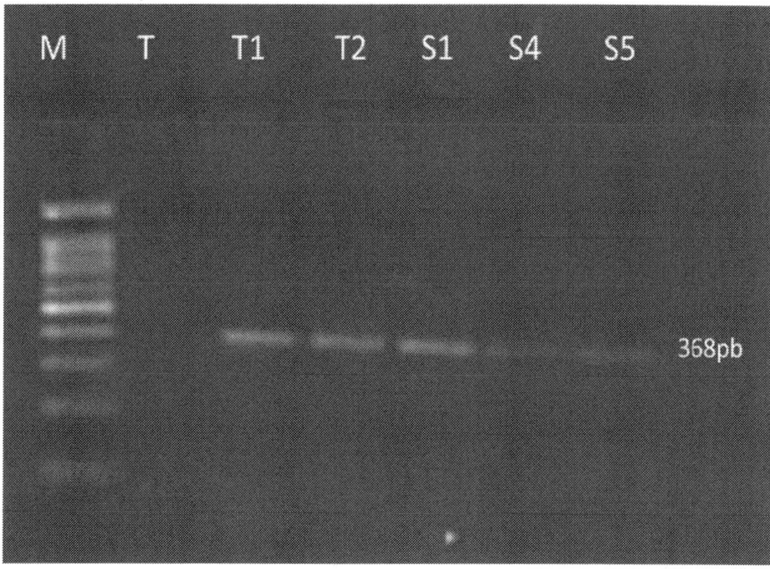

Figure 1: **Reaction (PCR).** M, molecular mass marker 100 bp; T, control; T1 and T2, controls + *V. parahaemolyticus* ATCC 43996 and ATCC 17802; S1, S4 and S5, environmental strains.

Characterization of *V. parahaemolyticus* isolates

In the present study, none of the strains showed beta haemolytic activity, but four strains were found to exhibit weak haemolysis, and all stains were negative for urease (Table 2). Most of the *Vibrio* isolates were producers of extracellular lipase, amylase, caseinase, DNase and gelatinase (Table 2). Experimentally infected fish showed external signs similar to those observed in outbreaks. During the determination of the LD50, all strains showing haemolytic activity were virulent and caused mortality during the infection tests. Pure culture of *V. parahaemolyticus*was isolated from liver and skin of moribund fish, and no moralities were detected in the control group. Signs of infection began to appear on the second day with strains S1 and S2, whereas for other strains, such signs were present only from the fourth day. The LD50 of the pathogenic strains ranged between 3.52×10^4 and 2.29×10^6 cfu g^{-1} fish (Table 2). Healthy fish infected with *V. parahaemolyticus* (S1 and S2) during virulence tests developed skin ulcers, similar to those observed on the surface of sea bass collected from initial outbreaks in fish farms (Figure 2). With S4 and S5 strains, the fish do not show any signs of infection during the rest of experiments.

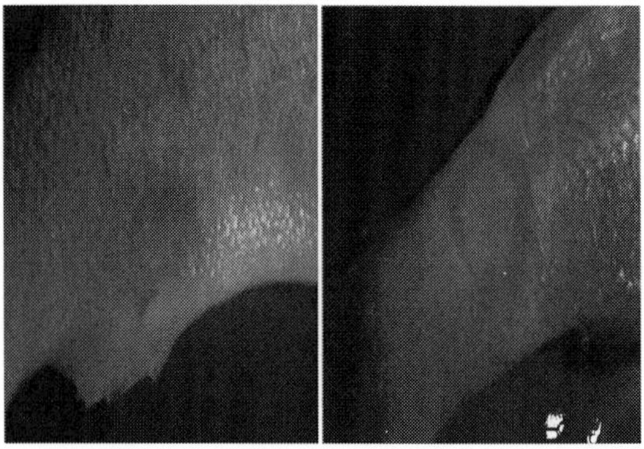

Figure 2: **Signs of infection observed on*D. labrax*larvae including.** Skin haemorrhages (left) and necroses (right) after infection tests.

V. parahaemolyticus survival in Fish Serum

Only two strains isolated from sea bass (S5 and S4) were found to be serum sensitive. The survival curves of *V. parahaemolyticus* strains in sea bass serum are shown in Figure 3. The two strains showed different profiles: (A) strains that resisted the bactericidal action of the serum and were capable of proliferating, and (B) other strains that showed sensitivity to the serum and of which the percentage of survival decreased.

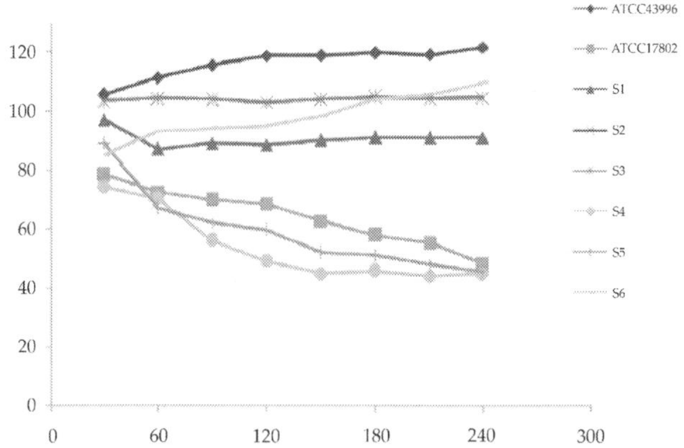

Figure 3: **Survival curves.** Six *V. parahaemolyticus* strains and the two reference strains (ATCC 43996 and ATCC 17802) in sea bass serum.

DISCUSSION

During the spring of 2010, a wide mortality occurred in two aquaculture farms, starting in the offshore farm and then the same sign and the same mortality occurred in the inshore farm. The clinical signs of vibriosis affecting sea bass (*D. labrax*) included darkened body colour, white nodular skin lesion and sudden death with haemorrhages in the skeletal muscle. The six strains isolated during our study were reported to exhibit variation in one or two biochemical reactions, inducing errors in species identification (Dileep et al. 2003). The atypical reactions in environmental *Vibrio* spices may pose difficulties in organism biochemical identification (Karunasagar et al. 1996). *B. gladioli* is indole negative and lysine decorboxylation negative bacteria, knowing that most of the Tunisian *Vibrio* isolates were negative for indole production (Ben Kahla-Nakbi et al. 2007). Identifying *V. parahaemolyticus* strains through PCR-based method which targets the conserved region of *V. parahaemolyticus* such as *gyrB* and *toxR* gene is more efficient, reliable and faster compared to identifying through biochemical tests (Kim et al. 1999). Although *V. parahaemolyticus* is widely distributed in the coastal environments all over the world, most of the environmental strains are not pathogenic to humans, and only 1 to 2% of the environmental strains have shown to be positive Kanagawa TDH (Hervio-Heath et al. 2002; Robert-Pillot et al. 2004). In the present study, four of the isolates were found to exhibit weak haemolysis, and none of them was found to exhibit positive amplification for the *tdh*gene; none of the weak haemolytic strains were found to exhibit urease activity. It is likely that the *tdh* and *trh* positive strains are present in very low numbers along with strains negative for these genes (Dileep et al. 2003). The weak haemolysis points towards the presence of other virulence factors other than *tdh* in *V. parahaemolyticus*. This is confirmed during the determination of LD_{50}. All strains showing haemolytic activity are virulent and cause mortality during the infection tests even with the absence *tdh* and *trh* genes. *V. parahaemolyticus* is considered as a human pathogen more than a fish pathogen. Several cases of fish mortality due to this germ are reported in Mediterranean coasts (Alcaide et al. 1999; Zorrilla et al. 2003). The organism has been associated with mortalities in Iberian toothcarp (*Aphanius iberus*) with the signs centred on external haemorrhages and tail rot (Austin and Austin 2007). Cultures regarded as intermediate between *V. alginolyticus* and *V. parahaemolyticus* were recovered from diseased milkfish (*Chanos chano* s) in the Philippines (Austin and Austin 2007). In *Penaeus monodon*, the organism has been implicated as a cause of red disease in India (Jayasree 2009). Pathogenicity has been established in *Solea senegalensis*, amberjack and eel by i.p. infections with LD50 dose of 6.3×10^5, 5×10^3 and 6.2×10^5 cfu fish^{-1}, respectively (Alcaide et al. 1999; Zorrilla et al. 2003). The corresponding value in abalone post-larvae was 3.5×10^5 cfu mL^{-1}, with the disease mirroring that of natural infections (Cai et al. 2007).

Many extracellular proteases are suggested to play important roles in the virulence of *Vibrio*spp. In a recent study (Lee et al. 2002), a protease was identified as the major virulence factor in a clinical *tdh*- and *trh*-negative *V. parahaemolyticus* isolate, and this enzyme showed cytotoxic activity on CHO and Vero cells (Ottaviani et al. 2005).

The enzymes› arsenal in these microorganisms gives them an edge in survival under adverse and stressful environmental conditions. Hydrolytic activities allow *Vibrio* strains the faculty to adhere to the epithelial cells of the juveniles of both *D. labrax* and *Sparus aurata*, to break the first barrier of natural defence and to colonize all internal organs, inducing external haemorrhages, necrotic eyes, deep ulcers, haemorrhagic liver, pale kidney and splenomegaly (Ben Kahla-Nakbi et al. 2009). In fact, virulence factors of non-toxigenic *V. parahaemolyticus*centre on proteases and adhesines. Serum plays an important role in the natural defence of fish against bacterial infections. The resistance to the serum is often associated with bacterial lipopolysaccharide (LPS). The high molecular weight portion of LPS is usually responsible for serum resistance in several bacterial pathogens (Amaro et al. 1995). Our results showed that virulent strains can survive the bactericidal action of the sea bass serum. Resistance to the bactericidal mechanisms of normal serum appears to be an important contributor to the virulence of fish-pathogenic *Vibrio* (Bordas et al. 1996).

However, the *V. parahaemolyticus* strains isolated during this outbreak were pathogenic to fish, but they did not possess any detectable virulence gene. The high diversity of strains belonging to the same species may explain the strain-specific expression of virulence factors and consequently the different data obtained from various geographic areas (Baffone et al. 2000).

CONCLUSIONS

This is the first report on the characterization and virulence of *V. parahaemolyticus* for sea bass in Tunisia and in Mediterranean coast. Vibriosis causes enormous economic loss in Tunisian aquaculture. The present study indicates that high abundance of extracellular hydrolytic enzyme profiles in the different strains isolated during this outbreak can be a main reason for the high mortality caused by this species. Knowing that, most of the fish larvae elevated in offshore farm are imported from several hatcheries in the Mediterranean Sea countries (France, Spain and Italy), and the inshore farm has its own hatchery. Moreover, genetic study is needed to view if strains responsible for such mortalities are imported with these larvae and introduced into Tunisian environment or they are already present in our environment (autochthon strains).

NOTES

Competing Interests

The authors declare that they have no competing interests.

REFERENCES

1. Alcaide E, Amaro C, Todoli R, Oltra R: Isolation and characterization of Vibrio parahaemolyticus causing infection in Iberian toothcarp Aphanius iberus . Dis Aquat Organ 1999,35(1):77–80. doi:10.3354/dao035077 doi:10.3354/dao035077
2. Alsina M, Blanch AR: A set of keys for biochemical identification of environmentalVibrio species. J Appl Bacteriol 1994,76(1):79–85. 10.1111/j.1365-2672.1994.tb04419.x
3. Amaro C, Biosca EG, Fouz B, Alcaide E, Esteve C: Evidence that water transmits Vibrio vulnificus biotype 2 infections to eels. Appl Environ Microbiol 1995,61(3):1133–1137.
4. Austin B, Austin DA: Bacterial fish pathogens, disease of farmed and wild fish. 4th edition. Godalming: Springer Praxis; 2007.
5. Baffone W, Pianetti A, Bruscolini F, Barbieri E, Citterio B: Occurrence and expression of virulence-related properties of Vibrio species isolated from widely consumed seafood products.

Int J Food Microbiol 2000,54(1–2):9–18. doi:10.1016/S0168–1605(99)00189–0 doi:10.1016/S0168-1605(99)00189-0 10.1016/S0168-1605(99)00189-0

6. Ben Kahla-Nakbi A, Besbes A, Bakhrouf A, Alcaide E: Characterization and virulence properties of Vibrio strains isolated from diseased gilthead sea bream (Sparus aurata) cultured in Tunisia. Bull Eur Assoc Fish Pathol 2007, 27: 90–99.
7. Ben Kahla-Nakbi A, Chaieb K, Bakhrouf A: Investigation of several virulence properties among Vibrio alginolyticus strains isolated from diseased cultured fish in Tunisia. Dis Aquat Organ 2009,86(1):21–28.
8. Bordas MA, Balebona MC, Zorrilla I, Borrego JJ, Morinigo MA: Kinetics of adhesion of selected fish-pathogenic Vibrio strains of skin mucus of gilt-head sea bream (Sparus aurata L .). Appl Environ Microbiol 1996,62(10):3650–3654.
9. Cai J, Li J, Thompson KD, Li C, Han H: Isolation and characterization of pathogenicVibrio parahaemolyticus from diseased post-larvae of abalone Haliotis diversicolor supertexta . J Basic Microbiol 2007,47(1):84–86. doi:10.1002/jobm.200610192 doi:10.1002/jobm.200610192 10.1002/jobm.200610192
10. Christensen WB: Urea decomposition as a means of differentiating Proteus and paracolon cultures from each other and from Salmonella and Shigella types. J Bacteriol 1946,52(4):461–466.
11. Croci L, Suffredini E, Cozzi L, Toti L, Ottaviani D, Pruzzo C, Serratore P, Fischetti R, Goffredo E, Loffredo G, Mioni R: Comparison of different biochemical and molecular methods for the identification of Vibrio parahaemolyticus . J Appl Microbiol 2006,102(1):229–237. doi:10.1111/j.1365–2672.2006.03046.x doi:10.1111/j.1365-2672.2006.03046.x
12. DePaola A, Nordstrom JL, Bowers JC, Wells JG, Cook DW: Seasonal abundance of total and pathogenic Vibrio parahaemolyticus in Alabama oysters. Appl Environ Microbiol 2003,69(3):1521–1526. 10.1128/AEM.69.3.1521-1526.2003
13. Dileep V, Kumar HS, Kumar Y, Nishibuchi M, Karunasagar I: Application of polymerase chain reaction for detection of Vibrio parahaemolyticus associated with tropical seafoods and

coastal environment. Lett Appl Microbiol 2003,36(6):423–427. doi:10.1046/j.1472–765X.2003.01333.x doi:10.1046/j.1472-765X.2003.01333.x 10.1046/j.1472-765X.2003.01333.x

14. Food and Drug Administration: Bacteriological analytical manual. 2nd edition. Washington DC: Association of Official Analytical Chemists; 1992.
15. Hammack TS, Amaguana RM, Andrews WH: An improved method for the recovery ofSalmonella serovars from orange juice using universal preenrichment broth. J Food Prot 2001,64(5):659–663.
16. Hara-Kudo Y, Sugiyama K, Nishina T, Saitoh A, Nakagawa H, Ichihara T, Konuma H, Hasegawa J, Kumagai S: Detection of TDH-producing Vibrio parahaemolyticusO3:K6 from naturally contaminated shellfish using an immunomagnetic separation method and chromogenic agar medium. Kansenshogaku Zasshi2001,75(11):955–960.
17. Hervio-Heath D, Colwell RR, Derrien A, Robert-Pillot A, Fournier JM, Pommepuy M: Occurrence ofpathogenic Vibrios in coastal areas of France. J Appl Microbiol2002,92(6):1123–1135. doi:10.1046/j.1365–2672.2002.01663.x doi:10.1046/j.1365-2672.2002.01663.x 10.1046/j.1365-2672.2002.01663.x
18. Hormansdorfer S, Wentges H, Neugebaur-Buchler K, Bauer J: Isolation of Vibrio alginolyticus from seawater aquaria. Int J Hyg Environ Health 2000,203(2):169–175. 10.1078/S1438-4639(04)70024-3
19. Jayasree S: Identification of immune cells interacting with Vibrio spp. and its in vitro post-phagocytic killing mechanism of haemocytes in the penaeid shrimp,Penaeus indicus H. Milne Edwards. J Fish Dis 2009,32(4):359–365. doi:10.1111/j.1365–2761.2009.01018.x. doi:10.1111/j.1365-2761.2009.01018.x. 10.1111/j.1365-2761.2009.01018.x
20. Karunasagar I, Sugumar G, Reilly PJ: Rapid polymerase chain reaction method for detection of Kanagawa positive Vibrio parahaemolyticus in seafoods. Int J Food Microbiol 1996,31(1–3):317–323.
21. Kim YB, Okuda J, Matsumoto C, Takahashi N, Hashimoto S, Nishibuchi M: Identification ofVibrio parahaemolyticus strains at

the species level by PCR targeted to the toxRgene. J Clin Microbiol 1999,37(4):1173–1177.

22. Lee CY, Cheng MF, Yu MS, Pan MJ: Purification and characterization of a putative virulence factor, serine protease, from Vibrio parahaemolyticus . FEMS Microbiol Lett 2002,209(1):31–37. doi:10.1016/S0378–1097(02)00477–9 doi:10.1016/S0378-1097(02)00477-9 10.1111/j.1574-6968.2002.tb11105.x
23. Martinez-Urtaza J, Simental L, Velasco D, DePaola A, Ishibashi M, Nakaguchi Y, Nishibuchi M, Carrera-Flores D, Rey-Alvarez C, Pousa A: Pandemic Vibrio parahaemolyticus O3:K6, Europe. Emerg Infect Dis 2005,11(8):1319–1320. 10.3201/eid1108.050322
24. Matsumoto C, Okuda J, Ishibashi M, Iwanaga M, Garg P, Rammamurthy T, Wong HC, Depaola A, Kim YB, Albert MJ, Nishibuchi M: Pandemic spread of an O3:K6 clone of Vibrio parahaemolyticus and emergence of related strains evidenced by arbitrarily primed PCR and toxRS sequence analyses. J Clin Microbiol 2000,38(2):578–585.
25. Nair GB, Ramamurthy T, Bhattacharya SK, Dutta B, Takeda Y, Sack DA: Global dissemination of Vibrio parahaemolyticus serotype O3:K6 and its serovariants.Clin Microbiol Rev 2007,20(1):39–48. doi:10.1128/CMR.00025–06 doi:10.1128/CMR.00025-06
26. Nishibuchi M, Kaper JB: Thermostable direct hemolysin gene of Vibrio parahaemolyticus : a virulence gene acquired by a marine bacterium. Infect Immun1995,63(6):2093–2099.
27. Okuda J, Ishibashi M, Abbott SL, Janda JM, Nishibuchi M: Analysis of the thermostable direct hemolysin (tdh) gene and the tdh -related hemolysin (trh) genes in urease-positive strains of Vibrio parahaemolyticus isolated on the West Coast of the United States. J Clin Microbiol 1997,35(8):1965–1971.
28. Ottaviani D, Santarelli S, Bacchiocchi S, Masini L, Ghittino C, Bacchiocchi I: Presence of pathogenic Vibrio parahaemolyticus strains in mussels from the Adriatic Sea, Italy. Food Microbiol 2005, 22: 585–590. 10.1016/j.fm.2005.01.005
29. Quilici ML, Robert-Pillot A, Picart J, Fournier JM: Pandemic Vibrio parahaemolyticusO3:K6 spread, France. Emerg Infect Dis 2005,11(7):1148–1149. 10.3201/eid1107.041008

30. Quindos G, Salesa R, Carrillo-Munoz AJ, Lipperheide V, Jaudenes L, San Millan R, Torres-Rodriguez JM, Ponton J: Multicenter evaluation of ATB fungus: a standardized micromethod for yeast susceptibility testing. Chemotherapy 1994,40(4):245–251. 10.1159/000239200
31. Reed M, Muench H: A simple method for estimating fifty per cent endpoints. Am J Hyg 1938, 27: 493–497.
32. Robert-Pillot A, Guenole A, Lesne J, Delesmont R, Fournier JM, Quilici ML: Occurrence of the tdh and trh genes in Vibrio parahaemolyticus isolates from waters and raw shellfish collected in two French coastal areas and from seafood imported into France. Int J Food Microbiol 2004,91(3):319–325. doi:10.1016/j.ijfoodmicro.2003.07.006 doi:10.1016/j.ijfoodmicro.2003.07.006 10.1016/j.ijfoodmicro.2003.07.006
33. Zhang XH, Austin B: Haemolysins in Vibrio species. J Appl Microbiol 2005,98(5):1011–1019. doi:10.1111/j.1365–2672.2005.02583.x doi:10.1111/j.1365-2672.2005.02583.x 10.1111/j.1365-2672.2005.02583.x
34. Zorrilla I, Arijo S, Chabrillon M, Diaz P, Martinez-Manzanares E, Balebona MC, Morinigo MA: Vibrio species isolated from diseased farmed sole, Solea senegalensis (Kaup), and evaluation of the potential virulence role of their extracellular products. J Fish Dis2003,26(2):103–108. 10.1046/j.1365-2761.2003.00437.x

CHAPTER

6

MOLECULAR IDENTIFICATION AND TRANSMISSION STUDIES OF X-CELL PARASITES FROM ATLANTIC COD GADUS MORHUA(GADIFORMES: GADIDAE) AND THE NORTHERN BLACK FLOUNDER PSEUDOPLEURONECTES OBSCURUS (PLEURONECTIFORMES: PLEURONECTIDAE)

MA Freeman[1,2], M Eydal[3], M Yoshimizu[5], K Watanabe[6], AP Shinn[2], K Miura[7], K Ogawa[4]

[1]Institute of Ocean and Earth Sciences & Institute of Biological Sciences, University of Malaya, Kuala Lumpur, Malaysia

[2]Institute of Aquaculture, University of Stirling, Scotland, UK

[3]Institute for Experimental Pathology, University of Iceland, Keldur, Iceland

[4]Laboratory of Fish Diseases, The University of Tokyo, Tokyo, Japan

[5]Laboratory of Microbiology, Hokkaido University, Hakodate, Japan

[6]Department of Aquatic Bioscience and Industry Faculty of Bio-Industry, Tokyo University of Agriculture, Abashiri, Japan

[7]Hokkaido Central Fisheries Experiment Station, Yoichi, Japan

ABSTRACT

Background

Epidermal pseudotumours from *Hippoglossoides dubius* and *Acanthogobius flavimanus* in Japan and gill lesions in *Limanda limanda* from the UK have been shown to be caused by phylogenetically related protozoan parasites, known collectively as X-cells. However, the phylogenetic position of the X-cell group is not well supported within any of the existing protozoan phyla and they are currently thought to be members of the Alveolata.

Ultrastructural features of X-cells in fish pseudotumours are somewhat limited and no typical environmental stages, such as spores or flagellated cells, have been observed. The life cycles for these parasites have not been demonstrated and it remains unknown how transmission to a new host occurs.

In the present study, pseudobranchial pseudotumours from Atlantic cod, *Gadus morhua*, in Iceland and epidermal pseudotumours from the northern black flounder, *Pseudopleuronectes obscurus*, in Japan were used in experimental transmission studies to establish whether direct transmission of the parasite is achievable. In addition, X-cells from Atlantic cod were sequenced to confirm whether they are phylogenetically related to other X-cells and epidermal pseudotumours from the northern black flounder were analysed to establish whether the same parasite is responsible for infecting different flatfish species in Japan.

Results

Phylogenetic analyses of small subunit ribosomal DNA (SSU rDNA) sequence data from Atlantic cod X-cells show that they are a related parasite that occupies a basal position to the clade containing other X-cell parasites. The X-cell parasite causing epidermal pseudotumours in *P. obscurus* is the same parasite that causes pseudotumours in *H. dubius*. Direct, fish to fish, transmission of the X-cell parasites used in this study, via oral feeding or injection, was not achieved. Non-amoeboid X-cells are contained within discrete sac-like structures that are loosely attached to epidermal pseudotumours in flatfish; these X-cells are able to tolerate exposure to seawater.

A sensitive nested PCR assay was developed for the sub clinical detection of both parasites and to assist in future life cycle studies. PCR revealed that the parasite in *P. obscurus* was detectable in non-pseudotumourous areas of fish that had pseudotumours present in other areas of the body.

Conclusions

The inability to successfully transmit both parasites in this study suggests that either host detachment combined with a period of independent development or an alternate host is required to complete the life cycle for X-cell parasites. Phylogenetic analyses of SSU rDNA confirm a monophyletic grouping for all sequenced X-cell parasites, but do not robustly support their placement within any established protist phylum. Analysis of SSU rDNA from X-cells in Japanese flatfish reveals that the same parasite can infect more than one species of fish.

BACKGROUND

X-cell disease in fish typically develops either as epidermal pseudotumours, gill filament lesions or pseudobranchial swellings in various marine species [1]. X-cells associated with epidermal pseudotumours in the flathead flounder, *Hippoglossoides dubius* Schmidt, 1904 and the yellowfin goby *Acanthogobius flavimanus* (Temminck et Schlegel, 1845) from northern Japan, have been shown, using small subunit ribosomal DNA (SSU rDNA) sequence data, to be related protozoan parasites that have an unresolved taxonomic identity [2]. Freeman [1] further demonstrated that the X-cell parasite causing gill filament lesions in the European dab, *Limanda limanda* (L., 1758), is related to the two Japanese X-cell parasites, and suggested they belong in the alveolate group and that they are basal members of the Myzozoa. Pseudobranchial X-cell pseudotumours occur in gadoid fish from the Pacific and Atlantic Oceans [3], but thus far have not been studied phylogenetically.

In the coastal waters of Hokkaido, seven species of pleuronectid flatfish have been reported to have epidermal pseudotumours containing X-cells [4]. Of these seven species, only X-cells from *H. dubius* have been characterised using SSU rDNA analyses [5], and it is not known how host specific X-cell parasites are, and whether the same X-cell parasite is responsible for causing epidermal pseudotumours in more than one flatfish species.

Experimental transmission of X-cell disease between fish has been attempted, but has never convincingly been achieved. However, most transmission studies were based on the assumption, at the time, that the X-cell condition had a viral aetiology and some studies may not have been suitable for the successful experimental transmission of protozoan parasites. A cell-free homogenate of epidermal pseudotumour tissue from the yellowfin goby, *A. flavimanus*, was subcutaneously inoculated into uninfected individuals, but no pseudotumour growth was observed during the trial [6]. Gill lesion regression was observed in European dab, *L. limanda*, that were being maintained in the laboratory, and subsequent attempts to transmit the X-cell condition to uninfected fish using an inoculum derived from X-cell material were not successful [7]. Cohabitation experiments with Atlantic cod, *Gadus morhua* L. 1758, were conducted by Morrison *et al.* [3], but were inconclusive due to the high mortalities of wild-caught X-cell infected fish under experimental conditions and the uncertainty that visibly uninfected wild-caught cod were truly naïve at the start of the experiment. However, a single uninfected fish did develop a large unilateral lesion after two months cohabitation with an X-cell infected cod.

In the present study we sampled pseudobranchial pseudotumours from Atlantic cod, *G. morhua*, from Iceland and epidermal pseudotumours from the northern black flounder, *Pseudopleuronectes obscurus*, (Herzenstein, 1890) from northern Japan. Experimental transmissions of X-cell parasites from Atlantic cod and northern black flounder to naïve fish were attempted and a sensitive PCR assay was developed for their sub-clinical detection. In addition, SSU rDNA analyses were utilised in order to confirm whether the same parasite is responsible for infecting different flatfish species in northern Japan, and whether the X-cell parasite causing pseudobranchial pseudotumours in Atlantic cod is phylogenetically related to those causing epidermal tumours and gill filament lesions in flatfish and gobies.

METHODS

Sampling and Transmission Studies of X-cell Parasites from Cod in Iceland

Pseudobranchial pseudotumours were excised from naturally infected young Atlantic cod selected from a land based cod farm in North West Iceland that is stocked with a local population of wild juvenile cod caught in an adjacent fjord for on-rearing. A sub-sample of pseudotumour tissue from each fish was fixed in 10% buffered formalin and 95% ethanol for histology and DNA analyses respectively; the remainder was briefly stored at 4°C until required. Pseudobranchial tissues from cod with no pseudobranchial swellings were fixed in 95% ethanol for use as negative control tissue in the DNA study. The pseudotumour tissue was gently homogenised in PBS with a sterile pestle and mortar and diluted with PBS to allow sufficient homogenate for the experiment. The homogenate was viewed under a compound microscope to ensure that intact X-cells were present. Disease free hatchery-reared juvenile cod, used in the experiment, came from the hatchery station of the Marine Research Institute in Iceland at Staður, Grindavík and the trial was conducted at the Sandgerði Marine Centre, Iceland. Fish were maintained in 1.5 m diameter rearing tanks and supplied with clean bore-hole sea water at a constant temperature of 9°C.

Experimental fish were split into three groups and kept in separate tanks. One group was orally intubated (oral group, n = 13) with 1 ml of pseudotumour homogenate using a flexible 3 ml plastic pipette; a second group was given an intracoelomic (IC) injection (IC group, n = 15) of 100 μl of pseudotumour homogenate; and a control group (n = 10) received no treatment. The fish were observed and fed daily and the trial ran for 17 weeks. At the start of trial fish had a mean weight of 65 g and a mean length of 19 cm.

Three fish from each group were sacrificed and examined for signs of pseudotumour development nine weeks into the trial, and the remaining fish were examined on week 17 at the end of the trial (oral group, n = 10; IC group, n = 4; control group, n = 6). All pseudobranchial tissues were removed, longitudinally bisected and fixed in either 95% ethanol for DNA analyses or 10% buffered formalin for histology. Nine fish that died during the trial (weeks 1-9) were frozen until week 17 and examined with those that survived the duration of the experiment.

Sampling and Transmission Studies of X-cell Parasites from Flatfish in Japan

Pseudopleuronectes obscurus with typical X-cell epidermal pseudotumours were taken as a by-catch by shrimp fishermen during the summer shrimp fishing season in Notsuke bay in Eastern Hokkaido, Japan. Infected fish were maintained in tanks until required for the experimental infections. Pseudotumours were dissected from the underlying tissues in recently culled fish. Some pseudotumour tissue was fixed in 10% buffered formalin and 95% ethanol for histology and DNA analyses respectively, the remainder was minced with scissors, half being set aside for the feeding experiment and the rest gently homogenised in physiological saline using a pestle and mortar and a ground-glass tissue homogeniser. The resulting homogenate was passed through a fine gauze cloth to remove large pieces of connective tissue and centrifuged at 1800 g for 5 min. The pellet was resuspended in 3 ml of saline, examined microscopically for the presence of intact X-cells and used immediately as an inoculum in the fin base injection experiment. Negative control skin samples, for the DNA study, were also taken from uninfected areas of X-cell infected fish and from fish with no visible signs of pseudotumours.

X-cells from discrete sac-like structures, found at the extremities of epidermal pseudotumours, were also examined microscopically and placed in seawater in 24-well tissue culture plates and maintained at 15°C to observe any amoeboid forms and monitor their ability to adhere to plastic over a 24 hour period.

Fish used in the transmission experiments were hatchery reared and assumed to be free from disease. Two species of flatfish were used in the experimental infections: barfin flounder, *Verasper moseri* (Jordan et Gilbert, 1898), family Pleuronectidae (righteye flounders); and Japanese flounder, *Paralichthys olivaceus* (Temminck et Schlegel, 1846), a lefteye flounder, family Paralichthyidae (Large-tooth flounders). In total, 150 fish of each species were used; fifty as control fish, fifty for the feeding of infected material and fifty for the fin base injection experiment.

For the feeding experiment, small pieces of pseudotumour tissue approximately 1 mm^3 in size were fed directly to fish that had been starved for two days and their feeding behaviour was observed to ensure that fish fed uniformly. For the subcutaneous fin base injection experiment, fish were anaesthetised and 25 μl of inoculum injected with a fine needle at the base of the dorsal fin rays. 25 μl was considered the largest volume to safely inject into fish smaller than 50 mm total length. Control fish received no treatment. Experimental fish were kept in six separate tanks in a flow-through system at 15°C in full strength seawater, and observed for three months. Total length measurements of fish at the beginning of the experiment were: barfin flounder 35-49 mm (mean 41.9 mm; n = 20) and Japanese flounder 23-37 mm (mean 30.8 mm; n = 20). At the end of the experiment, fish were examined using a dissecting microscope for the presence of epithelial pseudotumours on the skin and fins. PCR was not performed on tissue samples from the fin base injection sites for the flatfish group.

Histological Examination of Pseudotumours

Fresh pseudotumours were fixed in 10% buffered formalin for 48 h and transferred to 70% ethanol for processing in an automatic tissue processor. After embedding in wax, blocks were trimmed and sections of 5 μm were cut on a Reichert-Jung Biocut microtome before being stretched on a water-bath at 45°C and floated on to slides. Slides were dried overnight in an oven at 60°C prior to staining with haematoxylin and eosin. The sac-like structures from *P. obscurus* were carefully removed, from live fish, using forceps, and fixed in 2.5% glutaraldehyde in 0.1 M Sorensen›s phosphate buffer (pH 7.4) at 4°C for 4 h. Fixed tissues were then rinsed in 0.1 M Sorensen›s phosphate buffer (pH 7.4) at 4°C overnight before being post-fixed in 1% osmium tetroxide for 1 h, dehydrated through an ethanol series, embedded in Spurr›s resin and polymerised at 60°C for 48 h. Semi-thin sections of 0.5 μm thickness were cut using a glass knife on a Reichert Ultracut E ultramicrotome and stained with 1% Azur II followed by 1% methylene blue in 1% borax (50:50).

DNA Extraction, PCR Amplification and SSU rDNA Sequencing of X-cell Parasites

Ethanol fixed pseudotumour material and negative control tissues were homogenized with a sterile Eppendorf pestle and digested overnight at 56°C in 0.4 ml high concentration urea buffer containing 100 μg/ml proteinase K. DNA was extracted using a QIAamp DNA Mini Kit (QIAGEN Inc.) following the manufacturer›s tissue protocol and used as template DNA for subsequent PCR reactions.

Partial SSU rDNA for both X-cell parasites was first amplified using the primer pair: H-F1 M 5› gttctttcttgattctatrag 3› and H-R3 M 5› taggaattcctcgttcaagacg 3› that were modified from degenerate haplosporidian primers [8], and require a 48°C annealing temperature. Universal primers 18e [9] and 606f/r and 18gM [10] were used in combination with the above primers and with more specific X-cell primers designed from initial sequence reads (Table 1). All PCRs were performed in 20 μl volumes containing ~10 ng of genomic DNA, 15 pmol of each primer, 0.25 mM of dNTP, PCR buffer with a final $MgCl_2$ concentration of 2 mM and 0.5 units of Taq DNA polymerase. After an initial denaturation at 95°C for 5 min, samples were subjected to 35 cycles of amplification (denaturation at 95°C for 30 s, primer annealing at 55°C for 30 s (unless otherwise stated), and extension at 72°C for 1 min), followed by a 7 min terminal extension at 72°C. PCR amplicons were purified using a PCR purification kit (QIAGEN Inc) and used directly in sequencing reactions. The sequencing reactions were performed using BigDye®Terminator Cycle Sequencing chemistry utilising the same oligonucleotide primers that were used for the PCRs. DNA sequencing was completed on amplicons from four infected fish for each species. Sense and anti-sense strands were sequenced for all PCR products and contiguous sequences constructed manually using CLUSTAL_X [11] and BioEdit [12]. CLUSTAL_X was used for the sequence alignments with the settings for gap opening/extension penalties being adjusted manually to achieve optimum alignments. Regions of ambiguous sequence alignments were manually edited using the BioEdit sequence alignment editor and final alignments of specific taxa were generated using CLUSTAL_X.

Table 1: PCR primers used for the nested amplification of X-cell SSU rDNA.

nested **primer pairs**	**fwd primer (5'-3')**	**rev primer (5'-3')**	**annealing temp °C**	**size (bp)**
cod				
i) 305f[s]/H-R3M[x]	tgacctatcatgctgtgatgg	taggaattcctcgttcaagacg	54	1350
ii) 390Xf[x]/1400r[s]	agagggagcctgagagacg	agcaagcccgtatggagaagacg	52	1000
flounder				
i) 280f[s]/H-R3M[x]	atccatcagccatcgacgc	taggaattcctcgttcaagacg	55	1290
ii) 390Xf[x]/1000r[s]	agagggagcctgagagacg	tcgtccgatcctcagtcgg	55	645

[s] X-cell species-specific primer

[x] X-cell universal primer

Appropriate taxa were chosen for the phylogenetic analyses by performing nucleotide BLAST searches with the X-cell sequences [13] and reviewing previous phylogenetic analyses that included X-cell sequence data [1, 2, 5]. Additional taxa were chosen to further represent the major protist phyla (see Additional file 1).

Phylogenetic analyses were conducted using maximum parsimony (MP) methodologies in PAUP*4.0 beta10 [14]. MP analysis was done using a heuristic search with tree bisection-reconnection (TBR) branch swapping, 10 random taxon addition replicates, using the accelerated transformation (ACCTRAN) option. Gaps were treated as missing data and clade support was assessed using bootstrapping with 1000 replicates. Bayesian inference (BI) analyses were conducted using MrBayes v. 3.0 [15]. Models of nucleotide substitution were evaluated for the data using MrModeltest v. 2.2 [16]. The most parameter-rich evolutionary model based on the AIC was the general time-reversible, GTR+I+G model of evolution. Therefore, the settings used for the analysis were nst = 6, with the gamma-distributed rate variation across sites and a proportion of invariable sites (rates = invgamma). The priors on state frequency were left at the default setting (Prset statefreqpr = dirichlet (1,1,1,1)). Posterior probability distributions were generated using the Markov Chain Monte Carlo (MCMC) method with four chains being run simultaneously for 1,000,000 generations. Burn in was set at 2500 and trees were sampled every 100 generations making a total of 7500 trees used to compile the consensus trees.

Specific Nested PCR Detection System

A highly sensitive and specific nested PCR detection assay was developed for the X-cell parasites from Atlantic cod and the northern black flounder, using the newly acquired SSU rDNA sequence data. Oligonucleotide primers were designed that were universal for all currently known X-cell sequences and specific primers were also designed for each new X-cell parasite sequence. These primers were then used in the (universal/specific) combinations described in Table 1. Nested PCR reactions were conducted using the same PCR conditions described above, but with adjusted annealing temperatures (Table 1), the products of the first round reaction were used as template DNA for the second reaction.

In accordance with section 8.6 of the ICZN's International Code of Zoological Nomenclature, copies of this article are deposited at the following five publicly accessible libraries: Natural History Museum, London, UK; American Museum of Natural History, New York, USA; Museum National d'Histoire Naturelle, Paris,

France; Russian Academy of Sciences, Moscow, Russia; Academia Sinica, Taipei, Taiwan.

RESULTS

Atlantic cod with pseudobranchial pseudotumours were hand-selected from schooling juvenile fish in large indoor tanks. Infected cod appeared to be darker in colour and generally smaller than uninfected ones from the same year class (Figure 1a). Infected pseudobranchs were significantly enlarged and had a creamy-whitish appearance, whereas uninfected pseudobranchs had a normal gill-red colour (Figure 1b & 1c). The excised pseudobranchial pseudotumours were approximately 0.8 cm in length from fish that measured 10-12 cm in total length. The pseudobranchial pseudotumours from six bilaterally infected fish were used in the transmission experiment.

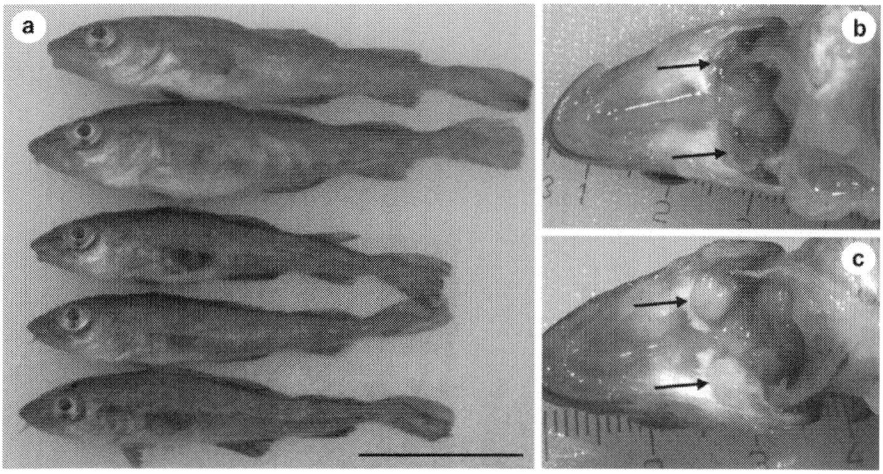

Figure 1: **Pseudobranchial pseudotumours in Atlantic cod*Gadus morhua*from Iceland**. a) Five cod from the same year-class; two uninfected fish (top) and three infected fish (bottom). b) Pseudobranchs have a gill-red colour in healthy fish and are enlarged and have a creamy-whitish appearance when infected (c) (black arrows). Scale bars a = 4 cm, b & c visible.

Pseudopleuronectes obscurus with epithelial pseudotumours caused by X-cell infection were caught in Notsuke Bay, Hokkaido (Figure 2a-d). Epithelial pseudotumours were typical of those described from numerous flatfish species from Hokkaido and were often large and significantly raised from the body surface, but did not penetrate the underlying musculature (Figure 2a & 2b), making them easy to remove from the fish to prepare the homogenate for the transmission experiment. Pseudotumours were always pigmented on the ocular (eyed) side, but could be either pigmented or non-pigmented on the abocular (non-eyed) side (Figure 2c & 2d) and when sampled, juvenile fish were also sometimes infected (Figure 2d). Pseudotumours often extended to the edge of the fins (Figure 2c & Figure 3a). Here, and at other margins of the pseudotumour, sac-like structures were observed that could be detached, intact, with mild pressure (Figure 3b). The sacs were relatively fragile, easy to rupture and contained X-cells and other cellular debris (Figure 3c). X-cells from ruptured sacs that had been maintained in seawater for 24 hrs showed no signs of adhering to the plastic tissue culture plates, had no pseudopodia and were not amoeboid in form. After 24 hrs in seawater they appeared similar to ones that were freshly removed from sacs and had not suffered any noticeable shrinkage (Figure 3d).

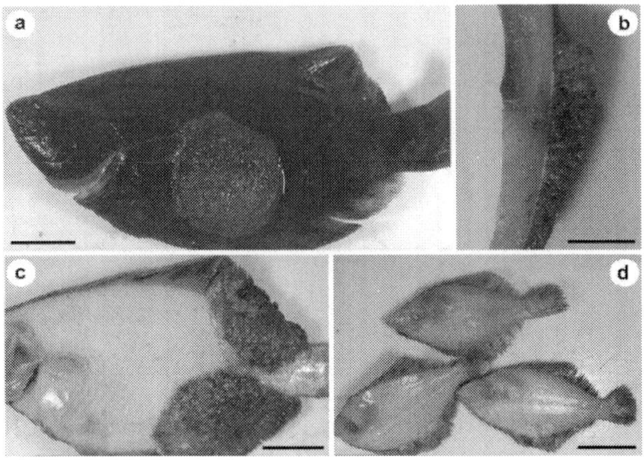

Figure 2: **Epidermal X-cell pseudotumours on formalin-fixed*Pseudopleuronectes obscurus*from Hokkaido, Japan**. a) A large central dorsal pseudotumour, seen in cross section (b). c) Pseudotumours can extend from the dorsal surface to the ventral surface and remain pigmented. d) Juvenile fish are also infected and ventral pseudotumours can also be unpigmented. Scale bars a, c & d = 3 cm, b = 10 mm.

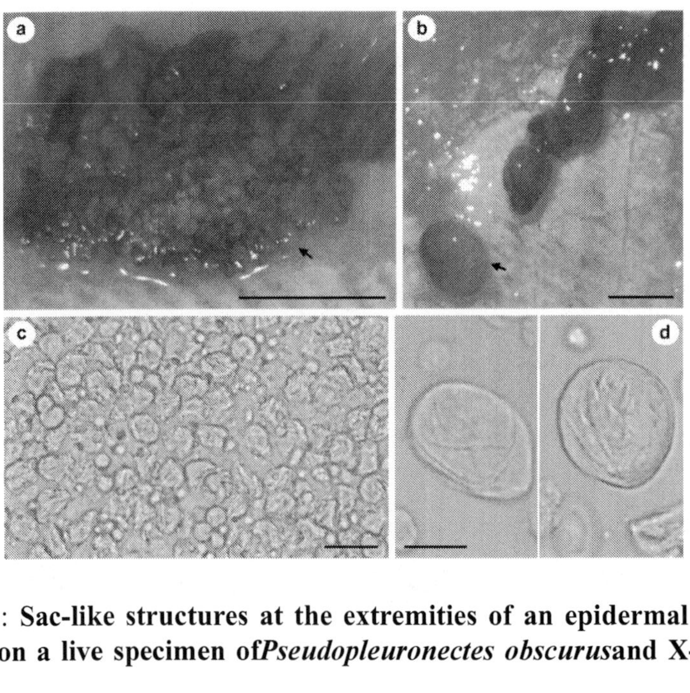

Figure 3: **Sac-like structures at the extremities of an epidermal pseudotumour on a live specimen of*Pseudopleuronectes obscurus*and X-cells re-**

leased from the sacs in tissue culture plates. a & b) Appearance of a fresh epidermal pseudotumour showing sac-like structures that can be separated from the main part of the pseudotumour intact, black arrows indicate the same sac attached and detached in a & b respectively. c) X-cells and host tissue debris immediately after being released from a sac-like structure. d) X-cells maintained in plastic tissue culture plates containing seawater, 24 hrs after being released from a sac-like structure. Scale bars a = 10 mm, b = 2 mm, c = 20 μm, d = 10 μm.

Histological analyses of the pseudotumours of cod and northern black flounder revealed that X-cells were present in large numbers in both pseudotumour types (Figure 4a-d). In *P. obscurus*, the X-cells were limited to the epidermal skin layer and were arranged in a regular pattern of folded tissues that contained numerous melanomacrophages and other unidentifiable host cells (Figure 4a & 4b). In cod the X-cells appeared to be adjacent to, but not intermingled with, pseudobranchial tissue (Figure 4c). However, pseudobranchial cartilages were also seen surrounded by X-cells deeper into the pseudotumour lesions (Figure 4d).

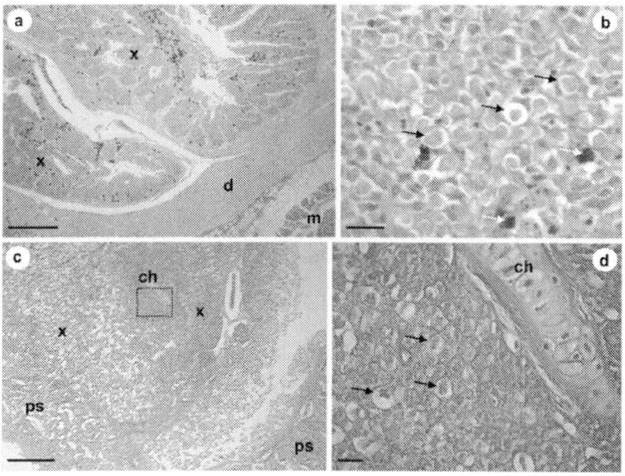

Figure 4: **Haematoxylin and eosin stained histological sections of X-cell pseudotumours**. a) Low magnification of a section through an epidermal pseudotumour in *P. obscurus*. Large numbers of X-cells (x) are in folds of infected host tissues. The dermis (d) and the underlying muscle (m) remain uninfected. b) High power magnification of X-cells in *P. obscurus* have a characteristic polygonal appearance with a lightly staining large nucleus (black arrows), melanomacrophages are present in high numbers (white arrows). c) Low magnification of a section through a pseudobranchial pseudotumour in juvenile Ice-

landic cod. X-cell masses (x) form adjacent to pseudobranchial tissues (ps) but are also found surrounding pseudobranchial cartilage containing chondrocytes (ch) deeper into the pseudotumour. d) High power magnification of the boxed section in Figure 3c; numerous X-cells (black arrows) form a distinctive mass that forms the bulk of the pseudotumour, pseudobranchial cartilage containing chondrocytes (ch) is enclosed by X-cells. Scale bars a & c = 400 μm, b & d = 20 μm.

Semi-thin sections of the sac-like structures show that in the central region of the sac, X-cells are found to be mostly restricted to large masses within the host epidermal tissue contained by a thick basal lamina-like membrane with underlying connective tissues. Some X-cells inside these large masses appear to be degenerate, whilst others were observed amongst the connective tissues outside of the membrane that contains the majority of the parasites (Figure 5a). At the border of the sac there is a less organised tissue structure observed. Aggregations of X-cells are no longer contained within membranous structures and they are less densely packed together with numerous X-cells being present at the peripheral margin of the sac and some appearing to be external to the sac (Figure 5b).

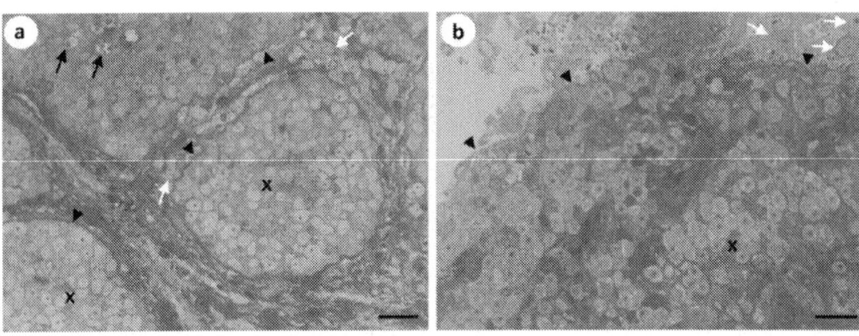

Figure 5: Semi-thin sections of the sac-like structures fromPseudopleuro-*nectes obscurus*. a) In the central region of the sac, X-cells (x) are found to be mostly restricted to large masses within the host epidermal tissue contained by a thick basal lamina-like membrane (black arrowheads), under which lies connective tissue. Some X-cells inside the large aggregations appear to be de-generate (black arrows), whilst others can be found amongst the connective tissues (white arrows) outside of the membrane that contains the majority of the X-cells. b) At the border of the sac there is a less organised tissue structure observed. Aggregations of X-cells (x) are no longer contained within membra-nous structures and numerous X-cells are present at the peripheral margin of the

sac (black arrowheads) and some appear to be independent from the sac (white arrows). Scale bars 20 μm.

Visual inspection of cod pseudobranchial tissues and microscopic inspection of both flatfish species at the end of the transmission experiments failed to detect the presence of abnormal tissue growth or signs of pseudotumours developing. Furthermore, PCR analyses of pseudobranchial DNA from experimental cod did not detect the presence of cod X-cell parasite DNA. PCR was not performed for the infection trials of the flatfish group.

Nested PCRs were successfully developed for both X-cell parasite species (Table 1), and no cross reactivity occurred between the two types of X-cells during PCR analyses. During the development of the nested PCR assay, negative control tissue from infected *P. obscurus* (skin samples taken >2 cm from pseudotumour sites), were sometimes found to be positive for the presence of X-cell DNA in the second round of the PCR; whereas skin samples from uninfected *P. obscurus* remained negative during both rounds of the PCR. Negative control tissues from visibly uninfected cod were not found to contain X-cell DNA with the nested PCR.

Partial SSU rDNA sequence data from the X-cell parasites infecting cod, *G. morhua*, and the northern black flounder, *P. obscurus*, were successfully obtained and have been deposited in GenBank (*G. morhua* X-cell, 1728 bp GU296508; *P. obscurus* X-cell, 1712 bp GU296509). A BLAST search with the contiguous X-cell sequence from *P. obscurus* showed a very high homology (> 99%) with the X-cell sequence reported from *H. dubius*[5]. A more detailed comparison with the X-cell sequence from *H. dubius* revealed a 99.76% similarity over 1692 bases of comparable data, confirming that the two isolates from different flatfish hosts should be considered conspecific.

Maximum parsimony and Bayesian phylogenetic analyses produced trees with a similar topology. The Bayesian inference tree (Figure 6) shows that all known X-cell sequences group together to form a monophyletic clade that is within the alveolates and is basal to the majority of the myzozoan taxa. The sequence from *P. obscurus* groups with the sequence from *H. dubius*and collectively they form a clade with the species infecting the yellowfin goby from Japan. More basal to the Japanese/Pacific grouping is the X-cell parasite from European dab and basal to this main X-cell group is the sequence for the Atlantic cod pseudobranchial parasite. This monophyletic grouping for the X-cell parasites was robustly supported with internal probabilities of 1.0 at all nodes within the group. However, it was less well supported from the main spine of the tree with a probability of 0.54 and did not group with other protist phyla.

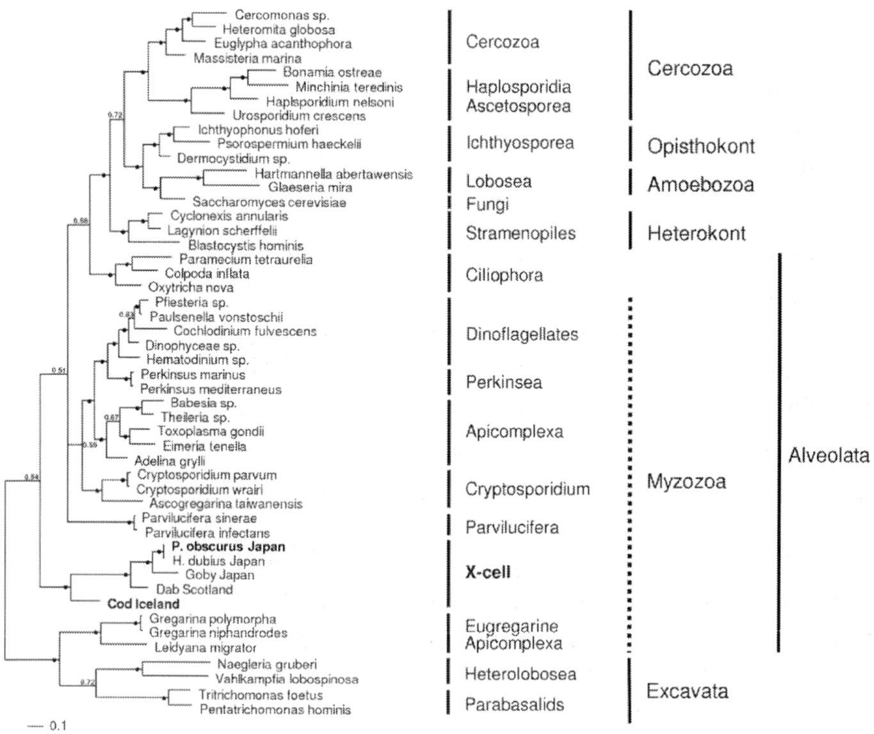

Figure 6: **Bayesian Inference phylogenetic tree**. Phylogenetic tree constructed using Bayesian Inference analyses using the general time-reversible (GTR+I+G) model of evolution. The numbers at the nodes represent the clade posterior probability, indicating the proportion of sampled trees that contained that branch. A solid circle at the node indicates a posterior probability of 0.9 and above. To the right of the tree are the currently accepted taxonomic rankings (order, class, phylum), at the far right are the higher scientific classifications (super-phylum).

DISCUSSION

The inability to successfully transmit both X-cell parasites in this study indicates that either the experimental design or conditions were not suitable for transmission to take place or that the X-cell stages seen in fish pseudotumours are not infective to new fish hosts. As comprehensive precautions, described in the methods, were taken in preparation for and throughout the transmission trials, we conclude that fish to fish transmission of X-cells is not readily achievable using feeding of infected tissues or by injection of an X-cell inoculum. Ideally we would have been able to use more fish and run the experiments for a longer period. However, the availability of infected tissues for feeding and homogenate preparations was limited.

We have shown that non-flagellate and non-amoeboid X-cells are easily shed from mature epidermal pseudotumours in discrete sac-like structures. These X-cells appear to lack a suitable apparatus with which to successfully infect a new fish host, unless direct transmission is via ingestion of the sacs. However, if simple ingestion of sloughed off sacs were sufficient then feeding of freshly minced tumour tissue should have resulted in transmission of the parasite. Therefore, we believe that it is more likely that the X-cells undergo further development in the environment or in an intermediate host before reinfection of the fish host can occur. The continued support for X-cells being phylogenetically related to the Myzozoa would also indicate that a flagellated life stage is present at some point in the life cycle which would be more capable of finding and infecting a new host fish.

Amoeboid X-cell Forms

As no environmental spore-like or flagellated stages have been described from X-cell pseudotumours in fish, in spite of numerous thorough histopathological and ultastructural studies [1], it is not immediately obvious how transmission to a new host might take place naturally. However, there have been several descriptions of amoeboid-like forms in advanced and ruptured X-cell pseudotumours which may represent the infective agent for a new host or a free-living stage leaving the fish host. Small amoeboid forms of X-cells have been described at the boundaries of epidermal pseudotumours in flatfish from Japan [17], free-floating amoeboid X-cells have been observed in advanced pseudobranchial pseudotumours in Atlantic cod [18] and free X-cells seen in the interlamellar spaces of ruptured gill lesions in the American plaice *H. platessoides*[19]. Freeman [1] recently suggested that transmission of X-cells from the fish host could be via these amoeboid stages and that this amoeboid phase to their life cycle would most likely be present in marine sediments and indeed most fish species known to become infected with X-cells are closely associated with the benthic environment [1]. However, recent molecular phylogenetic studies [[1, 2, 5], present study] do not support the hypothesis that X-cells are amphizoic amoebae having a free-living stage in coastal marine sediments. Furthermore, in the present study we have clearly demonstrated that the X-cells contained in sac-like structures are not amoeboid in form, do not adhere to plastic culture plates and do not form pseudopodia. We consider this sac stage to be the most likely form to naturally detach from the host fish as they are delicately attached to the main pseudotumour and they are seemingly a discrete unit. During the histological analysis of these structures, the margins of the sacs were not found to contain X-cells in the typical organised pattern that is observed in other parts of the pseudotumour. Rather, these X-cells are less densely packed together and some appear outside the sac membrane. This could signify that the sacs ultimately degenerate and release the X-cells into the environment. In addition, X-cells from the sacs appear to be able to withstand full salinity seawater, indicating that they are capable of surviving in the marine environment.

During the development of the diagnostic PCR, visibly uninfected areas of skin from infected *P. obscurus* were found to contain X-cell DNA. This supports the findings by Yamazaki et al. [17] who described the presence of very small amoeboid-like ‹wandering› X-cells, only 3-4 μm in diameter, at the boundaries of epidermal pseudotumours in flatfish from northern Japan. These data suggest that amoeboid forms of X-cells do exist in fish tissues and are probably responsible for autoinfection within the host, causing new pseudotumour growth once host invasion and initial development has occurred.

Experimental Design

It is possible that the methods used in the present study were not suitable for parasite transmission to occur between fish. Nevertheless, the transmission of fish parasites with monoxenous life cycles between fish using similar methods to the ones used in this study has been achieved for various fish parasites [20, 21]. However, successful parasite transmission can also be affected by environmental conditions such as salinity and temperature. Therefore, in order to reduce the likelihood of this occurring, in our experiments we endeavoured to use seawater whose temperature and salinity were the same as those of the seawater of the natural environment from where infected fish were taken.

If the release of X-cells from mature or ruptured pseudotumours is required for transmission to a new host, and transmission between fish is possible without an alternate host, then long-term cohabitation experiments in tanks with suitable substrate cover, allowing time for the development of an environmental stage, could prove to be more successful in future X-cell transmission studies. Furthermore, using fish with advanced pseudotumours, close to rupture, could be important for successful transmission to occur between experimental fish. However, both gadoid and pleuronectid fish with X-cell pseudotumours have very poor survival rates after capture [3, 22] which may hinder such long-term cohabitation experiments. In addition, lesion regression has also been observed in captive X-cell infected dab [22], suggesting that environmental conditions, such as temperature, may also be critical for the experimental transmission of X-cell parasites to take place and for pseudotumours to develop.

Geographical and Environmental Niches

X-cell pseudotumours in numerous wild fish populations have been reported to occur within extremely narrow geographic locations [4, 23], being found in high numbers at one location and almost absent in adjacent localities or bays. This irregular distribution pattern originally led to suggestions that the X-cell condition was caused by localised coastal pollution [24, 25]. A comprehensive study by Katsura et al. [4] of the distribution of X-cell infected fish in Hokkaido, however, revealed that the presence of X-cell pseudotumours was not related to coastal pollution, but had a strong correlation to the substrate type and amount of tidal water exchange. They found a very low frequency of infected fish (0.02-0.27%) on the north side of the Notsuke peninsula where a strong tide flows from north to south, but south of the peninsula, where the bay is protected from the strong current and the waters are 'stagnated and the bottom quite muddy', fish had a 12.3% prevalence of X-cell pseudotumours. Similar findings for the distribution patterns of populations of X-cell infected fish in Europe [23] suggest that certain environmental conditions such as substrate type and tidal flow potentially favour the survival and development of a free-living X-cell stage or can maintain an obligate alternate host that has a restricted zone for habitation.

Species Specificity

Katsura et al. [4] suggested that species specificity was present among X-cell infections in flatfish from Hokkaido, as some pleuronectid fish (*Liopsetta pinnifasciata* and *Limanda punctatissimus*) were collected in large numbers from numerous sites with muddy substrates with other X-cell infected pleuronectid fish being present, but they were never found to be infected with X-cells themselves. During this study we have shown that the same parasite causes X-cell pseudotumours in both *H. dubius* and *P. obscurus* from geographically distant locations in northern Japan, demonstrating that X-cell parasites are not species specific in some pleuronectid fish. It would be interesting to expand on this study and investigate whether the same parasite infects all seven pleuronectid fish known to be susceptible to epidermal X-cells infections in northern Japan. In our experimental transmission study, we used two fish species due to their availability from hatcheries; *Verasper moseri* (family Pleuronectidae) and *Paralichthys olivaceus* (family Paralichthyidae). *Verasper moseri*, sampled from the same location we sampled *P. obscurus* for this study, have been shown to be susceptible to infection with X-cell parasites causing characteristic epidermal pseudotumours [4], however it remains possible that the X-cells from *P. obscurus* are a different species to those in *V. moseri*. The Japanese flounder, *P. olivaceus*, has not previously been reported to be susceptible to infection with X-cell parasites and was included in this study due to its availability from hatcheries and its economic importance. We believe that *P. olivaceus* is probably not susceptible to the X-cell parasite infecting other flatfish species in northern Japan, as no reports are found in the literature and they are not pleuronectid fish. Its inclusion in this experiment could also have been valuable as a negative control had transmission been successful for *V. moseri*. In the gadoid transmission experiment we were able to use the same species for both donor and experimental fish, hence eliminating any species specificity concerns for the experiment.

Phylogenetic Relationships

Using SSU rDNA alone, Miwa et al. [2, 5] were able to demonstrate that X-cell parasites infecting different fish hosts (goby and flathead flounder) were clearly related, but were not able to demonstrate a stable phylogenetic relationship between the X-cell group and the other taxa used in their analyses. In a more comprehensive phylogenetic analysis, again limited to SSU rDNA, Freeman [1] confirmed monophyly for the X-cell group, but could only place the X-cell clade within the superphylum Alveolata and suggested that they were possibly related to both apicomplexans and dinoflagellates, and referred to them as basal myzozoans. The present study is the first to include a SSU rDNA sequence for the cod X-cell parasite, and our phylogenetic analyses confirm monophyly for the X-cell group, but again fail to locate the X-cell clade convincingly within any of the recognised alveolate groupings. However, the inclusion of the sequence from cod X-cells has altered the expected topology of the tree for the alveolate group. *Parvilucifera* spp. that normally group with other perkinsids at the base of the dinoflagellates clade now occupy an unresolved branch in the tree. It is evident from recent molecular studies [[1, 2, 5], present study] that SSU rDNA sequence data alone is not sufficiently informative to robustly place the X-cell group in phylogenetic studies and that additional more conserved gene regions should be studied to clarify their phylogenetic relationship with other alveolate taxa.

Suggested life Cycle for X-cell Parasites

To date, numerous histopathological and ultrastructural studies of X-cell parasites in fish have failed to reveal a developmental stage beyond the familiar X-cell, either locked in host tissue or as a free form in advanced pseudotumours. Here, we demonstrate that non-amoeboid X-cells are contained within discrete sac-like structures that are loosely attached to the fish host and that the X-cells from these sacs are able to tolerate seawater conditions. We believe that this represents good evidence that detachment from fish occurs naturally. Sac-like structures have not been observed or reported from pseudobranchial pseudotumours in gadoids or in X-cell gill lesions. However, free-floating X-cells have been observed in advanced pseudobranchial pseudotumours in Atlantic cod [18] and ruptured pseudotumours have been reported from X-cell gill lesions [19], which may serve as alternative release mechanisms for non-epidermal X-cells.

We do not believe that fish represent a dead-end or incidental host for X-cells. X-cells have evolved very specific tissue tropisms in the fish species they infect and have been reported from 5 teleost orders globally [1], suggesting a long-term and well-established host parasite relationship. Therefore, we can conclude that X-cell development in the fish host is purely a proliferative phase that leads to further development either outside of the fish host or in an alternate host. The lack of direct transmission for the X-cell parasites in this study supports this theory. Figure 7 shows four proposed stages in the life cycle for the X-cell parasite infecting flatfish in northern Japan.

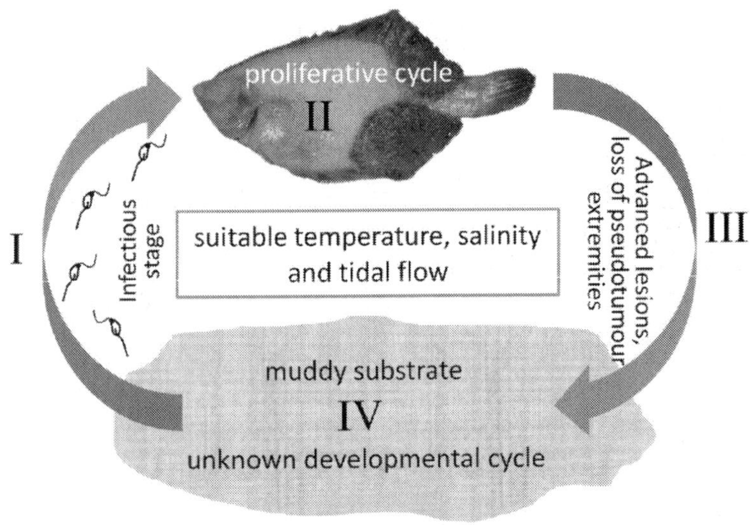

Figure 7: **Proposed life cycle for the X-cell parasite from flatfish in northern Japan**. The proposed life cycle consists of four stages. (I) An infective stage, either flagellated or equipped with suitable apparatus for host penetration/infection, which is released into the environment to infect a fish host. (II) A proliferative cycle starts in the fish which develops into large epidermal pseudotumours. (III) As the pseudotumours mature, sac-like extremities are lost from the host fish and fall to the seabed where the X-cells are released into the substrate. (IV)

In the substrate, either a free-living development occurs or the X-cells find an alternate host where development proceeds.

I. The infective stage to new fish hosts is either flagellated or equipped with suitable apparatus for host penetration and is released into the environment either from an alternate host or from a stage that developed during a free-living environmental phase in the life cycle.

II. A proliferative cycle starts in the fish which develops into large epidermal pseudotumours, each containing large quantities of X-cells.

III. As the pseudotumours mature, discrete sac-like structures are lost from the host fish and fall to the seabed where the X-cells are released into the substrate.

IV. If the substrate type is suitable, either free-living environmental development occurs or the X-cells infect and develop within an alternate host to produce infective stages for new fish hosts.

Understanding the life cycle of parasites that can infect commercially valuable fish species is important. In Europe, cod farming is an emerging industry and is still in the relatively early stages of development. However, pseudobranchial X-cell infections in Atlantic cod have already been shown to cause serious pathology and mortalities in farmed fish [26]. Identifying farm-sites that might favour the propagation of or allow X-cells parasites to become established in surrounding sediments will assist future aquaculture endeavours, such as gadoid farming.

CONCLUSIONS

The presence of discrete sac-like structures, filled with non-amoeboid X-cells, which are readily detached from the host fish and the lack of fish to fish transmission for both parasites in this study, suggests that other stages exist in the X-cell life cycle. This type of developmental cycle for X-cells may explain the lack of typical alveolate features seen in fish X-cells, which may be present in other stages of the life cycle. We suggest that X-cells have a second development phase either as free-living environmental organisms in the marine sediments or in an alternate host.

Analysis of SSU rDNA from X-cells infecting flatfish in Japan confirmed that the same X-cell parasite can infect more than one species of fish. Phylogenetic analyses of X-cell SSU rDNA places them as a monophyletic group in the alveolates. PCR analysis of infected flatfish showed that X-cell DNA is detectable in non-pseudotumour tissues, suggesting that some X-cells are motile in fish and may be responsible for autoinfection in the fish host.

NOTES

Acknowledgements

The Royal Society of Edinburgh (2× RSE International Exchange Programme: IEP Open - Outgoing Fellowships), The Sasakawa Foundation (grant No: 3034) and The Daiwa Foundation (grant No: 6211/6466) all provided initial funding for international travel with associated costs from the UK to Japan and Iceland and funding for work conducted at the University of Stirling. The University of Tokyo funded a visiting Associate Professorship for MAF to finalise the research in Japan during 2010.

Competing Interests

The authors declare that they have no competing interests.

Authors' contributions

ME and MAF collected material in Iceland and set up the transmission experiment, ME maintained the experiment for the duration. MY and KW organised sample collection in Japan, KO and MF set up the transmission experiment and KM maintained the experiment for the duration. MAF performed wax histology and DNA analyses, developed the nested PCR and drafted the manuscript. APS cut and viewed semi-thin plastic sections of the sac-like structures.

REFERENCES

1. Freeman MA: X-cell parasites in the European dab *Limanda limanda* are related to other X-cell organisms: a discussion on the potential identity of this new group of parasites. Parasitology. 2009,

10: 151-170.

2. Miwa S, Kamaishi T: Protistan X-cells in pseudotumors of yellowfin goby *Acanthogobius flavimanus* are a distinct organism from those in flathead flounder *Hippoglossoides dubius*. Dis Aquat Org. 2009, 85: 53-57. 10.3354/dao02058.
3. Morrison CM, Shum G, Appy RG, Odense P, Annand C: Histology and prevalence of X-cells lesions in Atlantic cod (*Gadus morhua*). Can J Fish Aquat Sci. 1982, 39: 1519-1530. 10.1139/f82-204.
4. Katsura K, Yamazaki F, Hamada K, Oishi K, Harada T, Shinkawa T: Geographic distribution and frequency of pseudotumourous fishes collected from the coastal waters of Hokkaido, Japan. B Jpn Soc Sci Fish. 1984, 50: 979-984.
5. Miwa S, Nakayasu C, Kamaishi T, Yoshiura Y: X-cells in fish pseudopseudotumours are parasitic protozoans. Dis Aquat Org. 2004, 58: 165-170. 10.3354/dao058165.
6. Ito Y, Kimura I, Miyake T: Histopathological and virological investigations of papillomas in soles and gobies in coastal waters of Japan. Prog Exp Pseudotumour Res. 1976, 20: 86-93.
7. Diamant A, Smail DA, McFarlane L, Thomson AM: An infectious pancreatic necrosis virus isolated from common dab *Limanda limanda* previously affected with X-cell disease, a disease apparently unrelated to the presence of the virus. Dis Aquat Org. 1988, 4: 223-227. 10.3354/dao004223.
8. Renault T, Stokes NA, Chollet B, Cochennec N, Berthe F, Gerard A, Burreso EM: Haplosporidiosis in the Pacific oyster *Crassostrea gigas* from the French Atlantic coast. Dis Aquat Org. 2000, 42: 207-214. 10.3354/dao042207.
9. Hillis DM, Dixon MT: Ribosomal DNA: molecular evolution and phylogenetic inference. Quart Rev Biol. 1991, 66: 411-453. 10.1086/417338.
10. Freeman MA, Yokoyama H, Ogawa K: Description and phylogeny of *Ceratomyxa anko* sp. n. and *Zschokkella lophii* sp. n. from the Japanese anglerfish, *Lophius litulon* (Jordan). J Fish Dis. 2008, 31: 921-930. 10.1111/j.1365-2761.2008.00965.x.
11. Thompson JD, Gibson TJ, Plewniak F, Jeanmougin F, Higgins DG: The CLUSTAL-X windows interface: flexible strategies for multiple

sequence alignment aided by quality analysis tools. Nucl Acids Res. 1997, 24: 4876-4882. 10.1093/nar/25.24.4876.

12. Hall TA: BioEdit: a user-friendly biological sequence alignment editor and analysis program for Windows 95/98/NT. Nucleic Acids Symp Ser. 1999, 41: 95-98.
13. Altschul SF, Gish W, Miller W, Myers EW, Lipman DJ: Basic local alignment search tool. J Mol Biol. 1990, 215: 403-410.
14. Swofford DL: PAUP* Phylogenetic analysis using parsimony (*and other methods), v. 4.0 beta10. 2002, Sinauer Associates, Sunderland, MA, USA
15. Ronquist F, Huelsenbeck JP: MrBayes 3 Bayesian phylogenetic inference under mixed models. Bioinformatics. 2003, 19: 1572-1574. 10.1093/bioinformatics/btg180.
16. Nylander JAA, Ronquist F, Huelsenbeck JP, Nieves-Aldrey JL: Bayesian phylogenetic analysis of combined data. Syst Biol. 2004, 53: 47-67. 10.1080/10635150490264699.
17. Yamazaki F, Hibino T, Oishi K, Harada T, Stich HF, Acton AB: X-cell morphology in the epidermal papillomas of flatfish collected from coastal waters of Hokkaido, Japan. B Jpn Soc Sci Fish. 1978, 44: 407-413.
18. Watermann B, Dethlefsen V: Histology of pseudobranchial pseudotumours in Atlantic cod (*Gadus morhua*) from the North Sea and the Baltic Sea. Helgolander Meeresun. 1982, 35: 231-242. 10.1007/BF01997554.
19. Maclean SA, Despres-Patanjo LI: Gill X-cell lesion in American plaice (*Hippoglossoides platessoides*) concurrent with lymphocystis infection. J Fish Biol. 1993, 43: 947-950. 10.1111/j.1095-8649.1993.tb01171.x.
20. Redondo MJ, Palenzuela O, Riaza A, Ángeles A, Álvarez-Pellitero P: Experimental transmission of *Enteromyxum scophthalmi* (Myxozoa) an enteric parasite of turbot *Scophthalmus maximus*. J Parasitol. 2002, 88: 482-488.
21. Lee SJ, Yokoyama H, Ogawa K: Modes of transmission of *Glugea plecoglossi* (Microspora) via the skin and digestive tract in an experimental infection model using rainbow trout, *Oncorhynchus*

mykiss (Walbaum). J Fish Dis. 2004, 27: 435-444. 10.1111/j.1365-2761.2004.00556.x.

22. Diamant A, McVicar AH: The effect of internal and external X-cell lesions on common dab, *Limanda limanda* L. Aquaculture. 1987, 67: 127-133. 10.1016/0044-8486(87)90017-2.
23. Diamant A, McVicar AH: Distribution of X-cell disease in common dab, *Limanda limanda*L., in the North Sea, and ultrastructural observations of previously undescribed developmental stages. J Fish Dis. 1989, 12: 25-37.
24. Stich HF, Acton AB, Forrester CR: Fish tumors and sub-lethal effects of pollutants. J Fish Res Board Can. 1976, 33: 1993-2001.
25. Wellings SR, McCain BB, Miller BS: Epidermal papillomas in Pleuronectidae of Puget Sound, Washington. Prog Exp Tumor Res. 1976, 20: 55-74.
26. Eydal M, Kristmundsson A, Bambir SH: Pseudobranchial X-cell pseudotumors in young wild and farmed Atlantic cod *Gadus morhua* in Iceland. Dis Aquat Org. 2010, 91: 83-88. 10.3354/dao02248.

CHAPTER

7

SEATROUT (SALMO TRUTTA) IS A NATURAL HOST FOR PARVICAPSULA PSEUDOBRANCHICOLA (MYXOZOA, MYXOSPOREA), AN IMPORTANT PATHOGEN OF FARMED ATLANTIC SALMON (SALMO SALAR)

Haakon Hansen[1], Trygve T Poppe[1,2], Turhan Markussen[1], Egil Karlsbakk[3]

[1]Norwegian Veterinary Institute, Oslo, Norway.

[2]Norwegian University of Life Sciences, School of Veterinary Medicine, Ås, Norway.

[3]Institute of Marine Research, 5817 Nordnes Bergen, Norway

ABSTRACT

Background

Parvicapsula pseudobranchicola (Myxozoa) causes widespread infections in farmed Atlantic salmon in northern Norway. Heavily infected salmon become runts, probably due to vision impairment or blindness. The salmon are likely infected by waterborne actinospores, released by an alternating annelid host, but the life cycle of *P. pseudobranchicola* is unknown. Seatrout and Arctic charr have been considered possible hosts for the parasite, but firm evidence has been lacking.

Findings

We show for the first time the presence of mature spores of *P. pseudobranchicola* in seatrout. The seatrout were infected with high intensities of *P. pseudobranchicola* in the pseudobranchs in early April. The presence of mature spores in early spring suggests that the fish had been infected late the previous year, a pattern of infection similar to that observed for farmed salmon stocked in autumn. Although heavily infected, the fish did not display any symptoms consistent with parvicapsulosis. The results suggest that the life cycle of *P. pseudobranchicola* is more adapted to seatrout, rather than to Atlantic salmon.

Conclusions

The presence of mature spores of *P. pseudobranchicola* in seatrout confirms that seatrout is a natural host for this myxosporean and this is also the first record of these spores in the pseudobranch of a wild salmonid. Furthermore, wild trout from non-farming areas may become heavily infected with *P. pseudobranchicola*, developing pseudobranch pathology resembling that of farmed Atlantic salmon suffering from parvicapsulosis.

Keywords: Natural host, Life-cycle, Salmonids, Marine Parasites, Parvicapsulidae, Norway

FINDINGS

Background

Infections with the myxosporean *Parvicapsula pseudobranchicola* are common in seawater farmed Atlantic salmon (*Salmo salar*) in Norway, especially in the three northernmost counties [1-4]. The known life cycles of parvicapsulids involve a polychaete alternate host [5-7]. However, the life cycle of *P. pseudobranchicola* is unknown.

Parvicapsula pseudobranchicola is most commonly diagnosed from the pseudobranch tissue in Atlantic salmon, and disporic trophozoites are produced interlamellary in this tissue [8]. In heavily infected pseudobranchs, there are few intact pseudobranch cells left [9,10]. Foci of infection have also been detected at other sites, but only in heavily infected salmon [9]. In severe infections with clinical parvicapsulosis, the pseudobranch is macroscopically observed to be swollen or papillate, or with a whitish "cheesy" matter sometimes covered with haemorrhages. Occasionally the pseudobranch tissue may be more or less ulcerated. Typical clinical signs include surfacing of a proportion of the fish in the pens; the fish swim disorganised, appear lethargic and may be unresponsive to visual challenge, as if blind. The eyes usually show crescent-shaped haemorrhage and cataract and exophthalmia may also occur. The fish go off feed, and tend to be cachectic and anaemic [9,11]. Parvicapsulosis in farmed Atlantic salmon emerged as a problem in 2002, roughly coinciding with a more widespread autumn stocking of smolts in northern Norway [11]. The discovery of *P. pseudobranchicola* infections in farmed Atlantic salmon suggested that the parasite could be present also in wild salmon, but also led to the assumption that other salmonids might host the parasite. Using real-time PCR, parasite DNA was detected in all three salmonid species present in Norwegian waters; wild Atlantic salmon, sea run Arctic charr (*Salvelinus alpinus*) and seatrout (*Salmo trutta*) [3]. However, the presence of myxosporean DNA in a fish does not provide clear evidence that this particular fish species is a suitable host supporting parasite sporogony. Sporoplasm entry may be unspecific, so unsuitable hosts may be still be PCR positive (see [12]). So far, the presence of mature spores has only been verified in farmed Atlantic salmon, hence the role of Arctic charr and seatrout as potential susceptible hosts is still unclear. Arctic charr are anadromous only in Northern Norway, and can therefore not be involved in the life cycle of the parasite in the southern parts of Norway. Wild Atlantic salmon smolts migrate to the sea during spring, usually spending 1-2 years in oceanic feeding areas before returning to the coast for spawning migration [13]. Since it has been shown from studies on farmed salmon that they develop parvicapsulosis and mature spores of *P. pseudobranchicola* spores in 4-7 months or less [1,4], it seems unlikely that wild salmon play an important role in the natural lifecycle of the parasite. Spores will be produced when the wild salmon is far out at sea and released over great depths, far from potential polychaete hosts. Seatrout, on the other hand, are present in much higher numbers, reside in coastal waters during summer and may overwinter in the estuarine waters [13]. Hence, seatrout may be an important, or even represent the principal vertebrate host for *P. pseudobranchicola*.

To investigate this hypothesis, we collected seatrout in a non-farming area in southern Norway and examined them for infection by *P. pseudobranchicola* and the presence of mature spores.

METHODS

Three seatrout, weighing approximately 1000 g, 600 g and 500 g, were sampled by angling near Rygge, Østfold County (59°18›32.82»N 10°44›31.82»E) on April 2nd 2014. The heads of the fish were collected and kept on ice for 12 hours prior to examination. The pseudobranchs were dissected out and squash preparations made from the pseudobranch. Each squash preparation was subsequently examined for the presence of spores under 400X-1000X magnification in a light microscope. The remaining pseudobranch tissue was cut in half. Half the tissue was fixed in 10% neutral phosphate buffered formalin (12 h) for histological examination. These samples were then transferred to 70% ethanol, embedded in paraffin wax and sectioned (3-5 μm) for histology and *in situ* hybridization. For histology, sections were stained with haematoxylin and eosin (HE) or May-Grünwald Giemsa. *In situ* hybridization was performed as previously described [10]. Images were captured using a Leica DM5000B microscope equipped with a Nikon DXM 1200 digital camera. The other half of the pseudobranchs were stored in 96% ethanol for molecular analyses. A pseudobranch from Atlantic salmon from a fish farm in Northern Norway experiencing parvicapsulosis was used as positive control in the molecular analyses.

DNA was extracted from pseudobranch tissue using the DNeasy Blood and Tissue kit on a QIAcube robot (Qiagen). The DNA extracts were then analysed for the presence of *P. pseudobranchicola* by real-time-PCR [3] and by conventional PCR amplifying 900bp of ribosomal 18S using the primer pair 3LinF and Myxgen4R [14]. PCR products from the latter assay were treated with ExoSAP-IT to remove unincorporated dNTPs and primers, following manufacturer instructions (Affymetrix). Sequencing of a 900 nucleotide 18S rDNA fragment was performed using both amplification primers and the BigDye®Terminator v3.1 Cycle Sequencing Kit. The resulting sequence data were proof-read and assembled with the Vector NTI 11 software (ver. 11.5) (Invitrogen) and subjected to a BLASTn search in GenBank.

RESULTS

The seatrout pseudobranchs examined in the present study appeared macroscopically normal. However, microscopic examination of squash preparations revealed a number of mature spores per field (no attempt to quantify the spores was done) and early developmental stages (trophozoites). The morphology and measurements (data not shown) of the most developed spores matched perfectly the description of *P. pseudobranchicola* (Figure 1A).

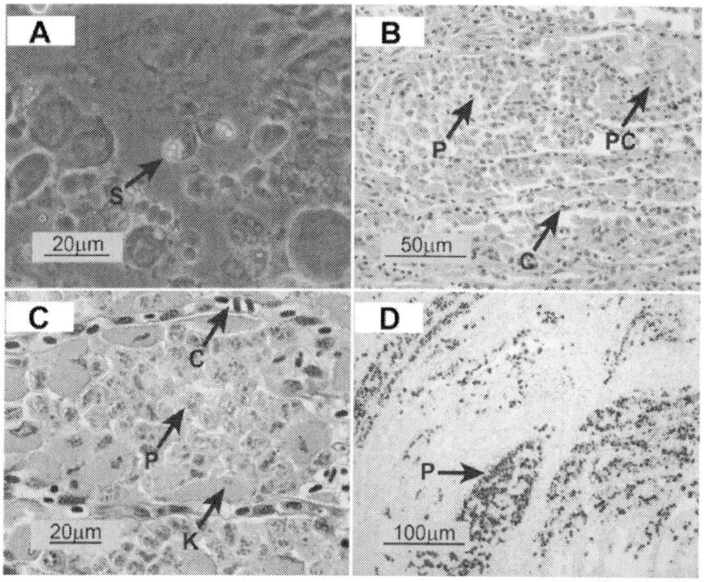

Figure 1: ***Parvicapsula pseudobranchicola*in pseudobranchs from sea trout (*Salmo trutta*). A)** Squash preparation showing mature spores (S). **B)** Histological section stained with May-Grünwald Giemsa showing heavily infected pseudobranch tissue. Proliferating spores (P) are clearly observed with few intact pseudobranch cells (PC) left. C=capillaries. **C)** Histological section stained with Haematoxylin-Eosin (HE) showing heavily infected pseudobranch region at higher magnification. Pseudobranch cells (PC) are observed, but the majority are either destroyed or nuclei show karyorheksis (K). C=capillaries, P=proliferating spores. **D)***In situ* hybridisation showing specific staining in areas with proliferating spores (P) surrounded by more intact pseudobranch tissue (not stained).

Although most of the pseudobranch was heavily affected by the infection with resultant disruption of the normal architecture, small, scattered islands of intact tissue with intact pseudobranch cells lining the lamellae could be identified (Figure 1B, C and D). In affected areas interlamellar spaces were filled with different developmental stages of the parasite, sometimes with associated haemorrhage from disrupted lamellar vessels. Remaining pseudobranch cells were pale and hypertrophic, many of them partly necrotic showing different stages of cell death with pyknosis and karyorhexis (Figure 1A and B). Inflammatory reaction was sparse in the interlamellar regions, while a mononuclear cell infiltrate was obvious near the base of the lamellae.

All three fish tested positive by real-time-PCR, producing Ct's from 12 to 15; the lowest value was obtained from the largest fish. By comparison, the positive control fish originating from an outbreak of parvicapsulosis produced a Ct-value of 16. Sequencing of samples from all three fish confirmed the presence of *P. pseudobranchicola* with 100% sequence identity (GenBank Acc. no. AY308481).

DISCUSSION

We show for the first time the presence of mature *P. pseudobranchicola* spores in seatrout. This is also the first record of spores from this parasite species in the pseudobranch of a wild salmonid. These observations confirm that seatrout is a host for this myxosporean. The life cycle of the parasite has been a focus of research since it was first described back in 2002 [11]. Since wild salmon leaves coastal waters on their feeding migrations, their role in the life cycle of the parasite is most likely limited. Ongoing studies on returning wild salmon, based on real-time PCR, suggests they contain low levels of parasite DNA with Ct's averaging 30 (unpublished data). This is much lower than obtained for seatrout in the present study. Although evidence so far has been lacking, seatrout has been suspected to host the parasite. In our study, we have shown that the parasite life cycle involve sea trout, a species with a different ecology than wild Atlantic salmon. This has implications for the search for the final host of *P. pseudobranchicola* and should result in an increased focus on examination of polychaetes from seatrout habitats in estuaries.

The present observations are at variance with the previously perceived natural infection dynamics of the parasite [4]. Spring stocked Atlantic salmon smolts (April-June) in northern Norway may become

PCR positive from July and develop spores in September-October, rarely developing clinical parvicapsulosis [4]. This could mimic a natural situation, as wild Atlantic salmon smolts enter the sea during spring. However, a farming situation with stocking of salmon smolts in the autumn often leads to clinical parvicapsulosis and development of spores during winter-spring (February–May). The seatrout examined in early April from the Oslofjord contained mature spores. These fish were relatively large (i.e. not first seawater sojourn), and may have resided in the sea during winter [13]. This suggests that some seatrout in southern Norway become infected during autumn or winter and release mature spores in the spring (April-May). Consequently, the infection dynamics suggested by our observations on wild seatrout resemble that associated with autumn stocked farmed Atlantic salmon. In previous studies on *P. pseudobranchicola* in wild salmonids, using the same real-time PCR assay as in the present study, high Ct values were observed but no myxospores were detected [3]. However, the results presented here verify that this parasite indeed does infect and develop mature spores in seatrout. Furthermore, this shows that wild trout from non-farming areas may become heavily infected with this parasite, developing pseudobranch pathology resembling that of farmed Atlantic salmon suffering from parvicapsulosis. The examined trout were apparently healthy judged by the fact that they were caught on angling equipment and no pseudobranch lesions were macroscopically evident. In contrast to this, a significant portion of the pseudobranch cells were clearly destroyed. The effect an infection with this myxosporean has on salmonids will depend on the function these principal cells have in this organ, which is presently unknown (see [15]). However, typical clinical signs of parvicapsulosis in farmed Atlantic salmon suggest impaired vision or blindness. Since the ocular blood supply is provided exclusively through the pseudobranchs, the infection may affect retinal oxygenation. In the examined seatrout, no vascular lesions could be found, and only the pseudobranch cells were found to be affected. Inflammation was mainly restricted to the areas close to the filaments. Possibly, inflammation is triggered at a later stage since the formation of exudative lesions may be necessary for spore release to the environment. A seasonal study of *P. pseudobranchicola*infection and development in a natural host, the seatrout, would provide valuable information on this economically important and enigmatic parasite.

NOTES

Acknowledgements

The study was funded by the Norwegian Research Council grant no. 207269 and The Norwegian Seafood Research Fund grant no. 900896. The authors wish to thank Tor Atle Mo and Sigurd Hytterød for assistance in sampling of seatrout, and Saima Nasrin Mohammad for assistance in PCR and sequencing.

Competing Interests

The authors declare that they have no competing interests.

Authors' Contributions

All authors contributed to the writing of the manuscript. HH sampled, made squash preparations and isolated material for further analyses. TM performed the *in situ* hybridisation analysis. TTP, EK and HH analysed histological data. HH did the molecular analyses. All authors read and approved the final manuscript.

REFERENCES

1. Nylund A, Karlsbakk E, Sæther PA, Koren C, Larsen T, Nielsen BD, et al. *Parvicapsula pseudobranchicola* (Myxosporea) in farmed Atlantic salmon *Salmo salar*. Tissue distribution, diagnosis and phylogeny. Dis Aquat Organ. 2005;63:197–204.
2. Hansen H, Brevik ØJ, Jørgensen A, Garseth Å, Nylund A, Karlsbakk E. The distribution of *Parvicapsula pseudobranchicola* in wild salmonids in Norway. In: 16th International Conference on Diseases of Fish and Shellfish 2013; Tampere, Finland.
3. Jørgensen A, Nylund A, Nikolaisen V, Alexandersen S, Karlsbakk E. Real-time PCR detection of *Parvicapsula pseudobranchicola* (Myxozoa: Myxosporea) in wild salmonids in Norway. J Fish Dis. 2011;34(5):365–71.
4. Karlsbakk E, Jørgensen A, Alexandersen S, Ottem KF, Nylund A. Parvicapsulosis in farmed Atlantic salmon. In: Fisken og Havet 2010 Havforskningsrapporten 2010) (in Norwegian), vol. Særnr. 1:

Havforskningsinstituttet; 2010. p. 105–06.

5. Køie M, Karlsbakk E, Nylund A. A new genus *Gadimyxa* with three new species (Myxozoa, Parvicapsulidae) parasitic in marine fish (Gadidae) and the two host life-cycle of *Gadimyxa atlantica* n. sp. J Parasitol. 2007;93(6):1459–67.
6. Køie M, Karlsbakk E, Einen A-CB, Nylund A. A parvicapsulid (Myxozoa) infecting *Sprattus sprattus* and *Clupea harengus* (Clupeidae) in the Northeast Atlantic uses *Hydroides norvegicus* (Serpulidae) as invertebrate host. Folia Parasitol. 2013;60(2):149–54.
7. Bartholomew JL, Atkinson SD, Hallett SL. Involvement of *Manayunkia speciosa*(Annelida: Polychaeta: Sabellidae) in the life cycle of *Parvicapsula minibicornis*, a myxozoan parasite of Pacific salmon. J Parasitol. 2006;92(4):742–8.
8. Karlsbakk E, Nylund A. *Parvicapsula pseudobranchicola*. In: Raynard R, Wahli T, Vatsos I, Mortensen S, editors. DIPNET - Review of disease interactions and pathogen exchange between farmed and wild finfish and shellfish in Europe. European Commission/ Veterinaermedisinsk Oppdragssenter; 2007. p. 119–20.
9. Sterud E, Simolin P, Kvellestad A. Infection by *Parvicapsula* sp. (Myxozoa) is associated with mortality in sea-caged Atlantic salmon *Salmo salar* in northern Norway. Dis Aquat Organ. 2003;54:259–63.
10. Markussen T, Agusti C, Karlsbakk E, Nylund A, Brevik ØJ, Hansen H. Detection of the myxosporean parasite *Parvicapsula pseudobranchicola* in Atlantic salmon (*Salmo salar*L.) using *in situ* hybridization (ISH). Parasit Vectors. 2015;8:105.
11. Karlsbakk E, Sæther PA, Høstlund C, Fjellsøy KR, Nylund A. *Parvicapsula pseudobranchicola* n.sp. (Myxozoa), a myxosporidian infecting the pseudobranch of cultured Atlantic salmon (*Salmo salar*) in Norway. Bull Eur Assoc Fish Pathol. 2002;22(6):381–7.
12. Holzer AS, Bartosova P, Peckova H, Tyml T, Atkinson S, Bartholomew J, et al. 'Who's who' in renal sphaerosporids (Bivalvulida: Myxozoa) from common carp, Prussian carp and goldfish–molecular identification of cryptic species, blood stages and new members of *Sphaerospora sensu stricto*. Parasitology. 2013;140(1):46–60.

13. Klemetsen A, Amundsen PA, Dempson JB, Jonsson B, Jonsson N, O'Connell MF, et al. Atlantic salmon *Salmo salar* L., brown trout *Salmo trutta* L. and Arctic charr *Salvelinus alpinus* (L.): a review of aspects of their life histories. Ecol Freshw Fish. 2003;12:1–59.
14. Kent ML, Khattra J, Hedrick RP, Devlin RH. *Tetracapsula renicola* n. sp. (Myxozoa:Saccosporidae); the PKX myxozoan—the cause of proliferative kidney disease of salmonis fishes. J Parasitol. 2000;86(1):103–11.
15. Mölich A, Waser W, Heisler N. The teleost pseudobranch: a role for preconditioning of ocular blood supply? Fish Physiol Biochem. 2009;35(2):273–86.

CHAPTER

8

RISK MAP AND SPATIAL DETERMINANTS OF PANCREAS DISEASE IN THE MARINE PHASE OF NORWEGIAN ATLANTIC SALMON FARMING SITES

Saraya Tavornpanich[1], Mathilde Paul[2], Hildegunn Viljugrein[1,3], David Abrial[4], Daniel Jimenez[1], Edgar Brun[1]

[1]Section for Epidemiology, Department of Health Surveillance, Norwegian Veterinary Institute, Sentrum, Norway

[2]UR AGIRs, Centre de coopération internationale en recherche agronomique pour le développement (CIRAD), Montpellier, France

[3]Centre for Ecological and Evolutionary Synthesis (CEES), Department of Biology, University of Oslo, Blindern, Norway

[4]UR 346Institut National de la Recherche Agronomique (INRA), Saint-Genès-Champanelle, France

ABSTRACT

Background

Outbreaks of pancreas disease (PD) greatly contribute to economic losses due to high mortality, control measures, interrupted production cycles, reduced feed conversion and flesh quality in the aquaculture industries in European salmon-producing countries. The overall objective of this study was to evaluate an effect of potential factors contributing to PD occurrence accounting for spatial congruity of neighboring infected sites, and then create quantitative risk maps for predicting PD occurrence. The study population included active Atlantic salmon farming sites located in the coastal area of 6 southern counties of Norway (where most of PD outbreaks have been reported so far) from 1 January 2009 to 31 December 2010.

Results

Using a Bayesian modeling approach, with and without spatial component, the final model included site latitude, site density, PD history, and local biomass density. Clearly, the PD infected sites were spatially clustered; however, the cluster was well explained by the covariates of the final model. Based on the final model, we produced a map presenting the predicted probability of the PD occurrence in the southern part of Norway. Subsequently, the predictive capacity of the final model was validated by comparing the predicted probabilities with the observed PD outbreaks in 2011.

Conclusions

The framework of the study could be applied for spatial studies of other infectious aquatic animal diseases.

Keywords: Pancreas disease, Aquatic epidemiology, Spatial analysis, Disease mapping, Bayesian modeling

BACKGROUND

Pancreas disease (PD) is a viral disease affecting Atlantic salmon (*Salmon salar L.*,) and rainbow trout (*Oncorhynchus mykiss*). A necessary cause of the disease is *salmon pancreas disease virus*, an alphavirus often referred as *salmonid alphavirus* (SAV). The onset of a PD outbreak is favored by unspecified influential environmental factors. Diagnostic criteria for PD outbreak were described by Taksdal et al., 2007 [1]. So far, 6 subtypes of SAV have been reported worldwide, whereas, only subtypes 2 and 3 have been detected in Norwegian salmonid farming sites [2, Hilde Sindre per communications]. The disease has been a significant problem in salmonid farming industries in Scotland, Ireland, and Norway [2]. An economic modeling study estimated that the direct cost of PD accounting for biological losses, disease controls and treatments, and insurance payment in a Norwegian salmonid site with an approximate of 500,000 smolts to be at least 14.4 million NOK in 2007 [3].

The characteristics of SAV and PD outbreaks have been thoroughly reviewed [4, 5, 6, 7, 8, 9, 10, 11]. In Norway, PD outbreaks have been reported only during the marine phase. The disease is regarded as endemic in the area along the western coast of Norway from Rogaland county to a firewall at Hustadvika (Figure 1) in Møre and Romsal with few sporadic outbreaks occurring north of this border. This firewall is a 10 nautical mile exposed costal area without any aquaculture activity. Veterinary authorities together with salmon industries have launched different prevention and control strategies to control the situation in the south and to keep the northern area free of PD [12]. All findings of epidemiological studies of PD outbreaks support the hypothesis that the virus can spread through the water media, causing a PD outbreak both with and without human intervention. This aspect of the disease spreading via water media is a crucial element of aquatic infectious diseases [9, 13]. Therefore, it is worthwhile to evaluate the spatial clustering of infected sites in epidemiological studies of PD outbreaks. In this study, we initially evaluated potential risk factors of PD in marine phase of Atlantic salmon farming sites, accounting for spatial congruity of infected sites. Then, the predicted probabilities of PD occurrence based on the model results were displayed in an interpolated map covering the southern part of Norwegian marine aquaculture.

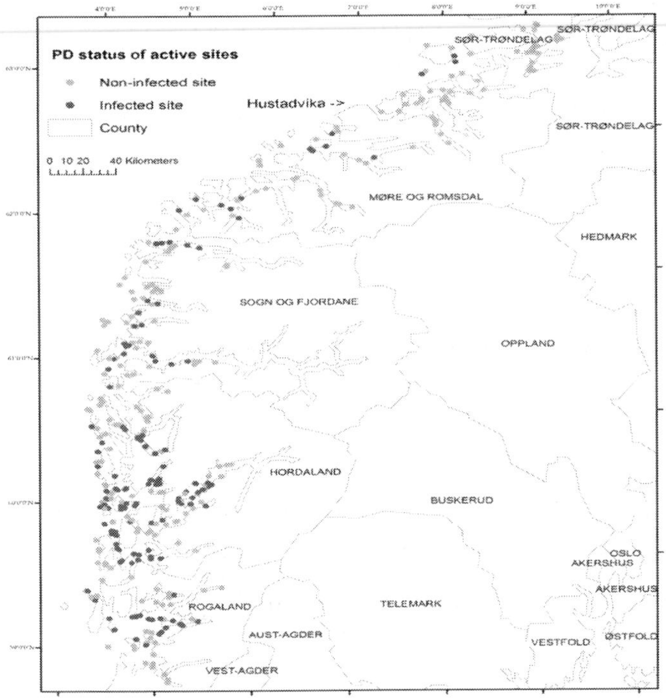

Figure 1: Geographical distribution of 359 active Atlantic salmon farming sites located near 6 counties of Norway (Vest-Agder, Rogaland, Hordaland, Sogn & Fjordane, Møre & Romsdal, SørTrøndelag) from 1 January 2009 to 31 December 2010.

METHODS

Data

In this study, we included Atlantic salmon farming sites located in 6 Norwegian counties (Vest-Agder, Rogaland, Hordaland, Sogn & Fjordane, Møre & Romsdal, SørTrøndelag), where most of the PD outbreaks were reported so far. To be considered as a PD at-risk site, the included sites were required to rear fish on the site for at least 6 consecutive months during the study period (1 January 2009 to 31 December 2010). All operators of Norwegian salmonid farming industries are required to register production statistical data to responsible authorities on a monthly basis. These statistics are linked to a farm site identity, which is georeferenced in the Directorate of Fisheries (www. fiskeridir.no). According to the inclusion criteria, 359 Atlantic salmon farming sites were qualified for model estimation of this study. The geographical distribution of the study sites is presented in Figure 1. In addition, a second dataset of active farming sites during a period of 2011 was created for validating the predictive capacity of the model. The second dataset contained 418 farming sites located throughout the study area.

PD Infection Status

All farming sites were monitored by the farmer for unusual increases in fish mortality and activities (e.g. runt, moribund). If PD was suspected, samples were routinely sent for laboratory confirmation using histopathological examination and PCR testing. The study sites were classified as PD positive if at least 1 PD outbreak were confirmed during the study period, and PD negative, if otherwise.

Site Latitude

In the present study, we evaluated if the site location, based on the latitude, had an influence on the PD occurrence. In this study, the latitudes of the study sites cover the area of 63°39'5,481"N (the farthest north) to 58°12'33,449"N (the farthest south). The value of site latitude was converted to meters prior to analyses.

Site Density

Site density for a given site was the number of farming sites (excluding the given site) that located within 10-km seaway distance of the given site. In this study, the site density ranged from 1 to14, had mean of 4, and a median of 4.5.

Smolt Cohort

A smolt cohort is the group of smolts that were put to sea and reared in sea cages until slaughtering. We identified the time when the smolt cohorts were put to sea. If the cohort was put to sea in the period of August-February it was classified as 'a cohort of autumn smolt'; otherwise, if the cohort was put to sea in the period of March-July, it was classified as 'a cohort of spring smolt'. Of the 359 study sites, 168 (46.8%) sites had only autumn cohorts; 170 (47.4%) sites had only spring cohorts; and 21 (5.8%) sites had mixed cohorts (both cohorts of autumn and spring smolts).

PD History in 2008

Previous recent history of PD in a nearby area is likely to have an impact on a future probability of PD occurrence. In this study, we evaluated this effect by classifying the study sites based on whether or not they were located within 10-km seaway distance of sites where PD was suspected or confirmed in 2008. The 10-km seaway distance was used as the cutoff because it was estimated that sites locating within 10-km seaway distance of PD infected sites had a large impact on PD risk [14]. Of 359 study sites, 225 (63%) sites were located within 10-km radius of PD suspected/confirmed sites in 2008. Furthermore, among these 225 sites, 110 (49%) sites were classified as PD confirmed cases during the study period.

Local Biomass Density

The local biomass density (LBD) indicates the potential of infection endangerment at a given site. This factor depends on average biomass of the surrounding sites to a given site (excluding LBD of the given site). The LBD data were estimated for all registered salmon fish farming sites in Norway and the estimation method was described by Jansen et al (2012) [15]. In the present study, we evaluated the effect of LBD on the probability of PD occurrence at each farming site. The LBD of the study sites ranged from 4.87e+02 to 4.62e+08; had a mean of 2.19e+08; and median of 2.11e+08. For the analysis purpose (facilitating convergence of the model), the LBD was transformed using log10, and then back transformed for interpretation.

Statistical Analysis

Univariate and Multivariate Analysis

All regression analyses were performed using WinBUGS (windows version of Bayesian inference using Gibbs Sampling) open source software [16]. First, we performed univariate logistic regression to individually evaluate the effect of each potential risk factor on PD occurrence. Records of PD confirmed cases from 1 January 2009 to 31 December 2010 were used as a dependent variable of the model. The evaluated risk factors were site latitude, site density within 10-km seaway distance, smolt cohort, PD history in 2008, and LBD. We explored the assumption of linear relationships with generalized additive models, using the mgcv package in R software. Variables with a 80% posterior probability interval including zero were excluded from further analyses. Then, multivariate logistic regression models were performed with and without a spatial component. A stepwise selection based on the Deviance Information Criterion (DIC) was used for selecting the combination of the covariates; the model with lower DIC is preferred to the model with larger DIC [17]. DIC was also used for evaluating whether adding spatial random effect would improve model fit in the final model. Residuals of all models were in turn checked for spatial dependency, as any pattern present in the residuals of a statistical model may indicate that results are biased [18]. To do so, we applied a Monte Carlo method [19, 20] using the geoRglm package of the R software. This consists in comparing the observed variogram with variogram 'envelopes' that were computed by simulating 999 permutations of the data values across locations.

Spatial Dependency Modeling

We grouped fish farming sites into hexagonal subunits which were included in the model as a spatial area-level random effect [21]. Nineteen hexagons (width of 80 km, corresponding to 5546.6 km^2) were created; they contained between 1 and 52 farming sites (mean=35). The area-level effect was modeled using a Conditional Autoregressive (CAR) prior structure, in which an adjacency matrix was specified with a weight of 1 given to adjacent hexagons and a weight of 0 given to nonadjacent hexagons.

Briefly, each site in this study was labeled with i=1...19 (the hexagon in which the site is localised), and j=1...359 (the unique site identifier). The observed case of PD outbreaks at each site ($y_{[ij]}$) during the study period followed a Bernoulli distribution. The model can be written as the following equations:

$$y_{[ij]} \sim \text{Bernoulli}(\pi_{[ij]}) \quad (1)$$

$$\text{logit}(\pi_{[ij]}))=\alpha+\beta x_{[ij]}+v_{[i]} \quad (2)$$

$v_{[i]} \sim \text{Normal}(\mu\delta_{[i]}, 1/\sigma^2)$ (3)

β is the vector of coefficients, $x_{[ij]}$ is the covariates matrix observed on site j, $v_{[i]}$ is the random area-level effect which is spatially structured according to the model of Besag et al (1991) [22]. The mean of this model is driven by the average risk level of the adjacent hexagon set $\delta_{[i]}$ of the hexagon i. The precision (σ^2) of $v_{[i]}$ follow a gamma distribution. Priors for model coefficients were based on a normal distribution (with a mean of 0.5 and a precision of 5e-04). We implemented the models in WinBUGS, and estimated the model parameters using Gibbs sampling, which is the method for simulating a new value for the model parameters from its full posterior conditional distribution, given the current values for the remaining parameters [23]. In this study, the model parameters were estimated based on 45000 effective iterations. An additional first 5000 iterations were discarded as a burn-in period. Geweke and Heidelberger-Welch tests were used to assess the convergence of the models [16].

Interpolated Map of PD Predicted Probability

The median value (50th) of the posterior distribution of the probability estimate of PD occurrence, obtained from the final Bayesian logistic model, was used for creating a PD risk map. First, the empirical semi-variogram of the point estimates was plotted. Several statistical models (exponential, Gaussian, spherical) were investigated to identify the best-fit variogram model. Those parameters were then used to interpolate values between all the 359 farming sites considering a fixed radius of 40 km and a cell size of 500 m. An ordinary spatial kriging was used and allowed to produce a raster map; the variance of output raster was also mapped. Kriging is a group of geostatistical techniques for interpolating the value of a random field at an unobserved location based on observed values at nearby locations; it has been widely used in veterinary epidemiology [20]. Kriging and mapping were performed with the gstat package of the R software and the spatial analyst extension of ArcGIS software v.9.3 (ESRI Inc. Redlands, CA USA).

Sensitivity Analysis

The threshold of 80 km, which was used to define the width of the spatial grid, was selected in order to have an acceptable number of farming sites per hexagon. To see whether the size of the grid would influence the model outputs, we carried out a sensitivity analysis by running the final model with a grid size of 40 km. As this size produced hexagons without any farming sites, the empty adjacent hexagons were merged. This resulted in a final grid of 39 polygons, which were used in the model. DIC and effect estimates of this model were compared to those of the model of originally selected grid size.

Model Validation

Relative operation characteristic (ROC) statistical methodology was used to validate the accuracy of the model. The approach has been widely used to access performance of diagnostic tests of animal diseases, and has been applied for validation of models predicting spatial distributions [24, 25, 26, 27]. For the present study, the ROC curve presents the plot of the "sensitivity" (true-positive value) versus "specificity" (true-negative value) at each plotted point. The sensitivity is the probability that the model predicts that a site is infected when that site is truly infected. The specificity is the probability that the model predicts that a site is not infected when that site is truly not infected. Each plotted point reflects the average probability that infection will be present on a site. We plotted ROC and calculated an area under the curve (AUC) to check the model fit by comparing the estimated probability of PD occurrence and the observed PD status during the study period. Also, we compared the predicted probabilities based on the final model with the observed PD status in 2011 in order to evaluate if this model could be used for predicting future outbreaks. If the PD occurrence is 100% predicted by the model, AUC of 1 is obtained [24]. All those performance measures were calculated using the pROC package of the R software [28].

RESULTS

Univariate and Multivariate Analyses

A total of 359 salmon fish farming sites were included in the present study based on the inclusion criteria. Of this study population, 127 sites (35%) were classified as PD cases during the study period. Geographical distribution of the study sites classified by PD status is presented in

Figure 1. DIC and coefficient estimates from the univariate logistic regression analysis are presented in Table 1. According to our significance criteria, all covariates except the smolt cohort were further evaluated in multivariate analysis. Table 2 presents the model covariates and DIC of various models depending on combinations of covariates.

Table 1: Univariate analysis using a Bayesian approach of potential risk factors for PD occurrence of 359 Atlantic salmon farming sites located in the southern part of Norway from 2009 to 2010

Model covariates	Median (80%PI)	DIC
intercept	-0.60 (-0.75, -0.46)	468.51
intercept + spatial component	NA	440.03
intercept + latitude	-0.62 (-0.80, -0.45)	442.67
intercept + site density	0.30 (0.19, 0.42)	459.14
intercept + smolt cohort	0.13 (-0.15, 0.41)	469.99
intercept + PD history	1.89 (1.53, 2.29)	417.79
intercept + LBD	2.40 (1.85, 3.02)	440.20

NA = Not available, *PI* = Probability interval.

Table 2: Multivariate analysis using a Bayesian approach of potential risk factors for PD occurrence of 359 Atlantic salmon farming sites located in the southern part of Norway from 2009 to 2010

Model covariates	DIC
intercept + latitude + site density + PD history + LBD + spatial component	402.79
intercept + latitude + site density + PD history + LBD ***	401.97
intercept + site density + PD history + LBD	408.05
intercept + latitude + PD history + LBD	403.19
intercept + PD history + LBD	410.71
intercept + LBD + spatial component	414.99
intercept + PD history + LBD + spatial component	409.78
intercept + site density + PD history + LBD + spatial component	408.36

*** final model.

There was no evidence of multicollinearity among the covariates. Linear correlation between LBD and site density was moderate (Pearson coefficient equals to 0.63 ($p<0.05$). A generalized additive model of LBD as a nonlinear function of site density showed a linear positive association for small site densities, the association leveling off for the higher site densities. The final model consisted of site latitude, site density, PD history, and LBD. Although there was an evidence of a spatial cluster of PD infected sites, which was indicated by a decrease of DIC after adding a spatial component to a model with an intercept alone (Table 1), the PD predicted probability was well explained by the covariates of the final model and the spatial component was finally removed. A plot of empirical semivariogram of model residuals and simulation envelopes (data not shown here) confirmed that the spatial pattern of PD cases was well accounted for by the final model. According to the final model, the site latitude alone had a negative effect on the PD probability (median=-0.42 with 95% probability interval between -0.73 and -0.15) meaning that the risk is increasing as one goes south. Having a history of surrounding sites with PD suspected/confirmed the previous year would increase the risk of getting PD (median=1.22, 95% probability interval between 0.58 and 1.89). Increasing LBD 10 times of an average LBD would increase the risk of getting PD by 1.24 (95% a probability interval between 0.04 and -2.59). At last, increasing site density by adding one more site in a 10-km seaway distance would slightly increase the risk of getting PD (median =0.08 with a 95% probability interval between -0.03 and 0.18).

Probability Estimates of PD Occurrence

Based on the results of the final model, probability estimates of PD occurrence of the study sites ranged from 0 to 0.75 with a median of 0.4, and about 34% of the study sites had the probability estimates of≥0.5. An interpolated map of the PD predicted probability based on the final model covering the southern part of Norway is presented in Figure 2. To improve the legibility of the map, we classified the predicted probability into 6 categories ranging from 0 to 1.0. The variance of output raster, which helps assessing the uncertainty of the PD predicted probability estimates produced by kriging, ranged from 1-5%. The predictions are more precise in the center of the circles than at their edges (Appendix 1).

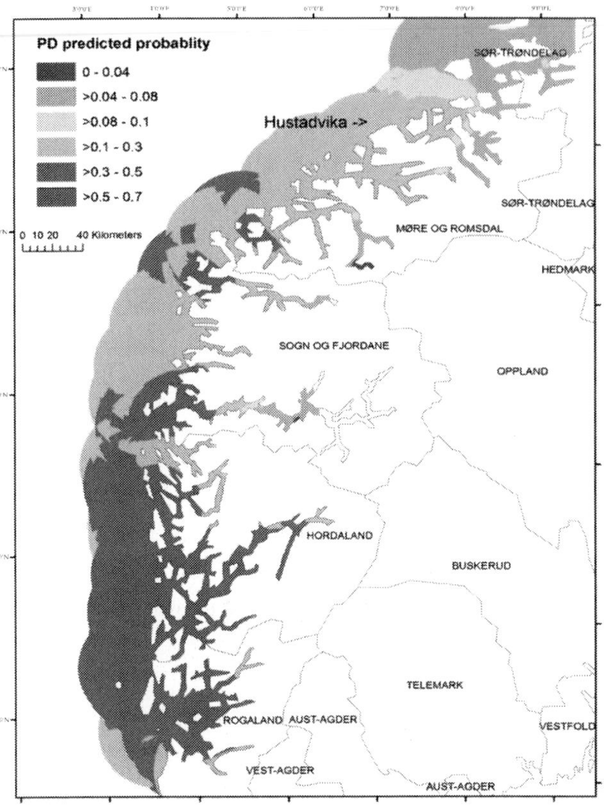

Figure 2: An interpolated map presenting PD predicted probability based on the final model of the present study.

Model Validation

Figure 3 shows the ROC curves for validating the capacity of the final model for: a) model estimation using observed PD data (359 sites) from 2009 to 2010, b) model prediction using observed PD data of 2011 (438 sites). A reasonable predictive capacity of the final model was observed. The AUC were 0.76 (95%CI: 0.71-0.81), and 0.72 (95%CI: 0.66-0.78) for the model estimation, and the model prediction, respectively.

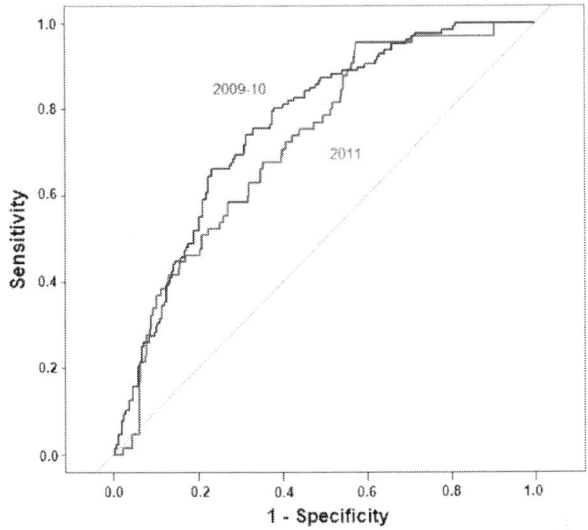

Figure 3: ROC curves for a) validation of the model estimates with observed PD occurrence in time period 2009-2010 (AUC=0.76, 95%CI: 0.71-0.81), b) validation of the model prediction with observed PD occurrence in 2011 (AUC=0.72, 95%CI: 0.66-0.78). Each plotted point reflects the average probability that infection will be present on a site.

Sensitivity Analysis

For the model with only a spatial component, changing the grid size yielded different estimate of model DICs. The models corresponding to a width of 40 km and 80 km yielded DICs of 424, and 440, respectively. However, changing grid size presented similar DICs and model estimates in the full model (Table 3). This indicates that the effect of spatial congruity on PD occurrence, if occurred, was not influenced by changing of grid size when other significant explanatory variables were included in the model.

Table 3: Sensitivity analysis of the effect of changing grid size on DIC and coefficient estimates of the final model

	40 km grid size (95% PI)	80 km grid size (95% PI)
intercept	-7.74 (-20.61, -1.24)	-9.22 (-19.5, -0.22)
latitude	-0.42 (-0.83, -0.10)	-0.42 (-0.73, -0.12)

site density	0.09 (-0.03, 0.20)	0.08 (-0.03, 0.19)
PD history	1.23 (0.58, 1.90)	1.23 (0.58, 1.89)
LBD	1.00 (0.21, 2.55)	1.19 (0.02, 2.42)
DIC	403.79	402.79

PI = Probability interval.

DISCUSSION

In the present study, we created quantitative PD risk maps of Norwegian Atlantic salmon farming sites accounting for influential factors of PD occurrence at site level. The study was based on the observed data from 1 January 2009 to 31 December 2010. The influential factors included site latitude, site density, PD history, and local biomass density. Subsequently, the probabilities of PD occurrence at site level were displayed in a map covering the southern part of Norwegian marine aquaculture, and validated with observed data of 2011.

The present study included sites located in the southern part of Norway where PD is endemic as well as sites in the southernmost part of the non-endemic area bordering the endemic zone. By including this non-endemic area, the strength of the model to predict the development of PD in this area could be observed. PD strategies in the PD endemic area mainly focus on control and prevention of the disease spreading outside of the affected sites and the endemic area. In the endemic area, with movement restriction, the PD-affected sites have been allowed to keep their fish until slaughtering time. Vaccination has been practiced in the majority of the sites in this PD endemic area; however, the efficacy and effectiveness of the currently used vaccine is debatable [29]. Conversely, vaccination is not a common practice in the PD non-endemic area (the northern part of Norway), but a stamping out policy is regulated and followed by a fallowing period [12]. So far, relatively few cases of PD have been detected in the non-endemic area; therefore, controlling PD outbreak in this area has been manageable financially. Nonetheless, the confining of PD outbreaks within the endemic area may not be feasible in the near future. A great number of financial supports and a strong commitment of fish farming industries would be needed if more stringent control strategies would be applied to prevent the disease spreading northward and to minimize the number of PD outbreaks in the endemic area. An economic model accessing a merit of new disease control strategies would be an assistant tool for decision makers.

The present study showed that the recent history of PD was a useful predictive factor of upcoming PD outbreaks. We found a positive association between the occurrence of PD outbreaks in 2009-2010 and the sites located within 10-km seaway distance of PD suspected/confirmed sites in 2008. This result supports the findings that increasing risk of getting PD is associated with a close distance to infected farming sites [7, 14]. We explored the influence of LBD and site density on PD occurrence, and found that these variables also yielded a positive association with increasing risk of getting PD after adjusting for the effects of the other variables. Both variables (LBD, site density) present a concentration of fish; however, they could yield different interpretations. A large site density could have a low LBD if the biomass of neighboring sites is low. The large site density means that there are many sites located in a certain area (in this case 10 km seaway distance). Therefore, frequent visits or multiple activities, such as movement of live fish by a well boat, leading to a higher chance of introducing and spreading the disease agent. On the other hand, a small number of sites each with a large biomass could result in a large overall value of LBD being likely to affect the spreading of virus more than introducing it.

The present study found a non-significant association between the time when smolt were put to sea and the PD occurrence. It has been hypothesized that the cohorts of autumn smolts were more susceptible to PD occurrence than cohorts of spring smolts due to a smaller starting weight of the fish, a lower seawater temperature and a shorter day length, etc [7]. The previous study used a fish cohort as a unit of analysis, and included the cohorts since 2002, which could introduce confounding due to the changes in site management and environment over a period of time. Further works are needed to explore in more details the effect of time when smolt cohorts are put to sea on PD occurrence.

Our results showed a significant effect of latitude, with the risk of getting PD increasing as one goes south. Interpretation of this result is sensitive as latitude can be correlated with several unmeasured variables related to PD risk, including seawater temperature. Exploring the effect of seasonality and seawater temperature on the PD occurrence would be of great interest. However, a cyclical pattern and large missing records of this variable in our study population made its integration into our spatial model challenging. An evaluation of this variable using a subset of the study population was attempted by assessing the proportion of months when the seawater temperature is suitable for survival of SAV [30]. Partly due to a similar pattern of seawater temperature of the subset of the study population we did not find this variable significant (data not shown here). A further evaluation of the effect of this variable on the PD occurrence would require a selection of farming sites with a difference in the pattern of seawater temperature, good-quality and comprehensive data, and would be facilitated by a more dynamic time-scale incorporated into the spatial analysis.

Other research groups have conducted epidemiological studies to determine risk factors of PD transmission [5, 31, 32]. A close proximity to infected sites, sharing the same water current network, sharing the same equipment with infected sites, sharing of farm ownerships, movement of fish and well boats, inadequate fallowing period, and abundance of sea lice were suggested as potential PD risk factor. This knowledge is invaluable; however, the accessibility and data quality for an extensive area is limited. Another challenging aspect is to clearly distinguish between the risk factors influencing SAV transmission and those inducing PD outbreak.

Peeler and Taylor (2011) published a thorough review emphasizing importance of applying epidemiology tools in a study of aquatic animal diseases, and mentioned a little use of tools describing a spread of infectious disease [33]. In the case of PD, a close proximity of PD-infected sites is clearly a risk factor for the disease transmission [7, 14], and sea currents can greatly contribute to spread of the virus between sites [9, 13]. Spatial autocorrelation often occurs in outbreaks of infectious diseases in which a farming site located closer to other infected sites is likely to experience an increased chance of acquiring the disease. If this spatial correlation exists and is not accounted for in the model, it could lead to biased parameter estimates and overly optimistic standard errors [20, 34]. In our case, despite the spatial clustering of PD infected sites, we did not find a significant improve by adding a spatial component to the final model. This indicates that the predictor variables appropriately accounted for the local spatial correlation.

The use of a Bayesian modeling approach in this work instead of a frequentist approach for estimating the probability of PD occurrence offers a few advantages. It is a flexible tool for accounting for hierarchical levels, such as spatial dependencies [35]. The method allows incorporating previous information providing an appropriate setting for complex models and missing data problem, and the approach yields results in a form of probability distribution that is interpretable intuitively. In our study, non-informative priors were used throughout due to uncertainty of previous information of the parameter estimates. However, the model estimates obtained from this present study could be later used as the prior knowledge to update the future prediction of PD occurrence. The outputs of the Bayesian method gave the posterior distribution of monitoring parameters. We extracted the median of the posterior distribution of the probability of PD occurrence, and displayed in a so-called "PD risk map". The raster map of PD risk that we produced made it possible to identify local "cold" (green color) and "hot" spots (red color) of PD risk in the area where PD outbreaks area commonly occurring. Based on the PD risk map, the PD predicted probability was high (> 0.5) in the area of Hordaland and some area of Rogaland [Figure 2. In contrast, the probability in Møre and Romsdal was moderate. Without considering the site density and LBD, one could mistakenly believe that there is a higher PD risk in Møre and Romsdal based solely on observed data.

Area under the curve indicated that the raster map produced from data collected in 2009-2010 had a reasonable capacity in discriminating infected farming sites from non-infected ones for the year 2011. This provided a considerable confidence that the risk map could be used as a tool to select areas with an acceptable probability of PD occurrence for salmon farming sites, and to focus surveillance and control measures on high-risk areas. The spatial congruity of infectious aquatic animal diseases is undeveloped and could be explored more in the future by including the use of time-dynamic spatial risk models, and the investigation of the spread of pathogenic agents (with no clinical outbreak) using intensive screening programs. The framework of the study could be applied for spatial studies of other infectious aquatic animal diseases.

CONCLUSIONS

This study presents quantitative risk maps of pancreas disease of Atlantic salmon in Norway. It also provides an insight into the relationship between the risk of PD outbreaks in farming sites and several spatial determinants, including latitude, site density, recent history of PD in a nearby area and local biomass density. The framework of the study could be applied for spatial studies of other infectious aquatic animal diseases.

APPENDIX 1

Variance of PD predicted probability.

NOTES

Acknowledgements

The study was financially supported by the Research Council of Norway (RCN): TRISAV and SIP RCN project number 190484 and 186907. We thank Anja B. Kristoffersen and Brit Bang Jensen for a critical review of the manuscript. We are also grateful to Matthew Farnsworth and Michael Ward for an intensive training in spatial epidemiology.

Competing Interests

The authors declare that they have no competing interests.

Authors' Contributions

1. ST coordinated the work, performed all statistical analyses of the present study including ROC/AUC, and drafted the manuscript. MP contributed to development of the Bayesian code, participated in the analyses of the spatial aspect of the study, and assisted in manuscript writing. HV and EB were involved in the study design, analyses of potential risk factors, and assisted in manuscript writing. DA contributed to development of the Bayesian approach. DJ carried out the handling of the raw data, and assisted in performing descriptive statistics of each variable evaluated in the study. All authors read and approved the final manuscript.
2.

3. REFERENCES

4. Taksdal T, Olsen AB, Bjerkås I, Hjortaas MJ, Dannevig BH, Graham DA, McLoughlin MF: Pancreas disease in farmed Atlantic salmon, Salmo salar L., and rainbow trout, Oncorhynchus mykiss (Walbaum) in Norway. J Fish Dis. 2007, 30: 545-558. 10.1111/j.1365-2761.2007.00845.x.
5. Graham DA, Frost P, McLoughlin K, Rowley HM, Gabestad I, Gordon A, McLoughlin MF: A comparative study of marine salmonid alphavirus subtype 1-6 using an experimental cohabitation challenge model. J Fish Dis. 2011, 33: 123-135.
6. Aunsmo A, Valle PS, Sandberg M, Midtlyng PJ, Bruheim T: Stochastic modeling of direct costs of pancreas disease (PD) in Norwegian farmed Atlantic salmon (Salmo salar L.). Prev Vet Med. 2010, 93: 233-241. 10.1016/j.prevetmed.2009.10.001.
7. Poppe T, Rimstad E, Hyllseth B: Pancreas disease of Atlantic salmon Salmo salar L. post-smolts infected with infectious pancreatic necrosis virus (IPNV). Bull Euro Asso Fish Pathol. 1989, 9: 83-85.
8. McLoughlin MF, Peeler EJ, Foyle KL, Rodger H, O'Ceallachain D, Geoghegan F: An epidemiological investigation of the re-emergence of pancreas disease in Irish farmed atlantic salmon (*Salmo salar* L.) in 2002. Marine Environment and Health Series. 2003, 14: 1-14.
9. McLoughlin MF, Graham DA: Alphavirus infections in salmonids – a review. J Fish Dis. 2007, 33: 511-531.
10. Kristoffersen AB, Viljugrein H, Kongtorp RT, Brun E, Jansen PA: Risk factors for pancreas disease (PD) outbreaks in farmed Atlantic salmon and rainbow trout in Norway during 2003–2007. Prev Vet Med. 2009, 90: 127-136. 10.1016/j.prevetmed.2009.04.003.
11. Fringuelli E, Rowley HM, Wilson JC, Hunter R, Rodger H, Graham DA: Phylogenetic analyses and molecular epidemiology of European salmonid alphaviruses (SAV) based on partial E2 and nsP3 gene nucleotide sequences. J Fish Dis. 2008, 31: 811-823. 10.1111/j.1365-2761.2008.00944.x.
12. Viljugrein H, Staalstrøm A, Molvær J, Urke HA, Jansen PA: Integration of hydrodynamics into a statistical model on the spread of pancreas disease (PD) in salmon farming. Dis Aquat Organ. 2009, 88: 35-44.

13. Kongtorp RT, Stene A, Andreassen PA, Aspehaug V, Graham DA, Lyngstad TM, Olsen AB, Olsen RS, Sandberg M, Santi N, Wallace C, Breck O: Lack of evidence for vertical transmission of SAV 3 using gametes of Atlantic salmon, Salmo salar L., exposed by natural and experimental routes. J Fish Dis. 2010, 33: 879-888. 10.1111/j.1365-2761.2010.01190.x.
14. Jansen MD, Taksdal T, Wasmuth MA, Gjerset B, Brun E, Olsen AB, Breck O, Sandberg M: Salmonid alphavirus (SAV) and pancreas disease (PD) in Atlantic salmon, Salmo salar L., in freshwater and seawater sites in Norway from 2006 to 2008. Prev Vet Med. 2010, 33: 391-402.
15. Anonymous. Norwegian PD regulations 2007-11-20 number 1315: regulations relating to zone to prevent infection and fight pancreas disease in aquaculture. Norwegian: PD regulations 2008-06-17 number 819: regulations relating to the sale of aquaculture animals and products of aquaculture animals; 2008.
16. Stene A, Viljugren H, Jansen P-A: Transmission of SAV3 by water transport? Current patterns and infection pressure between farms. Belfast: PD-trination meeting; 2011.
17. Aldrin M, Storvik B, Frigessi A, Viljugrein H, Jansen PA: A stochastic model for the assessment of the transmission pathways of heart and skeleton muscle inflammation, pancreas disease and infectious salmon anemia in marine fish farms in Norway. Prev Vet Med. 2010, 93: 51-61. 10.1016/j.prevetmed.2009.09.010.
18. Jansen PA, Kristoffersen AB, Viljugrein H, Jimenez D, Aldrin M, Stien A: Sea lice as a density dependent constraint to salmonid farming. Proc R S B. 2012, -10.1098/rspb.2012.0084. published online. 1-9
19. Lunn DJ, Thomas A, Best N, Spiegelhalter DJ: WinBUGS – a Bayesian modelling framework: concepts, structure, and extensibility. Stat Comput. 2000, 10: 325-337. 10.1023/A:1008929526011.
20. Spiegelhalter DJ, Best N, Carlin BP, Van Der Linde A: Bayesian measures of model complexity and fit. J R Stat Soc Series B Stat Methodol. 2002, 64: 583-639. 10.1111/1467-9868.00353.
21. Dormann CF, McPherson JM, Araujo MB, Bivand R, Bolliger J, Carl G, Davies RG, Hirzel A, Jetz W, Kissling WD, Kuhn I, Ohlemuller

R, Peres-Neto PR, Reineking B, Schroder B, Schurr FM, Wilson R: Methods to account for spatial autocorrelation in the analysis of species distribution data: a review. Ecography. 2007, 30: 609-628. 10.1111/j.2007.0906-7590.05171.x.

22. Diggle PJ, Ribeiro PJ: Model-based Geostatistics. Springer, New York, USA: Springer Series in Statistics; 2007.
23. Pfeiffer DU, Robinson TP, Stevenson M, Stevens KB, Rogers DJ, Clements ACA: Spatial Analysis in Epidemiology. New York: Oxford University Press; 2008.
24. Pullan RL, Kabatereine NB, Bukirwa H, Staedke SG, Brooker S: Heterogeneities and Consequences of *Plasmodium* Species and Hookworm Coinfection: a Population Based Study in Uganda. J Infect Dis. 2011, 203: 406-417. 10.1093/infdis/jiq063.
25. Besag J, York J, Mollié A: Bayesian image restoration, with two applications in spatial statistics. Ann Inst Stat Math. 1991, 43: 1-59. 10.1007/BF00116466.
26. Mollié A: From Bayesian and Empirical Bayes approaches to disease mapping. Disease mapping and risk assessment for public health. Edited by: Lawson A, Biggeri A, Bohning D, Lesaffre E, Viel JF, Bertollini R. Chichester, UK: John Wiley & Sons Ltd; 1999:15-29.
27. Greiner M, Pfeiffer D, Smith RD: Principles and practical application of the receiver-operating characteristic analysis for diagnostic tests. Prev Vet Med. 2000, 45: 23-41. 10.1016/S0167-5877(00)00115-X.
28. Fosgate GT, Adesiyun AA, Herd DW, Johnson WO, Hietala SK, Schurig GG, Ryan J: Estimation of receiver-operating characteristic curves to determine accuracy of a competitive enzyme-linked immunosorbent assay for the serodiagnosis of Brucella infection in domestic water buffalo (Bubalus bubalis) and cattle. Am J Vet Res. 2003, 64: 57-64. 10.2460/ajvr.2003.64.57.
29. Pontius RG, Schneider LC: Land-cover change model validation by an ROC method for the Ipswich watershed, Massachusetts, USA. Agric Ecosyst Environ. 2001, 85: 239-248. 10.1016/S0167-8809(01)00187-6.
30. Arekhi S: Modelling spatial pattern of deforestation using GIS and logistic regression: a case study of northern llam forests, llam province, Iran. Afr J Biotechnol. 2011, 10: 16236-16249.

31. Robin X, Turck N, Hainard A, Tiberti N, Lisacek F, Sanchez J-C, Müller M: pROC: an open-source package for R and S + to analyze and compare ROC curves. BMC Bioinforma. 2011, 12: 77-10.1186/1471-2105-12-77.

32. Karlsen M, Tingbø T, Solbakk I, Evensen Ø, Furevik A, Aas-Eng A: Efficacy and safety of an inactivated vaccine against Salmonid alphavirus (family Togaviridae). Journal of Vaccines. 2012, In press

33. Graham DA, Staples C, Wilson CJ, Jewhurst H, Cherry K, Gordon A, Rowley HM: Biophysical properties of salmonid alphaviruses: influence of temperature and pH on virus survival. J Fish Dis. 2007, 30: 533-543. 10.1111/j.1365-2761.2007.00811.x.

34. Wheatly SB, McLaughlin MF, Menzies FD, Goodall EA: Site management factors influencing mortality rates in Atlantic salmon (Salmo salar L.) during marine production. Aquaculture. 1995, 136: 195-207. 10.1016/0044-8486(95)01058-0.

35. Rodge H, Mitchell S: Epidemiological observations of pancreas disease of farmed Atlantic salmon, *Salmo salar* L., in Ireland. J Fish Dis. 2007, 30: 157-167. 10.1111/j.1365-2761.2007.00799.x.

36. Muller EJ, Taylor NG: The application of epidemiology in aquatic animal health-opportunities and challenges. Vet Res. 2011, 42: 94-10.1186/1297-9716-42-94.

37. Firestone SM, Christley RM, Ward MP, Dhand NK: Adding the spatial dimension to the social network analysis of an epidemic: investigation of the 2007 outbreak of equine influenza in Australia. Prev Vet Med. 2012, In press

38. Torabi M, Rosychuk RJ: Hierarchical Bayesian Spatiotemporal Analysis of Childhood Cancer Trends. Geogr Anal. 2012, 44: 109-120. 10.1111/j.1538-4632.2012.00839.x.

SECTION IV

FEED AND NUTRITION

CHAPTER

9

PROGRESS ON LARVAL AND JUVENILE NUTRITION TO IMPROVE THE QUALITY AND HEALTH OF SEAWATER FISH: A REVIEW

Toshio Takeuchi

Tokyo University of Marine Science and Technology, Tokyo, Japan

ABSTRACT

In seedling production of seawater fish, providing appropriate nutrition is a necessity for successful production of quality larvae and juveniles. Mass-produced live prey organisms, such as the rotifer *Brachionus plicatilis* species complex and brine shrimp *Artemia* spp., alone do not provide sufficient nutrition to the larvae and juveniles of seawater animals. This inadequacy has led to various problems related to fish quality and health, including increased incidence of morphological and behavioral abnormalities and mass mortalities. It is, therefore, important to identify the factors associated with these problems to improve seedling production techniques. This review collates the efforts made during the past two decades in larval nutrition–focusing on advances made in the use of certain nutrients, such as docosahexaenoic acid, vitamin A derivatives, and taurine that are important for the mass production of seawater fish larvae and juveniles–with an aim to improve the quality and health of fish.

Keywords: DHA, Juvenile, Larvae, Nutrition, Seed production, Taurine, VA

INTRODUCTION

In 2006, Worm et al. [1] demonstrated in an alarming article the prediction that world marine resources would be exhausted by 2048. Since then, declining populations of marine animals, such as bluefin tuna and eel, have been reported one after another, making the *Science* article more and more realistic. At the same time, the demand for marine products, as demonstrated by annual per capita consumption of seafood products, has been increasing in the United States, Europe, East Asia, and especially in China where the quantity supplied increased from 7 % in 1961 to 35 % in 2010 [2]. Given these circumstances, the global fishing industry has made a transition from fishing to farming. Consequently, the production of edible fish from aquaculture has reached 47 % of the total fish production in 2010. One of the techniques essential to supporting the aquaculture industry is artificial seedling production.

Although artificial seedling production in the freshwater aquaculture sector has been practiced since the first half of the 20th century, it was only sometime in the 1960s that the seawater sector started practicing artificial seedling production. Seedling production consists of fertilized egg collection and the three types of biological production: the breeding of larval and juvenile fish, cultivation of prey organisms, and cultivation of algae (as feed for prey animals). The difficulty in mass-producing prey animals for larval and juvenile fish in large supply and the low nutritional value of the mass-produced prey animals were the major reasons for the particularly late start of the seedling production of seawater fish. However, with marked advances in aquaculture technology in recent years, Japan currently produces 1–10 million of seawater fish seeds per hatchery.

A live prey schedule containing *Brachionus plicatilis* sp. complex (hereafter referred to as rotifer) and brine shrimp *Artemia* larva (hereafter referred to as *Artemia*) is used for the production of most seawater fish seeds. Thus, the use of bivalve larvae and copepods (such as *Tigriopus* or *Acartia*) is rare today, with the exception of the planktonic crustacean *Daphnia*, which is still frequently used in addition to the formulated diet in the production of freshwater fish seeds. Due to the discovery that baking yeast is an excellent feed for rotifers in culture, it became possible to increase the growth density of rotifers from 40 to 60 rotifers/ml of culture water, achieved previously with chlorella, to 400–600 rotifers/ml. Subsequently, the use of yeast in culturing rotifers became widespread in response to mass seedling production [3]. However, larval and juvenile fishes that were reared on a diet of rotifers fed yeast exhibited a slow reaction time to stimuli, and there were many reports of significant abdominal buoyancy as well as the complete mortality of seawater fish, such as red sea bream and parrot fish, about 10 days after initiation of feeding. Conversely, when only *Artemia* was continuous given as feed, mass mortality occurred in a much larger population of fishes, and even when no mortality occurred fishes showed a lack of energy, with different species affected to varying degrees. Comprehensive investigation to determine whether mass mortality in larval and juvenile fishes was due to nutritional deficiency revealed that many live feeds were ,indeed, nutritionally incomplete. In particular, the lack of n-3 highly unsaturated fatty acid (n-3 HUFA) with docosahexaenoic acid (DHA), both of which are essential fatty acids for seawater fishes, was the cause of mass mortality associated with the diets containing rotifers fed yeast or *Artemia* [3]. In addition, it became clear that the dietary requirement for DHA varies among seawater fish species. Consequently, feeds for seed production today are sampled prior to feeding to ensure adequate amounts of nutrients, particularly taurine and DHA.

Conventionally, the research and development of seedling production technology aimed to quantitatively increase seed production. However, along with this quantitative expansion, it became clear that when artificially produced seeds are used for fish stocking and farming, recovery rates suffered a decline and farming outcomes are affected unless fish health (physiological and morphological conditions) and fish quality (environmental adaptability to the place of stocking and farming) are improved [4]. These findings led to an increase in the demand for healthy and high-quality seeds and shifted the focus of the research and development of seedling production technology from quantitative expansion to quality improvement.

In this review, with the aim of improving the quality of seedling production, we summarize recent findings on DHA, vitamin A-related substances, and taurine, all of which are vital nutrients for fish.

DOCOSAHEXAENOIC ACID AND QUALITY OF LARVAE AND JUVENILES OF SEAWATER FISH

As we know, the fatty acid compositions of traditional live foods (e.g., rotifer and *Artemia*) reflect the food supply in their mass production. One of the disadvantages of these commercially produced live foods is the low content of n-3 highly unsaturated fatty acid (n-3 HUFA) [5, 6]. In particular, docosahexaenoic acid (DHA, 22:6n-3) is the most important n-3 HUFA for larvae and juveniles of seawater fish, and a high concentration of DHA in rotifer and *Artemia* has been shown to be essential for the seed production of yellowtail *Seriola quinqueradiata*, red sea bream *Pagrus major*, and striped jack *Pseudocaranx dentex* [7]. This section of the review discusses the importance of DHA.

Post-hatching Changes in DHA Content in Seawater Fish

The body fatty acid composition of seawater fish changes after hatching. In newly hatched fed larvae, the percentage of DHA (the composition ratio of DHA that originated from the egg relative to all fatty acids) decreases sharply over the first 10 days or so after hatching in yellowtail, red sea bream, flounder *Paralichthys olivaceus*, striped jack, and Pacific cod *Gadus macrocephalus* fed rotifer, as well as in common octopus *Octopus vulgaris* fed *Artemia* (Fig. 1) [8]. During this period, the body weight of fish sharply increases while the lipid content decreases. For example, in Pacific cod there is a dramatic decrease in the DHA content: the amount of DHA per gram wet weight on day 15 dropped to 1.2 mg/g, one-fifth of the initial amounts (6.1 mg/g on post-hatching day 1); and the absolute amounts down to 1.7 μg/fish, one–half of the initial amounts (3.7 μg/fish). We can rule out a possibility of the changes in DHA during post-hatching as a result of fasting condition in the hatchling. Usually in the fasted fish, the percentage of certain fatty acids such as arachidonic acid (20:4n-6) and DHA increase due to decreased body fat, whereas saturated and monounsaturated fatty acids decrease, even in an absolute amount, most prominently in neutral lipid, as they are consumed for energy [9, 10]. On account of the fact that the fatty acid compositions in newly hatched fed larvae reflect the food supply, the quick decrease of the DHA content in both polar and neutral lipids may suggest insufficient dietary DHA content, especially during the post-hatching period.

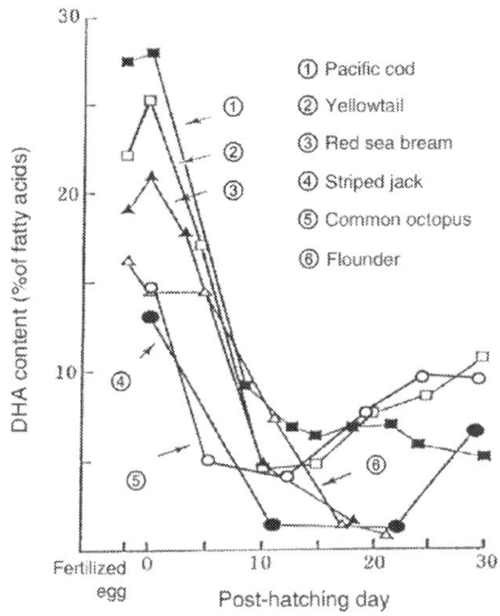

Figure. 1: Change in the DHA content in different seawater fish hatchlings [8].

Live Food Quality in Relation to DHA

Studies on the effect of the essential fatty acids eicosapentaenoic acid (EPA, 20:5n-3) and DHA during the *Artemia*-feeding period have revealed that yellowtail and striped jack have a higher DHA requirement than red sea bream and flounder [7]. Interestingly, in a "vitality test" where the larvae were exposed to the air for 5–20 s in a scoop net before being returned to the culture tank, only DHA-fed larvae showed a favorable survival rate (usually after 24 h). *Artemia*, regardless of origin, contain no DHA. In fish fed *Artemia* enriched with high-purity DHA ethyl ester, DHA is converted to docosapentaenoic acid (22:5n-3) [11] and EPA or used as an energy source. Consequently, the DHA-enrichment effect obtained with DHA ethyl ester is only one-third of that obtained with the same amount of EPA. These findings indicate that the low DHA content in live foods is likely to be one cause of the mass mortality of red sea bream larvae immediately after transfer to a net cage and for the low survival rate of yellowtail and striped jack during seed production.

The nutritional enrichment of live foods containing dried shark eggs is a possible solution for increasing the dietary content of DHA [12, 13, 14]. The fins of the spiny dogfish *Squalus acanthias* are used as food products, and the eggs from the captured sharks are obtained as a by-product. To ensure efficient by-product use, the eggs can be processed into dried powder and used to nutritionally enrich rotifer and *Artemia* with high concentrations of EPA and DHA. The mixture consists of roughly equal amounts of protein and lipid per dry weight and contains 3.9 % EPA, 7.6 % DHA [7], and 0.5 % taurine. Subsequent studies have demonstrated the utility of this shark egg powder not only as an enrichment material, but also as a paste feed for leptocephalus, the larvae of eel [15].

The DHA content required for live foods, especially *Artemia*, varies among fish species; the highest concentration is required by yellowtail and striped jack, followed by Pacific cod, red sea bream, and flounder [16]. In nutritionally enriched live foods, the DHA concentration varies substantially depending on the viability, strain, and enrichment procedures of the live foods such as water temperature, enrichment duration, enrichment concentration, and aeration intensity. Close attention is, therefore, needed when nutritionally enriching rotifer and *Artemia*for fish with a high DHA requirement.

Effects of Dietary DHA on Larvae and Juveniles of Seawater Fish

The relationship between the content of highly unsaturated fatty acids (e.g., DHA) in live foods and fish health (e.g., normal morphology rate) has been investigated in flatfish such as flounder [7], marbled sole [17], turbot [18], and brown sole *Pseudopleuronectes herzensteini* [19, 20]. A recent study on brown sole showed that an increased DHA content (from 0.6 % to 1.7–3.2 %) in rotifer between post-hatching days 15 and 25 (stages D to F) significantly improved survival and normal morphology rates, demonstrating that DHA supplementation is particularly beneficial during the metamorphic period known as the DHA-sensitive period (Fig. 2) [21]. The following details the investigation in some species of seawater fish (e.g., yellowtail and octopus).

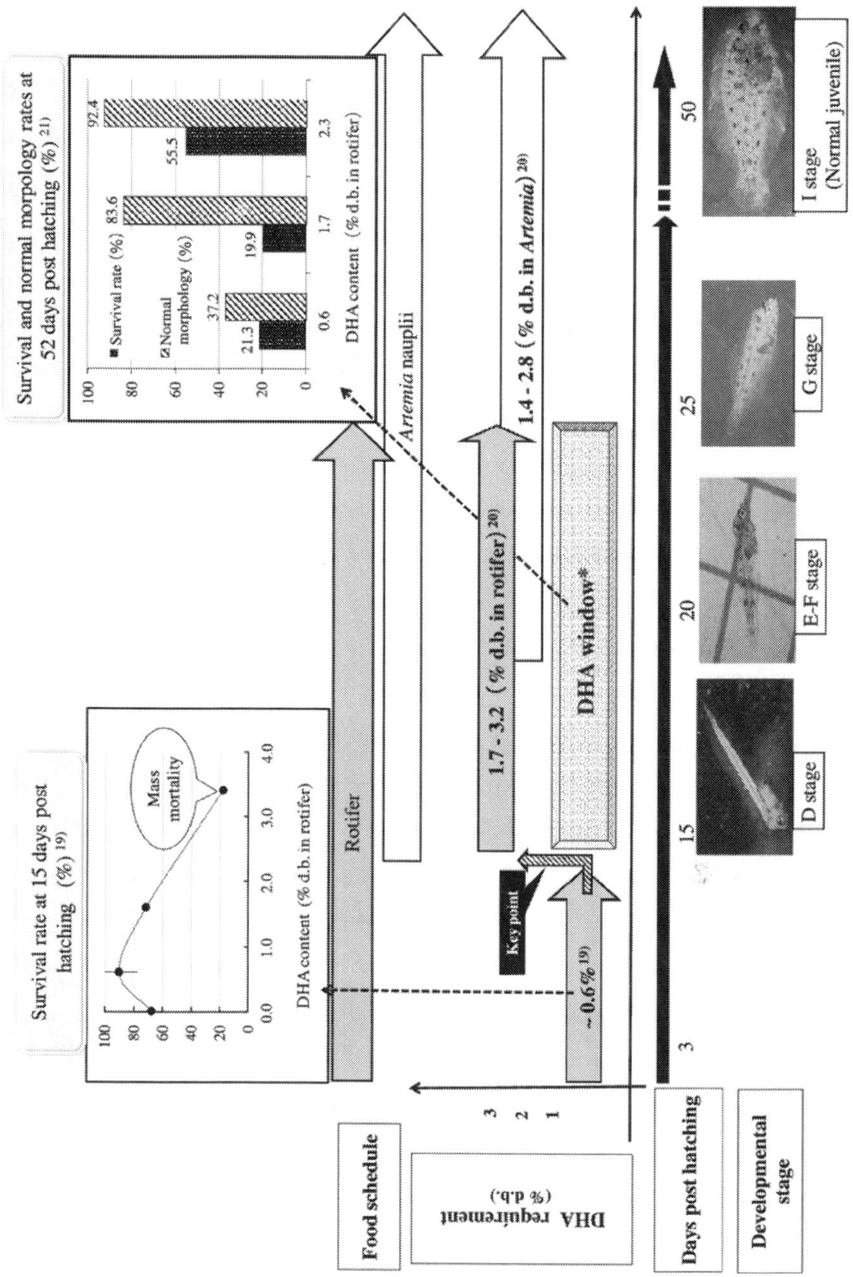

Figure. 2: Schematic representation of DHA nutritional requirements for larvae of brown sole *Pseudopleuronectes hersensteini*. *Asterisk* indicates the rearing period when DHA nutrition is considered critical [21].

Yellowtail

The reduced quality of released yellowtail juveniles is an ongoing problem. A previous study has shown that red sea bream reared in a net cage at a low population density do not exhibit schooling behavior but rather lying behavior as seen in wild individuals [22]. Another study has shown that wild and hatchery flounder exhibit distinct feeding behavior; the former show quick movement while the latter show slower off-bottom movement [23].

DHA has been shown to have a major effect on the vitality and behavior of larval and juvenile yellowtail. When yellowtail larvae and juveniles fed *Artemia* enriched with oleic acid (OA), EPA, or DHA, they were observed to exhibit dispersion, aggregation, and inherent schooling behavior, respectively [24, 25]. More specifically, the nearest neighbor angle decreased gradually only in the DHA-fed fish, indicating that the larvae were swimming in the same direction, while an increased nearest neighbor distance was observed in the OA-fed fish compared with the EPA-fed and DHA-fed fish, indicating a greater distance between individuals (Fig. 3).

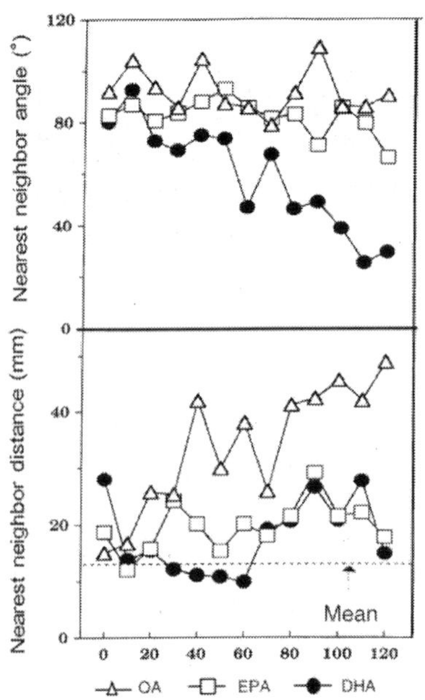

Figure. 3: Effect of dietary fatty acids including DHA on the orientation of free-swimming juvenile of yellowtail *Seriola quinqueradiata*. The fish positions are

mapped over a period of 120 min and presented in terms of distances or angle between adjacent juvenile. Nearest neighbor angle (NNA) indicates the angle formed by adjacent individuals; NNA = 0° indicates fish oriented in the same direction, NNA = 90° indicates they are swimming in different directions. Nearest neighbor distance indicates the distance between adjacent individuals, with a lower value indicating increased school formation.

A study to elucidate the reason for these behavioral differences revealed, from a comparison of the size of different parts of the brain, accelerated (retarded) development of the optic tectum and cerebellum in the DHA (OA)-fed fish [26]. Another study using [14] C-DHA to determine the distribution of DHA in the fish brain revealed an increased distribution of DHA in the periventricular gray and central white layers (nerve fiber layer) of the optic tectum and granular layer (nerve cell layer) of the cerebellum [27]. These observations suggest that dietary DHA intake by yellowtail larvae leads to the growth of nerve cells in the optic tectum and cerebellum, the extension of dendrites, and the creation of an information network through synapses at the tips of extended dendrites. Such development of the visual and motor centers might allow fish that would exhibit only the low-order behavior of taxis without supplementation to exhibit the higher-order behavior of schooling with it. Thus, DHA is involved not only in the development of vision, but also in the construction of higher-order neural networks that govern, for example, the development of swimming ability [28]. These results may reflect the essentiality of DHA to "better working of the brain", as demonstrated in rodents where DHA intake leads to an improved learning effect.

Common Octopus

Studies have revealed that insufficient dietary DHA intake is responsible for the initial mass depletion of planktonic common octopus larvae observed up to around 10 days after hatching (Fig. 4) [29]. Newly hatched octopus have short arms with 3–4 suckers and are planktonic, unlike adults. The number of suckers increases as the arms grow, and after 18 suckers have formed (28 days after hatching), the animals become benthic. During seed production of the common octopus, the number of newly hatched larvae starts to decrease rapidly around 1 week after hatching and most have died by around 15 days after hatching. At this time, adult *Artemia*is given as the initial feed; however, this feed is difficult to nutritionally enrich and its nutritional value quickly decreases [7]. Experiments have demonstrated the effectiveness of enriched foods by direct addition of nutrients such as shark eggs, lipid materials, and DHA algae to the culture tank and have also revealed a positive correlation between DHA content in *Artemia* and survival rate and the numbers of suckers [30]. These results have proven valuable in helping prevent the depletion of planktonic octopus paralarvae by maintaining a certain level of DHA in *Artemia*.

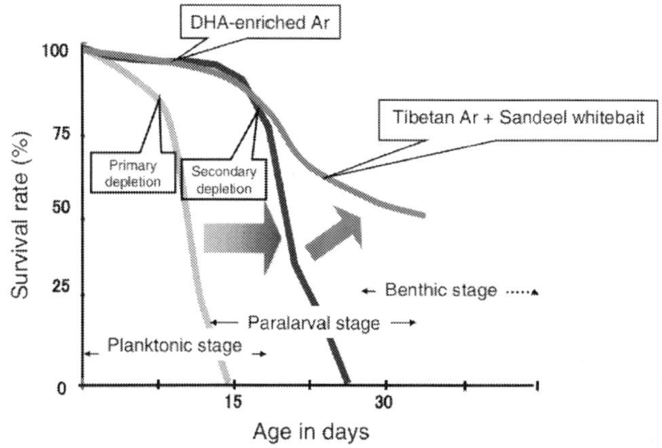

Figure. 4: Dietary DHA requirement to prevent primary and secondary losses of octopus *Octopus vulgaris* during their larval rearing. Note that larvae receiving DHA enriched *Artemia*has improved survival rate [29]. *Ar* indicates *Artemia.*

The common octopus suffers a secondary depletion, however, around 15–25 days after hatching, which must be prevented to obtain benthic juvenile octopus with ≥18 suckers (i.e., settling octopus). Although the experimental results above suggest the benefit of DHA supplementation, it is difficult to provide sufficient amounts of DHA through the nutritional enrichment of adult *Artemia*. Investigators have, therefore, looked at using sandeel whitebait instead, based on the knowledge that many wild common octopus paralarvae are captured with anchovy whitebait during seine boat fishing for whitebait, which suggests that the paralarvae fed on whitebait [31]. Subsequent studies have demonstrated better results with sandeel whitebait than with anchovy whitebait, and the addition of sandeel whitebait flakes containing large amounts of EPA and DHA directly to the culture tank has proven even more effective for rearing [32]. As another alternative, nutritionally enriched *Artemia* nauplii in place of adult *Artemia* were used. When common octopus paralarvae were fed Tibetan *Artemia*, North American *Artemia* (Utah strain produced from Great Salt Lake, US), which is routinely used in seed production), or Tianjin *Artemia* enriched with shark eggs and with or lacking sandeel whitebait, favorable results were obtained only in those animals fed the Tibetan *Artemia* plus sandeel whitebait. Thus, it would appear that only the combination of shark egg-enriched Tibetan *Artemia* plus sandeel whitebait is effective for preventing the secondary depletion [33]. However, the difficulties in routinely procuring Tibetan *Artemia* and in ensuring its sufficient quality (e.g., its hatching rate is variable) further complicates its use.

Working on the premise that if we can identify which ingredients of Tibetan *Artemia* are beneficial to common octopus larvae, it should be possible to sufficiently enrich *Artemia* of any origin, investigators began looking for these nutritional ingredients. It has been found that feeding with North American *Artemia* enriched with lipid, polar lipid in particular, results in an increased number of suckers, body weight, and survival rate, suggesting that polar lipids contain beneficial ingredients [34]. The identification and practical application of these ingredients should allow for mass seed production of the common octopus paralarvae.

DHA Overdose

Previous studies have shown that rainbow trout fed diets supplemented with high concentrations of essential fatty acids such as EPA and DHA showed reduced growth and feed efficiency [35]. Similar effects have also been observed in seawater fish larvae. In the seed production of Pacific cod and brown sole, rotifer containing 3 % or higher concentrations of DHA resulted in mass mortality [13, 19], presumably as a direct effect of DHA overdose. Moreover, in the seed production of *Scylla serrata*, a species of mud crab, excessive dietary levels of n-3HUFA and DHA have led to failed molting and morphological abnormalities, such as seta formation, in zoea [36, 37]. These observations suggest that an appropriate concentration of DHA in live foods is essential to improving the quality of seawater fish.

VITAMIN A AND LARVAL QUALITY

Flounder is one of the most popular fish species produced through seed production in Japan. Although established seed production procedures have enabled the mass production of up to one million flounders per facility, the occurrence of skeletal anomalies, pseudo albinism, and hypermelanosis on the blind side has reduced the survival of released individuals and the market price of caught individuals [23, 38]. More specifically, skeletal anomalies include malformations of the maxilla, mandible, caudal peduncle, and fin rays, and fused or wavy vertebrae. Pseudo albinism is the partial whitening of the body surface on the ocular side resulting from failed pigmentation at the end of metamorphosis, and hypermelanosis on the blind side is the partial or total pigmentation of the body surface on the normally white blind side.

History of Preventing Malformations

Since the 1980s, many reports have indicated the preventive and exterminating effects of early stage feeds on pseudo albinism in *P. olivaceus*. Early life stage feeding of *Artemia* and a diet of Brazilian *Artemia* has led to a high incidence of pseudo albinism, whereas feeding wild zooplankton, fertilized red sea bream eggs, and vitamin A-supplemented rotifer has reduced the incidence [39, 40, 41, 42, 43]. It was later reported that vitamin A, DHA, and phospholipids are necessary for normal pigmentation on the ocular side [17]. However, feeding vitamin A-enriched rotifer was found to lead to a high incidence of elongated caudal peduncles, partial loss of caudal fin rays, limited growth, and malformed caudal vertebrae. It thus became clear that vitamin A supplementation for preventing pseudo albinism causes skeletal anomalies [44, 45].

These abnormalities are thought to occur in certain sensitive stages in the development of flounder larvae [46]. According to Minami et al. [47], flounder have nine developmental stages (A–I), ranging from larvae to juveniles. This staging system has been adopted by many investigators as a useful tool for understanding the early stage life cycle and morphogenesis of flounder. In particular, the period between late stage D and stage E is important for the determination of pseudoalbinism [48].

Effects of Vitamin A Derivatives

Many studies conducted in the 1990s on vitamin A and vitamin A-related abnormalities suggested the impact of feeding concentration/stage and immersion concentration/stage (Fig. 5) [49]. Early studies reported rotifer enriched with vitamin A at a concentration of 50,000 IU/L in culture medium prevented pseudoalbinism but not skeletal malformations, which were observed even at a much lower vitamin A concentration of 10,000 IU/L [44, 45]. When *Artemia* was used, the addition of vitamin A at a concentration of ≥4,000 IU/L in culture medium caused vertebral anomalies, and the threshold dose of vitamin A to avoid vertebral anomalies was determined to be 2,000 IU/L [50]. These findings suggest that vitamin A-enriched *Artemia* is more likely to cause skeletal anomalies at lower vitamin A concentrations than vitamin A-enriched rotifer; we will return to the reason for this later in this section. Another study showed that feeding flounder larvae with *Artemia* enriched with 10,000 IU/L vitamin A for only 2–3 days caused vertebral anomalies [51]. Similarly, feeding stage E–H flounder larvae with *Artemia* enriched with vitamin A derivatives, such as ester-type vitamin A palmitate, retinol, retinoic acid, and vitamin A acetate, resulted in a high incidence of vertebral anomalies. In fact, retinoic acid, an active form of vitamin A, is present in all *Artemia* enriched with vitamin A derivatives [52].

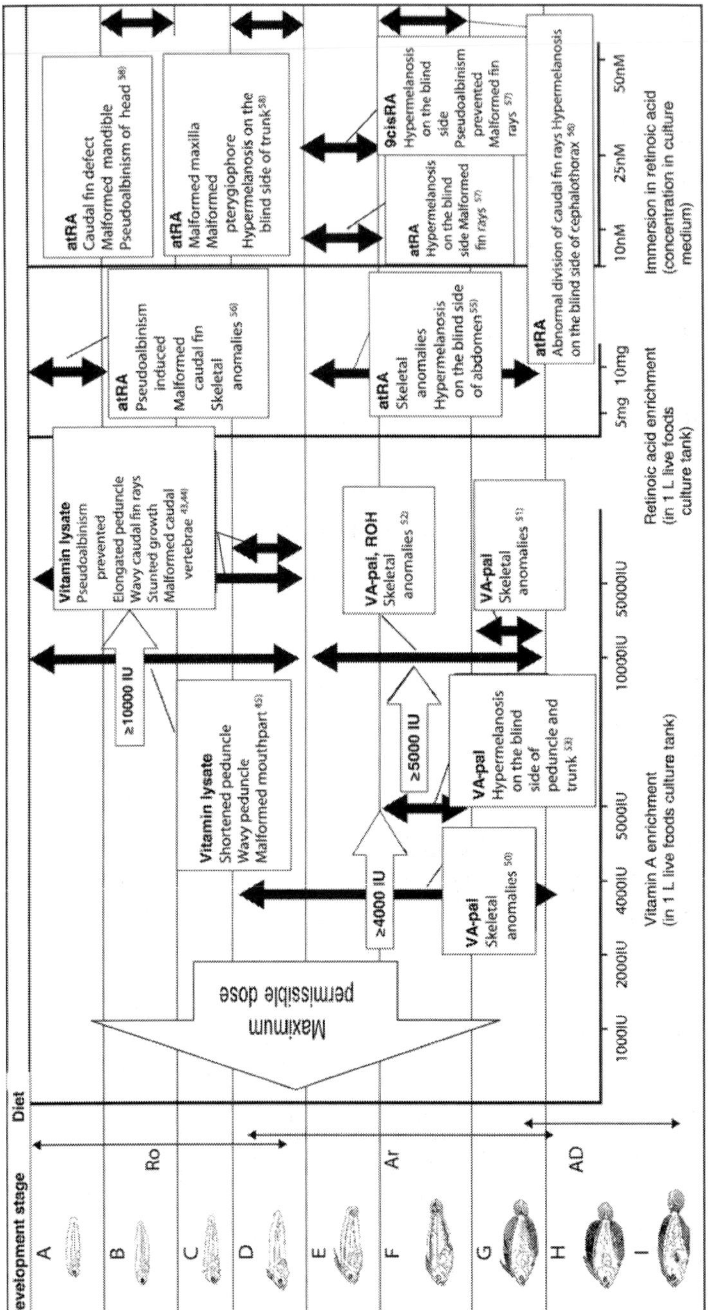

Figure. 5: Pictograph summarizing different studies that examined the effect of vitamin A and its derivatives on the development of flounder *Paralichthys*

olivaceus. The anomalies reported in the different studies are mentioned in the *different boxes*. The amount of vitamin provided for the fish are indicated on the *x* axis.

In regard to hypermelanosis on the blind side, flounder larvae fed *Artemia* enriched with ≥5,000 IU/L vitamin A palmitate only for 5 days during stages F-G developed the disorder. Retinoic acid was detected in these *Artemia* at a level of 7.2–20.5 μg/g, dry basis (d.b.) [53]. Analysis of the kinetics of vitamin A-enriched live foods revealed that vitamin A palmitate in rotifer- and *Artemia*-enriched feed containing 10,000 IU/L vitamin A palmitate was metabolized into retinol and then retinoic acid within around 18 h after enrichment. The retinoic acid content after 18 h of enrichment under 2,000 lux illumination was considerably higher in *Artemia* (25.6 μg/g, d.b.) than in rotifer (9.6 μg/g, d.b.). Thus, more retinoic acid is produced in *Artemia* after enrichment with the same amount of vitamin A. The previous observations that *Artemia* was more likely than rotifer to cause skeletal anomalies at a lower concentration of vitamin A appears to be due to the differential retinoic acid levels produced by metabolism in the two types of live foods. A possible contributing factor to this was found in the next study on the effect of photoirradiation during the culture of live foods: light shielding suppressed metabolism to retinoic acid in rotifer but had no effect in *Artemia* [54].

Flounder larvae fed retinoic acid-enriched *Artemia* for stages E–H showed a dose-dependent increase in the incidence of vertebral anomalies [55], while larvae fed retinoic acid-enriched rotifer for stages A–B developed pseudoalbinism and malformed caudal fins and vertebrae [56]. More clear-cut results on the effects of retinoic enrichment were obtained when premetamorphic flounder larvae (stages E–F) were immersed in water containing 10 or 25 nM retinoic acid: the larvae developed melanism (with a reduced incidence of albinism) and malformed fin rays [57]. When larvae at different stages of development were immersed in culture water containing 50 nM retinoic acid, larvae at stages B–C developed malformed mandibles and pseudoalbinism of the head, larvae at stages D–E developed malformed maxillae and pterygiophores in addition to hypermelanosis on the blind side of the trunk, and larvae at stages F–G showed excessive caudal fin rays and hypermelanosis on the blind side of the cephalothorax [58].

Based on these findings, the maximum recommended dose of vitamin A in live foods has been determined as 2,000 IU/L (Fig. 5). It appears that different anomalies are induced by vitamin A derivatives at different developmental stages: skeletal anomalies in stages A–G (from immediately after hatching to immediately before settling on the bottom), albinism in stages A–C, E, and F, and melanism in stages E–G. Thus, feeding diets containing a high concentration of vitamin A to larvae at any developmental stage can lead to anomalous development. Moreover, the effect of vitamin A appears to be more evident in immersion studies than in feeding studies. Unlike in immersion studies where larvae are continuously exposed to a consistent concentration of vitamin A, in feeding studies larvae shown a clear sharp increase and decrease in vitamin A content that reflects the dietary intake of vitamin A and its subsequent absorption and metabolism. This is supported by the observation that the concentration of all-trans retinoic acid peaks 3 h after flounder larvae are fed *Artemia* enriched with all-trans retinoic acid, before declining to about one-third of the peak concentration 24 h after feeding [59].

Taken together, the findings to date indicate that retinoic acid, which is produced by metabolizing vitamin A added at high concentration to live foods, is responsible for the occurrence of skeletal and pigmentation abnormalities in flounder.

Other Factors

Apart from vitamin A derivatives, hypermelanosis on the blind side can also be induced in larvae at stage G by a diet supplemented with vitamin D at a concentration of 20,000 IU/100 g [60]. However, rearing flounder after settlement in a culture tank with sand has been shown to be effective for preventing the development and progression of hypermelanosis on the blind side [61]. In addition, an analysis of the relationship between pseudoalbinism and the thyroid hormone thyroxin demonstrated that pseudoalbinism can be induced at a high rate simply by exposing flounder larvae at stages E–F to thyroxin by immersion [62].

Attempts have also been made to elucidate our understanding of the molecular mechanism underlying development of pseudoalbinism. Receptor analysis studies [63] are among the approaches taken so far. Also, a recent review by Hamre et al. [64] reviewed effects of dietary lipid-soluble vitamins on ontogeny of larvae of seawater fish. It is expected that further work will establish seed production techniques that do not rely purely on nutrients given to larval-stage fish but also incorporate research findings from various areas, such as the quality and genetic traits of parent fish.

TAURINE AND SEAWATER FISH QUALITY

Taurine, or aminoethylsulfonic acid ($C_2H_7NO_3S$; molecular weight 125.15), is a sulfur-containing amino acid-like compound that can be separated by the same extraction procedure used for regular free amino acids and quantified with an amino acid analyzer [65]. Unlike constituent amino acids, this amino acid-like compound possesses physiological functions, such as prevention of hypertension, improvement of liver function, maintenance of eye function, promotion of energy metabolism, and repair of radiation damage (e.g., suppression of leukopenia).

Taurine Biosynthesis

Taurine is usually biosynthesized from sulfur-containing amino acids, such as methionine, cysteine, and cystine (a cysteine dimer). The metabolic pathway from methionine to taurine is shown in Fig. 6. This pathway involves several enzymes, many of which are in the liver. Cysteine sulfuric acid, an intermediate metabolite, is decarboxylated by cysteine sulfuric acid decarboxylase (CSD) into the taurine precursor hypotaurine, which is then to taurine. Methionine, a source of taurine, is, therefore, regarded as an essential amino acid. When determining whether the required amount of essential amino acids is contained in a given diet, the combined amount of methionine and cystine is calculated. Althougvgh animals inherently possess this metabolic pathway, kittens and human infants are not able to synthesize a sufficient amount of taurine from methionine due to low CSD activity and, thus, require taurine as a conditionally essential nutrient. In recent studies, CSD activity was measured in various fish species and found to be high in rainbow trout and tilapia and low in flounder, red sea bream, yellowtail, and bluefin tuna *Thunnus orientalis* [66, 67]. It has also been demonstrated that flounder and red sea bream fed a cystine-supplemented diet do not have increased taurine content [68, 69].

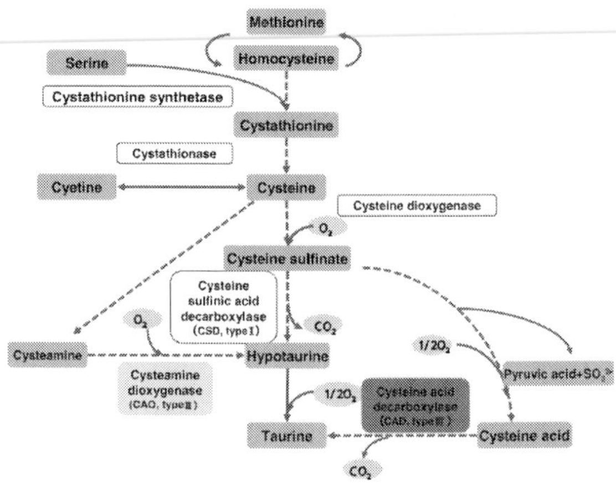

Figure. 6: Biosynthetic pathways of sulfur-containing amino acids in seawater fish.

When the free amino acid content in the body was measured in juvenile flounder, red sea bream, and yellowtail fed taurine-free diets, serine content was increased in the juveniles of all species and cystathionine content was additionally increased in flounder (Fig. 7). This suggests the insufficiency or lack of enzymes involved in the conversion from homocysteine to cystathionine and from cystathionine to cysteine.

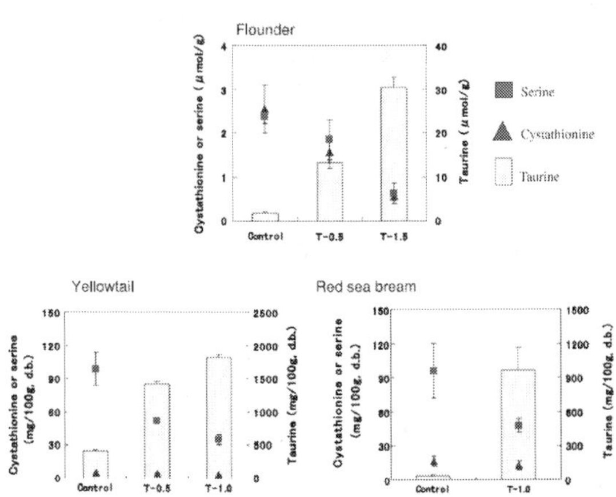

Figure. 7: Comparison of taurine (*yellow column*), serine (*red square*) and cystathionine (*blue triangle*) contents of seawater fish larvae (mean ± SD, $n \geq 3$). T-0, T-0.5, T-1.0, T-1.5 indicate 0, 0.5, 1.0 and 1.5 % of taurine supplementation in experimental diets, respectively.

Before 1990

Following the discovery of the taurine requirement of cats in the 1970s, the first study on the taurine requirement of fish was conducted in rainbow trout [70]. In the late 1980s, Sakaguchi and Murata reported a slight improvement in the growth in eel and guppy fed taurine-supplemented diets [71], though the news did not attract much attention.

Early 1990s

In 1992 Yokoyama and Nakazoe [72] concluded that taurine is not required for growth of rainbow trout as the fish can synthesize taurine, and they observed no effect of taurine supplementation in a feeding study.

Meanwhile, as part of the Development of Quality Seed Production Technology Project, a contract research project started in 1991 with the Fisheries Agency, comparative studies began to investigate the quality of hatchery fish and that of wild fish as a control. This project found that wild flounder achieved better growth when fed mysids (*Acanthomysis nakazatoi* and *A. kokuboi*) than with a formulated diet for flounder, and that feeding with live mysids resulted in higher n-3 HUFA and taurine levels and lower cystathionine levels in the fish body than feeding with a formulated diet [73, 74]. Given the fact that mysids are rich in taurine, various studies were then carried out to evaluate the effectiveness of taurine enrichment of live foods.

Late 1990s

Live mysids and dried mysid powder diets for juvenile flounder were produced to identify the diets' active ingredients and it was revealed that dietary supplementation of about 3.5 % taurine is needed to obtain a taurine content in the fish body comparable to that in juveniles immediately after becoming benthic (including wild juveniles). It was also found that after switching from a commercial formulated diet to live mysids, the content of DHA and taurine rapidly increased and reached a constant level within 5 days of switching [75]. These subsequent studies revealed that the active ingredients of mysids consist of free amino acids including taurine, and that the taurine content in commercial diets did not meet the taurine requirement of juvenile flounder [76]. Different patterns of behaviors are exhibited by wild and hatchery juvenile flounder; wild individuals rise up from the bottom to prey for a shorter period than hatchery individuals and are also more likely to turn around while swimming [77]. Of note, juvenile flounder fed a mysid powder diet exhibited a feeding behavior similar to that of wild juveniles [76].

After 2000

Research performed in the 1990s revealed that feeding mysids is effective for promoting the growth of juvenile flounder through the effect of taurine, and that taurine insufficiency is responsible for the development of green liver syndrome, characterized by green discoloration of the liver, in yellowtail and red sea bream fed fish meal-free diets. Following on from this, studies conducted at the start of the 21st century succeeded in determining the actual taurine requirements of juvenile flounder and red sea bream fed formulated diets.

These studies were able to obtain the optimal growth and feed efficiency at a taurine concentration of 15–20 mg/g diet [7]. This level is comparable to the taurine content in mysids (the preferred diet of flounder) and corresponds to more than 20 times the taurine requirement of cats (0.8 mg/g) [78]. Juvenile flounder fed low-taurine diets exhibit relatively slow feeding behavior; they did not rise quickly from the sandy seabed to catch prey and immediately settle back on the sandy bottom, as observed in wild juveniles, but rather after catching prey, they continue to swim near the water surface in search of more prey, much like carp in a pond [79]. Juvenile flounder do not exhibit normal behavior unless fed a diet containing at least 5 mg/g of taurine. A possible explanation for this abnormal behavior is that the lack of taurine, which is known to be involved in retinogenesis and also functions as a neurotransmitter, leads to difficulty in finding prey and improper transmission of behavioral control. When flounder that do not exhibit normal behavior are released to their natural habitat, they are immediately consumed by predators such as flathead fish (or Ishi-gani *Charybdis japonica* according to recent reports) and, thus, stocking efficiency decreases.

Juvenile yellowtail fed a fish meal-free diet containing concentrated soy protein as the main ingredient and supplemented with 4.5 % taurine for 39 weeks did not develop green liver syndrome but showed slightly reduced growth compared with those fed a fish meal-containing diet. Green liver syndrome is believed to be caused by failed conjugation of bilirubin and biliverdin by taurine, due to a lack of taurine in the fish body, and the accumulation of these two substances in the liver [80]. The formation of a taurine conjugate requires either glycine or taurine, but in fish only taurine is considered to be involved [81]. These findings indicate that taurine is also conditionally essential in yellowtail [82].

Red sea bream, as well as many other saltwater fish, do not typically eat purified diets where the main ingredient is a protein source used for freshwater fish, such as casein or egg albumin. Moreover, even when they eat these diets, a high growth rate and feed efficiency comparable to those achieved with fish meal-containing diets could not be obtained. The reason for this lack of effectiveness is unclear. Feeding red sea bream a purified taurine-containing casein diet was found to improve growth and feed efficiency, which peaked at a taurine concentration of ≥5 mg/g (0.5 %) [83]. Moreover, improved feeding behavior and pigmentation (i.e., brilliant pink without darkening, the natural color of the fish) were also observed in individuals fed a diet containing taurine (Fig. 8) [68]. These findings indicate that the previously observed poor growth of saltwater fish was due to the lack of taurine in purified diets and imply the need to review existing studies that have used taurine-supplemented purified diets to examine the deficiencies and requirements of various nutrients in saltwater fish.

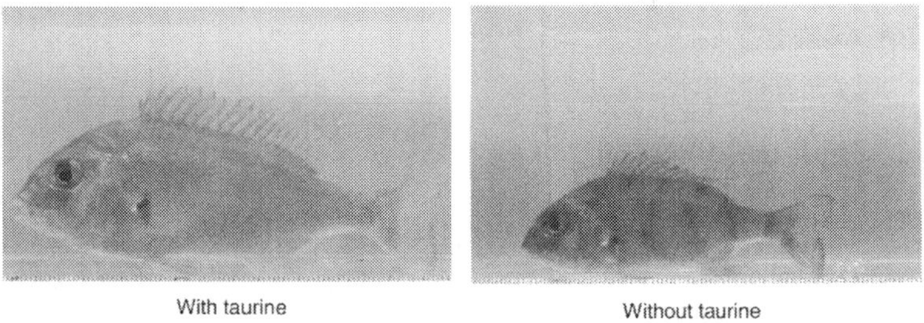

Figure. 8: Pictures of red sea bream *Pagrus major* juveniles showing differential pigmentation of fish that received diet with or without taurine.

Benefit of Taurine in Seed Production

There is some evidence to support the benefit of taurine in seed production, especially during the rotifer-feeding period. During the initial phase of seed production, larvae are usually fed rotifer. Since rotifer contain only a trace amount of taurine (<1 mg/g dry weight), a rapid decrease in taurine content in the fish body is observed in the hatched larvae of yellowtail, a similar trend to that observed for DHA. When taurine content was measured in hatchery yellowtail larvae and juveniles grown to a body size of up to 18 cm and in wild yellowtail larvae from 3 to 18 cm, wild larvae contained a taurine concentration 2–4 and 2 times (15 mg/g) higher than hatchery larvae, respectively; this appeared to be because wild larvae fed on taurine-rich organisms such as natural plankton (12 mg/g) and crustaceans (20–30 mg/g) [84]. With the aim of preventing a decrease in taurine content in the fish body, Takahashi et al. [85] attempted to enrich rotifers with taurine. They reported that taurine concentration in rotifer was in proportion to the amount of taurine added to a rotifer culture medium and the duration of enrichment. More specifically, the addition of taurine at a concentration of 400–800 mg/L to the rotifer culture medium resulted in the accumulation of taurine in rotifer at a concentration of 8–12 mg/g during an 18 h enrichment. This was the first successful attempt to introduce a free amino acid into live foods and was submitted for patent application [86]. Bivalve shellfish, such as oyster and scallop, have also been shown to directly absorb and accumulate taurine from water [87]. Feeding taurine-enriched rotifer to newly hatched larvae of flounder, red sea bream, bluefin tuna, amberjack, and Pacific cod was also found to be effective for improving growth and survival. Moreover, in a fasting experiment, red sea bream fed taurine-enriched rotifer survived 3–4 days longer than those fed non-taurine-enriched rotifer, demonstrating a clear survival benefit (i.e., starvation tolerance) [88]. These results suggest that taurine is an essential nutrient for seawater fish larvae in the initial stage of seed production.

In summary, taurine has been identified as an important factor for improving the health and quality of larval and juvenile fish in seed production. The inclusion of inexpensive synthesized taurine in the list of designated feed additives in June 2009 has allowed for more effective use of this compound [89]. The taurine content in live foods and the dietary taurine requirements of various species of fish are summarized in Table 1 [90, 91].

Table 1: Taurine content in live foods and dietary taurine requirements of various species of fish

Diet	mg/100 g	
	Wet weight	Dry weight
Rotifer	8–18	80–180
Taurine-enriched rotifer	30–125	300–1250
Artemia	100	650
Natural copepod	150	1200
Frozen copepod	60	460
Mysid *A. kokuboi*	380	2900
Formulated diet	–	400–460

Species	Requirement (mg/100 g)	Note
Saltwater fish		
Rotifer-feeding stage: Bluefin tuna/Amberjack [90]		Per dry weight rotifer
Flounder/Red sea bream/ Pacific cod	≥400	
Juvenile stage		
Flounder	1500–2000	Formulated diet
Red sea bream	≥500	Formulated diet
Yellowtail	≥1500	Formulated diet
Sea bass	≈800	Plant material-containing diet
Cobia	≥550	60 % yeast protein used
Dentex dentex	≥340	25 % soy protein concentrate used
Young/adult stage		
Red sea bream/Yellowtail	≥400	Formulated diet
Red sea bream	≥1000	High plant material-containing diet
Yellowtail	≈550	20 % fish meal + 46 % plant material
	≈4500	58 % soy protein concentrate used
Freshwater fish		
Rainbow trout/Carp/Tilapia	None	Formulated diet
Rainbow trout	Potential	100 % plant protein used
Juvenile eel	1500–2000	Formulated diet

Ayu	Potential	Formulated diet
Crustacean		
White shrimp [91]	1680	10 % fish meal + 73 % plant material
Mammal		
Kitten	≥80	Formulated diet

CONCLUSIONS

This review of the last two decades of research on improving the quality and health of fish in seed production has focused on the advances made in our knowledge and use of certain nutrients, such as DHA, vitamin A derivatives, and taurine, which are important for the mass production of seedling of seawater fish. These nutritional research findings will help in our efforts to solve problems encountered in the field of seed production and improve seed production techniques.

REFERENCES

1. Worm B, Barbier EB, Beaumont N, Duffy JE, Folke C, Halpern BS, Jackson JBC, Lotze HK, Micheli F, Palumbi SR, Sala E, Selkoe KA, Stachowicz JJ, Watson R (2006) Impacts of biodiversity loss on ocean ecosystem services. Science 314:787–790
2. FAO (2012) The States of World Fisheries and Aquaculture, p 209
3. Watanabe T (1980) Seed production of fish and live food organisms. In: Ogino C (ed) Nutrition and feeding in fish. Koseisha-Koseikaku, Tokyo, pp 81–110 (in Japanese)
4. Hino A (1994) Seed production. In: Publication Committee of the Japanese Society of Fisheries Science (ed) Recent advances in fisheries science. Koseisha-Koseikaku, Tokyo, pp 124–131 (in Japanese)
5. Watanabe T, Kitajima C, Arakawa T, Fukusho K, Fujita S (1978) Nutritional quality of rotifer *Brachionus plicatilis* as a living feed from the viewpoint of essential fatty acids for fish. Nippon Suisan Gakkaishi 44:1109–1114 (in Japanese with English abstract)
6. Watanabe T, Oowa F, Kitajima C, Fujita S (1978) Nutritional quality of brine shrimp, *Artemia salina*, as a living feed from the viewpoint of essential fatty acids for fish. Nippon Suisan Gakkaishi 44:1115–

1121 (in Japanese with English abstract)

7. Takeuchi T (2001) A review of feed development for early life stages of marine finfish in Japan. Aquaculture 200:203–222
8. Takeuchi T (1991) Variety of essential fatty acids requirement in fish. Kagaku Seibutu 29:571–580 (in Japanese)
9. Takeuchi T (1991) Nutritive deficiency and requirements in fish. In: Training project for sea farming in 1991—Textbooks of the basic theory course: v. development of fish in larval and juvenile stages series, No.4. Japan Sea Farming Association of Fisheries Agency, Tokyo, pp 1–68 (in Japanese)
10. Takeuchi T (2009) Nutrition in fish larvae and juvenile. In: Watanabe T (ed) Nutrition and feeding in fish and crustaceans. Koseisha-Koseikaku, Tokyo, pp 204–228 (in Japanese)
11. Takeuchi T (1994) Retroconversion of DHA in live food. In: Report of the Grant-in-Aid for scientific research (No. 04806029): general research C (the 1992–1993 academic years). Ministry of Education, Tokyo, pp 1–30 (in Japanese)
12. Takeuchi T, Sugiura M, Tei H, Yamamoto Y, Watanabe T (1992) Enrichment of live food organisms with shark roes. In: Abstracts of the annual meeting of the Japanese Society of Fisheries Sciences. Japanese Society of Fisheries Sciences, Shimozaki, p 51 (in Japanese)
13. Takeuchi T, Zheng F, Yoseda K, Hirokawa J, Watanahe T (1994) Nutritive value of DHA-enriched rotifer for larval cod. Nippon Suisan Gakkaishi 60:641–652 (in Japanese with English abstract)
14. Ishizaki Y, Takeuchi T, Watanabe T, Shimizu K, Imaizumi K (1996) Rearing test of larval yellowtail fed with rotifers enriched with different levels of Vitamin E and different types of freeze-dried powder of shark roes. Suisanzoshoku 44:527–535 (in Japanese with English abstract)
15. Tanaka S (1998) Growth of artificially hatched eel larvae via feeding. Yoshoku 35(7):40–44 (in Japanese)
16. Takeuchi T (1997) Nutritional requirement for improvement of rearing seed production. In: Japanese Society of Fisheries Science (ed) Biology and Stock Enhancement of Japanese Flounder (Fisheries Science Series 112). Koseisha-Koseikaku, Tokyo, pp

96–106 (in Japanese)

17. Kanazawa A (1993) Nutritional mechanisms involved in the occurrence of abnormal pigmentation in hatchery-reared flatfish. J World Aquac Soc 24:162–166
18. Reitan KI, Rainuzzo JR, Olsen Y (1994) Influence of lipid composition of turbot larvae. Aquac Int 2:33–48
19. Satoh N, Takeuchi T (2009) Docosahexaenoic acid requirement of larval brown sole *Pseudopleuronectes herzensteini*. Nippon Suisan Gakkaishi 75:28–37 (in Japanease with English abstract)
20. Satoh N, Takaya Y, Takeuchi T (2009) Docosahexaenoic acid requirement for the prevention of abnormal morphology in brown sole *Pseudopleuronectes herzensteini*during D-E larval stages. Fish Sci 75:1259–1266
21. Satoh N, Takeuchi T (2011) Proof test of the rearing technique based on DHA requirement during larval period to prevent the morphological abnormalities in brown sole *Pseudopleuronectes herzensteini*. Sci Rep Hokkaido Fish Res Inst 80:25–31 (in Japanease with English abstract)
22. Uchida K, Kuwada H, Tsukamoto K (1993) Tilting behaviour, a fear response to frightening stimuli, as a possible predictive index for stocking effectiveness in the juveniles of red sea bream *Pagrus major*. Nippon Suisan Gakkaishi 59:991–999 (in Japanese with English abstract)
23. Furuta S, Watanabe T, Yamada H (1998) Predation by fishes on hatchery-reared Japanese flounder *Paralichthys olivaceus* juveniles released in the coastal area of Tottori prefecture. Nippon Suisan Gakkaishi 64:1–7 (in Japanese with English abstract)
24. Masuda R, Takeuchi T, Tsukamoto K, Ishizaki Y, Kanematsu M, Imaizumi K (1998) Critical involvement of dietary docosahexaenoic acid in the ontogeny of schooling behaviour in the yellowtail. J Fish Biol 53:471–484
25. Takeuchi T, Ishizaki Y, Masuda R (2000) Effect of DHA on development of brain and behavior in larval yellowtail. Yoshoku 37(13):114–117
26. Ishizaki Y, Uematsu K, Takeuchi T (2000) A preliminary study of the effect of dietary docosahexaenoic acid on the volumetric growth

of the brain in larval yellowtail. Fish Sci 66:611–613

27. Ishizaki Y, Masuda R, Uematsu K, Shimizu K, Arimoto M, Takeuchi T (2001) The effect of dietary docosahexaenoic acid on schooling behaviour and brain development in larval yellowtail. J Fish Biol 58:1691–1703
28. Masuda R, Takeuchi T, Tsukamoto K, Sato H, Shimizu K, Imaizumi K (1999) Incorporation of dietary docosahexaenoic acid into the central nervous system of the yellowtail *Seriola quinqueradiata*. Brain Behav Evol 53:173–179
29. Takeuchi T (2008) Seed production of Japanese spiny lobster *Panulirus japonicus* and common octopus *Octopus vulgaris*. Food Packag 49:510–518 (in Japanese)
30. Hamasaki K, Takeuchi T (2001) Dietary value of *Artemia* enriched with the ω-Yeast or shark eggs as feed for planktonic larvae of *Octopus vulgaris*. Saibai Giken 28(2):65–68 (in Japanese with English abstract)
31. Shiraki M (2001) Incidence of common octopus *Octopus vulgaris* larvae caught by two-boat whitebait seine in the Naruto Straits of Tokushima Prefecture. Saibai Giken 29(1):55–57 (in Japanese with English abstract)
32. Okumura S, Kurihara A, Iwamoto A, Takeuchi T (2005) Improved survival and growth in *Octopus vulgaris* paralarvae by feeding large type *Artemia* and Pacific sandeel *Ammodytes personatus*. Aquaculture 244:147–157
33. Kurihara A, Okumura S, Iwamoto A, Takeuchi T (2006) Feeding Pacific sandeel enhances DHA level in common octopus paralarvae. Aquac Sci 54:413–420
34. Kurihara A (2008) Study on the searching effective nutrients in *Artemia* for common octopus *Octopus vulgaris* paralarvae. In: Doctoral thesis, Tokyo University of Marine Science and Technology, Tokyo (in Japanese)
35. Takeuchi T, Watanabe T (1979) Effect of excess amounts of essential fatty acids on growth of rainbow trout. Nippon Suisan Gakkaishi 45:1517–1519 (in Japanese with English abstract)
36. Hamasaki K, Suprayudi MA, Takeuchi T (2002) Effects of dietary n-3 HUFA on larval morphogenesis and metamorphosis

to megalops in the seed production of the mud crab, *Scylla serrate* (Brachyura:Portunidae). Suisanzoshoku 50:333–340

37. Suprayudi MA, Takeuchi T, Hamasaki K (2004) Effects of *Artemia* enriched with eicosapentaenoic and docosahexaenoic acid on survival and occurrence of melting failure in megalop larvae of the mud crab *Scylla serrata*. Fish Sci 70:650–658
38. Kaji S, Fukunaga T (1999) Results of a questionnaire on the recent status of seed production and market price of recaptured Japanese flounder *Paralichthys olivaceus*showing abnormal coloration. Saibai Giken 27:67–101 (in Japanese with English abstract)
39. Seikai T, Sinoda M (1981) Variations of frequency occurrence of color anormalies in hatchery-reared flounder *Paralichthys olivaceus* due to duration of *Artemia salina* fed. Bull Kyoto Inst Oceanic Fish Sci 5:29–37
40. Seikai T (1985) Influence of feeding periods of Brazilian *Artemia* during larval development of hatchery-reared flounder *Paralichthys olivaceus* on the appearance of albinism. Nippon Suisan Gakkaishi 51:521–527
41. Seikai T (1985) Reduction in occurrence frequency of albinism in juvenile flounder *Paralichthys olivaceus* hatchery-reared on wild zooplankton. Nippon Suisan Gakkaishi 51:1261–1267
42. Fukusho K, Nanba H, Yamamoto T, Yamasaki Y, Lee MT, Seikai T, Watanabe T (1987) Reduction of albinism in juvenile flounder *Paralichthys olivaceus* hatchery-reared on fertilized eggs of red sea bream *Pagrus major*, and its critical stage for the effective feeding. Bull Nat Res Inst Aquac 12:1–7 (in Japanese with English abstract)
43. Miki N, Taniguchi T, Hamakawa H (1989) Adequate vitamin level for reduction of albinism in hatchery-reared hirame *Paralichthys olivaceus* fed on rotifer enriched with fat-soluble vitamins. Suisanzoshoku 37:109–114 (in Japanese with English abstract)
44. Miki N, Taniguchi T, Hamakawa H, Yamada Y, Sakurai N (1990) Reduction of albinism in hatchery-reared flounder "Hirame" *Paralichthys olivaceus* by feeding on rotifer enriched with vitamin-A. Suisanzoshoku 38:147–155 (in Japanese with English abstract)

45. Yamamoto S, Oda M (1993) Effects of lipid-soluble vitamins on the reduction of colour abnormalities of juvenile flounders *Paralichthys olivaceus*. Bull Fish Expt Stn Okayama Pref 8:1–4 (in Japanese)
46. Takeuchi T, Haga Y (2000) Effect of vitamin on pigmentation and morphological abnormality in Japanese flounder *Paralichthys olivaceus*. Kagaku Seibutu 38:728–731 (in Japanese)
47. Minami T (1982) The early life history of a flounder *Paralichthys olivaceus*. Nippon Suisan Gakkaishi 48:1581–1588 (in Japanese with English abstract)
48. Seikai T, Shimozaki M, Watanabe T (1987) Estimation of larval stage determining the appearance of albinism in hatchery-reared juvenile flounder *Paralichthys olivaceus*. Nippon Suisan Gakkaishi 53:1107–1114
49. Takeuchi T, Tarui F (2005) Inducement of malformation to Japanese flounder by relative vitamin A compounds, especially retinoic acid. Aquabiology 27:400–403 (in Japanese)
50. Dedi J, Takeuchi T, Seikai T, Watanabe T (1995) Hypervitaminosis and safe levels of vitamin A for larval flounder *Paralichthys olivaceus* fed *Artemia* nauplii. Aquaculture 133:135–146
51. Dedi J, Takeuchi T, Seikai T, Watanabe T, Hosoya K (1997) Hypervitaminosis A during vertebral morphogenesis in larval Japanese flounder. Fish Sci 63:466–473
52. Takeuchi T, Dedi J, Haga Y, Seikai T, Watanabe T (1998) Effect of vitamin A compounds on bone deformity in larval Japanese flounder *Paralichthys olivaceus*. Aquaculture 169:155–165
53. Tarui F, Haga Y, Ohta K, Shima Y, Takeuchi T (2006) Effect of *Artemia* nauplii enriched with vitamin A palmitate on hypermelanosis on the blind side in juvenile Japanese flounder *Paralichthys olivaceus*. Fish Sci 72:256–262
54. Haga Y, Tarui F, Okumura K, Shima Y, Takeuchi T (2006) Effects of light irradiation on the dynamics of vitamin A compounds in rotifers and *Artemia*. Fish Sci 72:1020–1026
55. Haga Y, Takeuchi T, Seikai T (1999) Effect of retinoic acid on larval Japanese flounder *Paralichthys olivaceus* reared on *Artemia* nauplii. Suisanzoshoku 47:559–566

56. Haga Y, Takeuchi T, Seikai T (2002) Influence of all-*trans* retinoic acid on pigmentation and skeletal formation in larval Japanese flounder. Fish Sci 68:560–570
57. Miwa S, Yamano K (1999) Retinoic acid stimulates development of adult-type chromatophores in the flounder. J Exp Zool 284:317–324
58. Haga Y, Suzuki T, Takeuchi T (2002) Retinoids as potent teratogens on larval development of Japanese flounder *Paralichthys olivaceus*. Fish Sci 68:789–792
59. Haga Y, Takeuchi T, Seikai T (2004) Changes of retinoids contents in larval Japanese flounder *Paralichthys olivaceus* and *Artemia* nauplii enriched with a large dose of all-*trans* retinoic acid. Fish Sci 70:436–444
60. Hasegwa Y, Takeuchi T, Itagaki E, Fukunaga T (1998) Relationship between fat soluble vitamins in diet and the occurrence of colour abnormality on the blind side of juvenile Japanese flounder. Suisanzoshoku 46:279–286 (in Japanese with English abstract)
61. Yoo JH, Takeuchi T, Seikai T (2000) Sensitivity of the metamorphic events and morphogenesis of Japanese flounder *Paralichthys olivaceus* during larval development to thyroxine. Fish Sci 66:846–850
62. Murayama Y, Takeuchi T (2004) Effect of vitamin D enriched feed and bottom. In: Sea ranching technology series No.10. Fisheries Research Agency, Tokyo, pp 95–100 (in Japanese)
63. Suzuki T, Kurokawa T (2000) Organogenesis in flounder. Protein, nucleic acid and enzyme 45:86–92 (in Japanese)
64. Hamre K, Krossøy C, Lock EJ, Moren M (2010) Roles of lipid—soluble vitamins in the ontogeny of marine fish larvae. Aquac Res 41:745–750
65. Takeuchi T, Matunari H (2008) Taurine. In: Nishigawa K (ed) Science of functional food. Technology Research Service Center. Tokyo, pp 1134–1136 (in Japanese)
66. Yokoyama M, Takeuchi T, Park GS, Nakazoe J (2001) Hepatic cysteine sulphinate decarboxylase activity in fish. Aquac Res 32:216–220
67. Goto T (2002) Variety of taurine biosynthetic pathway in fish.

Kagaku Seibutu 40:635–637 (in Japanese)

68. Park GS, Takeuchi T, Yokoyama M, Seikai T (2002) Optimal dietary taurine level for growth of juvenile Japanese flounder *Paralichthys olivaceus*. Fish Sci 68:824–829
69. Matsunari H, Furuita H, Yamamoto T, Kim SK, Sakakura Y, Takeuchi T (2008) Effect of dietary taurine and cystine levels on growth performance and feeding behavior of red sea bream *Pagrus major*. Aquaculture 274:142–147
70. 7Walton MJ, Cowey CB, Adron JW (1982) Methionine metabolism in rainbow trout fed diets of differing methionine and cystine content. J Nutr 112:1525–1535
71. Sakaguchi M, Murata M (1988) Taurine. In: Sakaguchi M (ed) Extractive components of fish and shellfish. Koseisha-Koseikaku, Tokyo, pp 56–65 (in Japanese)
72. Yokoyama M, Nakazoe J (1992) Accumulation and excretion of taurine in rainbow trout *Oncorhynchus mykiss* fed diets supplemented with methionine, cystine and taurine. Comp Biochem Physiol 102A:565–568
73. Seikai T, Kinoshita I (1994) Comparative study on growth and development between the wild fish and the hatchery fish. In: Summary of research results of production techniques of healthy fish larvae in 1993. Fisheries Agency. Tokyo, pp 141–160 (in Japanese)
74. Seikai T, Takeuchi T, Park GS (1997) Comparison of growth, feed efficiency and chemical composition of juvenile flounder fed live mysids and formula feed under laboratory conditions. Fish Sci 63:520–524
75. Park GS, Takeuch T, Seikai T, Nakazoe J (1997) Nutritional study on mysids as food organisms for juvenile Japanese flounder. Suisanzoshoku 45:371–378 (in Japanese with English abstract)
76. Park GS, Takeuchi T, Seikai T, Yokoyama M (2000) The effects of residual salts and free amino acids in mysid meal on growth of juvenile Japanese flounder *Paralichthys olivaceus*. Nippon Suisan Gakkaishi 66:697–704 (in Japanese with English abstract)
77. Furuta S (1993) Releasing techniques and fry quality. In: Kitajima C (ed) Healthy fry for release, and their production techniques. Koseisha-Koseikaku, Tokyo, pp 94–101 (in Japanese)

78. Knopf K, Sturman JA, Armstrong M, Hayes KC (1978) Taurine: an essential nutrient for the cat. J Nutr 108:773–778

79. Kim SK, Takeuchi T, Yokoyama M, Murata Y, Kaneniwa M, Sakakura Y (2005) Effect of dietary taurine levels on growth and feeding behavior of juvenile Japanese flounder *Paralichthys olivaceus*. Aquaculture 250:765–774

80. Goto T, Takagi S, Ichiki T, Sakai T, Endo M, Yoshida T, Ukawa M, Murata H (2001) Studies on the green liver in cultured red sea bream fed low level and non-fish meal diets: relationship between hepatic taurine and biliverdin levels. Fish Sci 67:58–63

81. Kim SK, Matsunari H, Nomura K, Tanaka H, Yokoyama M, Murata Y, Ishihara K, Takeuchi T (2008) Effect of dietary taurine and lipid contents on conjugated bile acid composition and growth performance of juvenile Japanese flounder *Paralichthys olivaceus*. Fish Sci 74:875–881

82. Takagi S, Murata H, Goto T, Endo M, Yamashita H, Ukawa M (2008) Taurine is a essential nutrient for yellowtail *Seriola quinqueradiata* fed non-fish meal diets based on soy protein concentrate. Aquaculture 280:198–205

83. Matsunari H, Yamamoto T, Kim SK, Goto T, Takeuchi T (2008) Optimum dietary taurine level in casein-based diet for juvenile red sea bream *Pagrus major*. Fish Sci 74:347–353

84. Matsunari H, Takeuchi T, Tamura Y, Takahashi M, Ishibashi N, Chuda H, Arakawa T (2003) Changes in the taurine content during the early growth stages of artificially produced yellowtail compared with wild fish. Nippon Suisan Gakkaishi 69:757–763 (in Japanese with English abstract)

85. Takahashi T, Amano T, Takeuchi T (2005) Establishment of direct enrichment method of taurine to rotifer. Aquac Sci 53:121–126 (in Japanese with English abstract)

86. Takahashi T, Amano T, Takeuchi T (2003) Taurine directly enriched rotifer. J-Patent 2003-259760 (in Japanese)

87. Takeuchi T (2005) Recent development of on amino acids and peptides research in fish. In: Reports of the Research Committee of Essential Amino Acids (Japan); No. 174, pp 9–17 (in Japanese)

88. Chen JN, Takeuchi T, Takahashi T, Tomoda T, Koiso H, Kuwada H (2004) Effect of rotifers enriched with taurine on growth and survival activity of red sea bream *Pagrus major* larvae. Nippon Suisan Gakkaishi 70:542–547 (in Japanese with English abstract)

89. Takeuchi T (2010) Synthetic taurine is a good healthy supplement for fish. Nippon Suisan Gakkaishi 76:298–303 (in Japanese)

90. Matsunari H, Hashimoto H, Iwasaki T, Oda K, Masuda Y, Imaizumi H, Teruya K, Furuita H, Yamamoto T, Hamada K, Mushiake K (2013) Effect of feeding rotifers enriched with taurine on the growth and survival of larval amberjack *Seriola dumerili*. Fish Sci 79:815–821

91. Yue YR, Lin YJ, Tian LX, Yang HJ, Liang GY, He JT (2013) The effect of dietary taurine supplementation on growth performance, feed utilization and taurine contents in tissues of juvenile white shrimp *Litopenaeus vannamei* Boone, 1931, fed with low-fishmeal diets. Aquac Res 44:1317–1325

CHAPTER

10

SUBSTITUTION OF DIETARY FISH OIL WITH PLANT OILS IS ASSOCIATED WITH SHORTENED MID INTESTINAL FOLDS IN ATLANTIC SALMON (SALMO SALAR)

Torfinn Moldal[1] , Guro Løkka[2], Jannicke Wiik-Nielsen[1] , Lars Austbø[2], Bente E Torstensen[3] , Grethe Rosenlund[4] , Ole Bendik Dale[1] , Magne Kaldhusdal[1] and Erling Olaf Koppang[2]

[1]Norwegian Veterinary Institute, Sentrum, Norway

[2]Norwegian School of Veterinary Science, Norway

[3]National Institute of Nutrition and Seafood Research, Nordnes, Norway

[4]Skretting ARC, Stavanger, Norway

ABSTRACT

Background

Fish meal and fish oil are increasingly replaced by ingredients from terrestrial sources in the feeds for farmed salmonids due to expanding production and reduced availability of marine feed raw material. Fish oil that is rich in n-3 polyunsaturated fatty acids is considered beneficial to human health in general and to prevent intestinal inflammation and carcinogenesis in particular. In contrast, n-6 fatty acids that are present in many vegetable oils have been associated with increased risk of colitis and colon cancer in rodents and humans, as well as lowered transcription levels of certain stress and antioxidant-related genes in Atlantic salmon.

The aim of the present study was to investigate the intestinal health in Atlantic salmon fed with different vegetable oils as partial substitutes of fish oil in the diet. A feed trial lasting for 28 weeks included one reference diet containing fish oil as the sole lipid source and three diets where 80% of the fish oil was replaced by a plant oil blend with either olive oil, rapeseed oil or soybean oil as the main lipid source. These plant oils have intermediate or low n-3/n-6-ratios compared to fish oil having a high n-3/n-6-ratio. The protein and carbohydrate fractions were identical in all the feeds.

Results

Morphometric measurements showed significantly shorter folds in the mid intestine in all groups fed vegetable oils compared to the group fed fish oil. In the distal intestine, the complex folds were significantly shorter in the fish fed soybean oil compared to the fish fed rapeseed oil. Histological and immunohistochemical examination did not show clear difference in the degree of inflammation or proliferation of epithelial cells related to dietary groups, which was further confirmed by real-time RT-PCR which revealed only moderate alterations in the mRNA transcript levels of selected immune-related genes.

Conclusions

Shortened intestinal folds might be associated with reduced intestinal surface and impaired nutrient absorption and growth, but our results suggest that partial substitution of dietary fish oil with vegetable oils does not have any major negative impact on the intestinal health of Atlantic salmon.

Keywords: Atlantic salmon, Fatty acids, Fish oil, Inflammation, Intestine, Morphometric analyses, Plant oils, Real-time PCR

BACKGROUND

Salmonids are indigenous carnivores, and fish meal and fish oil have traditionally been the main ingredients in feed for farmed Atlantic salmon (*Salmo salar*) and rainbow trout (*Oncorhynchus mykiss*). Due to limited sources of marine raw materials, a fast growing aquaculture industry and increased focus on sustainability, the salmon farming industry has searched for alternative feeds [1, 2, 3]. The inclusion of plant-derived materials in fish feed may be beneficial from an economic and ecological point of view, but used without carefully considering the fish minimum requirements or upper tolerable levels of certain nutrients or anti-nutrients, it can unfortunately also cause adverse effects with regard to fish health, nutritive value and the consumers' acceptance [1].

In the early years of salmonid farming, proteins constituted more than half of the feed content, while the lipid fraction was as low as 10%. These figures have changed during the last decades as the protein level has decreased to less than 40%, while the lipid fraction has increased till about 35% [4]. Fish oil production requires 2-5 times more industrial fish than production of the same weight of fish meal [5], and replacement of fish oil with vegetable oils is hence of growing interest both from an economic and sustainability viewpoint.

Vegetable oils may contain high levels of n-6 polyunsaturated fatty acids (PUFAs) such as linoleic acid (LA, 18:2n-6) that can be further metabolized to arachidonic acid (AA; 20:4n-6). In contrast, fish oils are rich in n-3 PUFAs such as docosahexaenoic acid (DHA, 22:6n-3) and eicosapentaenoic acid (EPA, 20:5n-3) (Figure 1). In mammals, dietary n-3 fatty acids supplant the AA in inflammatory cell membranes [6] and therefore decrease the availability of the major precursor of pro-inflammatory eicosanoids as the same enzymes are involved in the metabolism of n-3 and n-6 fatty acids and further in the synthesis of eicosanoids [7]. Also in Atlantic salmon, dietary lipid has been shown to alter leucocyte phospholipid fatty acid composition and eicosanoid production [8], and increased n-6 levels in the feed gave increased n-6 fatty acids in leucocytes in an *ex vivo* study [9].

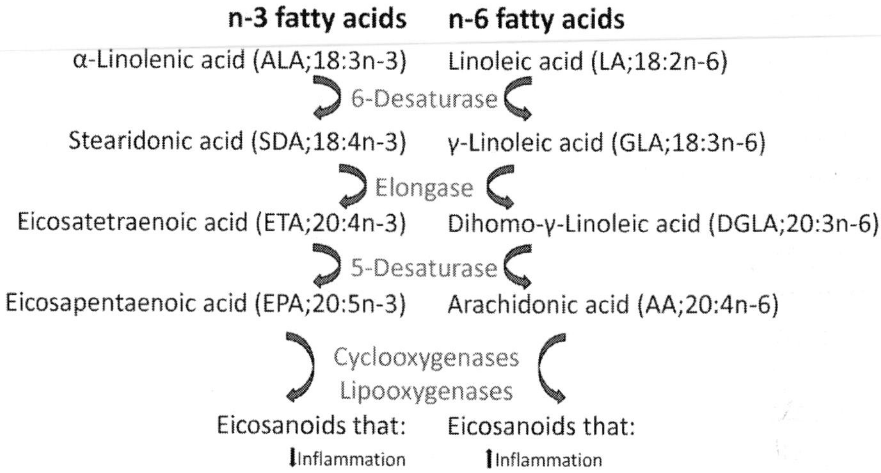

Figure 1: **The metabolism of n-3 and n-6 fatty acids and eicosanoids.** The same enzymes are involved in the metabolism of n-3 and n-6 fatty acids and synthesis of eicosanoids, but the biological properties of the eicosanoids are different.

In mammals, a diet rich in n-6 fatty acids has been associated with increased risk of ulcerative colitis [10] and promotion of intestinal carcinogenesis [6, 11], while a high intake of n-3 PUFAs is considered to be beneficial for health. Consumption of n-3 fatty acids has been shown to attenuate the dysbiosis and colitis caused by n-6 polyunsaturated fatty acid in mice [12] and to prevent and modulate a wide range of pathological conditions as cardiovascular diseases, diabetes and several inflammatory and neoplastic processes, including inflammatory bowel disease and colon cancer [13]. The n-3 fatty acids also inhibit the prostaglandin synthesizing enzyme cyclooxygenase-2 (COX-2) which is up-regulated during inflammation, the expression of the pro-inflammatory cytokines tumour necrosis factor-α (TNF-α) and interleukin-1 (IL-1) and the proliferation of lymphocytes as shown both *in vitro* and in rodent models [13, 14, 15].

Several studies have addressed the effects of vegetable oils as lipid sources in the feed on Atlantic salmon intestinal absorption [16], post-absorptive fates [17], feed uptake, growth rate, metabolism and nutrient content of the fish filet [2, 18, 19, 20, 21, 22]. Whereas many studies have addressed the intestinal health of the fish when fish meal is replaced by different plant-derived proteins, and both soybean meal and pea protein concentrate have been shown to induce enteritis [23, 24], there is a knowledge gap regarding the impact on intestinal health when fish oil is replaced by plant oils.

Complete substitution of fish oil with a plant oil blend containing rapeseed oil, palm oil and linseed oil in the feed induced lower transcription levels of certain stress and antioxidant-related genes in the intestine [25]. Another feed trial with the same oil blend partly substituting fish oil in combination with plant proteins at different inclusion levels demonstrated that in response to acute physiological stress, high levels of plant-derived dietary ingredients can enhance COX-2 induction and synthesis of pro-inflammatory eicosanoids in the intestine of salmon [26]. It has also been speculated whether inclusion of plant oils in the feed contributed to intestinal carcinogenesis in brood stock Atlantic salmon [27].

The aim of the present study was to investigate the morphology of the intestinal wall, the presence of antigen presenting cells and T lymphocytes, the proliferation pattern of epithelial cells, and the transcript levels of selected immune-related genes including relevant cytokines, major histocompatibility complex class II (MHC class II), cluster of differentiation 3ζ (CD3ζ), immunoglobulins, the intracellular receptor nucleotide-binding oligomerization domain-containing protein 2 (NOD2) and COX-2a in the intestine of Atlantic salmon when dietary fish oil was partially replaced by different vegetable oil blends with varying n-3/n-6-ratio.

METHODS

Animal Ethics, Fish and Feed

The feed trial was carried out at Skretting ARC Fish Trials Station that is approved by the Norwegian Animal Research Authority and was conducted according to current animal welfare regulations: FOR-1996-01-15-23 (Norway). Six hundred Atlantic salmon (*Salmo salar*) with mean initial weight of 815±28 g were equally distributed into 12 tanks, and triplicate groups of fish were fed a diet with either fish oil as the sole lipid source or a diet where 80% of the fish oil was replaced by one of three vegetable oil blends with olive oil, rapeseed oil or soybean oil as the main lipid source (Figure 2). The n-3/n-6 ratio in the feed with fish oil as the sole lipid source was 5.3, while the n-3/n-6 ratios were 0.7, 0.9 and 0.3 in the diets where fish oil was largely substituted with olive oil, rapeseed oil and soybean oil respectively. The composition of the protein fraction was identical in all diets with 30% fish meal and 70% plant protein. The proximate composition of all diets was similar with 332-341 g kg^{-1} fat and 406-413 g kg^{-1}protein.

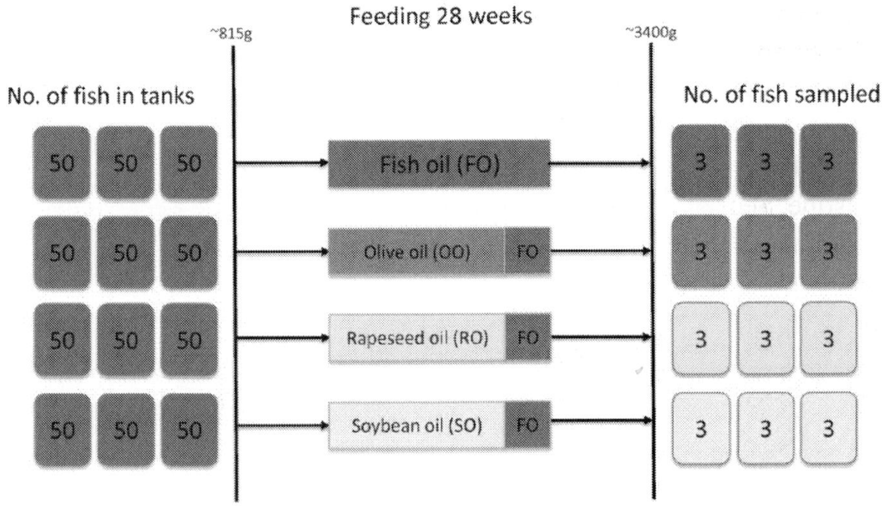

Figure 2: **The design of the feed trial.** Six hundred Atlantic salmon with mean initial weight of 815±28 g were equally distributed into 12 tanks, and triplicate groups of fish were fed a diet with either fish oil as the sole lipid source or a diet where 80% of the fish oil was replaced by one of three vegetable oil blends with olive oil (OO), rapeseed oil (RO) or soybean oil (SO) as the main lipid source. The protein fraction was identical in all diets, and 70% of the proteins were plant-derived. The trial lasted for 28 weeks, and the mean final weight of all dietary groups was 3399±76 g. Three fish in each tank were sampled by the end of the trial.

The fish were vaccinated intraperitonally (ALPHA JECT micro 6; PHARMAQ, Overhalla, Norway) to give protection towards furunculosis, vibriosis, cold-water vibriosis, winter ulcer disease and infectious pancreas necrosis one month prior to seawater transfer. The trial lasted for 28 weeks, and the mean final weight was 3399±76 g with no significant difference between dietary groups. The fish were anesthetized with MS222 at a concentration of 7 g L^{-1} and euthanized according to regulations (Forskrift om drift av akvakulturanlegg §34.

Avlivning av fisk). The design of the feed trial and feed composition is described in detail elsewhere [22].

Histology and Immunohistochemistry

For histological evaluation, tissues from the mid and distal intestine (Figure 3), also called first and second segment of mid intestine, respectively [28], from 24 fish in each dietary group (eight fish per tank) were fixed in 10% phosphate-buffered formalin for 24-48 hours. The tissues were routinely processed, embedded in paraffin and cut in 3 μm thick cross sections for the mid intestine and longitudinal sections (i.e. perpendicular to the macroscopically visible circular folds) for the distal intestine. Sections were routinely deparaffinized in xylene and rehydrated in graded alcohol baths before staining with hematoxylin and eosin (HE).

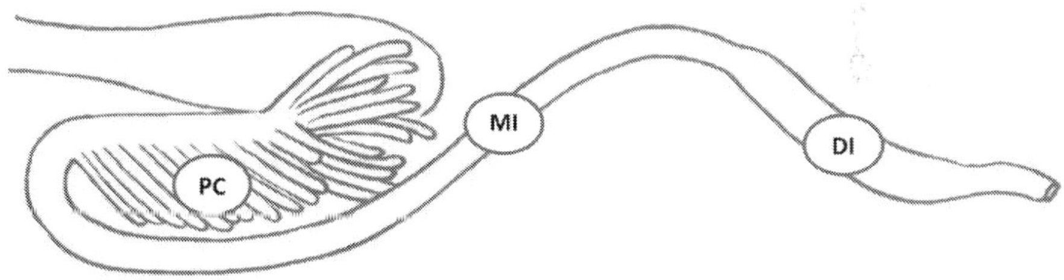

Figure 3: **Schematic drawing of the intestinal tract of Atlantic salmon.** The sampling sites are indicated as follows: PC=pyloric caeca, MI=mid intestine and DI=distal intestine.

Formalin-fixed and paraffin-embedded tissues from the mid intestine and the distal intestine from nine fish in each dietary groups (three fish in each tank) were prepared for detection of antigen presenting cells expressing MHC class II and T lymphocytes expressing CD3ε by immunohistochemistry with salmon specific polyclonal rabbit antisera as previously described by Koppang *et al.* [29, 30] with some modifications. Proliferating cells were recognized in sections from the mid intestine from nine fish in each dietary group (three fish in each tank) using a monoclonal mouse antibody against proliferating cell nuclear antigen (PCNA; M0879, Dako, Glostrup, Denmark). Sections were routinely deparaffinized in xylene and rehydrated in graded alcohol baths before they were transferred to distilled water. Antigen retrieval was performed by autoclaving the slides in 0.01 M citrate buffer at 121°C for 15 min, and the slides were cooled to room temperature and transferred to phosphate-buffered saline (PBS) before inhibition of endogenous peroxidase with 0.05% phenyl hydrazine (P26252, Sigma-Aldrich, Milwaukee, Wisconsin, US) in PBS at 37°C for 40 min. The slides were then incubated in goat serum diluted 1:50 in 5% bovine serum albumin (BSA) in Tris-buffered saline (TBS) for 20 min to prevent nonspecific binding. Antisera against MHC class II, CD3ϵ and PCNA were diluted 1:600, 1:400 and 1:3000, respectively, in 1% BSA/TBS before incubation for 30 min. The secondary antibody and substrate chromogen were provided from the EnVision® System kit (K4009, Dako). The sections were counterstained with hematoxylin added acetic acid (Mayer's hematoxylin) for 1 min and mounted with poly-vinyl alcohol mounting media (Ullevål Apotek, Oslo, Norway).

Morphometric Analysis

Micrographs of intestinal sections from nine fish in each dietary group (three fish per tank) were captured and morphometric measurements were performed in the software NIS-Elements D version 3 (Nikon, Tokyo, Japan) using Nikon digital sight camera configured with a Nikon eclipse 80i microscope. The measurements were performed as previously described by Løkka *et al.* [28]. The height of the folds was measured from the fold apex to the bottom of the epithelium at the base of the folds, and both simple and complex folds were measured in the distal intestine. The width of the folds was assessed at two points in every fold, and the thickness of the intestinal wall was measured from beneath the epithelium at the base of the folds (simple folds in distal intestine) to the serosa (Figure 4). Five measurements of the fold height and wall thickness and ten measurements of the fold width in both intestinal segments were recorded for each individual.

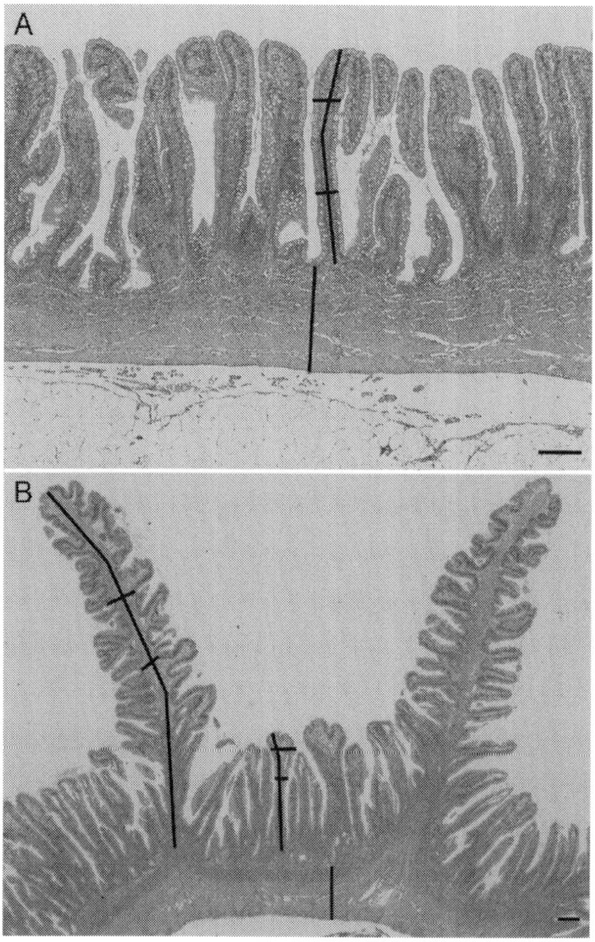

Figure 4: **Microphotographs showing measurements in sections of the mid (A) and distal (B) intestine.** The lines indicate how the height and width of the folds and the thickness of the wall were measured. Hematoxylin and eosin. Scale bar 200 μm.

Gene Transcription Analysis

Tissues from the pyloric caeca, mid intestine and distal intestine (Figure 3) from nine fish in each dietary group (three fish per tank) were frozen in liquid nitrogen and stored at -80°C. The samples from the mid intestine were homogenized in Trizol using zirconium beads (4 mm) in a Retsch MM 310 homogenizer (Retsch GmbH, Haan, Germany). Subsequent addition of chloroform separated RNA from proteins and DNA, and RNA was then precipitated from the water phase by adding isopropanol. Furthermore, the RNA pellet was cleansed twice in ethanol and dissolved in RNase free water. A DNase treatment with DNA-free™ (Applied biosystems, Foster City, CA, USA) was performed on the RNA extract.

The samples from the pyloric caeca and the distal intestine were homogenized in Buffer RLT added mercaptoetanol using stainless steelbeads (5 mm) in a Retsch MM 300 homogenizer (Retsch GmbH). RNA was extracted with RNeasy Mini kit (QIAGEN, Hilden, Germany) using the protocol "RNeasyMini Animal Tissues and Cells Standard V3" followed by the protocol "Cleanup RNeasyMini RNA Standard V3" in a QIAcube (QIAGEN).

The concentration of RNA was measured using a Biospec-Nano (Shimadzu Corporation, Kyoto, Japan) or Nanodrop ND-1000 UV-Vis Spectrophotometer (NanoDrop Technologies, Wilmington, DE, USA). To verify acceptable quality of the RNA, 24 random samples were selected and tested on an Agilent 2100 Bioanalyzer (Agilent Technologies, Palo Alto, CA, USA). Total RNA was stored at -80°C.

The cDNA synthesis from 1 μg of total RNA was prepared with oligo(dT), random hexamer primers, M-MLV Reverse transcriptase (Promega, Madison, WI, USA) and RNase inhibitor (Promega) to prevent RNA degradation. Real-time PCR was carried out in 13 μl reactions using TaqMan Gene Expression Master Mix (Applied Biosystems, Carlsbad, CA, USA) with cDNA template corresponding to 15 ng of RNA in each reaction in a 7900HT fast real-time PCR system (Applied Biosystems) according to the producer's instructions and running 40 cycles.

The following genes were analyzed by real-time RT-PCR: Cluster of differentiation 3ζ (CD3ζ), cyclooxygenase-2a (COX-2a), interleukin 1β (IL-1β), immunoglobulin M (IgM), immunoglobulin T (IgT), major histocompatibility class II (MHC class II), nucleotide-binding oligomerization domain-containing protein 2 (NOD2), transforming growth factor β (TGF-β), tumour necrosis factor α (TNF-α) and elongation factor $1A_B$ ($EF1A_B$) as the reference gene [31] (Table 1). When possible, primers and probe were designed to span across intron sections. All analyses were performed in triplicates, and a control lacking the template for each master mix was always included in the experiments. The data were analysed using Sequence Detection Systems Software v2.3 (Applied Biosystems).

Table 1: Sequences of primers and probes used in real-time RT-PCR analysis

Gene	Gene Sequence 5′→3′	GenBank accession no.	Pyloric caeca	Mid intestine	Distal intestine
EF1A$_B$	F-TGCCCCTCCAGGAT-GTCTAC	BG933853	18.19±0.09	19.74±0.28	18.90±0.15
	R-CACGGCCCACAGG-TACTG				
	P-FAM-AAATCGGCGG-TATTGG-MGB				
CD3ζ	F-AACAGGGATC-CAGAGAGTGCTG	BT060238	27.77±0.16	27.74±0.28	27.41±0.20
	R-AAGGGACGTGTA-AGTGTCGTCA				
	P-FAM-ACGGCACGC-GATAATCGCAGGA-BHQ				
COX-2a	F-CAGATCGCTG-GAAGGGTGG	AY848944	33.66±0.31	33.32±0.43	32.83±0.35
	R-TCATGTTGAAGC-GTTTCCTGTAG				
	P-FAM-AGCTA-AGGCCCTGGAGCA-CAGC-BHQ				
IgM	F-TGTAAAGAGAG-CAGACTGGGACAG	Y12456	25.60±0.55	24.89±0.60	24.63±0.45
	R-GAGACGGGTGCTG-CAGATATTC	Y12457			
	P-FAM-TGTTCCACGGC-GCATTCAAAGATTT-BHQ				
IgT	F-CAGCAGTCTGCT-GAAGGTC	GQ907004	29.91±0.29	30.29±0.50	28.00±0.30
	R-GGTTCTGTTTTGGA-GATCG	GQ907003			
	P-FAM-CTGCACCACA-CAGCTGTACTTGACC-BHQ				

IL-1β	F-GCTACCACAAAGTG-CATTTG	AY617117	34.91±0.42	33.91±0.64	31.98±0.33
	R-GAGGTTGGATCCCTT-TATGC				
	P-FAM-CCATTGAGAC-TAAAGCCAGACCTG-TAG-BHQ				
MHCII	F-CCACCTGGAGTACA-CACCCAG	X70165	20.95±0.25	20.34±0.35	20.49±0.18
	R-TTCCTCTCAGCCT-CAGGCAG				
	P-FAM-TCCTGCATGGTG-GAGCACATCAGC-BHQ				
NOD2	F-GCATCCAGTGTGAG-CACTTTCAG	EG915470	32.01 ± 0.19	31.55 ± 0.24	31.86 ± 0.23
	R-TTCATCTTCAGGAG-GTGAGCG				
	P-FAM-CAAGCTAACT-GATGCCTGCACAGAGT-GC-BHQ				
TGF-β	F-TGGAGCTGAGTGAG-GAGCAG	EU082211	34.04±0.29	34.24±0.41	33.46±0.32
	R-ACCGCATCTCAGA-CATGTTG				
	P-FAM-TGTG-GACCTCCTTTGCAAAG-TATGC-BHQ				
TNF-α	F-GCAGCTTTATGTGC-GGCAG	NM_001123589	36.38±0.30	NA	35.47±0.29
	R-TTTTGCACCAATGAG-TATCTCCAG	NM_001123590			
	P-FAM-TGGAAGACTG-GCAACGATGCAGGA-BHQ				

The mean Ct-values ± SEM for the fish oil group are given for each intestinal segment.

NA, Not applied.

Calculations and Statistical Analysis

Databases for the results for morphometric measurements and real-time RT-PCR were established in Excel® for Windows, and statistical calculations and graphical presentation of real-time PCR results were performed using Prism 6.0 software (GraphPad Software, San Diego, CA, USA). Morphometric data within each dietary group were pooled before calculating the group mean.

Data are given as mean ± standard error of the mean (SEM). The data were analyzed for normality using a Shapiro Wilk's test and for homogeneity of variance using a Brown-Forsythe's test. For data that were non-normal and/or with non-homogeneous residuals, a log transformation was performed prior to one way analysis of variance (ANOVA) followed by Tukey's multiple comparison test. The significance level was set to 0.05.

RESULTS

Histology and Immunohistochemistry

Histological examination of the mid intestine did not show any obvious pathological changes in any fish, while there was shortening, widening and fusion of the simple folds with leucocyte infiltration in the lamina propria in the distal intestine in a few individuals without any obvious association with the different dietary groups. Examination of sections stained with antibodies against MHC class II and CD3ϵ showed positive cells both in the epithelium and in the lamina propria and with similar density and distribution of antigen presenting cells and T lymphocytes regardless of diet (Figure 5). The proliferation pattern of epithelial cells in the mid intestine as demonstrated by immunohistochemical staining with an antibody against PCNA did not differ between the dietary groups.

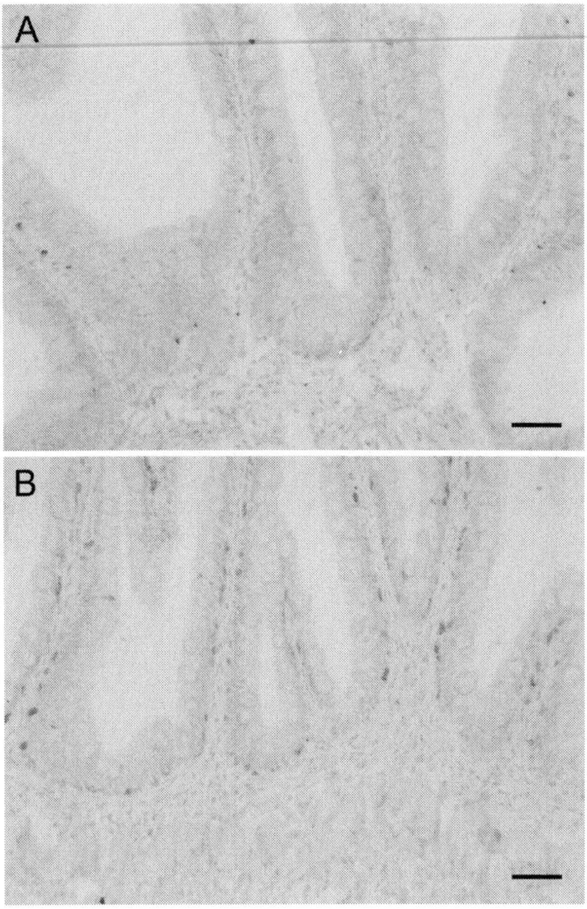

Figure 5: **Detection of antigen presenting cells and T lymphocytes in the mid intestine.**Microphotographs of mid intestinal sections from fish fed soybean oil stained with antisera against MHC class II **(A)** and CD3ϵ **(B)**. Positive cells visualized with red colour were found in the epithelium and only scattered cells were present in the lamina propria. The density and distribution of positive cells were independent of diet. Scale bar 50 μm.

Morphometric Analysis

Mid intestine: The folds of the mid intestine were tallest in the fish oil group (1393±36.4 μm), intermediate in the olive oil and rapeseed oil groups (1134±31.2 μm and 1131±42.4 μm respectively) and lowest in the soybean oil group (1012±24.8 μm). The differences were highly significant ($P<0.0001$) between the fish oil group and all vegetable oils groups. The folds of the mid intestine were widest in the soybean oil group (142.0±5.4 μm) and most slender in the olive oil group (122.6±4.3 μm) (P=0.0249). There was no significant difference between the dietary groups regarding the wall thickness. The results from the morphometric analyses of the mid intestine are shown in Table 2.

Table 2: The height and width of folds and wall thickness in mid and distal intestine (**μm**)

Mid intestine				
	FO	**OO**	**RO**	**SO**
Height of folds	**1393±36.4^{a}**	**1134±31.2^{b}**	**1131±42.4^{b}**	**1012±24.8^{b}**
Width of folds	135.8±4.8^{ab}	**122.6±4.3^{a}**	129.2±4.5^{ab}	**142.0±5.4^{b}**
Wall thickness	732.2±21.4^{a}	652.3±24.0^{a}	652.5±21.6^{a}	694.0±20.9^{a}
Distal intestine				
	FO	**OO**	**RO**	**SO**
Height of simple folds	1240±37.8^{a}	1204±39.7^{a}	1221±29.3^{a}	1242±35.7^{a}
Width of simple folds	138.9±5.3^{a}	127.1±5.5^{a}	131.3±5.9^{a}	131.8±5.3^{a}
Height of complex folds	3456±146.9^{ab}	3280±91.7^{ab}	**3598±100.2^{a}**	**3123±79.7^{b}**
Width of complex folds	265.2±10.8^{a}	267.1±10.7^{a}	234.1±7.8^{a}	238.7±10.1^{a}
Wall thickness	**705.2±20.4^{a}**	**616.8±22.8^{b}**	643.3±21.5^{ab}	**621.4±18.0^{b}**

FO, fish oil, *OO*, olive oil, *RO*, rapeseed oil, *SO*, soybean oil.

^{a,b}Significant differences (ANOVA, P<0.05) between dietary groups in the same intestinal segment are denoted by different superscript letters. The values that differ significantly are highlighted in bold text.

Distal intestine: The complex folds were tallest in the rapeseed oil group (3598±100.2 μm) and lowest in the soybean oil group (3123±79.7 μm) (P=0.0113), while the wall was thickest in the fish oil group (705.2±20.4 μm) and thinnest in the olive oil group (616.8±22.8 μm) (P=0.0080). There was no significant difference between the dietary groups regarding the height of the simple folds or the width of either simple or complex folds. The results from the morphometric analyses of the distal intestine are shown in Table 2.

Transcript Levels of Immune-Related Genes

Only small differences in the relative transcript levels of the various genes between the dietary groups were detected, and there was generally larger individual variation within a group than between the groups. Neither the transcript levels of the pro-inflammatory cytokines IL-1β and TNF-α, the intracellular receptor NOD2 nor the enzyme COX-2a that is involved in the synthesis of prostaglandins from fatty acids were significantly altered in any dietary group in any of the intestinal segments.

Pyloric caeca: In the pyloric caeca, the olive oil group had significantly higher transcript levels of CD3ζ and MHC class II (P=0.004 and 0.048) while the rapeseed oil group had a significantly higher transcript level of TGF-β (P=0.022) and the soybean oil group had significantly higher transcript levels of CD3ζ and TGF-β (P=0.033 and 0.017) compared to the fish oil group (Figure 6A).

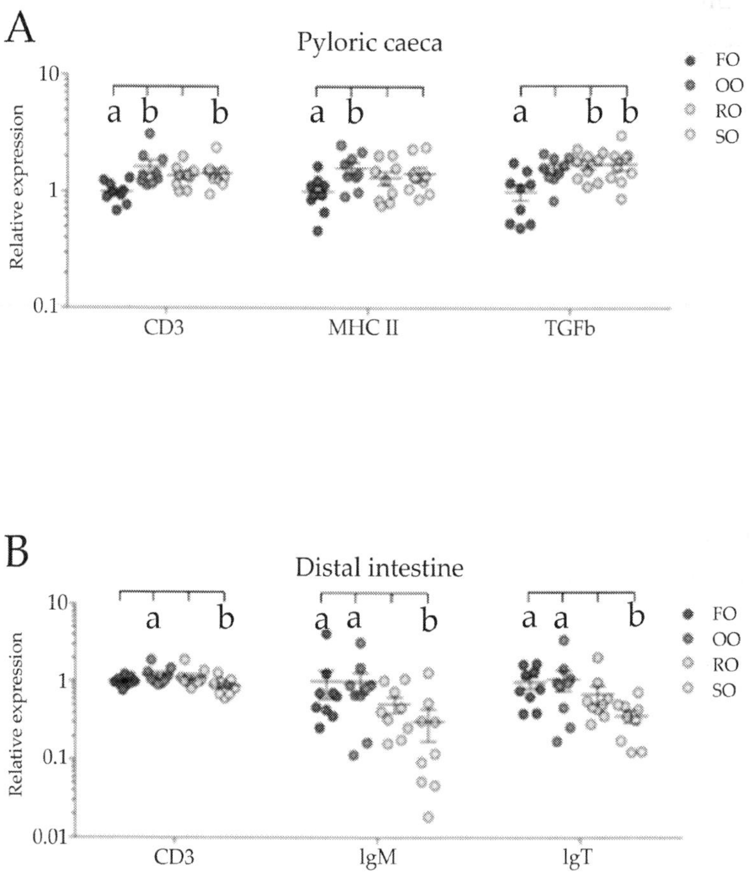

Figure 6: **Relative transcription levels of genes with significant differences between dietary groups.** The dot plots for CD3ζ, MHC class II and TGF-β in

the pyloric caeca **(A)** and CD3ζ, IgM and IgT in the distal intestine **(B)** show the transcript levels for each individual with mean for each group as obtained by real-time RT-PCR analysis. Transcript levels are presented relative to the mean of the fish oil group. Error bars represent standard error of the mean. [a,b]Significant differences (ANOVA, $P<0.05$) between dietary groups are denoted by different letters above the dot plots. FO=fish oil, OO=olive oil, RO=rapeseed oil and SO=soybean oil.

Mid intestine: The transcript levels of the selected genes appeared to be more uniform in the mid intestine, 7and no significant differences between any dietary groups were detected in this intestinal segment.

Distal intestine: In the distal intestine, the transcript level of CD3ζ was significantly higher in the olive oil group than the soybean oil group ($P=0.035$). The transcript levels of both IgM and IgT was significantly lower in the soybean oil group than in the fish oil group ($P=0.014$ and 0.016) and the olive oil group ($P=0.015$ and 0.038) (Figure 6B).

The underlying data for Figure 6A and B and the normalized transcript levels for all genes and dietary groups relative to the fish oil group are shown in Table 3. A disadvantage of calculating normalized transcription levels is that the differences between genes get lost, while the absolute Ct-values will reveal such differences. The Ct-values for MHC class II were relatively low (~18-20) in all intestinal segments of the fish in all dietary groups, indicative of a high transcript level of mRNA. The Ct-values for CD3ζ and the immunoglobulins were intermediate (~24-30), while the Ct-values for the selected cytokines, NOD2 and COX-2a were in general high (~31-37) and for some individuals beyond the detection limit. The mean Ct-values for all examined genes in the three intestinal segments for the fish oil group were included in Table 1.

Table 3: Relative transcription levels of the immune-related genes normalized to $EF1A_B$

	Pyloric caeca				Mid intestine				Distal intestine			
	FO	OO	RO	SO	FO	OO	RO	SO	FO	OO	RO	SO
CD3	**1±0.07[a]**	**1.63±0.21[b]**	1.38±0.10[ab]	**1.44±0.13[b]**	1±0.03[a]	1.22±0.29[a]	0.99±0.11[a]	0.96±0.06[a]	1±0.05[ab]	**1.21±0.11[a]**	1.13±0.11[ab]	**0.89±0.07[b]**
COX-2a	1±0.17[a]	1.56±0.29[a]	1.36±0.15[a]	1.28±0.22[a]	1±0.21[a]	0.98±0.10[a]	1.08±0.11[a]	1.20±0.17[a]	1±0.33[a]	0.85±0.17[a]	1.19±0.21[a]	0.85±0.24[a]
IgM	1±0.29[a]	1.37±0.38[a]	0.80±0.20[a]	0.97±0.28[a]	1±0.30[a]	0.93±0.28[a]	0.72±0.20[a]	0.95±0.17[a]	**1±0.40[a]**	**0.99±0.30[a]**	0.52±0.12[ab]	**0.31±0.14[b]**
IgT	1±0.17[a]	1.97±0.61[a]	0.85±0.12[a]	0.94±0.17[a]	1±0.22[a]	1.23±0.18[a]	0.89±0.22[a]	1.43±0.30[a]	**1±0.17[a]**	**1.09±0.33[a]**	0.71±0.18[ab]	**0.37±0.07[b]**
IL-1β	1±0.29[a]	0.94±0.23[a]	0.63±0.14[a]	0.91±0.22[a]	1±0.29[a]	1.51±0.68[a]	0.90±0.20[a]	0.89±0.29[a]	1±0.25[a]	1.26±0.68[a]	1.07±0.29[a]	0.92±0.52[a]
MHCII	**1±0.11[a]**	**1.59±0.18[b]**	1.32±0.17[ab]	1.44±0.19[ab]	1±0.11[a]	1.23±0.26[a]	0.92±0.15[a]	0.92±0.14[a]	1±0.14[a]	1.15±0.13[a]	1.48±0.18[a]	1.02±0.15[a]
NOD2	1±0.09[a]	1.31±0.12[a]	1.32±0.10[a]	1.21±0.09[a]	1±0.06[a]	0.98±0.09[a]	0.92±0.10[a]	0.87±0.06[a]	1±0.10[a]	1.23±0.16[a]	1.90±0.58[a]	0.96±0.15[a]
TGF-β	**1±0.15[a]**	1.59±0.13[ab]	**1.71±0.15[b]**	**1.74±0.21[b]**	1±0.16[a]	1.44±0.27[a]	1.19±0.21[a]	1.26±0.11[a]	1±0.16[a]	1.71±0.41[a]	1.54±0.27[a]	0.97±0.33[a]
TNF-α	1±0.14[a]	0.93±0.19[a]	0.68±0.11[a]	0.93±0.16[a]	NA	NA	NA	NA	1±0.13[a]	2.16±1.31[a]	1.44±0.28[a]	1.36±0.75[a]

The data are presented as mean ± SEM relative to the mean of the fish oil group.

FO, fish oil, *OO*, olive oil, *RO*, rapeseed oil, *SO*, soybean oil, *NA*, Not applied.

[a,b]Significant differences (ANOVA, P<0.05) between dietary groups in the same intestinal segment are denoted by different letters. The values that differ significantly are highlighted in bold text.

DISCUSSION

In this study, we have shown that fish fed diets where fish oil was largely replaced by three different vegetable oil blends had significantly shorter folds in the mid intestine compared with fish fed a diet with fish oil as the sole lipid source in a trial lasting for 28 weeks. The fold height decreased to a degree roughly corresponding to a decreasing n-3/n-6 fatty acid ratio of the feed, ie. the fish in the soybean oil group had the shortest folds. Additionally, the mid intestinal folds of the fish in the soybean oil group were significantly wider than of the fish in the olive oil group. In the distal intestine, the complex mucosal folds of the fish in the soybean oil group were significantly shorter than the folds of the fish in the rapeseed oil group, while the wall was significantly thicker for the fish in the fish oil group than the fish in the olive oil and soybean oil groups. Histological and immunohistochemical examination did not reveal any overt signs of inflammation in the lamina propria of the mid intestine. In the distal intestine, however, infiltration of inflammatory cells was observed in some individuals, but this could not be related to the dietary groups as fish in the fish oil group also were affected. Real time RT-PCR revealed only minor alterations in the transcript levels of the selected immune-related genes between dietary groups.

Dietary lipid sources have been reported to affect intestinal morphology in mammals. In weaning pigs, dietary supplementation with fish oil increased villus height in the small intestine combined with a decrease in transcript levels of inflammation related genes compared to a diet with corn oil [32]. Furthermore, dietary fatty acid composition has been reported to affect the height of intestinal villi in ileum in rats, the extent of the reductions increasing with increasing levels of n-6 fatty acids, ie. rats fed with fish oil had higher villi than those fed with olive oil and soybean oil [33]. Interestingly, rats fed with soybean oil had wider villi than the group given olive oil but not the group given fish oil, which is in agreement with our observations. The altered morphology was followed by a corresponding infiltration of mucosal lymphocytes [33].

In Atlantic salmon, shortening and widening of the simple mucosal folds of the distal intestine, in combination with infiltration of inflammatory cells in the lamina propria, has been repeatedly reported when feeding with soybean meal (SBM) and pea protein concentrate [23, 24, 34]. Starvation has also been described to mildly induce similar changes [23]. In humans, shortened intestinal villi and inflammation in the small intestine occurs in patients with coeliac disease caused by reaction to gluten proteins [35]. However, in the present study, histological investigation and immunohistochemical examination with antigen presenting cell and T lymphocyte markers did not show infiltration of inflammatory cells in the intestinal wall corresponding to the fold reduction pattern. Furthermore, there was no significant difference in the transcript levels of the investigated genes between either of the groups in the mid intestinal region. Combined, this indicates that the fold reductions observed in the mid intestinal region in the current study were not connected with a prolonged inflammatory response, but were probably caused by other factors.

In fish, the gastrointestinal microbiota is known to change with different feeding regimes [36, 37] and more specifically with different lipid levels and different vegetable oils [38]. Alterations in intestinal microbiota are hence not to be neglected as a possible explanatory factor for the altered morphology observed.

Significant reductions of mucosal folds in the mid intestine of all vegetable dietary groups were observed, in contrast to the mildly affected distal intestine. This finding might be related to the fact that long chain fatty acids (LCFAs) mainly are absorbed in the pyloric caeca and mid intestine and only to a limited extent in the distal parts of the intestine [16, 39]. Altering the composition of the LCFA in the feed can hence be speculated to cause most pronounced changes in the regions where these fatty acids are mainly absorbed.

Shortening of mucosal folds has been linked to altered proliferation pattern of the intestinal epithelium in Atlantic salmon; decrease in cell proliferation and apoptosis in smolt exposed to sublethal levels of inorganic mercury [40] and increase in cell proliferation in soybean meal induced enteropathy [36]. We did however not observe any differences in the proliferation pattern of epithelial cells in the mid intestine between any of the diet groups, even though lower turn-over in cell proliferation and apoptosis has been previously detected in Atlantic salmon fed with a diet where fish oil was completely replaced by plant oil [25].

Shortening of the mid intestinal folds probably reduces the total surface of the intestine and hence the absorption of nutrients, which may in turn influence the growth of the fish. A substantial proportion of starch and lipids is absorbed in the mid intestine [41]. Fish fed soybean oil were significantly shorter than the fish fed fish oil and somewhat (but not significantly) lighter than the other fish in the trial; results which were linked to reduced feed intake in the soybean oil group [22]. It cannot, however, be ruled out that the pronounced shortening of the folds in the mid intestine of fish fed soybean oil may be an additional factor contributing to the somewhat reduced growth in fish fed soybean oil. Furthermore, the difference in weight between the groups might have been more pronounced if not the fish fed fish oil as the sole lipid source had significantly reduced lipid digestibility due to high levels of dietary saturated fatty acids [22]. It cannot be ruled out that the reduced lipid digestibility in the fish oil group may have affected intestinal morphology as response to saturated fatty acids possibly being above an acceptable threshold level for Atlantic salmon. Overall, the shortened intestinal folds being most pronounced in the soybean oil group suggest that feeding with similar or higher levels of soybean oil as in the present trial might not be unproblematic for the production results.

In the pyloric caeca, the transcript levels of TGF-β and CD3ζ in the soybean oil group, the transcript level of TGF-β in the rapeseed oil group and the transcript levels of MHC class II and CD3ζ in the olive oil group was significantly higher than in the fish oil group. TGF-β is produced by cells of the innate immune system and by regulatory T lymphocytes and is both a pro- and anti-inflammatory cytokine that is involved in cell growth, migration, differentiation and apoptosis including inhibition of lymphocyte proliferation [42]. In SBM-induced inflammation in Atlantic salmon, transcription levels of TGF-β were reported up-regulated by 7-folds after 21 days, combined with a 20-folds up-regulation of IL-1β [43]. In the present trial, although significant, the differences observed in TGF-β transcript levels in the fish fed soybean oil and rapeseed oil compared to the fish fed fish oil was below two-fold, and there was no significant difference in IL-1β transcript levels between the groups; hence it is difficult to ascribe the differences in TGF-β transcript levels to an inflammatory process. The finding that neither the relative transcript levels of TNF-α, NOD2 nor COX-2a did vary significantly between the dietary groups in any of the intestinal segments also suggests that dietary lipids did not affect the degree of inflammation. The transcript levels of TNF-α and IL-1β in leucocytes from Atlantic salmon did not differ significantly between groups that were incubated in plasma with different n-3/n-6 ratio followed by stimulation with LPS, and a relative similar EPA/AA ratio in the cells was launched as an explanation to the lack of influence of fatty acid sources on inflammatory response [9]. A similar EPA/AA ratio in leucocytes may explain the relative stable transcript levels of cytokines regardless of diets in the current study too.

In contrast to mammals harboring mesenteric lymph nodes and distinct lymphoid follicles in the intestinal mucosa, the immune competent cells including antigen presenting cells as well as T and B lymphocytes are more diffusely spread in the intestinal tissue of teleosts like Atlantic salmon [44]. The moderately higher transcript levels of CD3ζ observed in the pyloric caeca of the olive and soybean oil group might indicate a slightly higher number of T lymphocytes as CD3ζ is part of the T cell receptor complex and expressed in all T lymphocytes [45]. MHC class II is in contrast expressed in antigen presenting cells that can present antigenic peptides to T lymphocytes and initiate the adaptive immunity [42]. In mammals, and presumably in teleosts, MHC class II is moreover expressed in intestinal enterocytes [46]. A higher transcript level of MHC class II might indicate a higher level of antigen presentation in the pyloric caeca of the olive oil group compared to the fish oil group. The differences in both CD3ζ and MHC class II transcript levels between the groups were however below two-fold and should be carefully interpreted. The higher transcript levels of certain genes in the plant oil groups compared to the fish oil group in the pyloric caeca could be linked to the high lipid absorption in this region [41]. However, this does not explain why we do not see a corresponding change in the transcript levels for these genes in the mid intestine as absorption rate for LCFAs are reported to be similarly high here [39].

The significantly lower transcript level of CD3ζ in the distal intestine of the soybean oil group compared to the olive oil group might in contrast to the pyloric caeca, indicate a lower number of T lymphocytes. Furthermore, the soybean oil group had significantly lower transcript levels of IgM and IgT than the fish fed both fish oil and olive oil. These immunoglobulins are expressed by different subpopulations of B lymphocytes in the teleost intestine [47] and are present both as membrane bound and secretory forms. Again, the differences were moderate, but summed together they suggest a difference in response to soybean oil between the pyloric cecea and distal intestine.

In mammals, replacement of n-3 fatty acids with n-6 polyunsaturated acids has been associated with intestinal inflammation and promotion of intestinal cancer [6, 10, 11]. Feed-induced intestinal carcinogenesis following inflammation has also been reported in brood stock Atlantic salmon [27], and it was speculated whether this partially could be related to replacement of fish oil with vegetable oils in the feed. Although the brood stock were exposed much longer to the commercial feed than the fish in the current 28 week-long trial, the results of the present study strongly suggest that partial replacement of fish oil with vegetable oils in the feed did not induce prolonged intestinal inflammation in Atlantic salmon.

The amount of fish oil still present in the feed might be of importance for the ability of the fish to cope with the increased amount of n-6 fatty acids. The regular feed composition used in the salmon industry has changed dramatically over the last decades as the lipid fraction has increased from 10% till about 35% [4]. This means that although vegetable oils constitute 80% of the lipid fraction as in the feeds of our study, the feed still contains approximately 1.4% EPA and DHA provided by the fish oil and fish meal included in the feed, which might be enough to sustain general intestinal health. It has been shown previously that 1% EPA and DHA in the feed is essential to attain good growth in fry [48]. A minimum proportion of EPA and DHA is considered to be required also for larger Atlantic salmon [49], however this has not yet been quantified.

CONCLUSIONS

The folds in the mid intestine were significantly shorter in all groups of Atlantic salmon fed vegetable oils compared to the group fed fish oil. In the distal intestine, the complex folds were significantly shorter in the fish fed soybean oil compared to the fish fed rapeseed oil. Histological examination did not reveal clear difference in degree of inflammation related to dietary groups, and this finding was confirmed by real-time RT-PCR that only revealed moderate alterations in the mRNA transcript levels of selected immune-related genes.

The shortening of the intestinal folds was most pronounced in the fish fed soybean oil and might be associated with reduced intestinal surface and impaired nutrient absorption and growth. Kept together with significantly higher transcript levels of TGF-β and CD3ζ in the pyloric caeca and significantly lower transcript levels of IgM and IgT in the distal intestine in the fish fed soybean oil compared to the fish fed fish oil, it can be concluded that inclusion of high levels of soybean oil in the feed for Atlantic salmon should be done with caution.

NOTES

Acknowledgements

We thank Thommy Holmvåg at Skretting ARC for excellent fish husbandry, PhD student Nina Liland at NIFES for providing RNA samples from the mid intestine and technical staff at the Norwegian Veterinary Institute for cutting sections and stainings. The study was funded by the Research Council of Norway through the projects 'SAFE FEED, SAFE AND HEALTHY SEAFOOD Contaminants, nutrients and health challenges in the novel production of farmed Atlantic salmon' (project number 199626; project leader Bente E. Torstensen) and 'InNoVacc' (grant number 183196/S40), Skretting ARC, the Norwegian Veterinary Institute and the Norwegian School of Veterinary Science.

Competing Interests

The authors declare that they have no competing interests.

Authors' Contributions

TM did the sampling, planned the experiments, performed the histological examination, morphometric measurements, immunohistochemical stainings, parts of the cDNA synthesis and PCR and statistical analyses and drafted the manuscript, GL planned the experiments, performed immunohistochemical stainings, examination of immunohistochemical stainings, parts of the cDNA synthesis and PCR and statistical analyses and drafted the manuscript, JWN performed parts of the RNA isolation, cDNA synthesis and PCR, LA planned the experiments, designed the probes and supervised the PCR work, BT and GR designed and supervised the dietary experiment and samplings, and OBD, EOK and MK planned the experiments and supervised the study. Everyone commented on the manuscript and approved the final version.

REFERENCES

1. Naylor RL, Hardy RW, Bureau DP, Chiu A, Elliott M, Farrell AP, Forster I, Gatlin DM, Goldburg RJ, Hua K, Nichols PD: Feeding aquaculture in an era of finite resources. Proc Natl Acad Sci USA. 2009, 106: 15103-15110. 10.1073/pnas.0905235106.
2. Torstensen BE, Espe M, Sanden M, Stubhaug I, Waagbø R, Hemre GI, Fontanillas R, Nordgarden U, Hevrøy EM, Olsvik P, Berntssen MHG: Novel production of Atlantic salmon (*Salmo salar*) protein based on combined replacement of fish meal and fish oil with plant meal and vegetable oil blends. Aquaculture. 2008, 285: 193-200. 10.1016/j.aquaculture.2008.08.025.
3. Powell K: Eat your veg. Nature. 2003, 426: 378-379. 10.1038/426378a.
4. Obach A: Trends and advances in feed formulation – towards lowest cost fish production. Practical Developments in Aquafeeds - Feed Advances to Boost Profitability: 2013. 2013, Nashville, Tennessee, USA: Nashville Convention Center
5. Sustainibility report 2012. http://www.skretting.com.
6. Chapkin RS, McMurray DN, Lupton JR: Colon cancer, fatty acids and anti-inflammatory compounds. Curr Opin Gastroenterol. 2007, 23: 48-54. 10.1097/MOG.0b013e32801145d7.
7. Calder PC: n-3 polyunsaturated fatty acids, inflammation, and inflammatory diseases. Am J Clin Nutr. 2006, 83: 1505S-1519S.
8. Bell JG, Ashton I, Secombes CJ, Weitzel BR, Dick JR, Sargent JR: Dietary lipid affects phospholipid fatty acid compositions, eicosanoid production and immune function in Atlantic salmon

(*Salmo salar*). Prostaglandins Leukot Essent Fatty Acids. 1996, 54: 173-182. 10.1016/S0952-3278(96)90013-7.

9. Seierstad SL, Haugland Ø, Larsen S, Waagbø R, Evensen Ø: Pro-inflammatory cytokine expression and respiratory burst activity following replacement of fish oil with rapeseed oil in the feed for Atlantic salmon (*Salmo salar* L.). Aquaculture. 2009, 289: 212-218. 10.1016/j.aquaculture.2008.12.004.
10. Tjønneland A, Overvad K, Bergmann MM, Nagel G, Linseisen J, Hallmans G, Palmqvist R, Sjödin H, Hägglund G, Berglund G, Lindgren S, Grip O, Palli D, Day NE, Khaw KT, Bingham S, Riboli E, Kennedy H, Hart A: Linoleic acid, a dietary n-6 polyunsaturated fatty acid, and the aetiology of ulcerative colitis: a nested case-control study within a European prospective cohort study. Gut. 2009, 58: 1606-1611.
11. Whelan J, McEntee MF: Dietary (n-6) PUFA and intestinal tumorigenesis. J Nutr. 2004, 134: 3421S-3426S.
12. Ghosh S, DeCoffe D, Brown K, Rajendiran E, Estaki M, Dai CB, Yip A, Gibson DL: Fish oil attenuates omega-6 polyunsaturated fatty acid-induced dysbiosis and infectious colitis but impairs LPS dephosphorylation activity causing sepsis. PLoS One. 2013, 8 (2): e55468-10.1371/journal.pone.0055468. doi:10.1371/journal.pone.0055468
13. Calder PC: Mechanisms of action of (n-3) fatty acids. J Nutr. 2012, 142: 592S-599S. 10.3945/jn.111.155259.
14. Calder PC, Yaqoob P, Thies F, Wallace FA, Miles EA: Fatty acids and lymphocyte functions. Br J Nutr. 2002, 87 (Suppl 1): 31-48.
15. Calder PC: Immunomodulation by omega-3 fatty acids. Prostaglandins Leukot Essent Fatty Acids. 2007, 77: 327-335. 10.1016/j.plefa.2007.10.015.
16. Denstadli V, Vegusdal A, Krogdahl Å, Bakke-McKellep AM, Berge GM, Holm H, Hillestad M, Ruyter B: Lipid absorption in different segments of the gastrointestinal tract of Atlantic salmon (*Salmo salar* L.). Aquaculture. 2004, 240: 385-398. 10.1016/j.aquaculture.2004.06.030.
17. Denstadli V, Bakke AM, Berge GM, Krogdahl Å, Hillestad M, Holm H, Ruyter B: Medium-chain and long-chain fatty acids have

different postabsorptive fates in atlantic salmon. J Nutr. 2011, 141: 1618-1625. 10.3945/jn.111.141820.

18. Torstensen BE, Lie Ø, Frøyland L: Lipid metabolism and tissue composition in Atlantic salmon (*Salmo salar* L.) - Effects of capelin oil, palm oil, and oleic acid-enriched sunflower oil as dietary lipid sources. Lipids. 2000, 35: 653-664. 10.1007/s11745-000-0570-6.
19. Torstensen BE, Frøyland L, Lie Ø: Replacing dietary fish oil with increasing levels of rapeseed oil and olive oil - effects on Atlantic salmon (*Salmo salar* L.) tissue and lipoprotein lipid composition and lipogenic enzyme activities. Aquac Nutr. 2004, 10: 175-192. 10.1111/j.1365-2095.2004.00289.x.
20. Torstensen BE, Frøyland L, Ørnsrud R, Lie Ø: Tailoring of a cardioprotective muscle fatty acid composition of Atlantic salmon (*Salmo salar*) fed vegetable oils. Food Chem. 2004, 87: 567-580. 10.1016/j.foodchem.2004.01.009.
21. Torstensen BE, Bell JG, Rosenlund G, Henderson RJ, Graff IE, Tocher DR, Lie Ø, Sargent JR: Tailoring of Atlantic salmon (*Salmo salar* L.) flesh lipid composition and sensory quality by replacing fish oil with a vegetable oil blend. J Agric Food Chem. 2005, 53: 10166-10178. 10.1021/jf051308i.
22. Liland NS, Rosenlund G, Berntssen MHG, Brattelid T, Madsen L, Torstensen BE: Net production of Atlantic salmon (FIFO, Fish in Fish out<1) with dietary plant proteins and vegetable oils. Aquac Nutr. 2013, 19: 289-300. 10.1111/j.1365-2095.2012.00958.x.
23. Bæverfjord G, Krogdahl Å: Development and regression of soybean meal induced enteritis in Atlantic salmon, *Salmo salar* L., distal intestine: a comparison with the intestines of fasted fish. J Fish Dis. 1996, 19: 375-387. 10.1111/j.1365-2761.1996.tb00376.x.
24. Penn MH, Bendiksen EA, Campbell P, Krogdahl Å: High level of dietary pea protein concentrate induces enteropathy in Atlantic salmon (*Salmo salar* L.). Aquaculture. 2011, 310: 267-273. 10.1016/j.aquaculture.2010.10.040.
25. Olsvik PA, Torstensen BE, Berntssen MHG: Effects of complete replacement of fish oil with plant oil on gastrointestinal cell death, proliferation and transcription of eight genes' encoding proteins responding to cellular stress in Atlantic salmon *Salmo salar* L. J

Fish Bio. 2007, 71: 550-568. 10.1111/j.1095-8649.2007.01521.x.

26. Oxley A, Jolly C, Eide T, Jordal AEO, Svardal A, Olsen RE: The combined impact of plant-derived dietary ingredients and acute stress on the intestinal arachidonic acid cascade in Atlantic salmon (*Salmo salar*). Br J Nutr. 2010, 103: 851-861. 10.1017/S0007114509992467.
27. Dale OB, Tørud B, Kvellestad A, Koppang HS, Koppang EO: From chronic feed-induced intestinal inflammation to adenocarcinoma with metastases in salmonid fish. Cancer Res. 2009, 69: 4355-4362. 10.1158/0008-5472.CAN-08-4877.
28. Løkka G, Austbø L, Falk K, Bjerkås I, Koppang EO: Intestinal morphology of the wild Atlantic salmon (*Salmo salar*). J Morphol. 2013, 274: 859-876. 10.1002/jmor.20142.
29. Koppang EO, Hordvik I, Bjerkås I, Torvund J, Aune L, Thevarajan J, Endresen C: Production of rabbit antisera against recombinant MHC class II beta chain and identification of immunoreactive cells in Atlantic salmon (*Salmo salar*). Fish Shellfish Immunol. 2003, 14: 115-132. 10.1006/fsim.2002.0424.
30. Koppang EO, Fischer U, Moore L, Tranulis MA, Dijkstra JM, Köllner B, Aune L, Jirillo E, Hordvik I: Salmonid T cells assemble in the thymus, spleen and in novel interbranchial lymphoid tissue. J Anat. 2010, 217: 728-739. 10.1111/j.1469-7580.2010.01305.x.
31. Olsvik PA, Lie KK, Jordal AEO, Nilsen TO, Hordvik I: Evaluation of potential reference genes in real-time RT-PCR studies of Atlantic salmon. BMC Mol Biol. 2005, 6: doi:10.1186/1471-2199-6-21
32. Liu YL, Chen F, Odle J, Lin X, Jacobi SK, Zhu HL, Wu ZF, Hou YQ: Fish oil enhances intestinal integrity and inhibits TLR4 and NOD2 signaling pathways in weaned pigs after LPS challenge. J Nutr. 2012, 142: 2017-2024. 10.3945/jn.112.164947.
33. Rosa DD, de Sales RL, Moraes LFD, Lourenco FC, Neves CA, Sabarense CM, Ribeiro SMR, Peluzio MDG: Flaxseed, olive and fish oil influence plasmatic lipids, lymphocyte migration and morphometry of the intestinal of Wistar rats. Acta Cir Bras. 2010, 25: 275-280. 10.1590/S0102-86502010000300010.
34. Van den Ingh TSGA, Krogdahl Å, Olli JJ, Hendriks HGCJ, Koninkx JGJF: Effects of soybean-containing diets on the proximal and distal

intestine in atlantic salmon (*Salmo salar*) - a morphological study. Aquaculture. 1991, 94: 297-305. 10.1016/0044-8486(91)90174-6.

35. Oberhuber G: Histopathology of celiac disease. Biomed Pharmacother. 2000, 54: 368-372.
36. Bakke-McKellep AM, Penn MH, Salas PM, Refstie S, Sperstad S, Landsverk T, Ringø E, Krogdahl Å: Effects of dietary soyabean meal, inulin and oxytetracycline on intestinal microbiota and epithelial cell stress, apoptosis and proliferation in the teleost Atlantic salmon (*Salmo salar* L.). Br J Nutr. 2007, 97: 699-713. 10.1017/S0007114507381397.
37. Merrifield DL, Dimitroglou A, Bradley G, Baker RTM, Davies SJ: Soybean meal alters autochthonous microbial populations, microvilli morphology and compromises intestinal enterocyte integrity of rainbow trout, *Oncorhynchus mykiss* (Walbaum). J Fish Dis. 2009, 32: 755-766. 10.1111/j.1365-2761.2009.01052.x.
38. Ringø E, Lødemel JB, Myklebust R, Jensen L, Lund V, Mayhew TM, Olsen RE: The effects of soybean, linseed and marine oils on aerobic gut microbiota of Arctic charr *Salvelinus alpinus* L. before and after challenge with *Aeromonas salmonicida* ssp. *salmonicida*. Aquac Res. 2002, 33: 591-606. 10.1046/j.1365-2109.2002.00697.x.
39. Røsjø C, Nordrum S, Olli JJ, Krogdahl Å, Ruyter B, Holm H: Lipid digestibility and metabolism in Atlantic salmon (*Salmo salar*) fed medium-chain triglycerides. Aquaculture. 2000, 190: 65-76. 10.1016/S0044-8486(00)00395-1.
40. Berntssen MHG, Hylland K, Julshamn K, Lundebye AK, Waagbø R: Maximum limits of organic and inorganic mercury in fish feed. Aquac Nutr. 2004, 10: 83-97. 10.1046/j.1365-2095.2003.00282.x.
41. Krogdahl Å, Nordrum S, Sørensen M, Brudeseth L, Røsjø C: Effects of diet composition on apparent nutrient absorption along the intestinal tract and of subsequent fasting on mucosal disaccharidase activities and plasma nutrient concentration in Atlantic salmon *Salmo salar* L. Aquac Nutr. 1999, 5: 121-133. 10.1046/j.1365-2095.1999.00095.x.
42. Zhu LY, Nie L, Zhu G, Xiang LX, Shao JZ: Advances in research of fish immune-relevant genes: A comparative overview of innate and

adaptive immunity in teleosts. Dev Comp Immunol. 2013, 39: 39-62. 10.1016/j.dci.2012.04.001.

43. Marjara IS, Chikwati EM, Valen EC, Krogdahl Å, Bakke AM: Transcriptional regulation of IL-17A and other inflammatory markers during the development of soybean meal-induced enteropathy in the distal intestine of Atlantic salmon (*Salmo salar* L.). Cytokine. 2012, 60: 186-196. 10.1016/j.cyto.2012.05.027.
44. Rombout J, Abelli L, Picchietti S, Scapigliati G, Kiron V: Teleost intestinal immunology. Fish Shellfish Immunol. 2011, 31: 616-626. 10.1016/j.fsi.2010.09.001.
45. Liu Y, Moore L, Koppang EO, Hordvik I: Characterization of the CD3 zeta, CD3 gamma delta and CD3 epsilon subunits of the T cell receptor complex in Atlantic salmon. Dev Comp Immunol. 2008, 32: 26-35. 10.1016/j.dci.2007.03.015.
46. Koppang EO, Lundin M, Press CM, Rønningen K, Lie Ø: Differing levels of Mhc class II beta chain expression in a range of tissues from vaccinated and non-vaccinated Atlantic salmon (*Salmo salar* L.). Fish Shellfish Immunol. 1998, 8: 183-196. 10.1006/fsim.1997.0136.
47. Salinas I, Zhang YA, Sunyer JO: Mucosal immunoglobulins and B cells of teleost fish. Dev Comp Immunol. 2011, 35: 1346-1365. 10.1016/j.dci.2011.11.009.
48. Ruyter B, Røsjø C, Einen O, Thomassen MS: Essential fatty acids in Atlantic salmon: effects of increasing dietary doses of n-6 and n-3 fatty acids on growth, survival and fatty acid composition of liver, blood and carcass. Aquac Nutr. 2000, 6: 119-127. 10.1046/j.1365-2095.2000.00137.x.
49. Sargent J, Bell G, McEvoy L, Tocher D, Estevez A: Recent developments in the essential fatty acid nutrition of fish. Aquaculture. 1999, 177: 191-199. 10.1016/S0044-8486(99)00083-6.

CHAPTER

11

TRANSCRIPTOMIC AND PHYSIOLOGICAL RESPONSES TO FISHMEAL SUBSTITUTION WITH PLANT PROTEINS IN FORMULATED FEED IN FARMED ATLANTIC SALMON (SALMO SALAR)

Luca Tacchi[1], Christopher J Secombes[1], Ralph Bickerdike[2], Michael A Adler[2], Claudia Venegas[3], Harald Takle[3,4] and Samuel AM Martin[1]

[1]Institute of Biological and Environmental Sciences, University of Aberdeen, Tillydrone Avenue, Aberdeen AB24 2TZ, UK.

[2]BioMar Ltd, Grangemouth Docks, Grangemouth FK3 8UL, UK.

[3]AVS Chile S.A., Casilla 300, Puerto Varas, Chile.

[4]Nofima, Aas, Norway.

ABSTRACT

Background

Aquaculture of piscivorous fish is in continual expansion resulting in a global requirement to reduce the dependence on wild caught fish for generation of fishmeal and fish oil. Plant proteins represent a suitable protein alternative to fish meal and are increasingly being used in fish feed. In this study, we examined the transcriptional response of Atlantic salmon (*Salmo salar*) to a high marine protein (MP) or low fishmeal, higher plant protein replacement diet (PP), formulated to the same nutritional specification within previously determined acceptable maximum levels of individual plant feed materials.

Results

After 77 days of feeding the fish in both groups doubled in weight, however neither growth performance, feed efficiency, condition factor nor organ indices were significantly different. Assessment of histopathological changes in the heart, intestine or liver did not reveal any negative effects of the PP diet. Transcriptomic analysis was performed in mid intestine, liver and skeletal muscle, using an Atlantic salmon oligonucleotide microarray (Salar_2, Agilent 4x44K). The dietary comparison revealed large alteration in gene expression in all the tissues studied between fish on the two diets. Gene ontology analysis showed, in the mid intestine of fish fed PP, higher expression of genes involved in enteritis, protein and energy metabolism, mitochondrial activity/kinases and transport, and a lower expression of genes involved in cell proliferation and apoptosis compared to fish fed MP. The liver of fish fed PP showed a lower expression of immune response genes but a higher expression of cell proliferation and apoptosis processes that may lead to cell reorganization in this tissue. The skeletal muscle of fish fed PP vs MP was characterized by a suppression of processes including immune response, energy and protein metabolism, cell proliferation and apoptosis which may reflect a more energy efficient tissue.

Conclusions

The PP diet resulted in significant effects on transcription in all the 3 tissues studied. Despite of these alterations, we demonstrated that high level of plant derived proteins in a salmon diet allowed fish to grow with equal efficiency as those on a high marine protein diet, and with no difference in biometric quality parameters.

BACKGROUND

Aquaculture production has been the fastest growing animal food-producing sector globally for over half a century, with production growing at an average rate of 8.1% per year since 1961[1], representing almost 50% of human consumed fish[1, 2]. Additionally, the capture from wild fisheries has plateaued with approximately three quarters of wild fisheries fished to capacity, overfished or recovering[3]. Atlantic salmon (*Salmo salar*) production alone now reaches over 1.5 million tonnes per year[4]. Total production in 2008 is estimated at 1.4 million tonnes, representing a 6% increase on 2007[5]. The positive growth trend of the industry is expected to continue, reflecting the rising demand for healthy human food products high in protein and marine oils.

Piscivorous fish, which include salmonids, require high protein diets and until recently this was almost exclusively derived from wild caught non food fish such as anchovies and sardines. It has become apparent that with increasing fish meal demand due to the expanding aquaculture industry the proportion of protein in diets cannot be sustained by use of fish meal alone. During 2003, 20% of total fish meal usage for aquaculture feed production was utilized by the salmon industry[6], but continued development of diets has reduced the inclusion of fish meal in salmon diets from 500 $g\,kg^{-1}$ to 350 $g\,kg^{-1}$ by replacement with alternative protein sources[7]. Plant proteins currently represent the only economic and sustainable protein alternative to fish meal and are increasingly being used in commercial fish feed, with the most common being soybean meal (SBM) or soybean protein concentrate (SPC) which have a high protein content and contain the majority of essential amino acids required by salmonids. All plants contain a number of anti nutritional factors (ANFs) as part of their inherent defence mechanisms, effected by lectins, saponins, phytic acid and proteinase inhibitors amongst others[8]. In early attempts to use plant proteins in fish diets ANFs were often co purified with the proteins, leading to metabolic dysfunction in fish liver[9], inflammation in the intestine[10, 11], reduced protein deposition[12], and an impaired immune response[13]. Currently used plant replacement diets use a combination of different plant proteins, with SPC often the major component[14]. Processing of soy protein to SPC with alcohol-extraction removes the majority of ANFs facilitating use of these products in fish feeds at a high inclusion level without causing enteritis or other gross morphological changes[15]. Replacement of up to 50% of the fish meal with a mixture of plant proteins is possible in rainbow trout without affecting fish growth, and even complete replacement of fish meal with SPC has been reported[16, 17]. Other successful plant proteins used regularly in aquaculture include wheat gluten at up to 40% of fish meal replacement in feed for salmon and trout (Hardy 1996), corn gluten and sunflower meal at up to 30% replacement, which show no adverse effects on salmonid growth performance[18, 19, 20]. Extensive studies investigating responses of Atlantic salmon to such diets indicate that a replacement of fish meal close to 100% can be used without negative effects on salmon growth providing the proteins are highly purified and the amino acid profile is well balanced[21]. High levels of pea protein concentrate do still cause enteropathy in salmon though[22].

In this paper we report a whole transcriptome based study on post-smolt Atlantic salmon fed on a low marine protein diet, where marine protein fish meal was partially replaced with a combination of plant derived proteins, compared to fish fed a diet high in fish meal. The high plant protein (PP) diet contained a balanced combination of soy protein concentrate, corn gluten and wheat gluten, and was supplemented with lysine and methionine to ensure the requirement for essential amino acids was met. The fish were fed for 77 days following which growth, feed efficiency and quality biometric parameters were measured and tissues were collected for histopathological changes and gene transcription analysis. The transcriptional response was examined using an Atlantic salmon"salar_2" microarray platform (Agilent)[23], enriched for genes involved in protein metabolism, lipid metabolism and immune function.

RESULTS

Feeding Trial

During the 77 day feeding trial fish were fed either a high marine protein (MP) diet or a plant protein replacement (PP) diet (Table1), and both groups doubled in weight over this period. No significant differences were seen between the groups in growth performance, feeding efficiency, condition factor or organ indexes (Table2). There were also no significant mortalities in any of the tanks.

Table 1: Dietary formulations

	PP	MP
Raw Materials (% of inclusion)		

Fish meal (%)	26.0	51.0
Fish oil (%)	19.2	17.3
Wheat (%)	6.3	11.9
Maize gluten (%)	10.0	3.0
Sunflower meal (%)	15.0	10.0
Rapeseed meal (%)	3.0	3.0
Wheat gluten (%)	1.3	
Soy protein concentrate(%)	18.0	3.0
Premix (%)	0.4	0.4
Monocalcium phosphate (%)	0.5	0.3
Lysine (%)	0.3	
Methionine(%)	0.1	
Proximate Analysis (% in feed)		
Protein (%)	44.8	44.3
Oil(%)	23.5	24
Ash(%)	6.8	8.9
Moisture(%)	7.7	8.5
Marine protein (as % of total Crude Protein)	39%	78%

Dietary formulation of high marine protein diet (MP) and plant protein diet (PP). Values represent percentage inclusion or content. Marine content calculated for total protein. The Fish oil is a commercial Chilean fish oil.

Table 2: Growth performance and somatic indexes

	PP		**MP**	
	Mean	**+/− SD**	**Mean**	**+/− SD**
Initial Weight (g)	110.6	0.8	110.8	0.4
Final Weight (g)	231.4	1.9	237.0	3.2
SGR (%)	1.09	0.02	1.12	0.02
FCR	0.87	0.01	0.90	0.03
SFR (%)	0.95	0.02	1.00	0.04
Condition factor	1.08	0.00	1.06	0.03
CSI	0.08	0.00	0.08	0.01
HSI	1.09	0.07	1.06	0.02
ISI	2.32	0.20	2.40	0.20

Growth performance and somatic indexes. SGR represents the specific growth rate of fish, FCR is the feed conversion ratio, SFR is the specific feeding rate, CSI is the cardiac somatic indice, HSI is the hepatic somatic indice and ISI is the intestinal somatic indice.

Histology

A number of histological changes were found, however different types of anomalies were observed in tissues of individuals from both treatment groups and such changes were observed in low frequency indicating that the fish were in a good condition (Table3). There was no significant dietary effect for any histological change except the frequency of perivascular and peri-biliar duct infiltrations in the liver, where 21% of fish fed the MP diet exhibited this change compared to 3.8% of fish fed the PP diet (Table3). A number of fish from both diets showed a mild fatty change in the liver, but this was not restricted to a particular dietary group (Table3). Equally, some anomalies in heart were observed at a low level in both dietary treatments. For the intestine, no changes were observed in pyloric caeca or mid intestine, whereas for the distal intestine an inflammatory infiltration in the submucosa and lamina propria was observed but at a low level in both diets, with no significant difference in occurrence between dietary groups (Table3).

Table 3: Histopathological observations/changes

Organ	Changes	PP	MP	*P*-value*
Liver	No significant findings	39.0%	33.3%	0.39
	Mild fatty change	38.1%	31.4%	0.312
	Mild to moderate fatty change	8.6%	7.6%	0.801
	Moderate fatty change	8.6%	8.6%	n.d.
	Moderate to severe fatty change	0.0%	1.9%	0.156
	Severe fatty change	0.0%	0.0%	n.d.
	Perivascular infiltration	2.9%	6.7%	0.196
	Peri biliar ducts infiltration	3.8%[a]	21.0%	<0.001
	Necrotic foci w/infiltration	1.0%	0.0%	0.317
	Mild congestion	0.0%	1.0%	0.317
	Focal necrosis	0.0%	2.9%	0.082
	Mild sinusoidal dilatation	0.0%	1.0%	0.317

	Biliar ducts esclerosis	0.0%	2.9%	0.082
	Biliar ducts proliferation	0.0%	0.0%	n.d.
	Foci of inflammatory cells	0.0%	1.0%	0.317
Heart (ventricle)	No significant findings	99.0%	98.1%	0.562
	Severe infiltration and myodegeneration	1.0%	1.0%	n.d.
	Infiltration, no myodegeneration	0.0%	1.0%	0.317
	Presence of melanin granules	0.0%	0.0%	n.d.
Mid-intestine and Pyloric caeca	No significant findings	100%	100%	n.d.
Distal intestine	No significant findings	97.1%	95.2%	0.472
	Infiltration of submucosa and lamina propria, widened of lamina propria within folds (simple and complex), prominent goblet cells/low number of supranuclear vacuoles	2.9%	4.8%	0.472

Histopathological observations/changes. Values (in percentage) represent occurrence where $n = 105$. Significantly different when $P < 0.05$, different superscript letters indicate significant differences within each row-wise comparison.

Transcriptomic Responses

Fish fed the PP diet showed changes in transcriptome in all 3 tissues examined compared to fish fed the marine protein rich diet MP. In total 8,151 oligo features were found to be expressed at different levels as a result of the dietary manipulation. For subsequent analysis only significant ($p < 0.05$) features with a fold difference of greater than 2 were analysed further (1961 mRNAs). The intestine exhibited the greatest differential gene expression with 1,236 genes significantly different in expression level between fish fed the PP diet and MP diet. Of these 615 genes were higher in expression and 621 lower in PP fed fish. In skeletal muscle 505 genes had an expression higher than two fold with 132 higher in PP fed fish and 373 lower and in liver 220 genes were found modified with 161 higher and 59 genes lower in PP fed fish vs MP fed fish (Table4).

Table 4: Genes significantly different in mid intestine of fish fed PP diet

Probe name[1]	ACC[2]	FC±SEM[3]	Identity[4]
Immune function and stress related			
Ssa#CK897125	CK897125	7.6±1.1	B-cell linker
Ssa#CL60Contig3	X70167	5.2±1.2	MHC class II antigen beta chain
Ssa#STIR13675	TC71772	5.0±1.3	CD209f
Ssa#NP9934055	NP9934055	4.8±1.4	T cell receptor alpha
Ssa#STIR21272	TC82967	4.3±1.2	Interferon inducible mx protein
Ssa#S35685629	S35685629	4.3±1.0	TCR-gamma
Ssa#S30239635	S30239635	2.8±1.2	IRF1
Ssa#STIR10385	TC67231	2.5±1.1	Vig-2 protein
Ssa#STIR15805	TC74805	2.4±1.0	cd3 epsilon
Ssa#STIR05606	BT056756	2.4±1.1	β2 microglobulin
Ssa#gi156446662	EF579742	2.0±1.2	MyD88
Ssa#CL81Contig1	BT072778	2.0±1.1	Virus induced TRIM protein
Ssa#S37438814	S37438814	2.0±1.0	CD3 gamma/delta
Ssa#STIR00071_2	DW580947	2.0±1.2	Interferon induced protein 35
Ssa#S18833713	S18833713	−7.5±1.2	Serum lectin 2
Ssa#CK882427	CK882427	−3.2±1.3	Serum amyloid A
Ssa#S35474845	S35474845	−3.0±1.1	HSP β-7
Ssa#S35677496	S35677496	−2.9±1.1	CCR3
Ssa#S31986130	S31986130	−2.7±1.2	Macrophage colony stimulating factor receptor
Ssa#S18892409	S18892409	−2.7±1.1	IgM
Ssa#STIR03818		−2.4±1.3	CXC13
Ssa#STIR19205	TC79827	−2.3±1.1	HSP 70a
Ssa#STIR13083	TC70912	−2.3±1.2	SAMHD1
Ssa#S35663823	S35663823	−2.2±1.2	Complement c1q

Ssa#S35583279	S35583279	−2.1 ± 1.0	IL-17D
Ssa#S35517748	S35517748	−2.0 ± 1.1	HSP 30
Ssa#S35536386	S35536386	−2.0 ± 1.1	HSP β-11
Omy#gi31087931	AY160984	−2.0 ± 1.2	IL-8
Ssa#TC106255	TC106255	−2.0 ± 1.1	Galectin-4
Cell proliferation and apoptosis			
Ssa#S35582566	S35582566	7.1 ± 1.2	Caspase-14
Ssa#CL201Ctg1		2.4 ± 1.0	Caspase-3
Ssa#S35693335	S35693335	2.2 ± 1.0	Caspase 8
Ssa#S32005165	S32005165	−9.4 ± 1.0	DNA topoisomerase 2-alpha
Omy#S34421775	S34421775	−6.6 ± 1.5	Replication factor C subunit 3
Omy#CA346576	CA346576	−4.5 ± 1.2	Tetraspanin-14
Ssa#CA038824	CA038824	−3.9 ± 1.1	Fgfr3 protein
Omy#S34313679	S34313679	−3.6 ± 1.4	Cyclin A1
Omy#S22901990	S22901990	−3.4 ± 1.0	Chromodomain-helicase-DNA-binding protein 7
Ssa#S47728937	S47728937	−3.1 ± 1.1	Tetraspanin-16
Ssa#S35496360	S35496360	−2.9 ± 1.0	DNA-repair protein complementing XP-A cells
Ssa#S35490761_S	S35490761	−2.7 ± 1.1	Transcription factor HES-1
Ssa#STIR38390	TC108636	−2.7 ± 1.3	RNA helicase
Ssa#S35531441	S35531441	−2.4 ± 1.1	Melanoma-derived growth regulatory protein
Ssa#STIR17200	TC76867	−2.4 ± 1.0	Tumor necrosisalpha-induced protein 2
Protein metabolism			
Ssa#S35499913	S35499913	6.5 ± 1.0	Titin
Ssa#S31974046	S31974046	4.2 ± 1.1	Cathepsin C
Ssa#S30293144	S30293144	3.9 ± 1.5	Proteasome subunit alpha type-5
Ssa#STIR05302	BT046757	3.2 ± 1.1	Proteasome subunit beta type-7
Ssa#CL233Ctg1	BT072668	3.0 ± 1.0	Cathepsin Z
Omy#S15290109	S15290109	3.0 ± 1.2	Keratin e1

Ssa#STIR25562	TC89420	3.0±1.1	Troponinskeletal muscle
Ssa#STIR20536	TC81827	2.9±1.1	Cathepsin A
Ssa#STIR24947	TC88495	2.8±1.2	Keratin 18
Ssa#DY714088	DY714088	2.7±1.0	Ribosomal protein S6 kinase b
Ssa#KSS4531	KSS4531	2.6±1.1	Ubiquitin-conjugating enzyme E2
Ssa#STIR08978	TC65424	2.2±1.0	Eukaryotic translation initiation factor 4e
Ssa#DY712052	DY712052	2.0±1.1	Gamma-tubulin complex component 4
Ssa#S30279979	S30279979	2.0±1.2	Eukaryotic translation initiation factor 2 subunit 1
Ssa#S30240560	S30240560	2.0±1.2	β-actin
Ssa#KSS1565	KSS1565	2.0±1.0	Proteasome subunit alpha type-6
Ssa#KSSb2668	KSSb2668	2.0±1.2	Tubulin beta-1 chain
Ssa#STIR39880	TC110797	2.0±1.1	Proteasome subunit alpha type-7
Ssa#S31996856	S31996856	2.0±1.1	E3 ubiquitin-protein ligase RNF128
Ssa#STIR19643	TC80514	−3.9±1.3	Myosin light chain 1–3
Ssa#TC76471	TC76471	−3.6±1.2	Tropomyosin-1 alpha chain
Ssa#AJ425777	AJ425777	−2.7±1.3	60 S ribosomal protein L34
Ssa#STIR11900	TC69277	−2.5±1.1	Myosin ic
Ssa#S35582711	S35582711	−2.3±1.0	Tropomyosin-1 alpha
Ssa#S35505113	S35505113	−2.2±1.1	60 S ribosomal protein L30
Ssa#FC072705	FC072705	−2.1±1.1	40 S ribosomal protein S10
Omy#CX150249	CX150249	−2.1±1.0	60 S ribosomal protein L36
Lipid metabolism			
Ssa#CB509140	CB509140	25.3±1.2	Fatty acid-binding protein
Ssa#S31963508	S31963508	14.5±1.5	Apolipoprotein A-I
Ssa#STIR00045_2	AY170327	5.2±1.0	PUFA elongase 5A

Ssa#STIR00100_2	CK887422	4.7± 1.0	Delta-6 fatty acyl desaturase
Omy#S18154618	S18154618	4.4± 1.2	Glycolipid transfer protein
Ssa#STIR00093_4	AF478472	4.3± 1.1	Delta-5 fatty acyl desaturase
Energy and mitochondrial activity/ kinases			
Ssa#STIR00144_4	TC64612	4.2± 1.0	Glyceraldehyde-3-phosphate dehydrogenase
Ssa#STIR39924	TC110855	2.6± 1.2	Cytochrome P450
Ssa#S32006874	S32006874	2.3± 1.1	Peroxisomal membrane protein 11 C
Ssa#S30263228	S30263228	2.4± 1.1	Succinate dehydrogenase
Ssa#S35661441	S35661441	2.0± 1.0	Fructose-1.6-bisphosphatase 1
Ssa#S32000347	S32000347	−2.5±1.2	Hemoglobin subunit beta-1
Cellular transport			
Ssa#S30284157	S30284157	23.6± 1.2	Solute carrier family 22
Ssa#STIR21606	C83482	9.1± 1.2	Solute carrier family 16
Ssa#DW575876	DW575876	7.2± 1.3	Solute carrier family 39
Ssa#STIR19539	TC80341	4.6± 1.0	Solute carrier family 31
Ssa#STIR19539	S35536215	2.8± 1.0	Solute carrier family 25
Ssa#S35454393	S35454393	2.1± 1.1	Solute carrier family 27
Omy#TC172518	TC172518	−2.2± 1.1	Beta globin
Ssa#CK991045	CK991045	−2.0± 1.0	Hemoglobin subunit alpha

There were no genes expressed at different levels in common between all the tissues examined indicating there was not a systematic response to the diets. However there was a co regulation of features between the mid intestine and skeletal muscle with 9 genes higher in both tissues of PP fed fish vs MP fed fish and 10 lower in both. Similarly in the liver and intestine 6 genes were found to be significantly different in common, with 5 genes higher in expression and 1 gene lower in the PP fed fish. To further investigate the biological significance of the differently expressed gene sets we used gene ontology analysis to help identify biological processes that were significantly different in tissues following the feeding trial.

Annotation of the microarray allowed 77% of the features to be allocated to a functional protein and 55% of these proteins were assigned a gene ontology (GO) identifier. This allowed statistical analysis for enrichment for GO biological processes to help interpret the changes in the transcriptome following the feeding trial (Figure1) to gain a holistic view of which biological processes were significantly modified following feeding the PP diet.

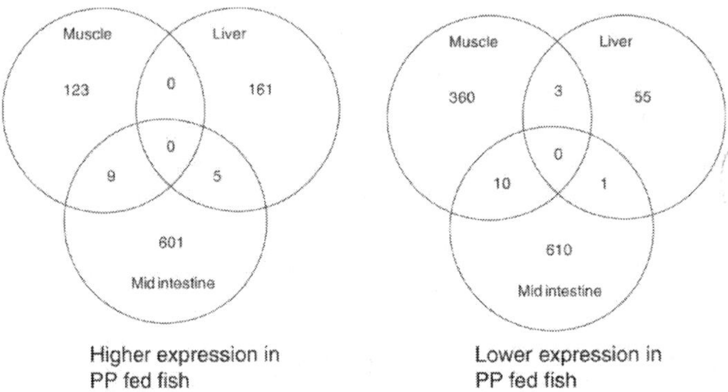

Figure 1: Venn diagram showing numbers of genes identified as expressed at different levels by microarray analysis. Summary of numbers of genes higher and lower expressed. The genes presented are all significantly different in expression ($p < 0.05$) with >2 fold change in expression.

Genes expressed at Different Levels in Mid Intestine

The intestine is the major tissue to come into direct contact with any feed components and for this reason may be very sensitive to dietary changes. This is shown here by the intestine having the greatest transcriptional response in terms of the number of genes expressed at different levels and the magnitude of different expression of these genes. Several biological processes were significantly different in the PP diet fed fish, related to intestinal functions including immune and stress related processes, protein metabolism, energy and mitochondrial activity, lipid metabolism and transport (Table4).

Immune and Stress Response

Genes encoding proteins related to both innate and acquired immune function were expressed at different levels by the dietary treatments.

For fish fed PP innate immune parameters were found to be both higher and lower in expression indicating a dynamic response. Modulators of the immune response were higher in PP fed fish, such as MyD88, a key transcription factor associated with induction of an inflammatory response and P105 subunit of an inhibitory protein that sequesters NF- B in the cytoplasm. Another transcription factor, interferon regulatory factor (IRF) 1 was also higher, and potentially increased expression of interferon responsive genes including Mx, Vig-2, interferon induced protein 35 and virus induced TRIM protein. Genes encoding cytokine receptors (IL-17R) and regulators of cytokine function including the suppressor of cytokine signalling (SOCS)-7) were also higher. A number of genes related to innate immune responses were also lower, and included IL-17D, MCSFR and chemokines such as CCL3, CXC13 and IL-8 (CXCL8). Other innate immune serum proteins were also lower in expression including serum amyloid A, and a number of lectins. Genes related to proteins in the acquired immune response were mostly higher in expression in fish fed the PP diet compared to the MP diet. For example, T cell receptor chains (α and γ) and their signalling subunits CD3 epsilon and gamma/delta were higher as were genes involved with antigen presentation such as β2 microglobulin and MHC class II). Related to antibody production by B cells, a B-cell linker protein that regulates B-cell function and development was higher in PP fed fish as was a B cell enhancing factor.

Regarding stress related genes, PP fed fish showed an elevated expression of glutathione-S-transferase and a thioredoxin interacting protein, involved in reducing oxidative stress, while heat shock proteins (HSP) β-7, β-11, 30 and 70a were all suppressed.

Cell Proliferation and Apoptosis

The intestinal transcription of genes encoding proteins related to cell proliferation (tumor necrosis alpha-induced protein 2, tetraspanins-14 and −16 and melanoma-derived growth regulatory protein) and the cell cycle (DNA topoisomerase 2-alpha, replication factor C subunit 3, FGFR3 protein, cyclin A1) were found to be higher in PP fed fish; whilst transcription of caspases 3, 8 and 14,which are directly involved in apoptosis, were found to be lower in expression relative to fish fed MP.

Protein Metabolism

Feeding the PP diet the expression of genes involved in protein metabolism was significantly different compared to MP. For example, protein synthesis related genes including elongation factor, translation initiation factors 2 and 4e and ribosomal protein S6 kinase b were higher, while a number of 60 S ribosomal protein encoding genes were lower relative to MP fed fish. Protein degradation encoding genes were more highly expressed as shown by lysosomal proteases, cathepsins A, C, and Z, ubiquitin proteasome route (UbP) of proteolysis, and proteasome subunits α5, α6, α7 and β7. Genes involved in targeting of proteins for degradation in the UbP were also higher including ubiquitin conjugating enzyme E2 and ubiquitin E3 ligases. Genes encoding the structural proteins troponin, tubulin, titin, actin and keratin were all found to have a higher expression level whereas a number of myosin mRNAs were lower in expression.

Lipid Metabolism

A higher transcript level of genes involved in lipid transport and metabolism was observed, with higher transcription of fatty acid-binding protein, apolipoprotein A-I, apolipoprotein B and delta-6 fatty acyl desaturase, delta-5 fatty acyl desaturase, PUFA elongase 5A and glycolipid transfer protein, respectively.

Energy and Mitochondrial Activity/Kinases

There was an induction of many genes related to energy metabolism in pathways such as glycolysis (glyceraldehyde-3 phosphate dehydrogenase), the tricarboxylic acid cycle (succinate dehydrogenase) and gluconeogenesis (fructose 1-6- bisphosphate) which were expressed at a higher level in PP fed fish relative to the MP fed fish.

Cellular Transport

Transport of solutes into and from a cell is an energy demanding process. Genes involved in membrane transport were higher in expression in fish fed the PP diet, and included a number of solute carrier family members such as SCF22 (organic cation transporter), SCF 27 (fatty acid transporter), SCF 25 (mitochondrialphosphate carrier), SCF 39 (zinc transporter), SCF

16 (monocarboxylic acid transporter), and SCF 31 (copper transporter). Two genes encoding haemoglobin subunits (hemoglobin subunit alpha-4 and hemoglobin subunit beta-1) and beta globin, however, were found to have a lower relative expression level.

Genes expressed at Different Levels in Liver

Several biological processes were significantly different in liver that can be related to immune parameters, stress responses, and protein and lipid metabolism, with key genes shown in Table5.

Table 5: Genes significantly different in liver of fish PP diet

Probe name[1]	ACC[2]	FC±SEM[3]	Identity[4]
Immune function and stress response			
Ssa#S35601811	S35601811	2.7±1.0	MPV17 protein
Ssa#gi156446662	EF579742	2.4±1.1	MyD88
Omy#S18150823	S18150823	2.2±1.1	Amine oxidase
Ssa#STIR13675	TC71772	2.0±1.0	CD209f
Ssa#CK874360	CK874360	2.0±1.1	Heat shock 70
Ssa#STIR36546	TC105929	−2.8±1.1	C-type MBL-2 protein
Ssa#NP797925	NP797925	−2.4±1.1	MHC class I alpha 2
Ssa#STIR04816	NM_001140849	−2.4±1.2	Hepcidin
Ssa#NP9934311	NP9934311	−2.2±1.1	T cell receptor alpha
Ssa#S48440415	S48440415	−2.2±1.3	SMAD3
Ssa#S35663823	S35663823	−2.1±1.1	Complement c1q
Cell proliferation and apoptosis			
Ssa#CA041082	CA041082	2.6±1.2	Transforming growth factor beta receptor
Ssa#S35486979	S35486979	2.5±1.0	Cell death activator CIDE-3
Ssa#S35559076	S35559076	2.3±1.0	Angiopoietin-related protein 4
Ssa#S30276711	S30276711	2.3±1.1	Activin receptor type-1B

Ssa#S35582821	S35582821	2.3 ± 1.0	Serine protease HTRA1
Ssa#STIR31305	TC98147	2.2 ± 1.1	Annexin A3
Omy#S18150823	S18150823	2.2 ± 1.0	Amine oxidase
Protein metabolism			
Ssa#DW574268	DW574268	2.4 ± 1.1	Ribosomal protein S18
Ssa#STIR03071	BT048999	2.3 ± 1.1	Proteasome beta type 3
Ssa#KSSb2684	KSSb2684	2.2 ± 1.2	Peptidyl-prolyl cis-trans isomerase
Ssa#STIR16404	TC75662	2.2 ± 1.1	Calpain
Ssa#STIR04151	NM_001141015	2.1 ± 1.0	Ribosomal protein l39

Immune and Stress Response

Genes encoding proteins involved in innate immunity such as complement c1q, C-type MBL-2 and a dendritic cell specific lectin CD209f which binds mannose carbohydrate molecules, and hepcidin a major liver associated antimicrobial peptide were lower in PP fed fish. A limited number of genes related to adaptive immunity were also found to be expressed at different levels, with MHC class I and SMAD 3 lower in expression level, whereas MyD88 was higher in fish fed the PP diet relative to MP fed fish. Moreover, genes involved in the oxidative response (MPV17 protein, amine oxidase and heat shock cognate 70 kDa protein) were also higher in fish fed the PP diet.

Cell Proliferation and Apoptosis

The PP diet stimulated/modified hepatic cell turnover as indicated by higher expression of serine protease HTRA1 and annexin A3 and the stimulation of apoptotic processes through genes involved in TGF beta pathways (cell death activator CIDE-3 and angiopoietin-related protein 4).

Protein Metabolism

Hepatic protein metabolism was clearly stimulated in PP fed fish since all genes significantly different in expression were higher in these fish, including ribosomal protein S18a, ribosomal protein l39 and peptidyl-prolyl cis-trans isomerise. Similarly genes involved in protein degradation were more highly expressed as seen with the non-lysosomal protein calpain and the proteasome subunit β3.

Genes expressed at Different Levels in Skeletal Muscle

The GO analysis showed a number of processes to be significantly different in skeletal muscle that can be related to protein metabolism, immune function and energy metabolism (Table6).

Table 6: Genes significantly different in skeletal muscle of fish PP diet

Probe name[1]	ACC[2]	FC±SEM[3]	Identity[4]
Immune function			
Ssa#STIR12634	TC70300	5.0±1.2	Vig-2
Ssa#STIR00067_2	U66477	3.2±1.3	Interferon inducible Mx protein
Ssa#CK894557	CK894557	−11.2±1.4	MHC class Ib antigen
Ssa#STIR00132_2	NM_001140254	−7.4±1.3	Tumor necrosis alpha-induced protein 2
Omy#NP565601	NP565601	−6.4±1.4	T-cell receptor beta
Ssa#S31982089	S31982089	−5.4±1.1	Platelet-activating factor receptor
Ssa#STIR12498	TC70105	−4.6±1.2	γ-ip (CXCL10)
Ssa#S48440415	S48440415	−3.4±1.1	SMAD3
Omy#S15331473	S15331473	−3.2±1.2	Interferon-inducible protein Gig2-like
Ssa#S35544087	S35544087	−3.1±1.3	B-cell CLL/lymphoma 7 protein family member B
Ssa#STIR13083	TC70912	−3.6±1.0	SAMHD1
Omy#S18153399	S18153399	−2.4±1.1	CD80
Ssa#KSS5035	KSS5035	−2.2±1.2	Interferon regulatory factor 2-binding protein
Ssa#STIR00071_3	DW580947	−2.3±1.1	Interferon -induced protein 35
Ssa#STIR08451	TC64790	−2.2±1.2	Heat shock protein 47
Ssa#S35480903	S35480903	−2.0±1.2	Heat shock protein 30
Ssa#STIR29454	TC95297	−2.1±1.1	Beta defensin 1

Cell proliferation and apoptosis			
Ssa#S35602638	S35602638	3.7±1.4	Tripartite motif 39
Ssa#TC110067	TC110067	3.5±1.5	Cell death inducing protein
Ssa#STIR08668	TC65039	3.1±1.1	Bh3 interacting domain death agonist
Ssa#STIR12507	TC70118	3.0±1.3	Syndecan 4
Ssa#STIR02208	BT049868	−38.6±1.2	Caspase 14
Ssa#S35559333	S35559333	−4.1±1.3	Ankyrin repeat domain-containing protein 54
Omy#TC151190	TC151190	−2.9±1.2	Nucleostemin
Ssa#S35550715	S35550715	−2.8±1.2	Growth hormone-inducible transmembrane protein
Ssa#S35585784	S35585784	−2.6±1.1	Caspase-8
Ssa#KSS5035	KSS5035	−2.4±1.2	Interferon regulatory factor 2-binding protein 2-B
Protein metabolism			
Ssa#S35682089	S35682089	4.2±1.3	Ubiquitin-like protein 1
Ssa#STIR17445	TC77227	2.6±1.3	Acta1 protein
Ssa#STIR21465	TC83266	2.5±1.2	Actin-binding
Ssa#STIR03710	BT048358	2.0±1.2	Proteasome subunit beta type-9
Ssa#STIR00115_4	BT045917	2.0±1.2	Tropomyosin-1 alpha chain
Ssa#S32008331	S32008331	−3.6±1.7	40 S ribosomal protein S16
Ssa#CL224Ctg1	NM_001140473	−2.8±1.5	Receptor-interacting serine/threonine-protein kinase 4
Omy#CX150460	CX150460	−2.7±1.1	60 S ribosomal protein L27
Omy#S15320960	S15320960	−2.6±1.1	Cathepsin D
Ssa#TC111443	TC111443	−2.5±1.1	Serine/threonine-protein kinase 35
Ssa#FC072705	FC072705	−2.3±1.0	40 S ribosomal protein S10
Ssa#S35583213	S35583213	−2.1±1.2	Serine/threonine-protein kinase PINK1
Ssa#S35591236	S35591236	−2.1±1.1	Histidine triad nucleotide-binding protein 3
Energy metabolism			
Ssa#STIR39924	TC110855	2.5±1.1	Cytochrome P450

Ssa#S32006874	S32006874	2.2±1.1	Peroxisomal membrane protein 11 C
Ssa#KSS5038	KSS5038	−4.0±1.3	Phosphoglycerate kinase
Ssa#STIR19484	TC80264	−3.1±1.3	Cytochrome c oxidase polypeptide viii
Ssa#CL200Ctg1	BT059338	−2.8±1.3	L-lactate dehydrogenase B chain
Ssa#S30282925	S30282925	−2.7±1.1	Peroxisomal 3.2-trans-enoyl-CoA isomerase
Omy#CX150319	CX150319	−2.4±1.2	Cytochrome c oxidase subunit II
Ssa#STIR13627	TC71700	−2.1±1.2	Cox18 cytochrome c oxidase
Ssa#EL698167	EL698167	−2.0±1.1	Malate dehydrogenase 1

List of selected mRNAs found to be higher or slower in expression following the feeding trial in mid intestine (a), liver (b) or skeletal muscle (c) and grouped according to functional classes (shown in bold). The selection was based on manual assignment of function and genes with greatest fold differences in expression are presented, the genes that are lower in expression are denoted by (−) value. The genes shown were significant at $p<0.05$ following t- tests and greater than 2-fold change. [1]Indicates the unique code for the feature on the microarray, [2]Accession number of the cDNA sequence. [3]Fold-change for genes higher or lower expressed in fish fed PP relative to fish fed MP diet. [4]Identity of the probe target as determined by BLASTX and BLASTN searches.

Immune and Stress Response

Regarding the innate immune status, PP fed fish showed lower expression level of inflammatory related genes such as platelet-activating factor receptor, SAMHD1, tumor necrosis alpha-induced protein 2, and the antimicrobial peptide beta defensin 1. Likewise, fish fed the PP diet showed a suppression of genes involved in the adaptive immune response, i.e. TCR β, MHC class I, SMAD 3, CD80 (a T cell activator). Interferon γ induced proteins γ-ip (CXCL10), Gig2-like and both interferon-induced protein 35 and IRF2-binding protein were also lower in fish fed the PP diet. In contrast, Mx and Vig-2 which are responsive to type I interferon were higher expressed in PP compared to MP. Lastly, the stress induced genes HSP 30 and 47 were lower in expression in PP fed fish relative to MP fed fish.

Cell Proliferation and Apoptosis

Genes involved in cell proliferation were lower expressed in the PP fed fish, including ankyrin repeat domain-containing protein 54, nucleostemin and growth hormone-inducible transmembrane protein. Several genes involved in apoptosis were found to be higher such as cell death inducing protein and bh3 interacting domain death agonist although in contrast whereas both caspase 8 and 14 showed lower transcript levels in the PP fed fish.

Protein Metabolism

Genes involved in protein synthesis such as several 60 S and 40 S ribosomal proteins were lower in expression in skeletal muscle of PP fed fish. A number of enzyme genes that modify essential amino acids (eg serine/threonine-protein kinase PINK1, serine/threonine-protein kinase 35, receptor-interacting serine/threonine-protein kinase 4 amongst others) were also lower in expression. Likewise, a major lysosomal peptidase, cathepsin D was lower in expression in PP fed fish, however, two components of the UbP were higher expressed in PP (ubiquitin-like protein precursor and proteasome subunit beta type-9).

Genes encoding proteins involved in muscle structure and physiology (acta1 protein, actin-binding homolog 1a and tropomyosin-1 alpha chain) were also more highly expressed in skeletal muscle of fish fed the PP diet.

Energy Metabolism

Genes related to energy metabolism were transcriptionally modified with peroxisomal membrane protein 11 C and cytochrome P450 being higher in expression whereas cytochrome c oxidases (COX2, 8 and 18) were found to be lower in PP fed fish compared to MP fed fish. In addition genes involved in glycolysis (phosphoglycerate kinase 1 and L-lactate dehydrogenase B chain) and gluconeogenesis (malate dehydrogenase 1) were found to be expressed lower in PP fed fish.

Co-regulated Genes

Only a small number of genes were co regulated between tissues following the feeding trial. No genes were significantly higher or lower regulated in all the tissues in this experiment. The greatest co regulation was observed between mid intestine and skeletal muscle, with 9 genes significantly higher and 10 genes significantly lower in PP compared to MP. Genes that were more highly expressed in mid intestine and liver of fish fed the PP diet compared to the MP diet play a role in energy metabolism, such as cytochrome P450, a membrane-associated protein located in the inner membrane of mitochondria and peroxisomal membrane protein 11 C that induces peroxisomal proliferation proteins involved in the pentose phosphate pathway. Genes that showed lower expression in mid intestine and liver have a role in protein synthesis (40 S ribosomal protein S10) and in the immune response, such as SAMHD1 a nuclease involved in innate immune responses by acting as a negative regulator of the cell-intrinsic antiviral response. Five genes were higher expressed in both intestine and liver of fish fed the PP diet, including immune related genes, such as CD209f and MyD88 and only 1 gene (complement c1q) showed lower expression in both tissues. Skeletal muscle and liver did not have any higher expressed gene in common and only 3 genes (membrane-bound O-acyltransferase domain-containing protein 7, SMAD 3 and the non annotated gene (**NP_998306**) were significantly lower between these two tissues (Figure2; full list of genes expressed at different levels is given in supplementary Table1).

a

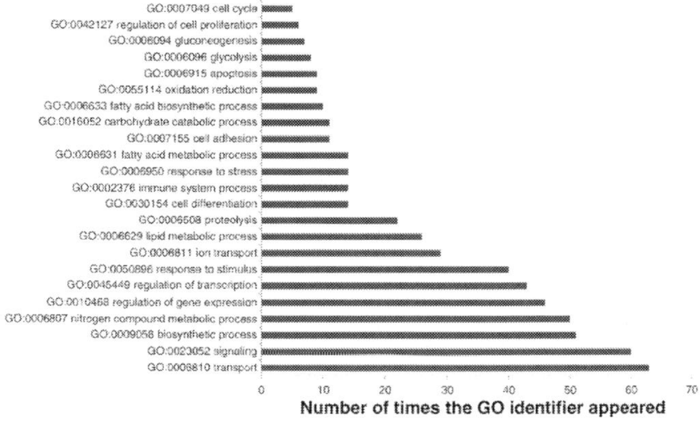

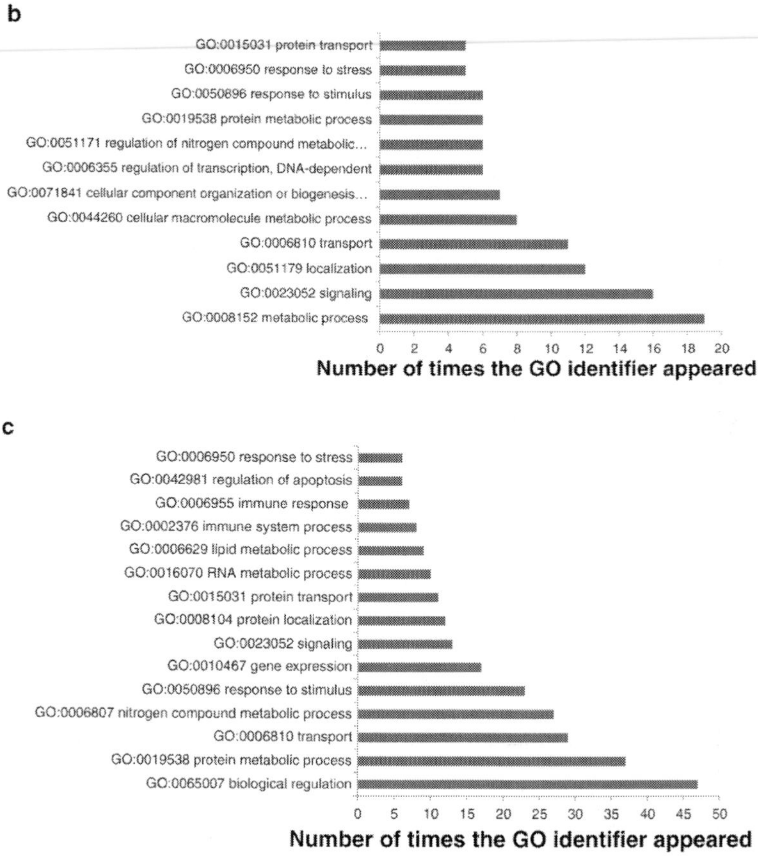

Figure 2: Gene ontology (biological processes) found to be significantly enriched following the feeding experiment in mid intestine (a), liver (b) and skeletal muscle (c). Only GO categories for which >3 genes were represented are included.

Confirmation of Expression by Real Time PCR

Real time PCR analysis was performed on a number of genes for each tissue to confirm microarray data (Table7). The real time PCR expression data was normalized with HPRT1 as the expression of this housekeeping gene was not found modulated by microarray analysis. For all genes the expression pattern showed the same trend between microarray and real time PCR analysis (Figure3).

Table 7: Primers used for real time PCR for expression analysis. Acc is accession number from NCBI

Gene name	Primer name	Primer sequence(5-prime to 3-prime)	Acc	Product size	Ann T	Tissue
Fabp2	FatF	GCTCTGTACTAGCTCTCCTCCC	CB509140	156 bp	55 °C	Mid intestine
	FatR	GGCGTACAGTTTGACTATGCAC				
Caspase-14	Cas14F	CGATTATACACCCGGACTATGG	S35582566	155p	55 °C	Mid intestine
	Cas14R	CCTATCAAGTGTGAATCCATGC				
TCR alpha	TCRaF	GGAAGACTCTGCTCTGTACCAC	U50991	147 bp	55 °C	Mid intestine
	TCRaR	GCTGTGGTATTTCTTGACTTC				
IgM heavy chain	IgMF	GCTTATAGCCATAGTACTACTG	S18892409	169 bp	55 °C	Mid intestine
	IgMR	GCATAGCTGCCCCATATCGC				
Tpm1	Trop1F	CGAAGATGAGAGAGATAAAGT-GC	TC76471	134 bp	55 °C	Mid intestine
	Trop1R	CCTCCTCAACCAGCTGGATACG				
RFC3	ReplF	GCTGACTCACTGCATTCCTCCTG	S34421775	163 bp	55 °C	Mid intestine
	ReplR	GAAGGCCTCTAGGTGGTA-AATGG				
HTRA1	HTRA1F	GGTCATCTTCATACAGAGAGG	S35582821	152 bp	55 °C	Liver
	HTRA1R	GCTTAGAGAATACCATCTTGC				

TGF beta receptor	TGBF	CCACAAGAAGCCAGCTGTCAG	gi|209735249	135 bp	55 °C	Liver
	TGFBR	CTAGCCAGGTATCTCTATCATGG				
MRPS18A	28 SF	CCATTGATTCAGTGAAGCCCATC	DW574268	158 bp	55 °C	Liver
	28 S R	CCTGCTGTGAGTTGACATGC				
Timd2	TcellF	CCATGGACAACCACACACACTG	CA368982	141 bp	55 °C	Liver
	TcellR	CCAGTAGAATGGACACCAG-GATC				
Hepcidin I	HepF	GCTTCTGCTGCAAATTCTGAGG	gi|209736931	157 bp	55 °C	Liver
	HepR	GTACAAGATTGAGGTTGTGCAG				
TCR alpha	TCRaF	GGAAGACTCTGCTCTGTACCAC	U50991	147 bp	55 °C	Liver
	TCRaR	GCTGTGGTATTTCTTGACTTC				
CDIP	CellF	CCATGTCTGAGACCTACTCTATG	TC110067	243 bp	55 °C	Skeletal muscle
	CellR	GATAGTCACTTGATGTCCAGTG				
acta1 protein	ActaF	CCTGTAAACTGTGAATGCGTC	TC77227	156 bp	55 °C	Skeletal muscle
	ActaR	CCAAAGTTTTATATCAGCTGC				
TGF beta 1	TGF1bF	GCTCGGAGTGTGAGA-CAGAACTG	S15319964	187 bp	55 °C	Skeletal muscle
	TGF1bR	CACTTGACGCAACAGAAA-CACTCC				
RT1-CE5	MHC1bF	GGAAAGATCTCCTGAAGACTT-GAG	CK894557	101 bp	55 °C	Skeletal muscle

	MHC1bR	CGTTTATGAGAAGTTCAGC				
60 S rib prot	60SF	GCTTCTTACCATGGTTCTTCAG	DR695852	140 bp	55 °C	Skeletal muscle
	60SR	GGTCAAGATCCCATCCACCATC				
Heat shock protein 30	HeatF	CCATCCAACCAGTCTCCTA-CAAGC	EG804126	303 bp	55 °C	Skeletal muscle
	HeatR	CCTCCTCAGCAGATAATGGATTC				
Hprt	HprtF	CCGCCTCAAGAGCTACTGTAAT	EG866745	255 bp	55 °C	All tis-sues
	HprtR	GTCTGGAACCTCAAACCCTATG				

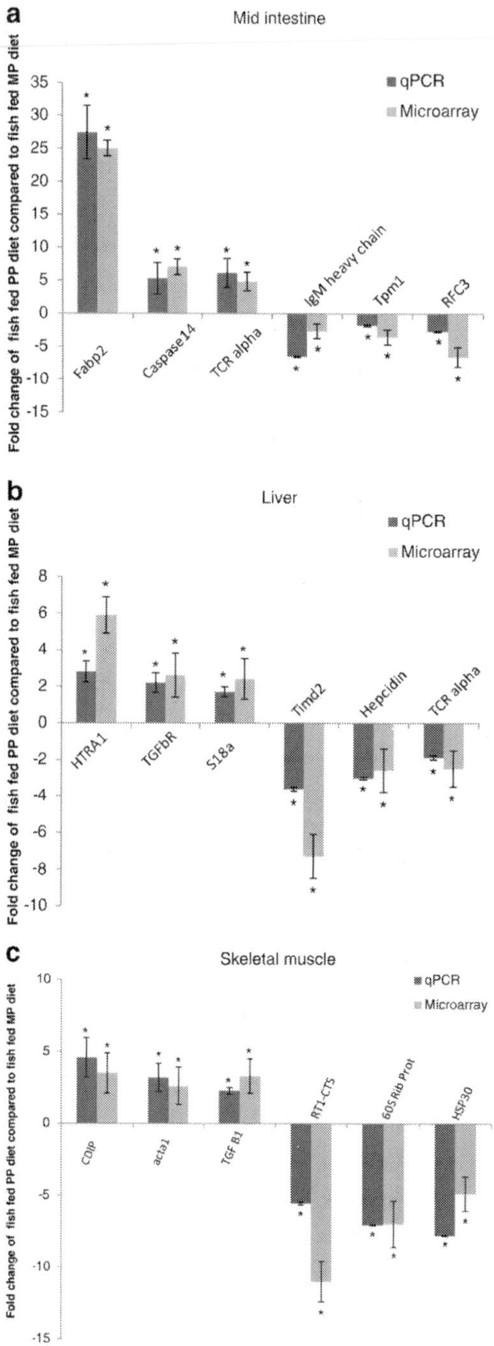

Figure 3: Quantitative real-time PCR confirmation of genes expressed at different levels in mid intestine (a), liver (b) and skeletal muscle (c) of fish fed

the PP diet compared to fish fed the MP diet for 6 genes identified by microarray analysis. Bars represent mean ± standard error of five fish, asterisks indicate significant ($p < 0.05$) differences. The genes chosen for the intestine were: Fatty acid-binding protein (Fabp2), Caspase-14 precursor, TCR alpha, IgM heavy chain, Tropomyosin-1 alpha chain (Tpm1) and Replication factor C subunit 3 (RFC3). The genes selected for the liver were: Serine protease HTRA1, TGF beta receptor, 28 S Ribosomal protein S18a (MRPS18A), T-cell immunoglobulin and mucin domain (Timd2), hepcidin I and TRC alpha. The gene chosen for the skeletal muscle were: cell death inducing protein (CDIP), acta1 protein, Transforming growth factor beta-1 (TGF beta 1), MHC class Ib antigen (RT1-CE5), 60Sribosomal protein L6 (60 S rib prot) and Heat shock protein 30.

DISCUSSION

With the continual expansion of aquaculture of piscivorous fish there is a global requirement to reduce the dependence on wild caught fish for the generation of fishmeal and fish oil. There are a number of studies that have explored the transcriptional response to plant and vegetable oil replacements in fish diets, often when 100% replacement experimental diets are used results in reduced growth[24, 25]. When both proteins and oils were replaced[24] in rainbow trout microarray analysis of liver indicated processes such as protein metabolism and cell cycle being altered, however the reasons for these changes could either have been combined effect of lowered essential fatty acids or changes in amino acid profile. There is also growing body of work demonstrating the genotype diet interaction in rainbow trout[25, 26, 27] and Atlantic salmon[28], with the latter concluding transcription of key metabolic regulators that respond to plant based feeds depend on the fish's genotype. Commercial salmon feed formulations partially replace fish meal with plant derived proteins, with varying effects on fish physiology and performance including fish growth and food conversion[24, 29]. This paper examined the physiological and performance effects with transcriptional responses in mid intestine, liver and skeletal muscle in salmon to a post smolt feed containing high levels of plant derived proteins with low fishmeal content compared to a low plant protein-high fishmeal diet. The tissues examined play different physiological roles in the fish all of which may be affected by dietary changes as reflected in the different response profiles seen in these tissues. In common with a number of other transcriptomic studies on nutritional effects (where extreme dietary manipulations are avoided) the impact of the different diets appears quite subtle suggesting limited changes to physiological and metabolic pathways[30]. Although, in general, the scale of gene response is quite low, distinct changes in all tissues examined do indicate that there are responses occurring as a result of the dietary components.

When interpreting the findings it is important to take a comparative holistic approach whereby apparent down regulation of genes in fish fed one diet may in fact be the result of up regulation in fish fed the other dietary treatment and vice versa. In this paper the expression is analysed relative to the marine profile (MP) diet.

Histological Changes

The occurrence of histological changes was very similar for fish fed either diet, with low frequencies observed for the majority of parameters measured. The main histological observation was a mild fatty change in liver, which is considered normal under intensive rearing conditions. This anomaly appeared to have no clinical effect, and was not prevalent for either dietary group. Although there was a higher incidence of peri-biliar duct infiltration in liver of fish fed the MP diet no other histological changes were observed to indicate an inflammatory response. This reduced infiltration in the liver by the biliary ducts may be related to the altered immune gene expression in the liver. In mammals viral infections can result in biliary atresia[31] and specifically can be related to immune related injury to bile ducts following infiltration of $CD4^+$ T and production of interferon γ[32]. At this stage we are unable to tell if the reduction in the structures is a direct result of the nutritional components or due to the altered hepatic immune gene expression in between the two diets. The distal intestine is often the target organ for anti-nutriental factors, in particular those present in SBM and pea protein concentrate[22], inducing histological changes including shortening of simple and complex mucosal folds with widening of central stroma, inflammatory cell infiltration in the submucosa and lamina propria with a mixed leukocyte population[33, 34]. In this study both dietary groups presented fish with distal intestine anomalies but with low frequencies, and no significant difference between diets. These results together with those reported by Sanden *et al.* (2005)[35] indicate soybean products, including the alcohol soluble fraction, may be used within formulation constraints without inflicting gross histopathological anomalies in the intestine of Atlantic salmon.

Transcriptome Changes in the Mid Intestine

The fish intestine has multiple functions which will be the first to respond to changes in nutritional intake; particularly digestion and absorption of nutrients and immune responsiveness to ingested pathogens, antigens and new antigens generated by the gut flora via gut associated lymphoid tissue[36]. The intestine contains three distinct regions: the proximal intestine containing pyloric caeca, the mid and distal intestine. Nutrient absorption occurs in all regions via enterocytes, by passive and facilitated diffusion and active transport[37], with the highest rates of uptake in the proximal section[37, 38]. The intestine of piscivorous fish can be particularly sensitive to plant derived ANFs and non-starch polysaccharides (NSP) in feed, resulting in altered permeability in trout feed 44% 44% SBM[7, 39] and enteritis in the distal intestine of salmon given a high dietary inclusion of SBM[33]. The inflammation/enteritis may be similar to a hypersensitivity reaction[33, 40]. Often these effects are temporary and quickly disappear when the intestinal tract is no longer exposed to the ANFs[11, 33, 41].

Processing of plant products for fish feed is under continual improvement and some current plant derived protein concentrates have very low contaminating factors or botanical impurities. In addition knowledge of ANF containing plant feed materials has increased to the point where commercial feed formulations permit plant derived proteins at acceptable inclusion levels where no negative health effects or impacts on growth and performance occur. This was confirmed in the current study by the histology assessment where no gross morphological changes associated with plant ANFs were observed. In addition there was no difference in growth or feed utilisation efficiency during the feeding period, where a doubling of weight was achieved. However there were more subtle changes to biological processes that were not apparent during the classical physiological evaluation but were detected by global transcriptomic analysis. In this trial the mid intestine showed the greatest transcriptome response of the tissues studied, reflecting the sensitivity of the intestinal cells to dietary factors. Processes modified in the intestine were related to immune parameters, cell proliferation, apoptosis, protein metabolism, energy metabolism, transport and lipid metabolism.

Fish fed the PP diet showed higher expression of genes involved in inflammation suggesting a possible dysfunction in immune regulation. Our findings support previous studies on gut intraepithelial and systemic T cells in fish which showed rainbow trout IELs are rich in T cells[10, 42]. Additionally the expression of TSC22D3, a regulator of T cell receptor mediated cell death, was found higher in PP, this protein may be induced by glucocorticoids[43, 44] activated by components in the PP diet. Together these results support previous reports that trace levels of substances with allergenic properties may cause expression of genes indicative of a hypersensitivity reaction in the intestine[45].

Interestingly genes involved in the inflammatory response were both higher and lower expressed in PP compared to MP. In particular genes involved in NF-κ B pathway were induced such as the signalling adaptor molecule MyD88 and the inhibitory proteins of the IkB family, NF-kB1 p105 which sequesters NFκ B in an inactive form in the cytoplasm[46]; inhibition of this pathway results in the production of proinflammatory cytokines[47]. MyD88 is also part of the signalling pathway that induces type I IFNs through the interaction of the MyD88–TRAF6–IRF7 complex[48]. IRF1[49] had higher expression in fish fed the PP diet, this transcription factor may have activated interferon responsive genes in PP fed fish including Mx[50], virus induced gene (vig) -2[51] and a virus induced TRIM protein[52]. PP fed fish also had higher expression of (SOCS)-7, which functions to reduce inflammation[53], potentially counteracting the expression of genes related to the inflammatory response in the intestine. Other genes involved in the innate immune response were expressed at lower level in fish fed PP. IL-8 is a chemokine that attracts neutrophils to a site of inflammation[54], whereas IL-17D coordinates the clearance of extracellular bacteria and contributes to the pathology of many autoimmune and allergic conditions in Atlantic salmon[55]. The lower expression of these genes in fish fed the PP diet may indicate there was not a proinflammatory response to the PP diet compared to MP diet.

Several antioxidant genes were expressed higher in PP fed fish mid intestine indicating protection against oxidative damage. The free radicals could be either endogenously produced by immune cells or present in the diet. Alternatively a potential lower concentration of antioxidants in the

PP feed, due to the lower fishmeal content, may require the antioxidant system within the fish to be increased accordingly. The overall low expression of heat shock protein mRNAs, suggests a limited stress response in the intestine. Additionally, two genes directly involved in apoptosis process (caspase-3 and14) were expressed at higher level in PP potentially indicating increased apoptotic activity of mid intestinal cells in these fish compared to MP fed fish.

The intestine has an extremely high rate of cellular turnover and hence generally high levels of protein synthesis and protein degradation. Protein metabolism genes relating to both synthesis and degradation were found generally to be higher in PP fed fish suggesting an increase in intestinal protein turnover. Genes related to both transcription and translations were at a higher level such as translation initiation factors, elongation factors and the ribosomal protein S6 kinase. Interestingly, a number of mRNAs encoding ribosomal proteins were at a lower expression in PP. This may relate to the stability of the mRNAs or multiple use of the ribosomal subunits during translation. In parallel to general increase in synthesis genes related to protein degradation were also at a higher level in PP such as cathepsins and components of the ubiqutin proteasome pathway[56, 57]. The higher protein turnover is likely to be energy demanding and this is related to an increase in genes encoding proteins involved in oxidative energy metabolism. Together these changes in expression suggest modulation in control of protein turnover in the intestine with components of both synthesis and degradation being altered which may reflect an increased activity of the intestine in fish fed the PP compared to the MP diet.

Cellular membrane transport related genes were higher expressed in PP, which could suggest that salmonids are able to adaptively modulate the densities of transporters to match changes in diet composition. Lipid metabolism and transport were also affected even though the PP diet contained the same fish oil content as the MP diet. mRNAs encoding two apolipoproteins were higher in expression in PP fed fish reflecting a greater mobilization and transport of cholesterol and fatty acids in the intestine, possibly an adaptive response to the lower dietary cholesterol content in PP compared to MP diet. Fatty acid metabolism genes were higher expressed in the PP diet including both FAD5 and 6, a PUFA elongase and other genes related to cholesterol metabolism. FADs are critical enzymes in the biosynthesis of long-chain highly unsaturated fatty acids (HUFA) from shorter chain PUFA[58, 59]. These lipid metabolism differences between PP and MP are of interest as the intestine is often over looked regarding these processes and cholesterol, even if the essential fatty acids are present at required levels, other factors including cholesterol may change, revealing the complex nature of the early digestion and modifications of nutrients in the mid intestine.

Transcriptome Changes in the Liver

The liver receives nutrients and compounds from the intestine and needs to respond to any substances that may have detrimental effects on the fish.

Transcriptome and proteomic studies on salmonids show nutritionally related modifications in both liver mRNAs and proteins due to feeding status[60, 61] and diet composition[9, 62, 63, 64].

The PP fed fish showed a significant difference to fish fed MP for genes related to immune function with a lower expression of innate factors such as lectins and hepcidin. Acquired immune system components were also found at lower level including T cell receptors, MHC I and II and components of the TGF-β pathway. TGF-β has an important role in the maintenance of T-cell[65] and B-cell homeostasis[66] by regulating cell proliferation process and apoptosis in these cells. This result, with the higher expression of genes involved in cell death such as CIDE-3 and angiopoietin-related protein 4, indicates that apoptosis may be a mechanism induced by the PP diet salmon liver. This is not surprising as apoptosis plays a central role in the differentiation and maintenance of the liver[67]. A balancing effect on the apoptotic TGF-β pathway is seen in the induction of several genes encoding proteins related to cell proliferation (such as HTRA1 serine protease and annexin A3). In particular, HTRA1 serine protease inhibits signalling mediated by TGF-β family proteins[68], playing an important role in contrast to cell death, whereas annexin A3 has a role in the signalling cascade during liver regeneration[69]. Other researchers have found the immune system to be altered following vegetable oil replacement in salmon[28] and in sea bass[70].

Genes involved in oxidative stress response (MPV17 protein, amine oxidase and HSP 70 kDa protein) were higher in liver of fish fed PP compared to MP. This is interesting to note as increases in antioxidant genes were also noted in a salmon diets that had marine oil replaced with vegetable oils[28] in the liver. In rainbow trout HSP expression in liver was increased following SBM rich replacement diets[9, 62, 63, 71], the induction of these genes may indicate a diet-induced stress response in fish fed the PP diet. Moreover, during general high protein turnover to deal with misfolded proteins[72] as may be the case of fish fed PP diet.

It is interesting to note that few lipid related metabolic genes were found significantly different in the liver. Vigilin, a protein implicated in both biosynthesis and metabolism of lipids and steroids, facilitates removal of excess cholesterol from cells[73] and secondly apolipoprotein A IV which facilitates transport of cholesterol to the liver were both expressed at higher level in fish fed PP compared to MP. The high expression of apolipoproteins in fish fed diets containing high levels of plant proteins has previously been observed[9, 74] and is most likely indicates reduced cellular cholesterol in fish fed the PP diet due to decreased dietary cholesterol, associated with low fishmeal content, and/or in response to trace levels of phytoestrogens[75] and phorbol esters[76] co purified with the plant proteins.

Transcriptome Changes in the Skeletal Muscle

Genes related to processes such as protein metabolism, energy metabolism, cell proliferation, apoptosis and immune function were all significantly different in PP fed fish compared to MP. Transcripts related to protein metabolism such as ribosomal protein mRNAs, transcription and translation initiation factors were generally lower in PP fed fish compared to MP fed fish indicating lower protein synthesis. In parallel, protein degradation related genes were also less, for example the ubiquitin proteasome pathway and lysosomal peptidase proteins, together these would suggest a lower protein turnover activity in the muscle tissue in PP group. Both protein synthesis and degradation are highly energy demanding processes[77] and the indication of lower protein turnover, may suggest reduced energy wastage[61, 78, 79]. This idea is strengthened by lower expression of COX2 and COX8 and other genes encoding proteins involved in glycolysis and gluconeogenesis in fish fed PP. Genes involved in cell proliferation were also expressed at lower levels in PP fed fish indicating further energy saving. Together these changes in biological processes may indicate efficient metabolic activity following feeding of the plant protein enriched diet.

Biological tissues with high metabolic rate require mechanisms to deal with free oxygen radicals, on the other hand those tissues where the metabolic rate is reduced, for example when protein turnover is decreased a reduction in oxidative stress response may be observed as was observed in this study. Additionally a number of HSPs 30, 47 and a heat shock transcription factor 1a were all at lower levels in PP reflecting the reduced protein turnover and requirement of stabilising many newly translated proteins. Genes involved in cell proliferation and related to cell death including two caspases (caspase-8 and 14) were also lower expressed in fish fed PP compared to the MP diet. The induction of bh3 interacting domain death agonist, a pro-apoptotic member of the Bcl-2 protein family[80] and the suppression of caspases is in accordance with apoptosis of skeletal muscle in mammals where the Bcl2/bax system was found crucial for muscle apoptosis, whilst the caspase activity appeared inhibited[81].

Differential expression of a number of immune related genes, particularly a decrease in interferon responsive genes including γ-ip[82], Gig2-like[83] and interferon-induced protein 35[84] were also expressed at lower level in PP. Other pro-inflammatory agents including platelet-activating factor receptor was also at a lower level PP, which regulates several pro-inflammatory functions such as chemokine and eicosanoid receptors[85]. An antimicrobial peptide beta defensin 1which is a central component of the non-specific defences[86] was also found lower in PP. Relating to the adaptive immune factors, TCRβ, MHCI and CD80 were all at lower levels in PP. Although we have observed differences in genes related to immune function in skeletal muscle of fish fed PP, the low level of inflammation and the subsequent immune response observed in the intestine did not cause a large immune shift in the muscle tissue. Instead, the lower expression of such genes, may allow fish fed PP to spend less energy resources on immune function for use in growth[61].

CONCLUSIONS

In conclusion, the present study confirm that a moderate level of plant protein derived proteins in a salmon diet allowed fish to grow with equal efficiency as those on a high marine protein diet, and with no difference in biometric quality parameters. The PP diet formulated with higher levels of soy protein concentrate, corn gluten, sunflower meal and wheat gluten resulted in significant effects on transcription in the mid intestine, liver, and skeletal muscle. The PP diet induced tissue specific changes in gene expression, with the mid intestine showing activation of the adaptive immune response indicating potential for hypersensitivity and an increase in protein turnover, although no difference in histopathological changes were observed in the proximate, mid or distal intestine. In liver cell proliferation and apoptosis indicate cellular reorganization and the general suppression of processes such as immune response. In contrast skeletal muscle tissue showed reduced protein metabolism and decrease in immune gene expression suggesting less energy expenditure in this tissue. The presence of only few genes in common between tissues may be due to the relatively mild changes that occurred and the complex nature of studying gene expression between tissue types on conservative dietary changes. These results improve the understanding of mechanisms and pathways activated by fishmeal replacement; in particular substitution by plant derived proteins and suggests that such diets can function well in Atlantic salmon aquaculture, hence to some extent reducing the burden on wild caught fish for fish meal. Additionally these results can assist in selection of molecular biomarkers useful for the development of new alternative feeds in salmonid aquaculture.

METHODS

Fish Husbandry and Sampling

One hundred and five juvenile mixed sex Atlantic salmon of approximately 100 g were maintained in 6 replicate 1 m^3 tanks at SGS Chile Ltd., Puerto Mont, Chile, at 10.1°C and 27.6% salinity. Fish were fed a plant protein diet (PP) or a high marine protein diet (MP), both of which were formulated to the same digestible protein and energy content within formulation constraints of a commercial feed specification (CPK 100, 3 mm, 24/44 lipid: protein, BioMar S.A., Chile) (Table1). The feeding trial was conducted in triplicate tanks per dietary treatment and lasted for 77 days. At the end of the feeding period all fish in each tank were bulk weighed and 35 fish from each tank were killed by percussive stunning for individual biometric measurements of round weight, fork length, gutted weight, and weight of liver, heart or intestine for calculation of specific growth rate (SGR), feed conversion ratio (FCR), specific feeding rate (SFR), condition factor, and somatic indices of hepatic (HSI), cardiac (CSI) or intestinal (ISI) organs. Liver, heart and intestine was excised from distinct anatomical regions of the same individuals and fixed in 10% buffered formalin for histological assessment. All fish were sampled 30 minutes following the final meal to ensure there were no differences in postprandial gene expression. An additional 12 fish from each tank were sampled for mid intestine, skeletal muscle and liver and immediately stabilised in RNAlater (Ambion) at 4°C overnight then stored at −20°C until RNA extraction for gene expression analysis.

Growth/feed Parameters Were Calculated As Follows

FCR = feed consumed (g)/biomass increase (g) SGR (% body wt d^{-1}) = $[(\ln W_2 - \ln W_1)/days] \times 100$; W_1 = start weight (g), W_2 = final weight (g), *days* = days in the growth periodSFR (%) = SGR × FCRHSI, CSI, ISI (%) = [organ weight (g)/round weight (g)] × 100 Condition factor = [round weight (g) × 100]/[fork length (cm)]^{3}Biometric parameters were analyzed for significant differences by Anova using InfoStat v 2009 software (University of Córdoba, Argentina), with $p < 0.05$ considered significant. Data is presented as means ± standard deviation for each dietary treatment.

Histology Assessment

Fixed tissues were submitted to dehydration process following paraffin embedding (Aquagestión, Puerto Montt, Chile). Sections (5 μm) were stained with hematoxylin and eosin (H & E) and examined under a light microscope. Micrographs were examined "blind" by the same experienced pathologist. Liver sections were evaluated for lipid degeneration level (fatty change) and the integrity of the whole organ. Heart tissue was evaluated for the presence of inflammatory infiltrates, myodegeneration and other possible abnormalities. Intestinal morphology was evaluated according to the following criteria: a) widening and shortening of intestinal folds, 2) loss of supranuclear vacuolization in enterocytes in the intestinal epithelium, 3) widening of central lamina propria within the intestinal folds, and 4) infiltration of (mixed) leukocytes in the lamina propria and submucosa[33]. The occurrence of histological changes for each dietary treatment were analysed by Kruskal-Wallis non-parametric

Anova with InfoStat v 2009 (University of Córdoba, Argentina), with $p<0.05$ considered significant.

RNA Isolation

RNA was extracted from 100 mg of tissue by homogenisation in 1 ml TRIZol (Invitrogen) using tungsten carbide beads (3 mm, Qiagen) and shaking (300 times per min) following the manufacturer's instructions. The RNA pellet was washed in 500 µl 80% ethanol, air dried and resuspended in RNase free H_2O. The concentration was determined by spectrophotometry (Nanodrop ND1000, LabTech) and the integrity of the RNA was determined by electrophoresis (Agilent Bioanalyser 2100). The RNA was then stored at −80° until required.

Microarray Analysis

Microarray platform:

Microarray experiments were performed using a custom-designed, Agilent-based microarray platform with 4×44 K probes per slide (Salar_2; Agilent Design ID:025520). The array contained primarily an Atlantic salmon resource with 34,441 features from Atlantic salmon coding sequences but additionally a further 9,111 features from rainbow trout (*Oncorhynchus mykiss*) sequences - the latter being selected when no homologues appeared to be available within *Salmo salar* datasets. Full details of the microarray platform and design are shown in Tacchi et al. 2011[23].

Hybridization and Analysis

For microarray analysis, 4 pools of RNA were produced for each tissue from fish fed PP and MP diets. Each RNA pool was an equimolar RNA mix from four different fish chosen randomly from each group. The microarray hybridisation was performed using a reference design, using a reference RNA sample, which comprised an equimolar mix of RNA extracted from all individual fish and tissue samples. Each experimental sample (labelled with Cy3™) was hybridised against this reference sample (labelled with Cy5™) in a 2-colour experiment. mRNA amplification, labelling and hybridization was performed as follows: mRNA was amplified using a MessageAMP™ aRNA Amplification Kit (Ambion). Briefly, 2 μg total RNA was reverse transcribed and the cDNA was used as a template for *in vitro* transcription in the presence of amino allyl modified dUTP, which allowed the generation of amplified antisense RNA (aRNA). For labelling, aRNA (3 μg) was denatured at 70°C for 2 min in a volume of 10 μl to which 3 μl of 0.5 M $NaHCO_3$ and 2 μl Cy dye (dye Cy3™ or Cy5™ mono-reactive dye pack, Amersham) was added. Incorporation of dyes was performed for 1 h in the dark, and after excess label was removed using a DyeEx tm 2.0 spin kit (QIAGEN) the amount incorporated was checked with a Nanodrop ND1000 (LabTech) spectrophotometer. Prior to hybridisation, 825 ng of each labelled template was fragmented in the presence of 11 μl of 10X blocking agent, 2.2 μl of 25X Fragmentation buffer (Agilent), and made up to a final volume of 20 μl with nuclease-free dH_2O. The solution was then incubated in the dark at 60°C for 30 min, after which 57 μl of 2X GEx Hybridisation buffer (Agilent) was added to each sample and 103 μl of each hybridisation solution was dispensed on the Agilent 4x44K Atlantic salmon "*Salmo salar2*" oligo array (Agilent array design, 025520, Array express platform A-MEXP-1940). The hybridisations were performed in a Microarray Hybridisation Oven (Agilent) overnight (18 h) at 65°C. Following hybridisation, the slides were washed in Gene Expression Wash buffers 1 and 2 (Agilent) following the manufacturer's instructions. The slides were then scanned using a GenePix personal 4100A Scanner (Axon Instruments) at a resolution of 5 μm and saved as *. TIF files. Images were extracted and initial analysis was performed by Feature extraction v9.5.3 (Agilent) performing background correction of feature intensities (within the software). A Lowess normalisation of background corrected data was next conducted and all intensity values <1.0 were set to 1.0. Statistical analysis of the arrays was performed using Genespring GX analysis platform (version 9.5; Agilent Technologies). Quality control of the data was performed within Genespring and included removal of saturated probe features, non-uniform features, population outliers and those features showing intensities not significantly different from background in the Cy3 or Cy5 channels. After these relatively stringent procedures, 20,095 of the original 43,730 array features were maintained for subsequent analyses. The experimental hybridisations are at European Bioinformatics Institute archived under accession number E-MEXP: E-TABM-1207.

Significant differential expression between fish fed the PP diet and fish fed the control diet was established by *t*-test analysis ($p<0.05$). Further filtering on fold change was conducted, and only transcripts showing more than two-fold change in expression were further characterised.

Analysis of Gene Ontology

Enrichment for gene ontology (GO) biological processes was performed on all cDNA features that had GO identifiers associated using the GOEAST program[87]. Fisher's exact test was performed within the GOEAST program to determine if GO identifiers occurred more often in a group than would appear by chance. For GO analysis only biological process GO identifiers were considered that occurred more than 3 times.

Real Time PCR

The cDNA was synthesized using 2 μg of total RNA as previously described[88]. Real time PCR was performed on a number of genes to confirm the microarray analysis results using the same RNA samples. For cDNA synthesis, 2 μg of total RNA was denatured (70°C, 3 min) in the presence of 1 μl of oligo-dT$_{17}$ (500 ng μl^{-1}), left at room temperature for 5 min to allow annealing, then stored on ice. The resulting cDNA was diluted to a final volume of 50μl in RNA/DNA free water (Sigma). For real time PCR, 3μl of cDNA was used as template with gene specific primers (Table7). A 2x iQ SYBR Green supermix (Bio-Rad) was used for qPCR, which was performed in a 96-well plate using the DNA Engine OpticonTM system (MJ Research, Inc.) with the following program: 95°C for 5 min, then 35 cycles of 94°C for 30° s, 55°C for 30 s and 72°C for 30 s, with a final extension of 72°C for 5 min. A negative control (no template) reaction was also performed for each primer pair. A sample from the serial dilution was run on a 2% agarose gel and stained with ethidium bromide and viewed under UV light to confirm a band of the correct size was amplified. A melting curve for each PCR was determined by reading fluorescence every degree between 72°C and 95°C to ensure only a single product had been amplified. Atlantic salmon hypoxanthine phosphoribosyl transferase 1 (HPRT1)[89] was used as control for normalization of expression since this gene was found not to be modulated by the diet treatments from the microarray analysis. The relative expression level of the genes was determined using the Pfaffl method[90]. Efficiency of the amplification was determined for each primer pair using serial 10 fold dilutions of pooled cDNA, performed on the same plate as the experimental samples. The efficiency was calculated as $E=10^{(-1/s)}$ where s is the slope generated from the serial dilutions, when Log dilution is plotted against ΔCT (threshold cycle number). Primers were design to have a Tm of 55°C, and where possible, to cross an exon-exon junction sites to avoid amplification of genomic DNA. Exon-intorn junction sites were determined comparing *Salmo salar* cDNA with genomic sequences for orthologous genes from *Danio rerio*, *Gasterosteus aculeatus*, *Oryzias latipes*, *Takifugu rubripes* and *Tetraodon nigroviridis* obtained from Ensembl (http://ensembl.org/).

The results obtained by real time PCR were analyzed using the Pfaffl method[90]. The qPCR measurements were analyzed by *T*-test, performed using R software, with $p<0.05$ considered significant. The expression data is presented as means ± standard error.

AUTHOR'S CONTRIBUTIONS

LT performed microarray experiments, analyzed the data, carried out the real time PCR and wrote the manuscript. MAC, CV and HT were involved in fish maintenance, feeding trial and sampling. SAMM, CJS, RB and MAC were involved in the experimental design and drafting of the manuscript. All the authors read and approved the final manuscript.

DECLARATIONS

Acknowledgements

The research was supported by an industrial studentship between the University of Aberdeen and BioMar Ltd (for L. Tacchi). The authors are grateful to AVS Chile S.A. and the staff at SGS Chile for conducting the feeding experiment and assistance during sampling.

Competing interests

The authors declare that they have no competing interests.

REFERENCES

1. Tacon AGJ: Aquaculture: A catch for all?. OECD Obs. 2010, 278: 32-
2. Tacon AGJ: FAO Fisheries Circular (FAO), no. 881 / FAO. Feed ingredients for carnivorous fish species: alternatives to fishmeal and other fishery resources. 1994, Fishery Resources and Environment Div, Rome (Italy), 39-
3. F.A.O: State of the World Fisheries and Aquaculture. 2007, Rome: FAO
4. ICES: Report of the Working Group on North Atlantic Salmon (WGNAS),13 April 2006,ICES Headquarters.ICES Headquarters 2006: ICES CM 2006/ACFM. 2023
5. ICES:*Report of the Working Group on North Atlantic Salmon (WGNAS),30 March-8 April 2009,Copenahagen, Denmark.* ICES CM 2009/ACOM:06. 2009,
6. F.A.O: FAO fisheries technical paper. 2005, FAO, Rome, 235-Review of the state of world marine fishery resources, 457
7. Gatlin DM, Barrows FT, Brown P, Dabrowski K, Gaylord TG, Hardy RW, Herman E, Hu G, Krogdahl A, Nelson R, et al: Expanding the utilization of sustainable plant products in aquafeeds: A review. Aquac Res. 2007, 38 (6): 551-579.
8. Krogdahl Å, Penn M, Thorsen J, Refstie S, Bakke AM: Important antinutrients in plant feedstuffs for aquaculture: An update on recent findings regarding responses in salmonids. Aquac Res. 2010, 41 (3): 333-344.
9. Martin SAM, Vilhelmsson O, Médale F, Watt P, Kaushik S, Houlihan DF: Proteomic sensitivity to dietary manipulations in rainbow trout. Biochim Biophys Acta Proteins Proteomics. 2003, 1651 (1–2): 17-29.
10. Bakke-McKellep AM, Frøystad MK, Lilleeng E, Dapra F, Refstie S, Krogdahl Å, Landsverk T: Response to soy: T-cell-like reactivity in the intestine of Atlantic salmon, *Salmo salar* L. J Fish Dis. 2007, 30 (1): 13-25.
11. Urán PA, Aydin R, Schrama JW, Verreth JAJ, Rombout JHWM: Soybean meal-induced uptake block in Atlantic salmon *Salmo*

salar distal enterocytes. J Fish Biol. 2008, 73 (10): 2571-2579.

12. Gómez-Requeni P, Mingarro M, Kirchner S, Calduch-Giner JA, Médale F, Corraze G, Panserat S, Martin SAM, Houlihan DF, Kaushik SJ, et al: Effects of dietary amino acid profile on growth performance, key metabolic enzymes and somatotropic axis responsiveness of gilthead sea bream (*Sparus aurata*). Aquaculture. 2003, 220 (1–4): 749-767.
13. Francis G, Makkar HPS, Becker K: Antinutritional factors present in plant-derived alternate fish feed ingredients and their effects in fish. Aquaculture. 2001, 199 (3–4): 197-227.
14. Zhao H, Jiang R, Xue M, Xie S, Wu X, Guo L: Fishmeal can be completely replaced by soy protein concentrate by increasing feeding frequency in Nile tilapia (*Oreochromis niloticus GIFT strain*) less than 2 g. Aquac Nutr. 2010, 16 (6): 648-653.
15. Escaffre AM, Kaushik S, Mambrini M: Morphometric evaluation of changes in the digestive tract of rainbow trout (*Oncorhynchus mykiss*) due to fish meal replacement with soy protein concentrate. Aquaculture. 2007, 273 (1): 127-138.
16. Kaushik SJ, Cravedi JP, Lalles JP, Sumpter J, Fauconneau B, Laroche M: Partial or total replacement of fish meal by soybean protein on growth, protein utilization, potential estrogenic or antigenic effects, cholesterolemia and flesh quality in rainbow trout, *Oncorhynchus mykiss*. Aquaculture. 1995, 133 (3–4): 257-274.
17. Gaylord TG, Teague AM, Barrows FT: Taurine supplementation of all-plant protein diets for rainbow trout (*Oncorhynchus mykiss*). J World Aquaculture Soc. 2006, 37 (4): 509-517.
18. Gill N, Higgs DA, Skura BJ, Rowshandeli M, Dosanjh BS, Mann J, Gannam AL: Nutritive value of partially dehulled and extruded sunflower meal for post-smolt Atlantic salmon (*Salmo salar* L.) in sea water. Aquac Res. 2006, 37 (13): 1348-1359.
19. Mente E, Deguara S, Begoña Santos M, Houlihan D: White muscle free amino acid concentrations following feeding a maize gluten dietary protein in Atlantic salmon (*Salmo salar* L.). Aquaculture. 2003, 225 (1–4): 133-147.
20. Thiessen DL, Campbell GL, Adelizi PD: Digestibility and growth performance of juvenile rainbow trout (*Oncorhynchus mykiss*) fed

with pea and canola products. Aquac Nutr. 2003, 9 (2): 67-75.

21. Espe M, Lemme A, Petri A, El-Mowafi A: Assessment of lysine requirement for maximal protein accretion in Atlantic salmon using plant protein diets. Aquaculture. 2007, 263 (1–4): 168-178.
22. Penn MH, Bendiksen EA, Campbell P, Krogdahl AS: High level of dietary pea protein concentrate induces enteropathy in Atlantic salmon (*Salmo salar* L.). Aquaculture. 2011, 310 (3–4): 267-273.
23. Tacchi L, Bron JE, Taggart JB, Secombes CJ, Bickerdike R, Adler MA, Takle H, Martin SAM: Multiple tissue transcriptomic responses to *Piscirickettsia salmonis* in Atlantic salmon (*Salmo salar*). Physiol Genom. 2011, 43 (21): 1241-1254.
24. Panserat S, Hortopan GA, Plagnes-Juan E, Kolditz C, Lansard M, Skiba-Cassy S, Esquerré D, Geurden I, Médale F, Kaushik S, et al: Differential gene expression after total replacement of dietary fish meal and fish oil by plant products in rainbow trout (*Oncorhynchus mykiss*) liver. Aquaculture. 2009, 294 (1–2): 123-131.
25. Le Boucher R, Quillet E, Vandeputte M, Lecalvez JM, Goardon L, Chatain B, Médale F, Dupont-Nivet M: Plant-based diet in rainbow trout (*Oncorhynchus mykiss*): Are there genotype-diet interactions for main production traits when fish are fed marine vs. plant-based diets from the first meal?. Aquaculture. 2011, 321 (1–2): 41-48.
26. Pierce LR, Palti Y, Silverstein JT, Barrows FT, Hallerman EM, Parsons JE: Family growth response to fishmeal and plant-based diets shows genotype x diet interaction in rainbow trout (*Oncorhynchus mykiss*). Aquaculture. 2008, 278 (1–4): 37-42.
27. Kolditz CI, Plagnes-Juan E, Quillet E, Lefèvre F, Médale F: Changes in white muscle transcriptome induced by dietary energy levels in two lines of rainbow trout (*Oncorhynchus mykiss*) selected for muscle fat content. Br J Nutr. 2010, 103 (5): 629-642.
28. Morais S, Pratoomyot J, Taggart J, Bron J, Guy D, Bell JG, Tocher D: Genotype-specific responses in Atlantic salmon (*Salmo salar*) subject to dietary fish oil replacement by vegetable oil: a liver transcriptomic analysis. BMC Genom. 2011, 12 (1): 255-
29. Tacon AGJ, Metian M: Global overview on the use of fish meal and fish oil in industrially compounded aquafeeds: Trends and future prospects. Aquaculture. 2008, 285 (1–4): 146-158.

30. Morais S, Pratoomyot J, Taggart J, Bron J, Guy D, Bell J, Tocher D: Genotype-specific responses in Atlantic salmon (*Salmo salar*) subject to dietary fish oil replacement by vegetable oil: a liver transcriptomic analysis. BMC Genom. 2011, 12 (1): 255-
31. Jafri M, Donnelly B, Bondoc A, Allen S, Tiao G: Cholangiocyte secretion of chemokines in experimental biliary atresia. J Pediatr Surg. 2009, 44 (3): 500-507.
32. Mack CL, Tucker RM, Sokol RJ, Kotzin BL: Armed CD4+ Th1 effector cells and activated macrophages participate in bile duct injury in murine biliary atresia. Clin Immunol. 2005, 115 (2): 200-209.
33. Baeverfjord G, Krogdahl Å: Development and regression of soybean meal induced enteritis in Atlantic salmon, *Salmo salar* L., distal intestine: A comparison with the intestines of fasted fish. J Fish Dis. 1996, 19 (5): 375-387.
34. van den Ingh TSGAM, Krogdahl Å, Olli JJ, Hendriks HGCJM, Koninkx JGJF: Effects of soybean-containing diets on the proximal and distal intestine in Atlantic salmon (*Salmo salar*): a morphological study. Aquaculture. 1991, 94 (4): 297-305.
35. Sanden M, Berntssen MHG, Krogdahl Å, Hemre GI, Bakke-McKellep AM: An examination of the intestinal tract of Atlantic salmon, Salmo salar L., parr fed different varieties of soy and maize. J Fish Dis. 2005, 28 (6): 317-330.
36. Salinas I, Meseguer J, Esteban MÁ: Assessment of different protocols for the isolation and purification of gut associated lymphoid cells from the gilthead seabream (*Sparus aurata* L.). Biol Proced Online. 2007, 9 (1): 43-55.
37. Collie NL, Ferraris RP: Chapter 9 Nutrient fluxes and regulation in fish intestine. Biochemistry and Molecular Biology of Fishes. 1995, 4: 221-239.
38. Bakke-McKellep AM, Nordrum S, Krogdahl Å, Buddington RK: Absorption of glucose, amino acids, and dipeptides by the intestines of Atlantic salmon (*Salmo salar* L.). Fish Physiol Biochem. 2000, 22 (1): 33-44.
39. Bureau DP, Harris AM, Young Cho C: The effects of purified alcohol extracts from soy products on feed intake and growth of

chinook salmon (*Oncorhynchus tshawytscha*) and rainbow trout (*Oncorhynchus mykiss*). Aquaculture. 1998, 161 (1–4): 27-43.

40. Rumsey GL, Siwicki AK, Anderson DP, Bowser PR: Effect of soybean protein on serological response, non-specific defense mechanisms, growth, and protein utilization in rainbow trout. Vet Immunol Immunopathol. 1994, 41 (3–4): 323-339.
41. Krogdahl A, Bakke-McKellep AM, Baeverfjord G: Effects of graded levels of standard soybean meal on intestinal structure, mucosal enzyme activities, and pancreatic response in Atlantic salmon (Salmo solar L.). Aquac Nutr. 2003, 9 (6): 361-371.
42. Bernard D, Six A, Rigottier-Gois L, Messiaen S, Chilmonczyk S, Quillet E, Boudinot P, Benmansour A: Phenotypic and functional similarity of gut intraepithelial and systemic T cells in a teleost fish. J Immunol. 2006, 176 (7): 3942-3949.
43. Riccardi C, Cifone MG, Migliorati G: Glucocorticoid hormone-induced modulation of gene expression and regulation of T-cell death: Role of GITR and GILZ, two dexamethasone-induced genes. Cell Death Differentiation. 1999, 6 (12): 1182-1189.
44. Delfino DV, Agostini M, Spinicelli S, Vacca C, Riccardi C: Inhibited cell death, NF-κB activity and increased IL-10 in TCR-triggered thymocytes of transgenic mice overexpressing the glucocorticoid-induced protein GILZ. Int Immunopharmacol. 2006, 6 (7): 1126-1134.
45. Wilson S, Blaschek K, Gonzalez De Mejia E: Allergenic proteins in soybean: Processing and reduction of P34 allergenicity. Nutr Rev. 2005, 63 (2): 47-58.
46. Jacobs MD, Harrison SC: Structure of an IκBα/NF-κB complex. Cell. 1998, 95 (6): 749-758.
47. Takeda K, Kaisho T, Akira S: Toll-like receptors. Annu Rev Immunol. 2003, 21: 335-376.
48. Honda K, Yanai H, Negishi H, Asagiri M, Sato M, Mizutani T, Shimada N, Ohba Y, Takaoka A, Yoshida N, et al: IRF-7 is the master regulator of type-I interferon-dependent immune responses. Nature. 2005, 434 (7034): 772-777.
49. Sun B, Skjaeveland I, Svingerud T, Zou J, Jorgensen J, Robertsen B: Antiviral Activity of Salmonid Gamma Interferon against

Infectious Pancreatic Necrosis Virus and Salmonid Alphavirus and Its Dependency on Type I Interferon. J Virol. 2011, 85 (17): 9188-9198.

50. Robertsen B, Trobridge G, Leongt J-A: Molecular cloning of double-stranded RNA inducible MX genes from atlantic salmon (Salmo salar l.). Dev Comp Immunol. 1998, 21 (5): 397-412.
51. Boudinot P, Salhi S, Blanco M, Benmansour A: Viral haemorrhagic septicaemia virus induces vig −2, a new interferon-responsive gene in rainbow trout. Fish Shellfish Immunol. 2001, 11 (5): 383-397.
52. van der Aa LM, Levraud JP, Yahmi M, Lauret E, Briolat V, Herbomel P, Benmansour A, Boudinot P: A large new subset of TRIM genes highly diversified by duplication and positive selection in teleost fish. BMC Biol. 2009, 7: 7-
53. Martens N, Uzan G, Wery M, Hooghe R, Hooghe-Peters EL, Gertler A: Suppressor of Cytokine Signaling 7 Inhibits Prolactin, Growth Hormone, and Leptin Signaling by Interacting with STAT5 or STAT3 and Attenuating Their Nuclear Translocation. J Biol Chem. 2005, 280 (14): 13817-13823.
54. Jimenez N, Coll J, Salguero FJ, Tafalla C: Co-injection of interleukin 8 with the glycoprotein gene from viral haemorrhagic septicemia virus (VHSV) modulates the cytokine response in rainbow trout (*Oncorhynchus mykiss*). Vaccine. 2006, 24 (27–28): 5615-5626.
55. Kumari J, Larsen AN, Bogwald J, Dalmo RA: Interleukin-17D in Atlantic salmon (*Salmo salar*): Molecular characterization, 3D modelling and promoter analysis. Fish Shellfish Immunol. 2009, 27 (5): 647-659.
56. Godiksen H, Morzel M, Hyldig G, Jessen F: Contribution of cathepsins B, L and D to muscle protein profiles correlated with texture in rainbow trout (*Oncorhynchus mykiss*). Food Chem. 2009, 113 (4): 889-896.
57. Tacchi L, Bickerdike R, Secombes CJ, Martin SAM: Muscle-Specific RING Finger (MuRF) cDNAs in Atlantic Salmon (*Salmo salar*) and Their Role as Regulators of Muscle Protein Degradation. Mar Biotechnol. 2012, 14 (1): 35-45.
58. Zheng X, Torstensen BE, Tocher DR, Dick JR, Henderson RJ, Bell JG: Environmental and dietary influences on highly unsaturated

fatty acid biosynthesis and expression of fatty acyl desaturase and elongase genes in liver of Atlantic salmon (*Salmo salar*). Biochim Biophys Acta Mol Cell Biol Lipids. 2005, 1734 (1): 13-24.

59. Hastings N, Agaba MK, Tocher DR, Zheng X, Dickson CA, Dick JR, Teale AJ: Molecular cloning and functional characterization of fatty acyl desaturase and elongase cDNAs involved in the production of eicosapentaenoic and docosahexaenoic acids from α-linolenic acid in Atlantic salmon (*Salmo salar*). Mar Biotechnol. 2004, 6 (5): 463-474.
60. Martin SAM, Cash P, Blaney S, Houlihan DF: Proteome analysis of rainbow trout (*Oncorhynchus mykiss*) liver proteins during short term starvation. Fish Physiol Biochem. 2001, 24 (3): 259-270.
61. Tacchi L, Bickerdike R, Douglas A, Secombes CJ, Martin SAM: Transcriptomic responses to functional feeds in Atlantic salmon (*Salmo salar*). Fish Shellfish Immunol. 2011, 31 (5): 704-715.
62. Vilhelmson OT, Martin SAM, Médale F, Kaushik SJ, Houlihan DF: Dietary plant-protein substitution affects hepatic metabolism in rainbow trout (*Oncorhynchus mykiss*). Br J Nutr. 2004, 92 (1): 71-80.
63. Taggart JB, Bron JE, Martin SAM, Seear PJ, Høyheim B, Talbot R, Carmichael SN, Villeneuve LAN, Sweeney GE, Houlihan DF, et al: A description of the origins, design and performance of the TRAITS-SGP Atlantic salmon *Salmo salar* L. cDNA microarray. J Fish Biol. 2008, 72 (9): 2071-2094.
64. Martin SAM, Douglas A, Houlihan D, Secombes C: Starvation alters the liver transcriptome of the innate immune response in Atlantic salmon (*Salmo salar*). BMC Genom. 2010, 11 (1): 418-
65. Lucas PJ, Kim SJ, Melby SJ, Gress RE: Disruption of T cell homeostasis in mice expressing a T cell-specific dominant negative transforming growth factor β II receptor. J Exp Med. 2000, 191 (7): 1187-1196.
66. Cazac BB, Roes J: TGF-β receptor controls B cell responsiveness and induction of IgA in vivo. Immunity. 2000, 13 (4): 443-451.
67. Schuster N, Krieglstein K: Mechanisms of TGF-β-mediated apoptosis. Cell Tissue Res. 2002, 307 (1): 1-14.
68. Oka C, Tsujimoto R, Kajikawa M, Koshiba-Takeuchi K, Ina

J, Yano M, Tsuchiya A, Ueta Y, Soma A, Kanda H, et al: HtrA1 serine protease inhibits signaling mediated by Tgfβ family proteins. Development. 2004, 131 (5): 1041-1053.

69. Harashima M, Harada K, Ito Y, Hyuga M, Seki T, Ariga T, Yamaguchi T, Niimi S: Annexin A3 expression increases in hepatocytes and is regulated by hepatocyte growth factor in rat liver regeneration. J Biochem. 2008, 143 (4): 537-545.

70. Geay F, Ferraresso S, Zambonino-Infante J, Bargelloni L, Quentel C, Vandeputte M, Kaushik S, Cahu C, Mazurais D: Effects of the total replacement of fish-based diet with plant-based diet on the hepatic transcriptome of two European sea bass (*Dicentrarchus labrax*) half-sibfamilies showing different growth rates with the plant-based diet. BMC Genom. 2011, 12 (1): 522-

71. Sagstad A, Sanden M, Krogdahl Å, Bakke-Mckellep AM, FrØystad M, Hemre GI: Organs development, gene expression and health of Atlantic salmon (*Salmo salar* L.) fed genetically modified soybeans compared to the near-isogenic non-modified parental line. Aquac Nutr. 2008, 14 (6): 556-572.

72. Freeman ML, Borrelli MJ, Meredith MJ, Lepock JR: On the path to the heat shock response: Destabilization and formation of partially folded protein intermediates, a consequence of protein thiol modification. Free Radic Biol Med. 1999, 26 (5–6): 737-745.

73. McKnight GL, Reasoner J, Gilbert T, Sundquist KO, Hokland B, McKernan PA, Champagne J, Johnson CJ, Bailey MC, Holly R, et al: Cloning and expression of a cellular high density lipoprotein-binding protein that is up-regulated by cholesterol loading of cells. J Biol Chem. 1992, 267 (17): 12131-12141.

74. Bouraoui L, SÁNchez-Gurmaches J, Cruz-Garcia L, GutiÉRrez J, Benedito-Palos L, PÉRez-SÁNchez J, Navarro I: Effect of dietary fish meal and fish oil replacement on lipogenic and lipoprotein lipase activities and plasma insulin in gilthead sea bream (*Sparus aurata*). Aquac Nutr. 2011, 17 (1): 54-63.

75. Lamon-Fava S: Genistein Activates Apolipoprotein A-I Gene Expression in the Human Hepatoma Cell Line Hep G2. J Nutr. 2000, 130 (10): 2489-2492.

76. Vandenbrouck Y, Lambert G, Janvier B, Girlich D, Bereziat G, Mangeney-Andreani M: Transcriptional regulation of apolipoprotein A-I expression in Hep G2 cells by phorbol ester. FEBS Lett. 1995, 376 (1–2): 99-102.

77. Houlihan DF, Carter CG, McCarthy ID: Nitrogen Metabolism and Excretion. Protein Turnover in Animals. Edited by: Walsh PJ, Wright P. 1995, CRC Press, Boca Raton, United States of America, 1-32.

78. Martin SAM, Blaney S, Bowman AS, Houlihan DF: Ubiquitin-proteasome-dependent proteolysis in rainbow trout (*Oncorhynchus mykiss*): Effect of food deprivation. Pflugers Arch Eur J Physiol. 2002, 445 (2): 257-266.

79. Dobly A, Martin SAM, Blaney SC, Houlihan DF: Protein growth rate in rainbow trout (*Oncorhynchus mykiss*) is negatively correlated to liver 20 S proteasome activity. Comp Biochem Physiol Mol Integr Physiol. 2004, 137 (1): 75-85.

80. Wang K, Yin XM, Chao DT, Milliman CL, Korsmeyer SJ: BID: A novel BH3 domain-only death agonist. Genes Dev. 1996, 10 (22): 2859-2869.

81. Sandri M, Rossini K, Podhorska-Okolow M, Carraro U: Role of Apoptosis in Muscle Disorders. Basic Appl Myol. 1999, 9 (6): 301-310.

82. Laing KJ, Bols N, Secombes CJ: A CXC chemokine sequence isolated from the rainbow trout *Oncorhynchus mykiss*resembles the closely related interferon-γ-inducible chemokines CXCL9, CXCL10 and CXCL11. Eur Cytokine Netw. 2002, 13 (4): 462-473.

83. Jiang J, Zhang Y-B, Li S, Yu F-F, Sun F, Gui J-F: Expression regulation and functional characterization of a novel interferon inducible gene Gig2 and its promoter. Mol Immunol. 2009, 46 (15): 3131-3140.

84. Bange FC, Vogel U, Flohr T, Kiekenbeck M, Denecke B, Böttger EC: IFP 35 is an interferon-induced leucine zipper protein that undergoes interferon-regulated cellular redistribution. J Biol Chem. 1994, 269 (2): 1091-1098.

85. Honda ZI, Ishii S, Shimizu T: Platelet-activating factor receptor. J Biochem. 2002, 131 (6): 773-779.

86. Casadei E, Wang T, Zou J, González Vecino JL, Wadsworth S, Secombes CJ: Characterization of three novel β-defensin antimicrobial peptides in rainbow trout (*Oncorhynchus mykiss*). Mol Immunol. 2009, 46 (16): 3358-3366.

87. Zheng Q, Wang XJ: GOEAST: a web-based software toolkit for Gene Ontology enrichment analysis. Nucleic Acids Res. 2008, 36: W358-W363. Web Server issue

88. Tacchi L, Bickerdike R, Secombes CJ, Pooley NJ, Urquhart KL, Collet B, Martin SAM: Ubiquitin E3 ligase atrogin-1 (Fbox-32) in Atlantic salmon (*Salmo salar*): Sequence analysis, genomic structure and modulation of expression. Comp Biochem Physiol Biochem Mol Biol. 2010, 157 (4): 364-373.

89. Bower NI, Johnston IA: Selection of reference genes for expression studies with fish myogenic cell cultures. BMC Mol Biol. 2009, 10: 80-

90. Pfaffl MW: A new mathematical model for relative quantification in real-time RT-PCR. Nucleic Acids Res. 2001, 29 (9): e45-

SECTION V

GENETIC IMPROVEMENT

CHAPTER

12

AN INTEGRATED APPROACH TO GENE DISCOVERY AND MARKER DEVELOPMENT IN ATLANTIC COD (GADUS MORHUA)

Sharen Bowman[1], Sophie Hubert[1,2], Brent Higgins[1], Cynthia Stone[1], Jennifer Kimball[1], Tudor Borza[1], Jillian Tarrant Bussey[1,3], Gary Simpson[1], Catherine Kozera[1,3], Bruce A. Curtis[1,8], Jennifer R. Hall[4], Tiago S. Hori[4], Charles Y. Feng[4], Marlies Rise[4], Marije Booman[4], A. Kurt Gamperl[4], Edward Trippel[4], Jane Symonds[6], Stewart C. Johnson[7], Matthew L. Rise[4]

[1]The Atlantic Genome Centre, Halifax, Canada

[2]Department des Sciences, Universite Sainte Anne, Pointe-de-l'Eglise, Canada

[3]The Atlantic Genome Centre, NRC Institute for Marine Biosciences, National Research Council, Halifax, Canada

[4]Ocean Sciences Centre, Memorial University of Newfoundland, St. John's, Canada

[5]Fisheries and Oceans Canada, St. Andrews, Canada

[6]National Institute of Water and Atmospheric Research, Ruakaka, New Zealand

[7]Pacific Biological Station, Fisheries and Oceans Canada, Nanaimo, Canada

[8]Department of Biochemistry and Molecular Biology, Dalhousie University, Halifax, Canada

ABSTRACT

Atlantic cod is a species that has been overexploited by the capture fishery. Programs to domesticate this species are underway in several countries, including Canada, to provide an alternative route for production. Selective breeding programs have been successfully applied in the domestication of other species, with genomics-based approaches used to augment conventional methods of animal production in recent years. Genomics tools, such as gene sequences and sets of variable markers, also have the potential to enhance and accelerate selective breeding programs in aquaculture, and to provide better monitoring tools to ensure that wild cod populations are well managed. We describe the generation of significant genomics resources for Atlantic cod through an integrated genomics/selective breeding approach. These include 158,877 expressed sequence tags (ESTs), a set of annotated putative transcripts and several thousand single nucleotide polymorphism markers that were developed from, and have been shown to be highly variable in, fish enrolled in two selective breeding programs. Our EST collection was generated from various tissues and life cycle stages. In some cases, tissues from which libraries were generated were isolated from fish exposed to stressors, including elevated temperature, or antigen stimulation (bacterial and viral) to enrich for transcripts that are involved in these response pathways. The genomics resources described here support the developing aquaculture industry, enabling the application of molecular markers within selective breeding programs. Marker sets should also find widespread application in fisheries management.

Keywords: Atlantic cod, ESTs, Markers, SNPs, Aquaculture

INTRODUCTION

The depletion of Atlantic cod stocks in the North Atlantic has resulted in reduced harvests of this species by the capture fishery (Rose 2007; Myers et al. 1997), thus providing an incentive to develop commercial farming for Atlantic cod. Initiatives to domesticate this species are underway in several countries including Canada and Norway (Rosenlund and Skretting 2006; Brown et al. 2003) using breeding to select for commercially desirable phenotypes such as high food conversion efficiencies, increased stress tolerance and enhanced disease resistance. The aquaculture community needs to overcome several challenges before a commercially viable industry is achieved for Atlantic cod. Infectious agents causing significant losses in production of this species in captivity include nodavirus (Patel et al. 2007), *Aeromonas salmonicida*subspecies (including atypical strains), and *Vibrio* species (Rodger and Colquhoun 2008), with emerging pathogens frequently being identified, such as the microsporidian *Loma* sp. (Khan 2005; Rodriguez-Tovar et al. 2003). Atlantic cod can also experience variations in temperature at mariculture sites. Fish confined within sea cages cannot avoid these non-optimal temperatures and thus experience stress (Gollock et al. 2006). Another stressor of fish grown in captivity is handling stress, which is encountered through routine aquaculture practices such as fish transfers, grading and tagging (Brown et al. 2003). In addition, issues such as early maturation can have a significant effect on production, as spawning fish undergo considerable weight reduction and loss of condition (Hansen et al. 2001).

Genomics tools have the potential to enhance and accelerate selective breeding programs through the identification of quantitative trait loci (QTL) and the incorporation of marker assisted selection (MAS). Many expressed sequence tag (EST) sequencing programs in commercially important species have already been completed (Rise et al. 2004; Gorodkin et al. 2007; Park et al. 2005; Koop et al. 2008). These sequence collections are regularly mined for marker development (He et al. 2003; Snelling et al. 2005; Hayes et al. 2007; Ramsey et al. 2007; Moen et al. 2008) with the aim of using these markers to enhance desirable characteristics through MAS. However, often sequencing initiatives have not been designed to maximise the opportunities for marker identification within breeding programs; frequently ESTs have been generated from varieties or individuals which have not been included in mapping populations or breeding programs. This necessitates an additional phase of directed re-sequencing to identify single nucleotide polymorphisms (SNPs) or microsatellite length variants that are present in the target individuals or populations (Labate and Baldo 2005; Lijavetzky et al. 2007).

We have attempted to address this issue by initiating an integrated genomics (Bowman et al. 2007) and broodstock development program (Symonds et al. 2007), the Atlantic Cod Genomics and Broodstock Development Project (CGP), with the generation of genomics resources designed to be applied directly in two family-based breeding programs located in New Brunswick (NB) and Newfoundland (NL), Canada. The rationale for undertaking separate province-based breeding programs has its basis in the observation of genetic differences among cod stocks in the

Northwest Atlantic (Hutchinson et al. 2001; Ruzzante et al. 1996). This has resulted in a desire to minimise the potential for inter-breeding of resident wild stocks with captive stock (either as a result of escapees, or via the release of eggs or sperm from fish in sea cages) that might be ill adapted to the local environmental conditions by using local stocks for breeding. ESTs have been generated from fish enrolled in the breeding programs, allowing identification of genetic markers present in family fish as a direct outcome from the sequencing component of the project, enabling genetic mapping, analysis of QTL and ultimately MAS.

This study describes the construction of cDNA libraries and the generation of a large EST set for Atlantic cod. The majority of the sequencing effort focused on normalized libraries to maximise gene identification (Bonaldo et al. 1996; Zhulidov et al. 2004), with sequences generated from multiple tissues and developmental stages. Several libraries were produced from tissues sampled after exposure to stressors or antigen stimulation to ensure that transcripts important in aquaculture-relevant traits such as disease resistance and temperature susceptibility/tolerance were represented in our collection. Several suppression subtractive hybridisation (SSH) libraries (Diatchenko et al. 1996) were also generated for targeted gene discovery i.e. for identification of genes that are induced or suppressed in physiologically relevant tissues following exposure to a viral pathogen or the viral mimic polyriboinosinic polyribocytidylic acid (pIC) (Rise et al. 2008), bacterial antigens (Feng et al. 2009), or heat-shock (Hori et al. 2010). A small number of these sequences, notably several of the SSH libraries, have been analysed separately (Rise et al. 2008; Feng et al. 2009; Hori et al. 2010), however approximately 130,000 of the ESTs presented here have not been described previously, and this is the first overall analysis of the entire data set. A set of novel microsatellite markers has already been identified from this dataset (Higgins et al. 2009). Putative SNP markers identified in the sequence data were tested by genotyping and found to be variable in Canadian and North European populations and in fish enrolled within the two family-based breeding programs from which the sequence resource was developed.

MATERIALS AND METHODS

Tissue Sampling

Fish for sampling were held at the St. Andrews Biological Station, St. Andrews, New Brunswick, Canada, at the Ocean Sciences Centre, Memorial University, Newfoundland, Canada, or at Great Bay Aquaculture, Newington, New Hampshire, USA. Tissue samples (approximately 0.5 cm^3) were taken from fish that had been euthanized using a lethal dose of anaesthetic. Samples were incubated in RNAlater (Ambion Inc., Austin, TX) overnight at 4°C, or flash frozen in liquid nitrogen, prior to storage at -80°C. The heat shock and immunogenic challenge procedures used in the production of a subset of libraries have been described in detail in Rise et al. 2008, Feng et al. 2009 and Hori et al. 2010.

Experimental manipulations and sampling of the fish were carried out in accordance with the Canadian Council on Animal Care Guidelines and approved by animal care committees at Dalhousie University, The Institute of Marine Biosciences, Halifax, and Memorial University of Newfoundland.

Production of Normalized cDNA and SSH Libraries

Tissue samples from multiple fish (80–100 mg $individual^{-1}$) were pooled prior to RNA extraction in the construction of normalized libraries. Total RNA was extracted from tissue pools by lysis using TRIzol reagent (Invitrogen, Burlington, ON) followed by chloroform extraction and precipitation with isopropanol as specified in the manufacturer's protocol. Normalized cDNA library generation was performed essentially as described in Bowman et al. 2007. cDNA production and library construction were carried out using the Creator SMART kit (Clontech, Mountain View, CA). cDNA was normalized using the Trimmer kit method (Evrogen JSC, Moscow, Russia) (Zhulidov et al. 2004) prior to directionally cloning into the SfiI sites of pDNR-LIB (Clontech). Methods utilized to construct reciprocal SSH libraries, including total RNA isolation, mRNA isolation, suppression subtractive hybridization, cloning, and initial evaluation of SSH library insert size and complexity, have been previously published (Rise et al. 2008; Feng et al. 2009; Diatchenko et al. 1996; Hori et al. 2010).

Normalized library names were generated using a series of two letter codes to identify production variables. For example: gmnbhkic is identified as a library created from *Gadus morhua* (gm) fish, from the New Brunswick Breeding Program (nb), head kidney (hk) was the tissue used, and the fish had undergone a treatment with pIC (ic). The SSH libraries follow similar nomenclature rules with respect to fish species, fish breeding program, and fish treatment; however, in addition, an "f" or an "r" was used in this naming convention to denote the "forward" or "reverse" subtracted libraries (i.e. enriched for transcripts that were up-regulated or down-regulated by a stressor, respectively), and only one letter was used to identify tissue type instead of two. For example: gmnlkfic is identified as a SSH library created from *G. morhua* (gm) fish, from the Newfoundland Breeding Program (nl), head kidney (k) was the tissue used, it was a forward SSH library (f), and the fish had undergone a treatment with pIC (ic).

DNA Sequencing

DNA sequencing was performed as described in Bowman et al. 2007. Briefly, individual bacterial clones were inoculated into 384 well plates containing LB/glycerol, plus an appropriate antibiotic to select for growth of transformed clones, using a QPIX robotic clone picker. After incubation overnight, clones were stored at −80°C prior to sequencing. DNA was prepared from each clone using templiphi (GE Healthcare, Piscataway, NJ), with sequencing reactions performed using ET terminator chemistry (GE Healthcare). After purification to remove excess fluorescent terminator reagents, samples were separated by electrophoresis on MegaBACE capillary sequencers (GE Healthcare).

EST Clustering and Annotation

Sequence trace files were basecalled using Phred (Ewing et al. 1998; Ewing and Green 1998), with poor quality sequences trimmed using Phred and vector sequences masked using Paracel Transcript Assembler (PTA). The resulting ESTs were clustered using PTA. Initially, normalized libraries were sequenced from the 3′ end. Clustering of these 3′ sequences was performed on a per-library or per-tissue basis with an additional clustering run of all 3′ ESTs. The contigs and singletons generated from clustering of 3′ sequences from all ESTs were used to select a set of unique clones for additional 5′ sequencing. This involved the clones for all singletons being selected for 5′ sequencing with, in addition, one clone from each contig also being selected (the clone representing the extreme 5′ end of that contig). Selected clones were rearrayed using a QPIX robot to inoculate wells from the original plates into new 384 well plates containing LB/glycerol/antibiotic. Sequencing and trace processing were carried out as described earlier. Additional clustering runs of 3′ and 5′ sequences were then performed for each normalized library, and a clustering run of all EST data (3′, 5′ and SSH) was also generated. ESTs have been deposited in the dbEST subsection of GenBank under accession numbers shown in Supplemental Table S1.

Sequences and contig consensi were processed through an automated annotation pipeline based on AutoFACT (Koski et al. 2005). This program processes the output of multiple database searches and generates a consensus annotation. Databases used in AutoFACT included UniRef90 (Uniprot), nr (NCBI), Kyoto Encyclopedia of Genes and Genomes (KEGG), Clusters of Orthologous Groups (COG), PFAM, LSU and SSU. BLAST hits were classified as significant in AutoFACT if they had a bit score of 40 or higher. Annotation of the complete dataset can be accessed via the project database Codgene (http://ri.imb.nrc.ca/codgene). Codgene displays information on all libraries constructed, and includes annotation summaries, access to relevant sequences by keyword searching, BLAST searching of selected datasets and access to statistics generated for libraries, sequences and clustering runs.

SNP Identification and Genotyping

SNPs were identified from clusters of four or more 3′ sequences with a predicted minor allele frequency greater than 25% as described in Hubert et al. 2009. Two GoldenGate panels (Illumina, San Diego, CA) of 1536 predicted SNPs were constructed, with high-throughput genotyping carried out at the Genome Quebec and McGill University Innovation Centre. DNA for genotyping was prepared from fin clips taken from fish collected around the Atlantic Canadian coast, including Cape Sable (NB

YC1), Georges Bank (NB YC2), Bay Bulls (NL YC2) and Smith Sound (NL YC3), and from Northern Europe, including Iceland, Ireland and Norway using the Qiagen DNeasy tissue kit (Mississauga, ON) according to the manufacturer's instructions.

RESULTS

EST Sequencing

In total, we generated 158,877 sequences for Atlantic cod. Of these, 138,760 were generated from 23 normalized libraries, and 20,117 from 19 SSH libraries (Table 1). These libraries were generated from 12 tissue types, and four developmental stages, including embryos, larvae, juveniles and adults. To maximise identification of genes relevant to desirable traits for aquaculture, some samples were taken from fish subjected to stressors (e.g. net or heat shock) (Hori et al. 2010) or exposures to immunogenic stimuli (e.g. pIC (Rise et al. 2008) or formalin-killed, atypical *A. salmonicida*(Feng et al. 2009)). Several SSH-identified transcripts have been shown by quantitative reverse transcription polymerase chain reaction to be responsive to immunogenic stimuli (Feng et al. 2009; Rise et al. 2008) or heat shock (Hori et al. 2010), indicating that our targeted gene discovery approach was effective for the identification of genes important in the cellular responses to such stimuli.

Table 1: List of CGP libraries

Library Name	Library Type	Breeding Program	Tissue Type	For./ Rev.	Tissue Treatment	No. of sequences
gmnsil	N	N/A	noda. infected liver	N/A	none	381
gmnsul	N	N/A	uninfected liver	N/A	none	136
gmnbbr	N	NB	brain	N/A	none	9322
gmnbgi	N	NB	gill	N/A	none	2224
gmapht	N	NB,NL,NH	heart	N/A	none	16983
gmapov	N	NB,NL,NH	ovary	N/A	none	11555
gmnbbrts	N	NB	brain	N/A	thermal stress	3358
gmnblits	N	NB	liver	N/A	thermal stress	15475
gmnbhkas	N	NB	head kidney	N/A	*A. sal.*	6057

gmnbhkic	N	NB	head kidney	N/A	pIC	7510
gmnbmd	N	NB	mixed digestive	N/A	none	6151
gmnbmu	N	NB	muscle	N/A	none	4600
gmapte	N	NB,NL,NH	testis	N/A	none	4225
gmnlskic	N	NL	spleen/head kidney	N/A	pIC	4665
gmlbgits	N	NB,NL	gill	N/A	thermal stress	1168
gmnbpcic	N	NB	pyloric caeca	N/A	pIC	4561
gmnbspic	N	NB	spleen	N/A	pIC	4544
gmnl2pia	N	NL	blood	N/A	pIC & *A. sal.*	2538
gmnlem	N	NL	embryo	N/A	none	12135
gmnlla	N	NL	larvae	N/A	none	20493
gmnlpbas	N	NL	blood	N/A	*A. sal.*	71
gmnlpbia	N	NL	blood	N/A	pIC & *A. sal.*	513
gmnlpbic	N	NL	blood	N/A	pIC	95
gmnlkfic	SSH	NL	head kidney	F	pIC	1005
gmnlkric	SSH	NL	head kidney	R	pIC	135
gmnlsfic	SSH	NL	spleen	F	pIC	3005
gmnlsric	SSH	NL	spleen	R	pIC	869
gmnlkfta	SSH	NL	head kidney	F	heat shock	1451
gmnlkrta	SSH	NL	head kidney	R	heat shock	93
gmnllfta	SSH	NL	liver	F	heat shock	1524
gmnllrta	SSH	NL	liver	R	heat shock	1586
gmnlmfta	SSH	NL	skeletal muscle	F	heat shock	1419
gmnlbfic	SSH	NL	brain	F	pIC	2067
gmnlbric	SSH	NL	brain	R	pIC	2111
gmnlkfas	SSH	NL	head kidney	F	*A. sal.*	1033
gmnlkras	SSH	NL	head kidney	R	*A. sal.*	986
gmnlpfas	SSH	NL	blood	F	*A. sal.*	81
gmnlpras	SSH	NL	blood	R	*A. sal.*	85
gmnlpfic	SSH	NL	blood	F	pIC	442
gmnlpric	SSH	NL	blood	R	pIC	90
gmnlsfas	SSH	NL	spleen	F	*A. sal.*	1048
gmnlsras	SSH	NL	spleen	R	*A. sal.*	1087

For each library, the type, breeding program, tissue(s) used and any treatments used on fish prior to tissue collection are shown, together with the number of sequences generated. Libraries produced were generated using either normalized library (N) or suppression subtractive hybridization (SSH) protocols, with the latter being divided into forward (F) or reverse (R) libraries. Fish were collected for the New Brunswick (NB), Newfoundland (NL) or New Hampshire (NH) breeding programs, or originated from a pre-project collection (N/A). The scheme used for library nomenclature is described in the Methods section. N/A = not applicable, noda. = nodavirus. For details on the heat shock challenge, see Hori et al 2010. For details on the formalin-killed, atypical *Aeromonas salmonicida* (*A. sal.*) challenge, see Feng et al. 2009. For details on the viral mimic (pIC) challenge, see Rise et al. 2008.

To facilitate marker development (e.g. SNP discovery), normalized cDNA libraries were generated from tissue samples isolated from numerous fish so that sequence variation between individuals could be identified. The number of individuals from which tissue was pooled prior to RNA extraction ranged from three to 340, with an average of 41. RNA used for SSH library generation was pooled after extraction. Individuals sampled were taken either from the same populations as the fish used for the production of families in NB and NL, or from fish enrolled in the family programs (for example parents of families or F1 offspring) to ensure that the markers identified were present in the family programs. New Brunswick parent fish contributing to the sequence data were collected from Northwest Atlantic Fisheries Organisation (NAFO) subdivision 4X (Myers et al. 1997) (Cape Sable in south west Nova Scotia) and were used to generate NB year class 1 families in 2006 (NB YC1). Newfoundland fish contributing to the sequence data were collected from NAFO subdivision 3L (Bay Bulls, NL) and were used the following year to generate NL YC2 families (Tables 1, 2 and 3).

Table 2: Normalized cDNA libraries created for *Gadus morhua* contributing sequences used for SNP identification

Library Name	No. of Fish	Breeding Program	Age	Fish Population	No. of sequences
gmnsil	3	N/A	Adult	Pre-project collection from NS waters	178
gmnsul	3	N/A	Adult	Pre-project collection from NS waters	67

gmnbbr	10	NB program	Adult	Same population as NB YC1 parents	5992
gmnbgi	10	NB program	Adult	Same population as NB YC1 parents	1346
gmapht	20	NB, NL, NH program	Adult	Same population as NB YC1 and NL YC2 parents	12374
gmapov	20	NB, NL, NH program	Adult	Same population as NB YC1 and NL YC2 parents	7683
gmnbbrts	14	NB program	Juvenile	NB YC1 F1 progeny	1959
gmnblits	14	NB program	Juvenile	NB YC1 F1 progeny	10650
gmnbhkas	14	NB program	Juvenile	NB YC1 F1 progeny	4665
gmnbhkic	18	NB program	Juvenile	NB YC1 F1 progeny	5460
gmnbmd	10	NB program	Adult	NB YC1 parents	4148
gmnbmu	10	NB program	Adult	NB YC1 parents	3720
gmnbspic	18	NB program	Juvenile	NB YC1 F1 progeny	3833
gmnbpcic	14	NB program	Juvenile	NB YC1 F1 progeny	3858
gmapte	17	NB, NL, NH program	Adult	Same population as NB YC1 and NL YC2 parents	3279
gmnlskic	21	NL program	Juvenile	NL YC1 F1 progeny	3795
gmnlem	340	NL program	Embryo	NL YC2 F1 progeny	8958
gmnlla	290	NL program	Larvae	NL YC2 F1 progeny	15550
gmlbgits	12	NB, NL program	Juvenile	NB YC1 F1 progeny, NL YC1 F1 progeny	241
gmnlpbia	26	NL program	Juvenile blood	NL YC1 F1 progeny	240

Libraries were generated by pooling one (or more) tissue types from multiple fish, from one (or more) breeding programs. YC1 and YC2 denote Year Class 1 and Year Class 2 fish, respectively.

Table 3: SSH libraries created for *Gadus morhua*

Library name	No. of fish treated	No. of fish control	Family
gmnlsfic	18	12	NL YC1 F1 family 32
gmnlsric	18	12	NL YC1 F1 family 32
gmnlkfic	18	12	NL YC1 F1 family 32
gmnlkric	18	12	NL YC1 F1 family 32
gmnlkfta	32	32	NL YC1 F1 family 4
gmnlkrta	32	32	NL YC1 F1 family 4
gmnllfta	32	32	NL YC1 F1 family 4
gmnllrta	32	32	NL YC1 F1 family 4
gmnlmfta	32	32	NL YC1 F1 family 4
gmnlbfic	24	16	NL YC1 F1 family 32
gmnlbric	24	16	NL YC1 F1 family 32
gmnlkfas	20	20	NL YC1 F1 family 32
gmnlkras	20	20	NL YC1 F1 family 32
gmnlpfas	20	20	NL YC1 F1 family 32
gmnlpras	20	20	NL YC1 F1 family 32
gmnlpfic	20	20	NL YC1 F1 family 32
gmnlpric	20	20	NL YC1 F1 family 32
gmnlsfas	20	20	NL YC1 F1 family 32
gmnlsras	20	20	NL YC1 F1 family 32

The numbers of fish contributing to each library are shown, together with the program and family details. In the "forward" SSH libraries (i.e. enriched for transcripts that were up-regulated by the treatment), the treated fish samples were used as the tester and the control samples were used as the driver. In the "reverse" SSH libraries (i.e. enriched for transcripts that were down-regulated by the treatment), the control fish samples were used as the tester and the treated samples were used as the driver as previously described (Rise et al. 2008). All fish were juveniles enrolled in the NL CGP breeding program.

Normalized libraries were initially sequenced from the 3′ direction, which facilitated clustering and ensured access to the 3′ untranslated region (3′UTR), where we anticipated finding a lower level of splicing (Kan et al. 2000) and a higher level of polymorphism for the purpose of marker identification. The disadvantages of this approach include reduced sequence quality, more sample failures, shorter read lengths and fewer informative hits to sequence databases. To allow better informative annotation of the sequence generated, a clone representing each singleton, or a single clone selected from the extreme 5′ end of each unique contig, was also sequenced from the 5′ direction. SSH libraries were cloned non-directionally and had relatively short insert lengths (200–500 base pairs on average); frequently single-pass sequencing was sufficient to generate the entire insert sequence. EST clustering was carried out for individual libraries, individual tissues, for all reads from the 3′ direction and for the entire sequence collection (all 3′, 5′ and SSH reads). Data for the clustering runs on sequences for individual libraries and tissues can be accessed from the CGP database (http://ri/imb.nrc.ca/codgene/index.php), with statistics from the clustering runs of 3′-only and the entire collection shown in Table 4.

Table 4: EST assembly summary statistics

	All Version 2.0	**3′ reads only**
Number of good sequences	154,142	97,976
Average trimmed EST length (bp)	563	591
Number of contigs	23,838	13,448
Number of singletons	27,976	21,746
Number of putative transcripts	51,814	35,194
Maximum no. of ESTs per contig	154	83
Average no. of ESTs per contig	5.27	5.66
Number of putative transcripts with		
Significant BLASTX hits	15,873	8,628
No significant BLAST hits	35,941	26,566
Percentage with no significant BLAST hits	69.37	75.48
Number of contigs containing		
2 ESTs	9,618	4,415
3 ESTs	3,993	2,310
4 ESTs	2,423	1,444
5-10 ESTs	5,194	3,558
11-20 ESTs	1,886	1,309
21-30 ESTs	452	279

31-50 ESTs	221	121
>50 ESTs	51	12

The number of putative transcripts is defined as the number of contigs plus the number of singletons. A score of 1 × e-5 was used as the *E* value cut-off for BLAST

A total of 154,142 reads from the set of 158,877 sequences were clustered due to additional screening criteria applied by the clustering software (PTA (Paracel 2002)). Clustering of the entire sequence set (All Version 2.0) resulted in 23,838 contigs and 27,976 singletons. The largest number of ESTs in a contig was 154 (all_v2.0.14.C13, best hit interferon stimulated gene 15, *G. morhua*), with an overall average of 5.27 ESTs per contig. The total number of putative transcripts (number of contigs plus number of singletons) identified in All Version 2.0 was 51,814. The average trimmed length for ESTs in this clustering run was 563 bp. This is slightly lower than that for the clustering of 3′ reads alone (591 bp average length) due to the shorter trimmed read lengths for sequences generated from the SSH libraries which were not included in the 3′ clustering. For 3′ reads only, 97,976 reads were included in EST clustering, resulting in 13,448 contigs, 21,746 singletons and a total of 35,194 for the number of putative transcripts. This lower estimate for the number of transcripts in Atlantic cod is probably a better estimate of the actual number of genes in this species, as the overall assembly contains many non-overlapping sequences for clones and transcripts, including both 5′ and 3′ reads from sequencing clones, but also potential internal regions of transcripts generated through SSH sequencing.

Annotation of Sequences

The number of putative transcripts which have a significant BLASTX hit (using 1 × e-5 as the *E*value cut-off) with an existing entry in the NCBI nr database was 15,873 (30.6%) from the complete sequence set (All Version 2.0) and 8,628 (24.5%) from the 3′ reads only. Both datasets are biased for the 3′ region of transcripts, with the low number of significantly similar sequences likely to be due to the high representation of 3′ UTRs within the sequence generated.

The complete sequence set was annotated using an automated pipeline based on AutoFACT (Koski et al. 2005). This performs sequence searches across multiple databases, combining the results to give a consensus annotation. AutoFACT uses a pipeline for automated annotation that attempts to use corroborating information from multiple sources before assigning an annotation to a sequence. It will also give informative hits greater priority than non-informative hits in the final annotation produced. All singletons and contigs were annotated using this approach. The 3′-only data were used primarily for SNP identification. A bit score of 40 was used as the cut-off value for AutoFACT annotation, allowing the comparison of equivalent hits from different databases. AutoFACT associated informative annotation to 16,258 putative transcripts, with a further 2,090 putative transcripts matching database entries that have no functional annotation attached.

In total, 2,287 putative transcripts can be assigned to KEGG pathways, with the most frequent assigned name being "ribosome" (312 sequences), "oxidative phosphorylation" (145) and "ubiquitin mediated proteolysis" (114). Similar results are obtained after comparison to the COG dataset, with 2,315 sequences assigned to COG categories, the most frequent meaningful categories being "translation, ribosomal structure and biogenesis" (382 sequences), "posttranslational modification, protein turnover, chaperones" (344) and "energy production and conversion" (136). Sequences have also been assigned to Gene Ontology (GO) categories as follows: 1,047 in the cellular component category with the most frequent being "membrane" (161), "intracellular" (161) and "nucleus" (123); 2,801 in the molecular function category, with "protein binding" (139), "catalytic activity" (131) and "zinc ion binding" (120) the most frequent; and 1,366 for the biological process category, with "transport" (92), "immune response" (91) and "proteolysis" (84) being the most frequent. Annotated sequences can be accessed from the project database (http://ri.imb.nrc.ca/codgene/index.php).

SNP Detection and Analysis

Contigs generated from clustering of 3′ reads only were used for SNP identification to reduce the likelihood of designing amplicons for SNP assays in spliced regions, and to prevent the over-sampling from a single clone that would occur if both 3′ and 5′ sequences had regions of overlap within a contig. A SNP identification pipeline incorporating PolyPhred (Nickerson et al. 1997) was developed, with custom Perl scripts used to extract information regarding contig coverage, predicted SNPs and the proportion of contributing sequences harbouring each variant (Hubert et al. 2009). SNPs identified in this process were grouped into categories based on the ratio of sequences having each variant, making the assumption that sequences from two independent clones in a contig that originated from the same allele of the same individual would represent a rare event. An ideal situation for a predicted frequent SNP (PF-SNP) was expected to consist of a contig with many contributing sequences over the region analysed where the two different versions of the SNP were present in a 50:50 ratio. Ultimately, the criteria chosen for selection of the SNPs included a minimum of four read coverage at the position of the SNP within a contig, and a minor allele frequency greater than 25%, which ensured that each version of the SNP was sampled at least twice. This led to the identification of 4753 PF-SNPs. PF-SNPs were chosen for SNP assay development as they were likely to be more frequently variable within family crosses in the breeding program, thus allowing their placement on a genetic map and their potential association with QTL. SNPs were identified as predicted rare SNPs (PR-SNPs) if the minor allele was observed infrequently within a contig, often a single read with a sequence variant, and these have been excluded from SNP assay development at this stage. We also treated PR-SNPs as potential artefacts as there is an increased likelihood that they arose from polymerase errors or sequence mis-calls rather than representing real variation. A small subset of PF-SNPs was confirmed as variable using a low throughput pipeline (Hubert et al. 2009) with a larger set subsequently used for high throughput genotyping using the Illumina GoldenGate platform.

The sequences used for SNP identification originated from two year classes of fish enrolled in the selective breeding programs, NB YC1 and NL YC2. Fish from which tissues were extracted for cDNA library construction included adults which contributed eggs and sperm for the generation of F1 individuals, F1 progeny (embryos, larvae and juveniles) and adults from the same populations as the wild fish used as broodstock, but which did not make a genetic contribution to the breeding programs. The assembly of the 3′ reads used for SNP discovery consisted of approximately 58% sequences originating from NB YC1 and 41% from sequences originating from NL YC2 (Table 2), with 1% of sequences from other sources. Because we had identified SNPs based on their frequency in sequences generated from NB YC1 and NL YC2, we expected our SNP set to be most variable in these populations and also to show the highest values for observed heterozygosity and minor allele frequency. Table 5 shows the results of high throughput GoldenGate genotyping for our selected SNPs on several populations of fish, collected from Canada and Europe. The fish genotyped from NB YC1 and NL YC2 were from the same populations as the fish that were used as parents in those year classes, but were not used as parents themselves so did not contribute to the sequence data. Table 6 shows values for observed heterozygosity and minor allele frequency for a small number of SNPs selected from the larger set in fish from Cape Sable and Bay Bulls populations.

Table 5: SNP characteristics

Population	No. of fish genotyped	Polymorphic SNPs	Sequence contribution of population to SNP assembly
NB YC1 (Cape Sable)	23	71%	58%
NB YC2 (Georges Bank)	24	71%	N/A
NL YC2 (Bay Bulls)	23	69%	41%
NL YC3 (Smith Sound)	23	70%	N/A
Iceland	26	61%	N/A
Ireland	15	53%	N/A
Norway	25	56%	N/A

The percentage of SNPs tested that were polymorphic in different populations of fish is shown. Only NB YC1 and NL YC2 fish provided samples from which the sequence collection was generated. N/A = not applicable.

Table 6: Examples of sequences which were used in microarray construction that also harbour a SNP

	Best Hit in NCBI nr database				**SNPs**		
Contig name	Accession number	Annotation	Species	*E*value	SNP name	Cape Sable	Bay Bulls
Ribosomal							
all_v2.0.609.C4	gb|ACQ58145.1	39S ribosomal protein L30, mitochondrial precursor	*Anoplopoma fimbria*	1.00E-11	cgpGmo-S198	0.09/0.04	0.30/0.15
all_v2.0.4842.C1	gb|ACO09691.1	39S ribosomal protein L32, mitochondrial precursor	*Osmerus mordax*	7.00E-46	cgpGmo-S967a	0.61/0.43	0.52/0.43
all_v2.0.9851.C1	gb|ACI67749.1	39S ribosomal protein L52, mitochondrial precursor	*Salmo salar*	2.00E-35	cgpGmo-S878	0.43/0.22	0.48/0.37
all_v2.0.560.C2	gb|ACH70774.1	ribosomal protein S7	*Salmo salar*	1.00E-101	cgpGmo-S1917	0.30/0.20	0.26/0.22
all_v2.0.3432.C1	gb|ACN10030.1	60S ribosomal protein L10	*Salmo salar*	1.00E-119	cgpGmo-S105	0.17/0.13	0.30/0.15
all_v2.0.1527.C1	gb|ACN10349.1	60S ribosomal protein L18a	*Salmo salar*	1.00E-90	cgpGmo-S1730	0.35/0.30	0.26/0.22
all_v2.0.4345.C1	gb|ACI67287.1	60S ribosomal protein L27	*Salmo salar*	9.00E-65	cgpGmo-S474	0.22/0.20	0.26/0.26
all_v2.0.14203.C1	gb|ACN10033.1	60S ribosomal protein L38	*Salmo salar*	5.00E-25	cgpGmo-S916	0.35/0.26	0.39/0.24
all_v2.0.983.C1	gb|ACO09602.1	60S ribosomal protein L8	*Osmerus mordax*	1.00E-140	cgpGmo-S357	0.43/0.22	0.39/0.32

all_v2.0.64.C1	dbj|BAF98661.1	Ribosomal protein L13a	*Solea senegalensis*	1.00E-91	cgpGmo-S1519	0.35/0.30	0.35/0.35
all_v2.0.603.C1	dbj|BAF45898.1	Ribosomal protein S10	*Solea senegalensis*	4.00E-69	cgpGmo-S2265	1.00/0.50	1.00/0.50[a]
all_v2.0.545.C7	gb|ACO09841.1	40S ribosomal protein S16	*Osmerus mordax*	5.00E-67	cgpGmo-S649a	0.30/0.20	0.52/0.26
all_v2.0.3134.C1	ref|NP_001134397.1	40S ribosomal protein SA	*Salmo salar*	1.00E-139	cgpGmo-S699b	0.22/0.11	0.09/0.04
Immune/Stress							
all_v2.0.2956.C2	gb|ACO09614.1	Complement component 1 Q subcomponent-binding protein	*Osmerus mordax*	1.00E-115	cgpGmo-S1200	0.52/0.48	0/0
all_v2.0.4174.C1	gb|ACO14444.1	Cystatin-F precursor	*Esox lucius*	1.00E-27	cgpGmo-S296	0.04/0.06	0.13/0.06
all_v2.0.2115.C1	gb|ACN10355.1	CXC chemokine receptor type 4	*Salmo salar*	1.00E-59	cgpGmo-S525	0.43/0.26	0.35/0.17
all_v2.0.2319.C4	gb|AAF72567.1	Immunoglobulin D heavy chain constant region variant a	*Gadus morhua*	3.00E-33	cgpGmo-S1542	1.00/0.50	1.00/0.50[a]
all_v2.0.4114.C1	ref|XP_001923855.1	Similar to interleukin 12 receptor beta 2.b	*Danio rerio*	1.00E-23	cgpGmo-S946	0.35/0.48	0.65/0.37
all_v2.0.1550.C1	gb|ACO14448.1	Macrophage migration inhibitory factor	*Esox lucius*	1.00E-41	cgpGmo-S2107	0.52/0.39	0.30/0.24
all_v2.0.8001.C1	ref|XP_001894282.1	T-cell receptor beta chain ANA 11	*Brugia malayi*	2.00E-07	cgpGmo-S1484	0.48/0.28	0.48/0.33
all_v2.0.682.C1	gb|ACN10899.1	T-complex protein 1 subunit alpha	*Salmo salar*	1.00E-106	cgpGmo-S772	0.30/0.28	0.52/0.43
all_v2.0.3543.C1	ref|NP_001133482.1	T-complex protein 1 subunit delta	*Salmo salar*	1.00E-107	cgpGmo-S2247	1.00/0.50	1.00/0.50[a]

The contig name in the All Version 2.0 assembly is given, together with the accession number and annotation of the best BLAST hit in the NCBI nr database. The SNP name for a SNP identified on that contig is shown, together with values for the observed heterozygosity and minor allele frequency for that SNP in fish collected from Cape Sable, Nova Scotia and Bay Bulls, Newfoundland.

[a]Indicates putative SNPs that may represent variation between different genes rather than different alleles of the same gene

The population used to produce NB YC1 showed the greatest number of polymorphic SNPs (71%). For the population used to generate NL YC2, 69% of SNPs were polymorphic. Two other Canadian populations were tested which contributed to two additional year classes enrolled in the CGP breeding programs, NB YC2 from Georges Bank in the Gulf of Maine, and NL YC3 from Smith Sound, NL. These populations also showed high levels of polymorphism (71% and 70% respectively). Fish isolated in European waters showed a slightly lower level of polymorphism, with fish from Iceland, Ireland and Norway having polymorphism levels of 61%, 53% and 56%, respectively.

Selection of Sequences for Microarray Fabrication

Putative transcripts have been selected from All Version 2.0 for inclusion on a 20,000 element, 50-mer oligonucleotide microarray. Initial criteria for selection attempted to maximise the inclusion of sequences with informative annotation (both contigs and singletons), followed by sequences with similarity to database entries of unknown function. The remaining sequences were selected based on criteria such as known sequence directionality (for design of sense oligonucleotides) and degree of representation, including contigs with a higher number of contributing sequences in preference to contigs with low depth of coverage, or singletons. A few of the deeper contigs containing SSH sequences alone (which were cloned non-directionally), but lacking significant BLAST hits, were included as both sense and anti-sense sequences, in the hope that they represented genes important in response to stimuli such as antigen injection or heat shock. All other contigs and singletons composed entirely of SSH-identified ESTs, but having significant sequence similarity (e.g. E values less than $1 \times e\text{-}5$), were included in the microarray probe design pipeline as they could be oriented using their BLAST hit. Selection of sequences included on the microarray will be discussed in detail elsewhere (M. Booman, personal communication). However, many of these sequences were also in the set used for SNP detection. A subset of oligonucleotides that were included on the microarray has been designed from sequences for which an informative SNP was also developed. In total, 1,391 oligonucleotides included on the 20,000 element microarray fall into this category, and some examples taken from the larger set are shown in Table 6. Of these, 605 oligonucleotides have associated AutoFACT annotation, with 567 having informative annotation and 38 hits annotated as "unassigned protein". The remaining 786 (56%) remain "unclassified" by AutoFACT as the sequences from which they were derived currently have no significant similarities in any of the databases searched. Many of the annotated sequences have similarities to structural or housekeeping proteins, but some have similarity to proteins involved in pathways of particular interest, such as immune response, thermal adaptation or reproduction. Two examples of functional categories identified from the larger SNP set, ribosomal subunits and proteins potentially involved in immunological or stress response pathways, are shown in Table 6.

DISCUSSION

Species-specific genomics projects have frequently been carried out in an iterative fashion, with an initial sequencing phase followed by a marker development phase, followed by the subsequent application of sequences and markers in downstream analyses such as microarray development and QTL identification (Rise et al. 2004; Snelling et al. 2005; Moen et al. 2008). The genomic information for a single species has often been developed by many groups of researchers without overall coordination, leading to the generation of sequences from strains or varieties that are not in current commercial use. Thus sequence information cannot be directly mined for markers relevant in commercial production, resulting in a requirement for additional re-sequencing to develop markers in varieties or strains used in industry. We have designed an integrated approach (Fig. 1) in developing genomics tools for Atlantic cod, such that the information generated is directly and immediately applicable in commercial cod breeding.

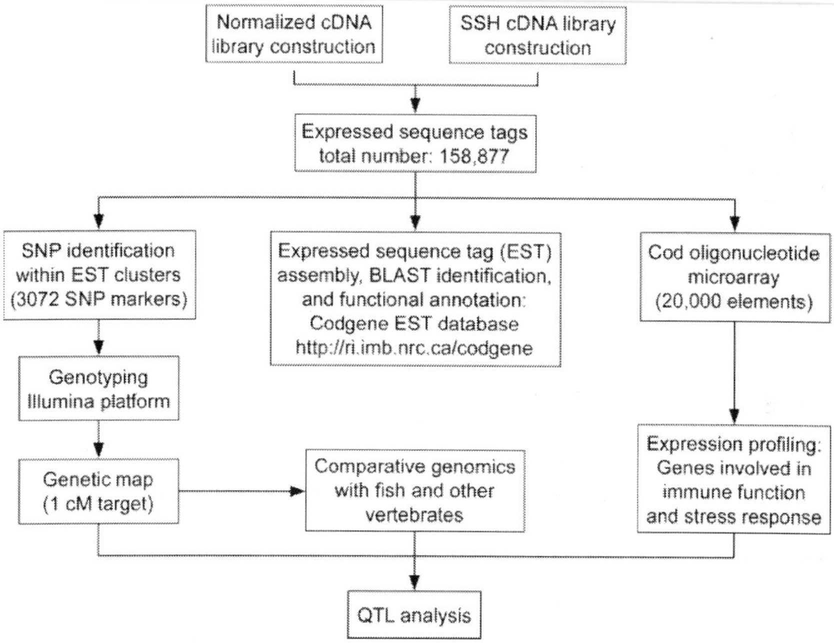

Figure. 1: An overview of the genomics workflow within the CGP. The sequence information has been generated from populations of fish enrolled in selective breeding and is being used to develop high-throughput molecular resources for Atlantic cod. These genomics tools (such as genetic markers and an oligonucleotide microarray) will be applied to dissection of QTL within the family-based breeding programs (adapted from Rise et al, in press).

Our sequencing approach was designed to maximise gene identification, to ensure that genes of relevance in aquaculture were present in our collection, and to allow identification of markers present in two CGP breeding programs for use in linkage mapping and QTL identification. Normalized cDNA libraries were constructed from major tissues, including liver, heart, brain, gill, skeletal muscle, digestive tissue (pyloric caecae, anterior stomach, anterior intestine and middle intestine), spleen, head kidney, blood, ovary and testis (Table 1). Four developmental stages were also used in library construction, including embryo, larvae, juvenile (F1 progeny within the breeding programs) and adult (either parents, or fish from the same populations as the parents enrolled in the breeding programs). Some of these libraries were selected to target genes involved in specific processes such as reproduction (liver, ovary, testis, embryo) and genes contributing to the immune response (head kidney, spleen, blood). In addition, several libraries were constructed from tissues collected from fish that had been subjected to biotic or abiotic stressors, including exposure to elevated temperature (liver, brain, gill), to the viral mimic pIC (head kidney, spleen, pyloric caecae, blood), to nodavirus (liver) and to the bacterial pathogen atypical *A. salmonicida* (head kidney, blood). Multiple individuals were used per library for the purpose of marker development, to ensure that the sequence contribution from a single individual was low, and that the variation captured originated from multiple fish.

As the ability to identify highly up-regulated transcripts is lost through use of normalization protocols, we also constructed a series of SSH libraries, with a focus on identifying genes involved in the immune response, or those that respond to temperature stress. Although these libraries were constructed from multiple individuals, each library was constructed from a single family of fish, and therefore any sequence variation captured represented the variation present in only two individuals i.e. the parents giving rise to that cross (Table 3). Although we would expect to find many differences between two unrelated individuals, it was possible that alleles would be identified in this set of sequences that were rare within the population as a whole. Therefore the SSH sequences were not included for marker discovery using the automated pipeline. The SSH library construction protocol uses restriction enzyme digestion with RsaI followed by non-directional cloning (Diatchenko et al. 1996), giving rise to sequences of shorter read length where directionality could only be inferred if they assembled with a sequence of known directionality from the collection of normalized reads, or if an informative sequence similarity was found after database searching. However, in general, similarities were identified at higher frequency for SSH libraries when compared to normalized libraries, as the former often represented an internal coding cDNA fragment, whereas the latter contained a high proportion of non-coding 3′ sequence.

A set of microsatellite markers was identified in this sequence at an early stage of production (Higgins et al. 2009), but it was necessary to accumulate sufficient depth of coverage within contigs before SNP detection could commence. Identification of putative SNPs was carried out on contigs generated through clustering and assembly of reads generated from the 3′ end of normalised libraries. The 5′ reads were not used to prevent over-representation of these clones in the sequence used for SNP discovery. SSH sequences were not used because of the limited genetic variation captured in these libraries as described above. The pipeline used for SNP identification has been described in detail elsewhere (Hubert et al. 2009). This study describes the SNPs with respect to the sequences and libraries from which they were generated, and in the context of the CGP breeding programs. Two year classes contributed to sequence generation, from which the SNPs were isolated; NB YC1 and NL YC2. The contribution of each year class to the EST assembly used for SNP discovery was 58% and 41%, respectively. With respect to NB YC1 and NL YC2, the individuals used for genotyping were taken from the same populations as the parents used for generation of those year classes, rather than the parents themselves to avoid over-estimating the levels of polymorphism in those populations. Because the SNPs had been selected based on their predicted high variability in those year classes, we expected to see the greatest levels of both polymorphism and minor allele frequencies in those populations, with NB YC1 predicted to have the higher value because of its larger contribution to the sequence data. This was found to be largely the case, although the population used in the generation of NB YC2 (Georges Bank) showed marginally higher levels of polymorphism than NL YC2. However, the two populations used for production of the NB year classes, Cape Sable and Georges Bank, are from the adjacent NAFO subdivisions 4X and 5Y, respectively, so this is not entirely unexpected. Again, the population used for NL YC3 was not used for sequence generation, and had intermediate values for polymorphism when compared with the other Canadian populations. In comparison, three European populations were tested, showing levels of polymorphism ranging from 53% to 61%. Therefore it will be possible to use the SNPs described here to analyse populations across the North Atlantic, although their utility will be greatest for Canadian populations.

Although it is possible that the differences in the levels of polymorphism observed between Canadian and European fish could be due to a lower level of genetic diversity for this species in European populations, it is unlikely, and the reduction observed is most probably due to ascertainment bias related to the origin of the sequence from which the SNPs were identified. It is therefore likely to result from the limited gene flow between remote populations. The polymorphic loci shared between all populations across the North Atlantic represent a valuable pool of potentially ancient, conserved polymorphisms that could find widespread application within population studies and breeding programs.

As we aimed to capture sequence diversity at the same time as maximising gene identification, parameters used for clustering and assembly allowed sequences with a small number of sequence differences to be incorporated within a contig. In addition, because a large number of individuals were being pooled in the construction of each library, we expected multiple alleles to be present for each gene within the sequence collection. However, this has also made it difficult to determine by bioinformatics means alone whether a contig contains different alleles of the same gene, multiple, closely related genes, or a combination of these two possibilities. We have tested this by genotyping the set of SNPs generated in several CGP families to determine whether SNPs are inherited in a manner consistent with Mendelian segregation. The observation that both parents and progeny from a single cross appear to be heterozygotes is diagnostic for inter-gene differences rather than allelic variants. From these results, we estimate that 4% of contigs represent multiple genes assembled together, rather than the desired result in our assembly such that one contig contains only alleles of a single gene. The ideal substrate for sequencing to facilitate correct gene and allele identification would be a doubled haploid fish (Streisinger et al. 1981). It should be possible to generate doubled haploid individuals in Atlantic cod that are homozygous at all loci, and this may be an appropriate approach to take in sequencing the genome of this species. Nevertheless, the incidence of multiple genes similar enough to assemble together in a single contig appears to be reasonably low in our assembly, and a more in-depth study of gene families can dissect out the different members and alleles by further sequencing and genotyping (Borza et al. 2009).

The CGP sequences have also been used to design oligonucleotides for spotting on a microarray (Fig. 1) for use in expression profiling. The focus of this effort has been to include sequences having informative annotation, whether they are represented once, or multiple times in our dataset. However, our selection strategy also incorporates sequences that are frequently represented in the CGP sequence collection, and this has ensured that the majority of the sequences used for SNP development are also present on the microarray. Because the sequences originated from tissues taken from fish which had been exposed to a variety of stressors such as immunogenic stimuli (e.g. exposure to bacterial antigens or a viral mimic) or thermal stress, or for several developmental stages, such as eggs, larvae and steps in the reproductive cycle in both sexes, genes involved in these processes are likely to have been expressed, and therefore will be present in our sequence collection. However, because the sequence generated is biased for the 3′UTR to facilitate clustering of individual transcripts and to aid in SNP identification, many of these sequences are currently without informative annotation. Nevertheless, we have been successful in identifying many transcripts where sequence similarity allows us to predict that they function in processes such as reproduction or the immune response. Many of these are represented on the microarray but, in addition, harbour a SNP that has been shown to exhibit polymorphism within the two populations used to generate fish for NB YC1 and NL YC2 (Cape Sable and Bay Bulls, respectively) within the CGP breeding programs. These transcripts will be extremely useful in connecting expression patterns with QTL, with the ultimate aim of identifying the causative genes for traits of interest. For example, a transcript up-regulated on stimulation with viral, bacterial and parasite antigens which is located in a QTL interval associated with disease resistance would be prioritised in further studies.

The integrated genomics tools described here, produced from a single sequence dataset that is tightly linked to fish breeding programs being developed for commercial aquaculture, should enable the rapid dissection of traits and the identification of linked markers which can be applied immediately within these programs to accelerate the development of aquaculture for Atlantic cod.

In conclusion, we have developed a significant sequence resource for Atlantic cod, a species previously under-represented in sequence databases. These sequences have been generated from fish involved in selective breeding and we have been successful in capturing sequence variation in individuals enrolled in these programs. Future work will focus on exploiting this information to identify markers for commercially relevant traits, and on analysis of data from the additional tools that have been developed from this sequence resource, such as a microarray and a high-density genetic map, to enable future high-throughput studies in Atlantic cod.

NOTES

Acknowledgements

This research was supported in part by Genome Canada, Genome Atlantic and the Atlantic Canada Opportunities Agency through the Atlantic Cod Genomics and Broodstock Development Project. A complete list of supporting partners can be found at www.codgene.ca/partners.php. We would like to thank fish husbandry teams at the Joe Brown Aquatic Research Building at the Ocean Sciences Centre of Memorial University, NL, at the Fisheries and Oceans Canada St. Andrews Biological Station, NB and at the Huntsman Marine Sciences Centre, NB, as well as all individuals who were involved in tissue sampling. We also thank Marie-Catherine Tessier and colleagues at the Genome Quebec and McGill University Innovation Centre for high-throughput genotyping.

Open Access

REFERENCES

1. Bonaldo MF, Lennon G, Soares MB (1996) Normalization and subtraction: two approaches to facilitate gene discovery. Genome Res 6:791–806
2. Borza T, Stone C, Gamperl AK, Bowman S (2009) Atlantic cod (*Gadus morhua*) hemoglobin genes: multiplicity and polymorphism. BMC Genet 10:51
3. Bowman S, Higgins B, Stone C, Kozera C, Curtis BA, Tarrant Bussey J, Kimball J, Verheul H, Johnson SC (2007) Generation of genomics resources for Atlantic cod (*Gadus morhua*): progress and plans. Bull Aquac Assoc Can 105:24–30

4. Brown JA, Minkoff G, Puvanendran V (2003) Larviculture of Atlantic cod (*Gadus morhua*): progress, protocols and problems. Aquaculture 227:357–372
5. Diatchenko L, Lau YF, Campbell AP, Chenchik A, Moqadam F, Huang B, Lukyanov S, Lukyanov K, Gurskaya N, Sverdlov ED, Siebert PD (1996) Suppression subtractive hybridization: a method for generating differentially regulated or tissue-specific cDNA probes and libraries. Proc Natl Acad Sci U S A 93:6025–6030
6. Ewing B, Green P (1998) Base-calling of automated sequencer traces using phred. II. Error probabilities. Genome Res 8:186–194
7. Ewing B, Hillier L, Wendl MC, Green P (1998) Base-calling of automated sequencer traces using phred. I. Accuracy assessment. Genome Res 8:175–185
8. Feng CY, Johnson SC, Hori TS, Rise M, Hall JR, Gamperl AK, Hubert S, Kimball J, Bowman S, Rise ML (2009) Identification and analysis of differentially expressed genes in immune tissues of Atlantic cod stimulated with formalin-killed atypical *Aeromonas salmonicida*. Physiological Genomics 37:149–163
9. Gollock MJ, Currie S, Petersen LH, Gamperl AK (2006) Cardiovascular and haematological responses of Atlantic cod (*Gadus morhua*) to acute temperature increase. J Exp Biol 209:2961–2970
10. Gorodkin J, Cirera S, Hedegaard J, Gilchrist MJ, Panitz F, Jorgensen C, Scheibye-Knudsen K, Arvin T, Lumholdt S, Sawera M, Green T, Nielsen BJ, Havgaard JH, Rosenkilde C, Wang J, Li H, Li R, Liu B, Hu S, Dong W, Li W, Yu J, Staefeldt HH, Wernersson R, Madsen LB, Thomsen B, Hornshoj H, Bujie Z, Wang X, Bolund L, Brunak S, Yang H, Bendixen C, Fredholm M (2007) Porcine transcriptome analysis based on 97 non-normalized cDNA libraries and assembly of 1, 021, 891 expressed sequence tags. Genome Biol 8:R45
11. Hansen T, Karlsen O, Taranger GL, Hemre G, Holm JC, Kjesbu OS (2001) Growth, gonadal development and spawning time of Atlantic cod (*Gadus morhua*) reared under different photoperiods. Aquaculture 203:51–67
12. Hayes BJ, Nilsen K, Berg PR, Grindflek E, Lien S (2007) SNP detection exploiting multiple sources of redundancy in large EST collections improves validation rates. Bioinformatics 23:1692–1693

13. He C, Chen L, Simmons M, Li P, Kim S, Liu ZJ (2003) Putative SNP discovery in interspecific hybrids of catfish by comparative EST analysis. Anim Genet 34:445–448

14. Higgins B, Hubert S, Simpson G, Stone C, Bowman S (2009) Characterization of 155 EST-derived microsatellites and validation for linkage mapping. Mol Ecol Resour 9:733–737

15. Hori TS, Kimball J, Johnson SC, Afonso LOB, Bowman S, Hubert S, Gamperl AK, Rise ML (2010) Heat-shock responsive genes identified and validated in Atlantic cod (*Gadus morhua*) using genomic techniques. BMC Genomics 11:72

16. Hubert S, Tarrant Bussey J, Higgins B, Curtis BA, Bowman S (2009) Development of single nucleotide polymorphism markers for Atlantic cod (*Gadus morhua*) using expressed sequences. Aquaculture 296:7–14

17. Hutchinson WF, Carvalho GR, Rogers SI (2001) Marked genetic structuring in localised spawning populations of cod (*Gadus morhua*) in the North Sea and adjoining waters, as revealed by microsatellites. Mar Ecol Prog Ser 223:251–260

18. Kan Z, Gish W, Rouchka E, Glasscock J, States DJ (2000) UTR reconstruction and analysis using genomically aligned EST sequences. Proc Int Conf Intell Syst Mol Biol 8:218–227

19. Khan RA (2005) Prevalence and influence of *Loma branchialis* (Microspora) on growth and mortality in Atlantic cod (*Gadus morhua*) in coastal Newfoundland. J Parasitol 91:1230–1232

20. Koop BF, von Schalburg KR, Leong J, Walker N, Lieph R, Cooper GA, Robb A, Beetz-Sargent M, Holt RA, Moore R, Brahmbhatt S, Rosner J, Rexroad CE 3rd, McGowan CR, Davidson WS (2008) A salmonid EST genomic study: genes, duplications, phylogeny and microarrays. BMC Genomics 9:545

21. Koski LB, Gray MW, Lang BF, Burger G (2005) AutoFACT: an automatic functional annotation and classification tool. BMC Bioinformatics 6:151

22. Labate JA, Baldo AM (2005) Tomato SNP discovery by EST mining and resequencing. Mol Breeding 16:343–349

23. Lijavetzky D, Cabezas JA, Ibanez A, Rodriguez V, Martinez-Zapater JM (2007) High throughput SNP discovery and genotyping in grapevine (*Vitis vinifera* L.) by combining a re-sequencing approach and SNPlex technology. BMC Genomics 8:424

24. Moen T, Hayes B, Baranski M, Berg PR, Kjoglum S, Koop BF, Davidson WS, Omholt SW, Lien S (2008) A linkage map of the Atlantic salmon (*Salmo salar*) based on EST-derived SNP markers. BMC Genomics 9:223

25. Myers RA, Hutchings JA, Barrowman NJ (1997) Why do fish stocks collapse? The example of cod in Atlantic Canada. Ecol Appl 7:91–106

26. Nickerson DA, Tobe VO, Taylor SL (1997) PolyPhred: automating the detection and genotyping of single nucleotide substitutions using fluorescence-based resequencing. Nucleic Acids Res 25:2745–2751

27. PARACEL (2002) PTA: Paracel transcript assembler user manual. Paracel, Inc, Pasadena, CA

28. Park KC, Osborne JA, Tsoi SC, Brown LL, Johnson SC (2005) Expressed sequence tags analysis of Atlantic halibut (*Hippoglossus hippoglossus*) liver, kidney and spleen tissues following vaccination against *Vibrio anguillarum* and *Aeromonas salmonicida*. Fish Shellfish Immunol 18:393–415

29. Patel S, Korsnes K, Bergh O, Vik-Mo F, Pedersen J, Nerland AH (2007) Nodavirus in farmed Atlantic cod *Gadus morhua* in Norway. Dis Aquat Organ 77:169–173

30. Ramsey JS, Wilson AC, De Vos M, Sun Q, Tamborindeguy C, Winfield A, Malloch G, Smith DM, Fenton B, Gray SM, Jander G (2007) Genomic resources for *Myzus persicae*: EST sequencing, SNP identification, and microarray design. BMC Genomics 8:423

31. Rise ML, Hall J, Rise M, Hori T, Gamperl AK, Kimball J, Hubert S, Bowman S, Johnson SC (2008) Functional genomic analysis of the response of Atlantic cod (*Gadus morhua*) spleen to the viral mimic polyriboinosinic polyribocytidylic acid (pIC). Dev Comp Immunol 32:916–931

32. Rise ML, von Schalburg KR, Brown GD, Mawer MA, Devlin RH, Kuipers N, Busby M, Beetz-Sargent M, Alberto R, Gibbs AR, Hunt P, Shukin R, Zeznik JA, Nelson C, Jones SR, Smailus DE, Jones SJ,

Schein JE, Marra MA, Butterfield YS, Stott JM, Ng SH, Davidson WS, Koop BF (2004) Development and application of a salmonid EST database and cDNA microarray: data mining and interspecific hybridization characteristics. Genome Res 14:478–490

33. Rise ML, Gamperl AK, Johnson SC, Robinson A, Culver K, Booman M, Borza T, Garber A, Hubert S, Tosh J, Afonso L, Boyce D, Hamoutene D, Lush L, Neil S, Symonds J, Butts I, Feng CY, Hori T, Pérez-Casanova J, Shine E, Hall JR, Higgins B, Kimball J, Simpson G, Stone C, Tarrant Bussey J, Elliot J, Moir J, Nardi G, Powell F, Walsh A, Rise M, Trippel EA, Bowman S (2009) The Atlantic Cod Genomics and Broodstock Development Project. Bull Aquac Assoc Can 107-3:21-30
34. Rodger HD, Colquhoun DJ (2008) Clinical vibriosis in farmed Atlantic cod (*Gadus morhua*) in Ireland. Vet Rec 162:94–95
35. Rodriguez-Tovar LE, Wadowska DW, Wright GM, Groman DB, Speare DJ, Whelan DS (2003) Ultrastructural evidence of autoinfection in the gills of Atlantic cod *Gadus morhua* infected with Loma sp. (phylum Microsporidia). Dis Aquat Organ 57:227–230
36. Rose GA (2007) Cod: The ecological history of the North Atlantic fisheries. Breakwater Books, St. John›s, Newfoundland
37. Rosenlund G, Skretting M (2006) Worldwide status and perspective on gadoid aquaculture. ICES J Mar Sci 63:194–197
38. Ruzzante DE, Taggart CT, Cook D, Goddard S (1996) Genetic differentiation between inshore and offshore Atlantic cod (*Gadus morhua*) off Newfoundland: Microsatellite DNA variation and antifreeze level. Can J Fish Aquat Sci 53:634–645
39. Snelling WM, Casas E, Stone RT, Keele JW, Harhay GP, Bennett GL, Smith TP (2005) Linkage mapping bovine EST-based SNP. BMC Genomics 6:74
40. Streisinger G, Walker C, Dower N, Knauber D, Singer F (1981) Production of clones of homozygous diploid zebra fish (*Brachydanio rerio*). Nature 291:293–296
41. Symonds J, Garber A, Puvanendran V, Robinson A, Neil S, Trippel E, Walker S, Boyce D, Gamperl K, Lush L, Nardi G, Powell F, Walsh

A, Bowman S (2007) Family-based Atlantic cod (*Gadus morhua*) broodstock development. Bull Aquac Assoc Can 105:39–46

42. Zhulidov PA, Bogdanova EA, Schcheglov AS, Vagner LL, Khaspekov GL, Kozhemyako VB, Matz MV, Meleshkevitch E, Moroz LL, Lukyanov SA, Shagin DA (2004) Simple cDNA normalization using kamchatcka crab duplex-specific nuclease. Nucleic Acids Res 32:e37

CHAPTER

13

EFFECTS OF GENOTYPE AND DIETARY FISH OIL REPLACEMENT WITH VEGETABLE OIL ON THE INTESTINAL TRANSCRIPTOME AND PROTEOME OF ATLANTIC SALMON (SALMO SALAR)

Sofia Morais[1], Tomé Silva[2], Odete Cordeiro[2], Pedro Rodrigues[2], Derrick R Guy[3], James E Bron[1], John B Taggart[1], J Gordon Bell[1] and Douglas R Tocher[1]

[1] Institute of Aquaculture, University of Stirling, Stirling FK9 4LA, Scotland, UK.

[2] Aquaculture Research Group, Centre of Marine Sciences (CCMAR), University of Algarve, 8005-139 Faro, Portugal.

[3] Landcatch Natural Selection Ltd., Cooperage Way Business Centre, Alloa FK10 3LP, Scotland, UK.

ABSTRACT

Background

Expansion of aquaculture requires alternative feeds and breeding strategies to reduce dependency on fish oil (FO) and better utilization of dietary vegetable oil (VO). Despite the central role of intestine in maintaining body homeostasis and health, its molecular response to replacement of dietary FO by VO has been little investigated. This study employed transcriptomic and proteomic analyses to study effects of dietary VO in two family groups of Atlantic salmon selected for flesh lipid content, ‘Lean’ or ‘Fat’.

Results

Metabolism, particularly of lipid and energy, was the functional category most affected by diet. Important effects were also measured in ribosomal proteins and signalling. The long-chain polyunsaturated fatty acid (LC-PUFA) biosynthesis pathway, assessed by fatty acid composition and gene expression, was influenced by genotype. Intestinal tissue contents of docosahexaenoic acid were equivalent in Lean salmon fed either a FO or VO diet and expression of LC-PUFA biosynthesis genes was up-regulated in VO-fed fish in Fat salmon. Dietary VO increased lipogenesis in Lean fish, assessed by expression of FAS, while no effect was observed on β-oxidation although transcripts of the mitochondrial respiratory chain were down-regulated, suggesting less active energetic metabolism in fish fed VO. In contrast, dietary VO up-regulated genes and proteins involved in detoxification, antioxidant defence and apoptosis, which could be associated with higher levels of polycyclic aromatic hydrocarbons in this diet. Regarding genotype, the following pathways were identified as being differentially affected: proteasomal proteolysis, response to oxidative and cellular stress (xenobiotic and oxidant metabolism and heat shock proteins), apoptosis and structural proteins particularly associated with tissue contractile properties. Genotype effects were accentuated by dietary VO.

Conclusions

Intestinal metabolism was affected by diet and genotype. Lean fish may have higher responsiveness to low dietary n-3 LC-PUFA, up-regulating the biosynthetic pathway when fed dietary VO. As global aquaculture searches for alternative oils for feeds, this study alerts to the potential of VO introducing contaminants and demonstrates the detoxifying role of intestine. Finally, data indicate genotype-specific responses in the intestinal transcriptome and proteome to dietary VO, including possibly structural properties of the intestinal layer and defence against cellular stress, with Lean fish being more susceptible to diet-induced oxidative stress.

BACKGROUND

Fish are important components of the human diet, being highly nutritious and valued as the main source of n-3 long-chain polyunsaturated fatty acids (LC-PUFA). These essential fatty acids, mainly eicosapentaenoic acid (EPA) and docosahexaenoic acid (DHA), have well-known health-promoting properties, including protection against a range of cardiovascular and inflammatory diseases, and neurological disorders [1]. With population growth and increasing awareness of the importance of fish consumption as part of a healthy diet, worldwide demand for seafood continues to grow. However, as traditional fisheries are largely in decline, aquaculture must meet this demand. Aquaculture is the fastest-growing food production sector with an average annual growth rate of 6.6%, accounting for 46% of total fish supply [2]. In the European and American continents, aquaculture production is largely dominated by salmonid species, mainly Atlantic salmon, and feeds for such carnivorous species have traditionally relied on fishmeal (FM) and fish oil (FO) from wild stocks. Recent estimates indicated that 88.5% of global production of FO was used by the aquaculture sector, with salmonid culture taking the largest share (56%) [3]. With ever increasing demands for aquafeeds and reduction in fisheries landings, the availability of FO and FM seriously limits the growth of aquaculture production and there is an urgent need to find more sustainable alternatives.

Vegetable oils (VO) can replace FO in salmon feeds without compromising fish growth or condition although, at high levels of replacement, tissue levels of n-3 LC-PUFA are significantly reduced [4, 5]. The effects of FO replacement by VO are becoming well characterized in the hepatic transcriptome of salmonids [6, 7, 8, 9, 10], and other species [11]. However, studies on intestinal transcriptome are few and restricted to effects of replacement of FM by plant proteins, particularly soybean meal [12, 13], given its potential to cause enteritis. It is now clear that intestine in salmonids is not simply a site for reacylation and packaging of dietary lipids but it also has important roles in fatty acid metabolism, including LC-PUFA biosynthesis [14, 15]. Furthermore, dietary VO can induce major histological changes in fish enterocytes, originating mostly from supranuclear lipid droplet formation, possibly due to altered reacylation mechanisms and decreased phospholipid synthesis [16, 17, 18]. In some cases, these accumulations were large enough to be deemed pathological [19, 20]. A recent study investigating effects of dietary FO replacement by VO on intestinal transcriptome in Atlantic cod indicated potential effects on lipid absorption and transport and suggested morphological and structural changes to the intestinal muscle layer [21]. Furthermore, both this and a previous study on Atlantic salmon showed significant effects on expression of genes involved in cell proliferation and apoptosis [22]. Therefore, there is indication that intestine may be affected by changes in lipid components of feed formulations. Given its crucial roles in nutrient absorption, protection against the entry of pathogens, and immune function [23], further attention is warranted and impacts of FO replacement require investigation in intestine, particularly in salmon where important changes in diet formulation are already being applied. This study is a large-scale analysis of the effects of replacement of dietary FO by VO on the transcriptome and proteome of Atlantic salmon intestine. Furthermore, given recent interest in evaluating genetic selection as a feasible strategy, in conjunction with changes in commercial feed formulation, to meet worldwide demand for farmed fish without compromising animal welfare or nutritional value [9, 10, 24], two groups of Atlantic salmon families, Lean and Fat, were studied to examine the potential effects of genetic background. This experiment was performed in parallel with another microarray study looking at effects in the hepatic transcriptome, analysing samples from the same individuals [9, 10], enabling a global and comprehensive assessment of the physiological and molecular effects of FO replacement by VO in Atlantic salmon, including potential interactions with genotype.

RESULTS

Microarray Analysis

Two way-ANOVA of the cDNA array dataset returned 1409, 1626 and 862 significant genes for the factors diet, genotype and diet × genotype interaction, respectively. Detailed analysis was restricted to the top 100 most significant features, which were categorised according to biological function, based on mammalian homolog genes. Metabolism, particularly of lipid and energy, was the functional category most affected by diet accounting for 39-41% of the top 100 annotated genes (Table 1), and showing highest diet × genotype interaction (Additional file 1). Diet also impacted translation (17%; mostly ribosomal proteins) and signalling (17%). In contrast, genotype affected less markedly metabolism (21%), whereas structural proteins (25%; mostly different types of collagen) and proteins involved in the regulation of transcription (21%) predominated (Table 2).

Table 1: Intestine transcripts corresponding to the top 100 most significant features exhibiting differential expression between diets

Accession no	Gene	VO/FO		p- value
		Lean	Fat	
Metabolism (39%)				
Lipid metabolism (9%)				
DW590668	Fatty aldehyde dehydrogenase	- 1.1	- 1.2	0.0003
CK888998	Ethanolamine kinase 1	1.2	1.2	0.0008
CK884623	Epidermis-type arachidonate lipoxygenase 3	1.4	1.1	0.0010
CK885725	Protein containing a beta-keto-acyl synthase domain	1.4	1.6	0.0017
AF478472	Delta-5 fatty acyl desaturase	1.2	2.5	0.0019
Energy metabolism/generation of precursor metabolites (11%)				
CK885194	Cytochrome c oxidase subunit 5A	- 1.2	- 1.1	0.0000
bra_snb_03E12	NADH dehydrogenase subunit 1	- 1.3	- 1.2	0.0000
EG649067	Ubiquinol-cytochrome c reductase core protein 1	- 1.2	- 1.1	0.0004
CK884638	Uncoupling protein	- 1.3	−1.3	0.0019
CK890974	Mitochondrial calcium-dependent solute carrier SCaMC-2	- 1.5	1.0	0.0021
EG648806	NADH dehydrogenase (ubiquinone) 1 beta subcomplex 9	- 1.1	- 1.1	0.0022
Protein and amino acid metabolism (9%)				
CK877521	Kunitz-type protease inhibitor 2 precursor	1.2	1.1	0.0002
DW590246	Betaine aldehyde dehydrogenase	- 1.4	- 1.0	0.0005
CK885180	Fumarylacetoacetate hydrolase domain-containing protein 2B	- 1.4	- 1.3	0.0007

CK897653	Ubiquitin-conjugating enzyme E2 A	1.2	1.2	0.0012
CK885145	Lysosomal protective protein precursor (Cathepsin A)	1.1	1.1	0.0014
Carbohydrate metabolism (5%)				
EG649263	Endosulfine alpha	- 1.3	- 1.0	0.0005
CK885545	Glucose transporter type 8	1.0	1.5	0.0013
Xenobiotic and oxidant metabolism (5%)				
CK885392	Catalase	1.6	1.1	0.0000
CO472279	Cytochrome P450 1[a]	1.2	1.4	0.0004
CK898307	Selenoprotein H-like protein	1.3	1.1	0.0013
Transport/ intracellular trafficking (5%)				
EG647578	Calcium-activated potassium channel subunit alpha 1	1.3	1.2	0.0005
EG648841	Sodium-coupled neutral amino acid transporter 2	- 1.2	- 1.2	0.0007
EG648303	Mitochondrial import receptor subunit TOM20 homolog	- 1.1	- 1.1	0.0011
Regulation of transcription (13%)				
CK898304	C-Maf	1.6	1.6	0.0000
CK890154	Butyrate response factor 2	1.6	1.2	0.0001
EG647831	Proliferation-associated 2G4 b	- 1.3	- 1.2	0.0003
AM041851	Enhancer trap locus homolog 1	- 1.1	- 1.2	0.0004
CK885179	Steroid receptor RNA activator 1	- 1.3	- 1.3	0.0007
CK884728	Butyrate response factor 1	1.1	1.4	0.0011
CK879187	Enhancer of polycomb homolog 1	1.4	1.1	0.0020
Translation (17%)				
CK878380	40S ribosomal protein S18	- 1.1	- 1.2	0.0003
DW591873	60S acidic ribosomal protein	- 1.3	- 1.2	0.0006
EG648644	60S ribosomal protein L29	- 1.1	- 1.2	0.0011
DW590580	40S ribosomal protein S5	- 1.2	- 1.2	0.0011
CK880046	40S ribosomal protein S28	- 1.1	- 1.2	0.0017
EG649167	60S ribosomal protein L37a	- 1.1	- 1.2	0.0017
EG648512	60S ribosomal protein L41	- 1.1	- 1.2	0.0017
CK886042	60S ribosomal protein L31	- 1.1	- 1.3	0.0021

CK895350	60S ribosomal protein L36	- 1.1	- 1.1	0.0021
Signalling/Signal transduction (17%)				
CK874863	Proto-oncogene serine/threonine-protein kinase	1.1	1.1	0.0002
CK881266	Mitogen-activated protein kinase kinase kinase kinase 1	1.2	1.2	0.0002
CK878590	Serine/threonine protein kinase (Tribbles homolog 2)	- 1.1	- 1.3	0.0004
AJ425351	Tensin-like C1 domain-containing	- 2.0	- 1.2	0.0006
CK885079	Sonic hedgehog-like protein	1.3	1.0	0.0008
CN181161	TGF-beta type II receptor	1.1	1.1	0.0009
EG649257	Serine/threonine-protein kinase VRK3	- 1.1	- 1.2	0.0009
CK886267	Guanylate cyclase 2C	1.1	1.1	0.0011
CK884467	Serine/threonine-protein phosphatase 2A regulatory subunit B	- 1.4	- 1.3	0.0020
Structural proteins (7%)				
mus_snm_13F09	Myosin heavy chain	- 1.2	- 1.1	0.0002
CK885118	Cardiac tropomyosin	1.0	1.1	0.0013
CK883354	Troponin C	1.2	1.2	0.0017
EG647866	Tubulin alpha chain	- 1.2	- 1.2	0.0020
Immune response (2%)				
BM414177	Liver-expressed antimicrobial peptide 2	1.5	1.4	0.0022
Miscellaneous				
CK884634	Ezrin-radixin-moesin binding phosphoprotein 50	1.1	1.1	0.0003
EG648821	Transposase	- 1.1	- 1.2	0.0004
CK883634	Poly [ADP-ribose] polymerase 12	1.3	1.1	0.0006
CK885090	Claudin 15	1.2	1.1	0.0008
DW589291	Transposase	- 1.3	- 1.3	0.0009
CK878888	Splicing factor, arginine/serine-rich 3	- 1.3	- 1.2	0.0013
CK885098	RNA-binding protein with multiple splicing 2	- 1.4	- 1.2	0.0013
CK891747	Occludin	- 1.2	- 1.2	0.0014

CK888908	All-trans-retinol 13,14-reductase precursor	1.1	1.2	0.0017
CK880622	Cyclin G1	- 1.1	- 1.2	0.0019
CK885604	Methylenetetrahydrofolate dehydrogenase (NADP+ dependent) 1-like	- 1.6	- 1.3	0.0020
CK891699	Barrier-to-autointegration factor	- 1.1	- 1.2	0.0020
CK879123	Hemoglobin subunit beta	1.5	1.0	0.0022

Annotated features (67% of all clones) are arranged by categories of biological function (with percentages of distribution of genes within each category being indicated) and, within these, by decreasing significance (assessed by two-way ANOVA). Also indicated are the GenBank accession numbers for each clone (or, when not available, the probe number) and the expression ratio between fish fed VO and those fed FO, for each genotype (Lean and Fat).

Table 2: Intestine transcripts corresponding to the top 100 most significant features exhibiting differential expression between family groups

Accession no	**Gene**	**Lean/Fat**		**p- value**
		FO	**VO**	
Metabolism (21%)				
Lipid metabolism (5%)				
CK874747	Acylglycerol kinase	- 1.0	- 1.1	0.0003
BI468033	ATP-binding cassette sub-family A member 1	- 1.1	- 1.6	0.0004
Energy metabolism/generation of precursor metabolites (2%)				
CK890974	Mitochondrial calcium-dependent solute carrier SCaMC-2	- 1.1	- 1.6	0.0000
Protein and amino acid metabolism (12%)				
EG647485	Proteasome subunit beta type-8 precursor	- 1.3	- 1.8	0.0000
DW591059	Heat shock protein Hsp-16	1.2	1.4	0.0001
AJ425553	26S proteasome non-ATPase regulatory subunit 9	- 1.2	- 1.1	0.0002

DW592187	26 proteasome complex subunit DSS1	1.5	1.3	0.0008
CK878109	Heat shock protein 60 kDa	1.2	1.4	0.0009
Xenobiotic and oxidant metabolism (2%)				
CO472279	Cytochrome P450 1A	- 1.2	- 1.5	0.0003
Transport/ intracellular trafficking (5%)				
CK886307	Peptide transporter PEPT2	- 1.4	- 1.3	0.0006
DW588251	Post-GPI attachment to proteins factor 2	1.1	1.3	0.0007
Regulation of transcription (21%)				
CK877881	Forkhead box Q1	- 1.4	- 1.7	0.0000
CK884548	Proline-rich nuclear receptor coactivator 2	- 1.1	- 1.2	0.0001
CK885013	Forkhead box protein F1	1.1	1.5	0.0003
CK891975	Y-box binding protein-2	- 1.1	- 1.1	0.0003
DW589341	Signal transducer/activator of transcription Stat1	- 1.2	- 1.2	0.0004
CN181282	Bromodomain containing protein 3	- 1.1	- 1.1	0.0007
CK884407	C-Myc-binding protein	- 1.3	- 1.1	0.0007
CK895950	Transcription factor CP2-like	- 1.5	- 1.2	0.0007
EG648473	BTB/POZ domain-containing protein	- 1.1	- 1.4	0.0009
Translation (7%)				
CN181322	40S ribosomal protein S4	1.1	1.1	0.0002
CK885979	Ribonuclease UK114	- 1.1	- 1.3	0.0005
EG648042	Eukaryotic translation initiation factor 3, subunit J	1.1	1.1	0.0009
Signalling/Signal transduction (16%)				
CO471793	Rho GDP-dissociation inhibitor 1	1.3	1.0	0.0000

CK885560	Protein tyrosine phosphatase non-receptor type 9	1.3	1.5	0.0004
CK885079	Sonic hedgehog-like protein	- 1.3	- 1.0	0.0004
EG648038	Kalirin, RhoGEF kinase	1.2	1.2	0.0005
DW589610	Chimaerin 1	1.2	1.2	0.0005
CK897590	Growth factor receptor-bound protein 2	1.1	1.1	0.0009
EG647545	Protein tyrosine phosphatase receptor type D	1.0	1.2	0.0009
Structural proteins (25%)				
CK882401	Troponin C	- 1.1	- 1.2	0.0000
CK889952	Transgelin 2	1.5	1.7	0.0000
EG648135	Osteonectin	1.3	1.5	0.0000
CK899058	Collagen alpha 3 type VI	1.5	1.5	0.0002
CK879405	Collagen alpha 2 type VI	1.3	1.5	0.0002
EG649361	Collagen alpha 2 type I	1.5	1.7	0.0005
EG649013	Collagen alpha 3 type I	1.4	1.5	0.0005
mus_mfo_15A10	Troponin-I isoform 3	1.4	1.6	0.0007
CK876833	Keratocan	- 1.0	1.6	0.0007
DW590534	Collagen alpha 2 type I	1.5	1.6	0.0010
CK892271	Collagen alpha 2 type V	1.2	1.4	0.00010
CK873441	Collagen alpha1 type VI	1.2	1.4	0.0010
Immune response (5%)				
AJ425527	T-cell immunoglobulin and mucin domain-containing protein 4	1.1	1.1	0.0001
EG649194	Mannose-binding protein C	1.4	1.4	0.0003
Miscellaneous				
AM041770	Preimplantation protein 3	- 1.3	- 1.2	0.0000
EG649106	Ornithine decarboxylase antizyme 2	1.4	1.1	0.0002
CK884355	Guanylin precursor	1.2	1.4	0.0002

CK879185	Aminolevulinate, delta-, synthetase 1	- 2.0	- 1.1	0.0002
CK885814	Similar to p53-associated parkin-like cytoplasmic protein	1.2	1.1	0.0002
CK886214	Septin-7	1.1	1.1	0.0003
BM414079	Tumor protein D52-like 2	- 1.1	- 1.3	0.0003
BM413818	Splicing factor, proline- and glutamine-rich	1.2	1.2	0.0003
CK874550	Follistatin-related protein 1	1.1	1.3	0.0004
DW589104	Nucleoside diphosphate kinase 6	1.1	1.1	0.0005
DW592207	Deoxyhypusine synthase	1.5	1.6	0.0007
CK892359	HCLS1-associated protein X-1	- 1.2	- 1.1	0.0007
BI468158	Nuclear protein 1	- 1.0	- 1.6	0.0009
AJ425003	Lamina-associated polypeptide 2	1.2	1.1	0.0010
CK899283	Small EDRK-rich factor 2	1.3	1.5	0.0010
CK885090	Claudin 15	- 1.2	- 1.1	0.0011

Annotated features (60% of all clones) are arranged by categories of biological function (with percentages of distribution of genes within each category being indicated) and, within these, by decreasing significance (assessed by two-way ANOVA). Also indicated are the GenBank accession numbers for each clone (or, when not available, the probe number) and the expression ratio between Lean and Fat fish fed either FO or VO.

Gene Ontology (GO) enrichment analysis was performed on the complete significant lists, enabling identification of GO terms significantly enriched in the input entity list, in comparison to the whole array, providing clues as to which biological processes might be particularly altered in the experimental conditions being compared. It revealed no significant enrichment of GO terms in the genotype list, while 20 and 7 GO terms were significantly enriched in the diet and interaction lists, respectively. GO terms enriched in the diet list included structural constituents of ribosome, structural molecule activity, cytosolic ribosome, cytosol, ribosomal subunit, translation, cellular biosynthetic process, gene expression, macromolecule and biopolymer biosynthetic process and other related terms. This was explained by the large number of ribosomal proteins, components of both the 40S and 60S subunits, which were down-regulated by dietary VO (Table 1). In contrast, several Δ6-desaturase (Δ6fad) clones showing a diet × genotype interaction (see Additional file 1) caused a significant enrichment of the GO terms oxidoreductase activity, stearoyl-CoA 9-desaturase activity, unsaturated fatty acid biosynthetic activity/metabolic processes and very long chain fatty acid (VLCFA) biosynthetic activity/metabolic processes.

RT-qPCR Analysis of Gene Expression

The expression of several genes significantly affected or related to processes affected by the two factors in the microarray analysis was determined by RT-qPCR (Table 3). For diet, a reasonably good match was found for Δ5 fatty acyl desaturase (Δ5fad), NADH dehydrogenase subunit 1 (NADH1), proliferation-associated 2G4b (PA2G4), 60S acidic ribosomal protein (RP60S), proliferating cell nuclear antigen (PCNA; also known as cyclin) and cytochrome P450 1A (CYP1A), particularly in the Fat group where fold-changes were generally more pronounced and significant. No change in expression of uncoupling protein 2 (UCP2) with diet was measured while, for myosin heavy chain (MYO) and methylenetetrahydrofolate dehydrogenase (NADP+ dependent) 1-like (MTHFD1), RT-qPCR indicated a change opposite to that suggested by microarray. Regarding genotype, a good match was obtained for CYP1A, proteasome subunit beta type-8 precursor (PSMB8) and alpha 2 type I collagen (COL1A2), while transgelin 2 (TAGLN) expression did not differ between family groups, and for ATP-binding cassette sub-family A member 1 (ABCA1) there was an inverse change in expression.

Table 3: Relative expression of genes assayed by RT-qPCR

Genes	VO/FO				Lean/Fat			
	Lean		Fat		FO		VO	
	Ratio	p value	Ratio	p value	Ratio	p value	Ratio	p value
Microarray validation								
Δ5fad (D)	1.72	ns	**4.34**	**0.005**	2.49	ns	−1.01	ns
NADH1 (D)	−1.03	ns	**−1.28**	**0.05**	1.00	ns	1.24	ns
UCP2 (D)	−1.10	ns	1.08	ns	−1.10	ns	**−1.28**	**0.003**
MTHFD1 (D)	1.20	ns	−1.09	ns	−1.08	ns	1.22	ns
PA2G4 (D)	−1.27	ns	**−1.23**	**0.032**	1.11	ns	1.1	ns
RP60S (D)	−1.02	ns	**−1.69**	**0.023**	1.13	ns	**1.9**	**0.006**
MYO (D)	**1.77**	**0.034**	1.22	ns	−1.35	ns	1.08	ns
PCNA (D)	−1.14	ns	−1.27	ns	1.08	ns	1.20	ns
CYP1A (D+G)	**1.71**	**0.004**	**1.81**	**0.023**	**−2.22**	**0.003**	**−2.38**	**0.001**
ABCA1 (G)	1.24	ns	1.17	ns	1.21	ns	**1.28**	**0.015**
PSMB8 (G)	−100.00	ns	**1.87**	**0.033**	**−16.67**	**0.026**	**<−1000**	**0.000**
TAGLN (G)	1.23	ns	−1.02	ns	−1.03	ns	1.22	ns
COL1A2 (G)	1.31	ns	−1.06	ns	1.47	ns	**2.06**	**0.001**
LC-PUFA biosynthesis and regulation								
Δ6fad_a	1.11	ns	**2.96**	**0.008**	1.76	ns	−1.52	ns
elovl5a	1.08	ns	1.2	ns	1.03	ns	−1.08	ns
elovl5b	−1.19	ns	**1.92**	**0.031**	1.54	ns	−1.47	ns
elovl2	1.2	ns	**14.74**	**0.003**	4.89	ns	−2.50	ns
Fatty acid synthesis								
FAS	**1.49**	**0.010**	1.20	ns	1.05	ns	1.31	ns
Fatty acid oxidation								
ACO	1.18	ns	1.28	ns	1.19	ns	1.09	ns
CPT1	1.23	ns	−1.09	ns	−1.16	ns	1.15	ns
Regulation of lipid metabolism								
PPARα	1.14	ns	1.38	ns	1.26	ns	1.04	ns
PPARβ	1.03	ns	1.03	ns	−1.02	ns	−1.02	ns
PPARγ	1.52	ns	**1.60**	**0.036**	1.21	ns	1.15	ns
Xenobiotic and oxidative stress								
GST	−1.30	ns	1.07	ns	1.21	ns	−1.14	ns
MT-A	−1.82	0.051*	−1.05	ns	−1.25	ns	**−2.13**	**0.007**
CAT	**1.57**	**0.036**	1.26	ns	−1.52	ns	−1.22	ns
SOD	−1.20	ns	−1.04	ns	1.21	ns	1.04	ns
Apoptosis								

CASP3B	**1.22**	**0.045**	1.22	ns	−1.04	ns	−1.04	ns
CASP6A/B	1.30	ns	1.37	ns	−1.14	ns	−1.20	ns

fad: Fatty acyl desaturase (Δ5 and Δ6 activities); NADH1: NADH dehydrogenase subunit 1; UCP2: Uncoupling protein 2; elovl: MTHFD1: Methylenetetrahydrofolate dehydrogenase (NADP+ dependent) 1-like; PA2G4: Proliferation-associated 2G4b; RP60S: 60S acidic ribosomal protein; MYO: Myosin heavy chain; PCNA: Proliferating cell nuclear antigen; CYP1A: Cytochrome P4501A; ABCA1: ATP-binding cassette sub-family A member 1; PSMB8: Proteasome subunit beta type-8 precursor; TAGLN: Transgelin 2; COL1A2: Collagen type I alpha 2; elovl2: Elongase 2; FAS: Fatty acid synthase; ACO: Acyl-CoA oxidase; CPT1: Carnitine palmitoyltransferase I; PPARs; Peroxisome proliferator-activated receptors; GST: glutathione S-transferase; MT-A: Metallothionein A; CAT: catalase; SOD: superoxide dismutase; CASP; caspase.

*Only marginally non-significant.

Genes correspond to features significantly affected by either diet (D) or genotype (G) in the microarray analysis or involved in biological processes being analysed in further detail. Values are normalized (by *cofilin-2* and *elf-1 α*) gene expression ratios between fish fed VO in relation to FO for each family group or of Lean fish in relation to the Fat group when fed either one of the diets. Values in bold are significantly different, at $p<0.05$ (REST 2008); ns - non-significant.

As well as validation above, RT-qPCR was used to further analyse genes/pathways identified by microarray and published data as potentially interesting, including lipid metabolism, xenobiotic and oxidative stress, and apoptosis. One was LC-PUFA biosynthesis, given that Δ5fad was significantly affected by diet in the microarray analysis, with a stronger response in Fat fish, whereas Δ6fad showed a significant diet × genotype interaction (supplementary file 1) confirmed by RT-qPCR. The Δ6fad transcript was only significantly up-regulated in Fat fish fed VO, compared to FO, and Lean fish showed higher levels of Δ6fad expression than Fat fish when fed FO, while the opposite trend was noted when fish were fed VO. Fatty acyl elongases (elovl's) were also quantified and their expression (except elovl5a) broadly followed that of fads: significantly up-regulated when dietary VO replaced FO in the Fat group. Additionally, elovl5b and, particularly, elovl2 showed a trend for increased expression (high fold-change) in Lean fish, compared to Fat fish, when fed FO, while an opposite trend was observed when salmon were fed VO. Although genes involved in fatty acid synthesis and oxidation showed few significant differences, expression of fatty acid synthase (FAS) was up-regulated in fish fed VO, but only in Lean fish. The expression of peroxisome proliferator-activated receptors (PPARs), involved in the regulation of multiple lipid metabolism genes, was determined but only PPARγ showed any significant change, being up-regulated in the Fat group when dietary VO replaced FO. Of the xenobiotic and oxidative metabolism genes assayed, apart from CYP1A, only catalase (CAT) was affected by diet (up-regulated in fish fed VO) and only significantly in the Lean family. In contrast, metallothionein A (MT-A) showed higher expression in Fat fish, but only when fed VO, while a marginal down-regulation was observed when comparing VO and FO-fed fish in the Lean group. Of genes related to apoptosis, CASP3B was up-regulated by VO in Lean fish whereas a similar fold-change was marginally non-significant in the Fat fish.

Intestine Fatty Acid Composition

The levels of most fatty acids in pyloric caeca were affected by diet, whereas genotype had no significant effect (Table 4). However, some fatty acids also showed a significant diet × genotype interaction, indicating that the effect of diet depended on the genetic background of the fish. For instance, interactions were observed for some LC-PUFA as a result of higher levels being found in the Lean group, compared to Fat, when fish were fed VO, while the reverse was observed when fed FO. Another unexpected result was that similar levels of DHA in FO- and VO-fed Lean fish meant that, in spite of substantial differences in Fat fish fed the two diets, the effect of diet on DHA was marginally non-significant. Similarly, levels of EPA and 22:5n-3 between FO- and VO-fed fish were noticeably closer in the Lean group.

Table 4: Fatty acid compositions (% of total fatty acids) and lipid contents (% of wet weight) of intestinal tissue of Lean and Fat family groups fed diets containing either FO or VO

	FO		VO		Two-way ANOVA		
	Fat	Lean	Fat	Lean	D	G	D×G
Fatty acid							
Total saturated	23.70 ± 1.21	22.82 ± 1.09	18.57 ± 0.73	20.08 ± 1.38	0.000	0.742	0.066
Total monoenes	38.25 ± 7.15	44.45 ± 5.31	44.98 ± 2.38	39.34 ± 4.49	0.717	0.928	0.046
18:2n-6	4.19 ± 0.67	4.51 ± 0.24	13.14 ± 0.59	11.28 ± 1.12	0.000	0.139	0.041
18:3n-6	0.10 ± 0.02	0.10 ± 0.02	0.28 ± 0.12	0.25 ± 0.13	0.003	0.611	0.611
20:2n-6	0.57 ± 0.06	0.58 ± 0.07	1.54 ± 0.40	1.65 ± 0.06	0.000	0.199	0.510
20:3n-6	0.17 ± 0.06	0.14 ± 0.02	1.01 ± 0.12	1.47 ± 0.26	0.000	0.056	0.019
20:4n-6	1.20 ± 0.60	0.85 ± 0.31	0.74 ± 0.18	1.40 ± 0.29	0.785	0.419	0.016
22:4n-6	0.08 ± 0.06	0.05 ± 0.06	0.00 ± 0.00	0.09 ± 0.07	0.337	0.337	0.038
22:5n-6	0.21 ± 0.06	0.16 ± 0.07	0.07 ± 0.02	0.12 ± 0.05	0.002	1.000	0.128
Total n-6 PUFA	6.51 ± 0.18	6.38 ± 0.36	16.79 ± 0.86	16.27 ± 0.67	0.000	0.410	0.781
18:3n-3	1.46 ± 0.38	1.67 ± 0.21	4.79 ± 0.53	3.71 ± 0.75	0.000	0.211	0.021
18:4n-3	1.36 ± 0.60	1.67 ± 0.66	0.85 ± 0.31	0.55 ± 0.20	0.002	0.712	0.127
20:3n-3	0.27 ± 0.05	0.33 ± 0.14	0.67 ± 0.18	0.62 ± 0.04	0.000	0.768	0.384
20:4n-3	1.77 ± 0.28	2.16 ± 0.46	0.97 ± 0.13	0.83 ± 0.10	0.000	0.504	0.061
20:5n-3	6.55 ± 0.55	5.79 ± 0.62	3.60 ± 0.90	4.40 ± 0.72	0.000	0.682	0.058
22:5n-3	2.82 ± 0.81	2.30 ± 0.74	1.57 ± 0.36	2.54 ± 0.77	0.157	0.530	0.052
22:6n-3	16.68 ± 6.68	11.65 ± 5.65	6.97 ± 1.58	11.50 ± 3.10	0.057	0.943	0.050
Total n-3 PUFA	30.90 ± 5.94	25.56 ± 5.19	19.41 ± 2.18	24.15 ± 3.76	0.015	0.962	0.042
Total PUFA	38.06 ± 5.96	32.73 ± 5.42	36.45 ± 2.08	40.59 ± 3.16	0.206	0.794	0.059
Total lipids	7.50 ± 3.75	9.56 ± 3.81	6.97 ± 1.07	4.59 ± 1.05	0.081	0.804	0.111

Results are means ± SD (n = 4) and p-values of two-way ANOVA are presented for diet (D), genotype (G) and interaction between both factors.

Proteomic Analysis

Of the protein spots identified as being differentially expressed between diets or genotypes (Additional files 2and 3), only 17 and 29 could be excised and, of these, only 9 and 20, respectively, returned reliable identifications by peptide fragment fingerprinting (Figure 1). Proteins significantly up-regulated by dietary VO are likely implicated in xenobiotic/drug metabolism (epoxide hydrolase 2, EPHX2), protection from oxidative stress (hemopexin-like protein, HPX and peroxiredoxin-1, PRDX1) and induction of apoptosis and inflammatory responses (galectin-2, LGALS2). Those proteins down-regulated by dietary VO included proteins responsible for protein folding and involved in signalling (2-peptidylprolyl isomerase A, PPIA), actin-based motility (myosin light chain, MYL) and DNA replication, repair or transcription (histone H2A, H2A) (Table 5). Proteins affected by genotype encompass a variety of pathways, of which only a few are related to metabolism, namely carbohydrate (alpha-enolase, ENO1; pancreatic alpha-amylase precursor, AMY2; triosephosphate isomerase 1b, TPI1), folate (dihydropteridine reductase, DHPR) or retinol (retinol-binding protein II, RBP2) metabolism (Table 6). Other proteins may have potentially multiple roles but can broadly be assigned roles in response to oxidative and cellular stress (HPX; PRDX; heat shock protein 70, HSP70), oxygen transport (alpha globin, HBA), signal transduction (calreticulin, CALR), transcription/RNA repair (heterogeneous nuclear ribonucleoprotein A0, HNRNPA0; histone H2A, H2A), apoptosis (CASP3), cellular transport, potentially also associated with apoptosis (voltage-dependent anion channel 2–2, VDAC2; annexin A4, ANX4), and proteolysis (proteasome beta 1 subunit, PSMB1). As with the microarray analysis, a few proteins with a more structural function and particularly associated with tissue contractile properties were affected by genotype, showing lower levels in Lean fish. These included calponin-1 (CNN1) and transgelin (TAGLN), the latter which was also found to be significantly affected by microarray, albeit up-regulated in Lean fish. Most proteins significantly affected by genotype showed lower levels of expression in the Lean group, with the exception of ENO1, HSP70, TPI1, H2A and HBA.

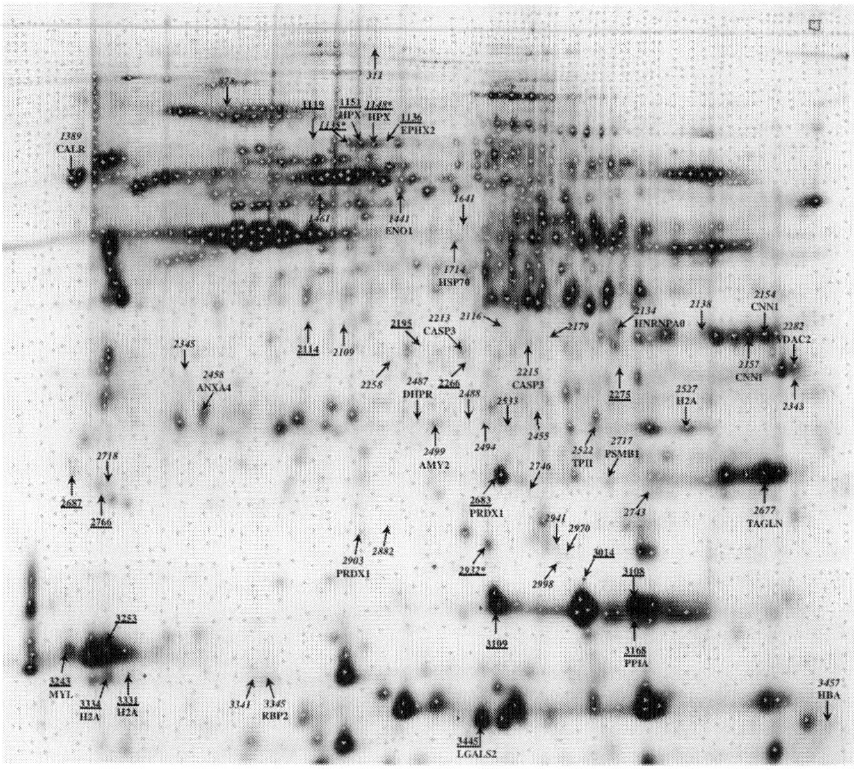

Figure 1: Proteins significantly affected by diet or genotype. Cropped image of a scanned 2-D polyacrylamide gel showing all spots identified as being differentially expressed ($P<0.05$, two-way ANOVA) in fish fed either FO or VO (underlined spot numbers), or between family groups (remaining spot numbers); the three underlined spot numbers containing an * were differentially affected by both diet and genotype. Spots that returned reliable identifications by peptide fragment fingerprinting are identified by their abbreviation.

Table 5: Proteins differentially regulated by diet

Spot No	Protein ID	Accession No. (NCBInr)	Theoretical Mw (KDa)/ pI	Protein Score[a]	No. Matched peptides (MS/MS)[b]	Best score peptide	VO/ FO Lean	VO/ FO Fat	p-value
1136	Epoxide hydrolase 2 (EPHX2)	gb|ACI33129.1|	54.2/5.52	84	1	GGLFVGLPDEIPR	1.23	1.40	0.0190
1148	Hemopexin-like protein* (HPX)	emb|CAA92147.1|	50.4/5.61	169	1	VHLDAITSDDAGNI-YAFR	1.32	1.56	0.0002
1151	Hemopexin-like protein* (HPX)	ref|NP_001104617.1|	51.0/6.18	161	1	VHLDAITSDDAGNI-YAFR	1.38	1.54	0.0001
2683	Peroxiredoxin 1 (PRDX1)	gb|ACI67145.1|	22.0/6.42	757	8	SISTDYGVLKEDE-GIAYR	1.14	1.09	0.0310
3168	2-peptidylprolyl isomer-ase A (PPIA)	ref|NP_001135161.1|	17.5/7.61	101	1	VYFDITIGDTPAGR	−1.46	−1.18	0.0170
3243	Myosin light chain smooth muscle isoform (MYL)	ref|NP_998803.1|	16.9/4.47	100	2	EAFLLFDR	−1.23	−1.11	0.0490
3331	similar to H2A histone family, member J (H2A)	ref|XP_001521566.1|	13.3/10.84	74	1	AGLQFPVGR	−1.58	−1.18	0 0250
3334	Histone H2A** (H2A)	emb|CAA25528.1|	13.7/10.88	37	1	AGLQFPVGR	−1.48	−1.18	0.0210
3445	Galectin 2 (LGALS2)	gb|ACN10131.1|	14.8/5.93	314	2	SGASSFSINIGHDS-DNYALHFNPR	1.17	1.15	0.0170

[a] The protein score probability limit (where $P < 0.05$) is 73.

[b] Peptides with confidence interval above 95% were considered.

* Equivalent to warm-temperature-acclimation-related-65 kDa-protein.

**Same protein identification obtained by MS/PMF (peptide mass fingerprinting): Score = 76; 7 peptides matched.

Only reliable identifications of Actinopterygii, obtained by mass spectrometry (MALDI-TOF-MS/MS) analysis and searches in MASCOT, NCBI and ExPASy (Mw/pI) databases are shown. Expression ratios between fish fed VO and FO diets are given for each family group, as well as p-value for diet (two-way ANOVA, DeCyder V7.0).

Table 6: Proteins differentially regulated by genotype

Spot No	Protein ID	Accession No. (NCBInr)	Theoretical Mw (KDa)/pI	Protein Score[a]	No. Matched peptides (MS/MS)[b]	Best score peptide	Lean/Fat FO	Lean/Fat VO	p-value
1148	Hemopexin-like protein* (HPX)	emb|CAA92147.1|	50.4/5.61	169	1	VHLDAITSD-DAGNIYAFR	−1.09	−1.29	0.0270
1389	Calreticulin precursor (CALR)	gb|ACI32936.1|	47.6/4.33	81	2	FEPFSNEGK	−1.28	−1.33	0.0026
1441	Alpha-enolase (ENO1)	ref|NP_001133366.1|	47.0/5.91	405	3	AAVPSGASTGI-YEALELR	1.28	1.2	0.0066
1714	Heat shock protein 70 (HSP70)	ref|NP_990334.1|	70.8/5.47	111	1	IINEPTAAAIAY-GLDKK	1.16	1.77	0.0260
2134	Heterogeneous nucle-ar ribonucleoprotein A0 (HNRNPA0)	gb|ACI67551.1|	28.8/9.10	235	2	LFVGGLN-VDTDDDGLRK	−1.19	−1.09	0.0460
2154	Calponin 1 (CNN1)	ref|NP_001139857.1|	33.2/8.56	565	4	KINTSPQN-WHQLENIGN-FVR	−1.34	−1.51	0.0350
2157	Calponin 1*** (CNN1)	ref|NP_001139857.1|	33.2/8.56	43	1	YDPQKEEELR	−1.34	−1.47	0.0240
2213	Caspase 3 (CASP3)	ref|NP_001133393.1|	31.0/5.97	254	2	IPVEADFLYAY-STAPGYYSWR	−1.15	−1.24	0.0300
2215	Caspase 3 (CASP3)	ref|NP_001133393.1|	31.0/5.97	279	2	VANDQTVQQ-IQQLLSK	−1.14	−1.31	0.0130

2282	Voltage-dependent anion channel 2–2 (VDAC2)	gb|ACH71030.1|	30.1/8.85	214	1	VNNNSLVGV-GYTQTLRPGVK	−1.24	−1.19	0.0160
2458	Annexin A4 (ANXA4)	gb|ACI69495.1|	28.4/5.22	218	3	NHLLQVFK	−1.14	−1.11	0.0300
2487	Dihydropteridine reductase (DHPR)	gb|ACI67281.1|	15.7/8.46	81	1	QSVWTSTISSH-LATR	−1.14	−1.11	0.0300
2499	Pancreatic alpha-amylase precursor (AMY2)	ref|NP_001036176.1|	57.4/6.89	86	1	ALVFVDNHD-NQR	−1.37	−1.27	0.0021
2522	Triosephosphate isomerase 1b (TPI1)	ref|NP_001133174.1|	26.6/7.63	336	2	LDPNTEVVC-GAPSIYLEFAR	1.26	1.16	0.0009
2527	Histone cluster 1 (H2A)	ref|NP_001086775.1|	14.0/10.88	94	1	AGLQFPVGR	1.12	1.11	0.0074
2677	Transgelin (TAGLN)	gb|ACM09025.1|	21.7/7.69	219	1	DGCVLSELIN-SLHK	−1.61	−1.35	0.0350
2717	Proteasome beta 1 subunit (PSMB1)	ref|NP_001003889.1|	26.1/6.32	87	1	GAVYSFDPVG-SYQR	−1.29	−1.12	0.0140
2903	Peroxiredoxin-1(PRDX1)	gb|ACI67145.1|	22.0/6.42	115	1	QITINDLPVGR	−1.62	−1.99	0.0069
3345	Retinol-binding protein II, cellular (RBP2)	ref|NP_001139954.1|	15.6/5.44	70	2	AIDIDFATR	−1.32	−1.1	0.0330
3457	Alpha globin (HBA)	emb|CAA65949.1|	15.1/9.19	170	2	TYFSHWADLSP-GSAPVK	1.03	2.09	0.0320

[a]The protein score probability limit (where $P < 0.05$) is 73.

[b] Peptides with confidence interval above 95% were considered.

* Equivalent to warm-temperature-acclimation-related-65 kDa-protein.

*** Same protein identification obtained by MS/PMF (peptide mass fingerprinting): Score = 193; 27 peptides matched.

Only reliable identifications of Actinopterygii, obtained by mass spectrometry (MALDI-TOF-MS/MS) analysis and searches in MASCOT, NCBI and ExPASy (Mw/pI) databases are shown. Expression ratios between Lean and Fat family groups are given for each dietary treatment, as well as p-value for genotype (two-way ANOVA, DeCyder V7.0).

DISCUSSION

Dietary plant ingredients can induce chronic intestinal inflammatory conditions in salmonids that can ultimately result in carcinogenesis [25]. This extreme reaction is rare and usually associated with soy protein at high levels [26]. Dietary n-3 LC-PUFA have important anti-inflammatory and anti-carcinogenic effects in mammalian intestine [23]. Therefore, use of feeds containing high percentages of plant proteins combined with replacement of FO by VO, as is now prevalent in the industry, requires studies on dietary effects on intestinal transcriptomes and proteomes. However, interpretation of the data was difficult as the effects on dietary treatments and/or family groups were subtle, as also observed in liver transcriptome [9], and is typical of this type of experiment [6, 8, 11, 27]. Partly as a consequence, validation of the microarray data gave variable results, from perfect match, to opposite changes in a few, although effects observed in the microarray, with fold changes as low as 1.2 (e.g., NADH1 and PA2G4) were validated by RT-qPCR. In view of the whole genome duplication event that occurred in salmonids [28], gene expression studies are often more challenging due to the presence of highly similar genes which may hybridize with cDNA probes presenting low specificity, further complicated if similar transcripts, corresponding to duplicated genes, are differentially regulated [10]. Nonetheless, the presence of several features related to specific processes in both the transcriptomic and proteomic analysis gave supporting evidence to the pathways likely differentially affected by dietary oil and genetic background related to flesh adiposity.

Intestinal LC-PUFA Biosynthesis Capacity Is Differentially Affected By Diet and Genotype

Considering whether genetic selection for fish families showing better adaptation to more sustainable feeds might be a viable approach to develop aquaculture, one outcome of this investigation was to establish if effects of diet on expression of LC-PUFA biosynthesis genes depended on genotype, as shown in the liver transcriptome of these fish [9]. This was not seen in the hepatic transcriptome of European sea bass families showing different growth rates when fed a vegetable diet but, in this, case similar LC-PUFA profiles were also noted in both genotypes in response to the vegetable diet [11]. In both salmon tissues, differences in n-3 LC-PUFA content between fish fed FO or VO were smaller in the Lean family group. This was due to higher levels of n-3 LC-PUFA in Lean salmon, compared to Fat, when fish were fed VO, but higher amounts in the Fat family group when fed FO. However in liver, up-regulation of LC-PUFA biosynthesis when fish were fed VO was much larger in the Lean family group, whereas in intestine the same individuals only showed significant up-regulation in Fat fish. This appears contradictory but can be explained by the differential tissue n-3 LC-PUFA contents. Although the difference was smaller in Lean fish compared to Fat fish in both tissues, in liver there was still a considerable difference in n-3 LC-PUFA levels between fish fed FO or VO, while in intestine levels were similar. PUFA have important activities on transcription factors, either as direct ligands or through effects on membrane composition [29], affecting transcription of many genes involved in lipid metabolism, including desaturases and elongases [30, 31]. In salmon, regulation of genes of LC-PUFA biosynthesis that are known to respond to dietary composition, i.e., Δ5fad, Δ6fad, elovl5b and elovl2 [32, 33, 34], appear to show high plasticity and are likely under feedback regulation by tissue n-3 LC-PUFA. Both studies suggest that the Lean family group may show an enhanced response to low dietary n-3 LC-PUFA, with greater up-regulation of biosynthesis when fed VO. In contrast to liver, this response was sufficient in intestine to maintain tissue n-3 LC-PUFA, particularly DHA, at similar levels to fish fed FO. Considering that differences in desaturase expression between the Fat and Lean fish were only significant when FO but not VO, was fed, suggests that the likely mechanism is through negative feedback by high levels of n-3 LC-PUFA rather than positive feedback from low levels of LC-PUFA and/or higher levels of shorter chain precursors.

Other Dietary Effects on Lipid Metabolism

Transcriptional regulation of desaturases and elongases by LC-PUFA may involve both PPARα and sterol regulatory element binding protein 1c (SREBP-1c) [30, 31]. In liver, expression of Δ5fad, Δ6fad, elovl2, PPARα, and possibly PPARβ, appeared co-ordinately regulated by diet depending on genotype, while PPARγ was not affected [9]. In intestine, however, expression of PPARα and PPARβ was not affected by either diet or genotype, while PPARγ was up-regulated by dietary VO, significantly in Fat fish. This suggests that dietary regulation of lipid metabolism genes in fish intestine might differ to mammals, where PPARα showed differential expression in response to dietary EPA and DHA in murine intestine [35]. Reasons for differential regulation of PPARs between salmon liver and intestine are unclear, but may be due to different patterns of tissue expression. In plaice and seabream, there was no nutritional regulation (fasting/feeding) of PPARs in the intestine, where PPARγ was the dominant isotype, in contrast to liver where PPARα was dominant [36]. PPARγ in both mammals and fish is predominantly expressed in adipose tissue and promotes adipocyte differentiation and lipid storage [36, 37, 38]. In mammals, PPARγ activates the expression of genes characteristic of mature adipocytes and adipogenesis, including FAS [37, 38] and hence the expression of PPARγ, up-regulated in salmon fed VO, might be related to increased expression of FAS. However, increased PPARγ expression was only significant in Fat fish whereas FAS was significantly up-regulated only in Lean salmon. As fish PPARγ is functionally the most different of the three isotypes compared to mammalian PPARs, and is expressed more widely in fish tissues that in mammals, other mechanisms and functions may underlie the observed regulation [35].

In this study, the hypotriglyceridemic effect of LC-PUFA, well established in mammals [39, 40], was also observed in salmon intestine. Lipogenesis was down-regulated in FO-fed fish, as demonstrated by decreased FAS expression (RT-qPCR) and the presence of a transcript containing a beta-ketoacyl synthase domain, a component of FAS (microarray). The differences in FAS expression were not as marked as in liver [9] and were only significant in Lean fish but, together with the LC-PUFA biosynthesis data, demonstrate the active role of salmon intestine in lipid metabolism [21]. However, despite up-regulation of lipogenesis by dietary VO, lipid accumulation in enterocytes was lower than in fish fed FO, contrary to previous reports of VO promoting lipid accumulation in enterocytes [16, 17, 18].

In contrast, the hypotriglyceridemic effect of LC-PUFA did not involve the typical increase in β-oxidation [39], reported in mice intestine [36, 40]. As in liver, no changes were observed in the expression of β-oxidation genes carnitine palmitoyltransferase I (CPT1) and acyl-CoA oxidase (ACO) [10]. Nonetheless, effects of dietary lipid on energy metabolism were observed in intestine. In particular, UCP and transcripts involved in the mitochondrial electron transport chain, including components of cytochrome c oxidase, NADH1 and ubiquinol-cytochrome c reductase complexes, and the mitochondrial metabolite transporter SCaMC-2, were slightly down-regulated by dietary VO, possibly suggesting reduced energetic metabolism. EPA may act as a mitochondrial proliferator in both rat and salmon liver [41, 42], which might also explain this result.

Vegetable Oils as Potential Sources of Contaminants in Alternative Diets for Aquaculture?

The use of FO in feeds has also been questioned in relation to levels of persistent organic pollutants, POPs [43], whereas VO are generally considered safer alternatives to dietary FO [44]. However, results suggested that VO had greater impact on xenobiotic metabolism in salmon intestine than FO, with the transcriptome and proteome both showing up-regulation of transcripts and proteins involved in detoxification and protection from oxidative stress in fish fed VO. This was surprising, considering expression of these genes has been associated with FO supplementation and higher levels of organic contaminants and/or increased peroxidative susceptibility of LC-PUFA [22, 45, 46]. In particular, and linked to a detoxification role, up-regulation of CYP1A transcript and epoxide hydrolase 2 (EPHX2) protein was found in fish fed VO. Cytochrome P450 1A metabolizes many exogenous and endogenous molecules, including pollutants (e.g., polychlorinated biphenyls-PCBs, polycyclic aromatic hydrocarbons-PAHs, dioxins and dibenzofurans) as well as several metabolic products (e.g., steroids, bile acids). Its expression is sensitive to low levels of contaminants and therefore it is a commonly used marker [47, 48]. Lipid peroxidation and antioxidant enzymes are also biomarkers for environmental xenobiotic contamination in fish, as the catalytic actions of detoxifying enzymes like CYP1A produce high levels of reactive oxygen species (ROS) [49, 50]. We observed up-regulation of CAT and of a selenoprotein transcript, as well as of HPX and PRDX1 proteins in salmon fed VO. CAT and PRDX1 catalyze the decomposition of hydrogen peroxide [51], hemopexin prevents heme-mediated oxidative stress [52], and selenoproteins have diverse roles, including selenium transport and antioxidant defense [53]. Expression of other genes involved in protection from oxidative stress like GST and SOD were not affected, but MT-A was down-regulated in the Lean family group fed VO. MT-A, besides having an antioxidant role, also protects against heavy metal toxicity and maintains physiological zinc homeostasis [54] and hence could be responding to contaminants more abundant in FO (e.g., heavy metals) [55].

Previous analysis of contaminants in comparable feeds using the same standard FO and a similar VO blend, showed that levels of several POPs, including organochlorine pesticides, and heavy metals were substantially lower, but polycyclic aromatic hydrocarbons (PAH) were 10-fold increased in the VO diet compared to the FO diet [55]. PAH derive from incomplete combustion of organic compounds and can be found in high concentrations in fats, including VO, through multiple routes of contamination, before, during or after oil processing [56]. They are metabolized by both CYP1A1 and EPHX2, among other enzymes, and CYP1A1 is induced by PAH in mammals [57]. Therefore, higher PAH levels in the VO diet might explain, at least partly, the results obtained in the present study although, unlike POPs, PAHs are not persistent and are readily eliminated from fish tissues [58].

High doses of PAH result in significant intestinal hyperplasia in fish, with an increase in cell proliferation and faster epithelial turnover [59]. Previous studies on intestinal gene expression in fish indicated a reduction in cell proliferation or differentiation associated with dietary FO replacement by VO [21, 22], possibly due to lower levels of membrane LC-PUFA and reduced oxidative stress [22]. In the present study, no major impact on cell proliferation was apparent in the intestinal transcriptome or proteome data. Two transcripts related to cell proliferation, PA2G4 and cyclin G1, were slightly down-regulated in fish fed VO, but in mammals these have opposing effects [60, 61] and, furthermore, two mammalian PA2G4 isoforms have been shown to have opposite effects in cellular proliferation [62] and hence results are inconclusive.

Previously, expression of caspases, effectors of controlled cell death or apoptosis, was affected by replacement of dietary FO by VO in fish [21, 22]. Apoptosis is particularly important in organs with high rates of cellular turnover such as intestine but, in addition to maintaining normal gut function, apoptosis may be affected by pathological or toxic conditions, including those induced by environmental chemical contaminants [63, 64]. In the present study, expression of CASP3B was up-regulated in salmon fed VO, particularly in the Lean family group and a similar, non-significant trend was observed for CASP6A/B. As ROS are important signalling molecules in apoptotic processes, these results could be linked to a cytotoxic effect causing increased oxidative stress in VO [64]. Relevant to the above was the up-regulation of galectin 2 (LGALS2) in the proteome of salmon fed VO. Galectins are pleiotropic regulators of immune functions and are up-regulated by injury and infectious conditions, have well-recognized modulatory roles in mammalian intestinal inflammatory diseases, and their mode of action involves induction of apoptosis [65, 66]. The lack of major effects on cell proliferation and only slight up-regulation of CASP3 and LGALS2 suggests that any contaminant doses experienced by the fish were unlikely to have caused any serious morphophysiological damage in the intestine. As similar trends were not seen in the hepatic transcriptome of these individuals [9], this may suggest intestine can potentially metabolize and detoxify xenobiotics present in the diet. Furthermore, there were no growth or general performance issues with these fish [5]. Therefore, the data do not imply abnormal gastrointestinal functions or effects on final product quality.

Effect of Genotype in Intestinal Transcriptome and Proteome

Contrary to diet, genotype did not have a major impact on metabolism genes, apart from transcripts related to the proteasomal degradation pathway including a strong down-regulation of PSMB8 in Lean fish, particularly fed VO. This gene has been recently found to have a molecular evolution history that suggests a very strong selective pressure for its functional dimorphism to be maintained in vertebrates [67]. Two different alleles, A-type and F-type, can be found in basal vertebrate species, including Atlantic salmon. The PSMB8F lineage was lost in common ancestors of higher teleosts and tetrapods but was then independently revived *de novo* through the appearance of F-type alleles within the PSMB8A lineage. In this study we did not find evidence of significant differences between families groups for the A-type allele in the transcriptomic analysis as the probe showing significant variation between families in the microarray corresponded to the PSMB8F allele. Hence, it was also the F-type transcript that was validated by RT-qPCR, using type-F specific primers. However, further studies would be required to confirm this and to assess the functional significance of this result. On the other hand, expression of PSMB1 was down-regulated in the intestine proteome of Lean fish. Proteolysis through this pathway is essential for many cellular processes, including the cell cycle, signalling, cellular defence and responses to oxidative stress [68]. Therefore, this response might be related to defence against cellular stress, as another difference between the two family groups was related to xenobiotic and oxidant metabolism. Apart from lower expression of a CYP1A transcript in Lean fish, two proteins with antioxidant roles, HPX and PRDX, were down-regulated in the proteome of Lean compared to Fat fish. Alpha globin, or haemoglobin alpha (HBA), a major component of blood and potent mediator of oxidative stress, can have both protective and damaging effects depending on complex interactions in H_2O_2-rich environments [69]. However, given its opposite regulation to HPX, whose main role is to scavenge heme and protect from its toxic effects [52], up-regulation of HBA in Lean fish may indicate heme-mediated oxidative stress.

The apoptotic pathway may be differentially affected by genotype, with down-regulation of CASP3, VDAC2 and ANXA4 in the Lean family group, the latter two transport proteins having well-recognized roles in apoptosis [70, 71]. In contrast, heat shock proteins that protect against environmental stresses were increased in the intestine transcriptome and proteome of Lean salmon. This response could be associated with the observed changes in the ubiquitin-proteasome degradation system, as the systems have been functionally coupled in mammals. Thus, moderate exposure to a heat shock can cause a transient increase in intracellular proteolysis by the ubiquitin-proteasome pathway, followed by a phase of slower or even inhibited protein degradation [72, 73]. Furthermore, Pirkkala et al. [74] demonstrated transcriptional induction of heat shock genes when proteasome activity was down-regulated. However, judging by the fold-changes, these effects are only relevant when fish were fed VO, and hence could be more related to dietary changes. Collectively, the data may indicate higher sensitivity of Lean fish to environmental or endogenous stresses due to replacement of dietary FO by VO.

The predominant influence of genotype was in the expression of intestinal transcripts of structural proteins, particularly collagen alpha chains, but also osteonectin, TAGLN, troponin I and keratocan, which were up-regulated in Lean fish, whereas troponin C was down-regulated. Furthermore, CNN1 and TAGLN were down-regulated in the intestinal proteome in Lean fish. Collagen, the main component of connective tissue, helps to maintain the structural integrity of tissues [75], while osteonectin is an extracellular matrix glycoprotein with high affinity towards collagen and whose expression has been associated with remodelling processes in tissues, including human intestine during development/morphogenesis and in diseased mucosa [76]. Troponin, TAGLN and CNN1 are all involved in actin binding, actin-myosin interaction and muscle contraction. The inverse regulation of troponins is not conflicting as they have different roles in actomyosin cross-bridge formation and contraction; binding of troponin C to Ca^{2+} induces conformational changes that counteract the inhibitory action of troponin I [77]. Expression of TAGLN transcript and protein showed opposite effects but a lack of correlation between transcriptomic and proteomic data is not unprecedented [78]. As discussed above, this result might also be explained by the presence of similar duplicated genes in Atlantic salmon that are regulated differently. Transcriptomic results were validated by RT-qPCR for COL1A2, although only significantly when fish were fed the VO diet, for which fold changes were higher. In addition, in the microarray results differences in expression of structural proteins between family groups were consistently more accentuated in fish fed VO which could suggest a cumulative effect of diet. Furthermore, MYO was up-regulated in fish fed VO compared to FO but only in Lean fish (RT-qPCR), and significant diet × genotype interactions were found for alpha-actinin 1, tubulin beta-2 chain and procollagen-lysine 2-oxoglutarate 5-dioxygenase 2 (Additional file 1), which were up-regulated in Lean salmon, compared to Fat, but only when fed VO. In cod, replacement of FO by VO resulted in changes in intestinal expression of structural genes with the potential to alter the structural and mechanical properties of the intestinal muscle layer, including a range of actin-binding transcripts [21].

The present study is the first investigation of the influence of genetic background of families with different flesh adiposity phenotypes on intestinal gene expression of a fish species. Effects were subtle and consequently their potential impacts difficult to fully assess. However, if genetic selection for families better adapted to alternative formulations is an approach taken in the future, the potential for genotype-specific differences being exacerbated when VO replaces dietary FO should be further examined to assess the consequences of these changes in intestinal gene expression.

CONCLUSIONS

Metabolic activity, particularly lipid and energy, of intestinal tissue responded to dietary lipid composition but was also affected by genotype. The LC-PUFA biosynthesis pathway, typically up-regulated when salmon are fed VO, was especially influenced by genetic background. The Lean fish showed an enhanced response to low dietary n-3 LC-PUFA and the expression of Δ5fad, Δ6fad, elovl5b and elovl2 in the intestine showed high plasticity and was reflected in tissue biochemical composition indicating that their transcriptional regulation might be under feedback control by n-3 LC-PUFA, mainly DHA. Lower n-3 LC-PUFA in VO increased lipogenesis in Lean salmon, assessed by expression of FAS, while β-oxidation appeared unaffected, although transcripts involved in mitochondrial respiratory or electron transport chains were down-regulated, suggesting reduced activity in fish fed VO. Higher expression of genes and proteins involved in xenobiotic metabolism (CYP1A and EPHX2), antioxidant defence (CAT, HPX and PRDX1), and apoptosis (Casp3B) were observed in VO-fed fish, suggesting they might be responding to higher levels of contaminants, particularly PAH, in the diet. However, the intestine appeared able to metabolize and detoxify xenobiotic substances potentially present in the diet without major deleterious effects. Nonetheless, the data suggest that further attention should be given to contaminants in VO in the future. On the other hand, the data indicate potential genotype-specific differences in the response of the intestinal transcriptome and proteome to dietary VO. These include potential changes in structural properties of the intestinal layer and defence against cellular stress suggesting the Lean group was more susceptible to diet-induced oxidative stress. Considering the possibility of selecting families better adapted to alternative diet formulations, and the central role of intestine as a major barrier to nutrients, contaminants and pathogens, greater attention should be given to this organ when evaluating the effects of diet and genotype.

METHODS

Feeding Trial and Sampling

A dietary trial was conducted using two genetically characterized groups of Atlantic salmon (*Salmo salar*) post-smolts comprising full-sib families selected from a breeding program (Landcatch Natural Selection Ltd., Argyll, Scotland). The choice of the family groups was based on estimated breeding values (EBVs) of the parents for high or low flesh adiposity, assessed by Torry Fatmeter (Distell Industries, West Lothian, UK), a trait that was found to have a heritability ranging from 0.17 to 0.39 in this dataset. The two groups were created from four unrelated full-sib families; two families from the extreme lower end of the EBV distribution for flesh lipid content (Lean) and two families from the extreme upper end of the distribution (Fat). The average EBV for the lipid content of the Fat families was 2.00 percentage units higher than that of the Lean families, representing a standardised selection differential of 2.33 standard deviations. Assessment of the flesh and visceral lipid contents at the end of the trial confirmed differences in adiposity between the groups [5].

Two thousand fish of each group were stocked into eight 12 x 5m^3 net pens at the Ardnish Fish Trials Unit (Marine Harvest Scotland, Lochailort, Scotland; 500 fish pen^{-1}). Duplicate pens of each group were fed one of two experimental diets (Skretting ARC, Stavanger, Norway) containing 25-32% fish meal, 40-45% plant meals and 27.5-30% oil (percentages varying according to pellet size) supplied either as standard northern FO or as a VO blend comprising rapeseed, palm and *Camelina* oils in a ratio of 5:3:2. Diets were formulated to fully satisfy the nutritional requirements of salmonid fish [79] and contained similar levels of PUFA but different n-3 and n-6 PUFA contents, 25.3% and 4.6% in the FO diet and 13.4% and 17.1% in the VO diet, respectively [5]. After 55 weeks, 25 fish per pen were sampled 24 h after the last meal. Fish were killed by a blow to the head following anaesthesia, and intestinal tissue (pyloric caeca) collected, immediately frozen in liquid nitrogen and stored at −70°C prior to analyses. Further details can be found in Bell et al. [5].

Lipid Extraction and Fatty Acid Analyses

Total lipid from 1 g of intestine of four fish per treatment was extracted and determined gravimetrically [80], and fatty acid methyl esters (FAME) prepared by acid-catalysed transesterification of total lipid [81]. FAME were separated and quantified by gas chromatography as described in detail previously [5]. Significant differences in intestinal fatty acid composition were determined by two-way ANOVA ($p<0.05$) using the SPSS 16.0 statistical package (SPSS Inc., Chicago IL, USA).

RNA Extraction and Purification

Intestinal tissue (0.2 g) from six individuals per experimental group was homogenised in 2mL TRI Reagent and total RNA isolated following manufacturer's instructions (Ambion, Applied Biosystems, Warrington, U.K.). RNA quantity and quality (integrity and purity) were assessed by gel electrophoresis and spectrophotometry (NanoDrop ND-1000, Thermo Scientific, Wilmington, U.S.A.), and 100 μg of total RNA from each sample further cleaned by mini spin-column purification (RNeasy Mini Kit, Qiagen, Crawley, UK).

Microarray Hybridizations, Image Processing and Statistical Analysis

The TRAITS/SGP (v.2.1) salmon 17k cDNA microarray described by Taggart et al. [33] was used (ArrayExpress accession: A-MEXP-1930). A dual-labelled experimental design was employed, with each sample being competitively hybridised against a common pooled-reference.

The experiment comprised 2 genotypes (Lean/Fat) × 2 diets (VO/FO) × 6 biological replicates. Indirect labelling was employed for preparing the microarray targets. Antisense amplified RNA (aRNA) was produced from 500 ng of purified total RNA per sample using the Amino Allyl MessageAmpTM II aRNA Amplification Kit as per manufacturer's instructions (Ambion, Applied Biosystems), followed by Cy3 (for samples) or Cy5 (for pooled reference) fluor (GE HealthCare, Buckinghamshire, U.K.) incorporation through dye-coupling reaction [10]. Microarray hybridizations were performed in a Lucidea semi-automated system (GE Healthcare) without pre-hybridization. For each array, every labelled biological replicate and corresponding pooled reference (40 pmol each dye, c. 150 ng aRNA) were combined and added to the hybridization solution. Two post-hybridization automatic washes followed by six manual washes to a final stringency of 0.1× SSC (EasyDipTM Slide staining system; Canemco Inc., Quebec, Canada) were performed before scanning.

Scanning was performed at 10 μm resolution using an Axon GenePix 4200AL Scanner (MDS Analytical Technologies, Wokingham, Berkshire, U.K.). Laser power was constant (80%) and "auto PMT" was enabled to adjust each channel at less than 0.1% feature saturation and Cy3/Cy5 mean intensity close to one. BlueFuse software (BlueGnome, Cambridge, U.K.) was used to identify features and extract fluorescence intensity values from TIF images. The resulting fluorescence intensity data (BlueFuse proprietary algorithm) and quality annotations for the 17,102 gene features were exported into the GeneSpring GX version 10.0.2 analysis platform (Agilent Technologies, Wokingham, Berkshire, U.K.) after undergoing block Lowess normalization. All control features were excluded from subsequent analyses. Data transformation and quality filtering were as in Morais et al. [9, 10]. This gave a final list of 15,498 genes that were eligible for statistical analysis. Experimental annotations complied fully with minimum information about a microarray experiment (MIAME) guidelines [82] and experimental hybridisations are archived on the EBI ArrayExpress database (http://www.ebi.ac.uk/arrayexpress/) under accession number E-TABM-1173. Hybridization data were analysed in GeneSpring by two-way ANOVA, which examined the explanatory power of the variable diet and genotype and the interaction between the two, followed by Gene Ontology (GO) enrichment analysis of the significant lists of features, at a significance level of 0.05. No multiple test correction was employed, as previous analyses, confirmed by RT-qPCR, indicated that such corrections are over-conservative for this type of data [9, 32].

RT-qPCR Gene Expression Analysis

Expression of selected genes, for microarray validation and to further examine biological processes of interest, was studied by reverse transcription quantitative real time PCR (RT-qPCR), with target qPCR primer sequences given in Additional file 2. In addition, amplification of two reference genes, *cofilin-2* and elongation factor-1α (*elf-1α*), was performed. One μg of column-purified total RNA per sample was reverse transcribed into cDNA using the VersoTM cDNA kit (ABgene, Surrey, U.K.) using a mixture of random hexamers (400ng/μL) and anchored oligo-dT (500ng/μL) at 3:1 (v/v). Negative controls (no enzyme) were performed to check for genomic DNA contamination. A similar amount of cDNA was pooled from all samples and the remaining cDNA diluted 20-fold with water. RT-qPCR analysis used relative quantification with the amplification efficiency of each primer pair assessed by serial dilutions of the cDNA pool. Amplifications were carried out in duplicate using a Quantica machine (Techne, Cambridge, U.K.) in a final volume of 20 μl containing 2–8 μl diluted cDNA, 0.5 μM of each primer and 10 μl AbsoluteTM QPCR SYBR® Green mix (ABgene), with a systematic negative control (NTC-non template control). The qPCR profiles contained an initial activation step at 95°C for 15 min, followed by 30–40 cycles: 15 s at 95°C, 15 s at the specific primer pair annealing temperature (Ta; Additional file 2) and 15 s at 72°C. After amplification, a melt curve was performed confirming a single product in each reaction, RT-qPCR product sizes checked by agarose gel electrophoresis, and identity of amplicons confirmed by sequencing. Gene expression was analysed using the relative expression software tool (REST 2008, http://www.gene-quantification.info/), employing a pair wise fixed reallocation randomisation test (10,000 randomisations) with efficiency correction [83].

Protein Extraction and Labelling

Six intestine samples (100 mg) per treatment were rapidly disrupted by homogenization and sonication on ice in 1 ml of DIGE lysis/labeling buffer (7M Urea, 2M Thiourea, 4% CHAPS and 30mM Tris, pH 8.5) in the presence of 10 μl of a protease inhibitor cocktail (ref. P8340, Sigma-Aldrich) and 4μl of 250 mM EDTA. After centrifugation at 12,000 × g for 20 min at 4°C, the supernatant was recovered and protein concentration determined (Quick Start Bradford Protein Assay Kit, ref. # 500–0202, Bio-Rad). Protein (500 μg) was purified by precipitation (ReadyPrepTM 2-D Cleanup kit, Bio-Rad) and the pellet re-suspended in DIGE lysis/labeling buffer at ~5μg/μl.

Samples were labelled using CyDye DIGE fluors (5nmol labelling kit; GE Healthcare), following manufacturer's instructions. Three of the experimental replicates of each treatment (50 μg protein with pH corrected to 8.0-8.5) were labelled individually with 400 pmol Cy3 and the remaining three with 400 pmol Cy5. In addition, equal amounts of all experimental samples were pooled and 600 μg of protein (50 μg per 2-D gel) were batch-labelled with Cy2. The three labelled samples, corresponding to two experimental samples and one internal reference pool, were then combined to have in each 2-D gel samples corresponding to fish fed either FO or VO within the same family group.

Two-Dimensional (2-D) Polyacrylamide Gel Electrophoresis

Rehydration buffer (ReadyPrep 2-D starter kit, Bio-Rad) containing 0.2% DTT was added to the pooled protein samples to a final volume of 450 μl, which were loaded onto Immobiline DryStrip pH 3–11 NL, 24 cm IPG strips (GE Healthcare) by passive rehydration at room temperature overnight in the dark. Proteins were separated in the first dimension by isoelectric focusing at 20°C (Ettan IPGphor, GE Healthcare), applying increasing voltage until 200 V for 4 h, increasing to 500 V over a period of 3 h, then keeping the applied tension at a constant 1000 V for 1 h, followed by a further increase to 8000 V over 90 min, maintaining this voltage for almost 9 h (total of 79,200 Vh). After isoelectric focusing the strips were equilibrated in two 40 min steps using 50mM Tris–HCl pH 8.8, 6M urea, 30% glycerol, 2% SDS buffer, to which 2 % DTT (w/v) and 2.8% iodoacetamide (GE Healthcare) were added to produce reducing and alkylating buffers, respectively. The strips were loaded onto a 12.5% acrylamide gel cast between low fluorescence glass cassettes (EttanDALTsix gel caster system, GE Healthcare). The strips were overlaid with ReadyPrep Overlay Agarose (Bio-Rad) and the six gel cassettes run in the EttanDALT system in two steps: at 60 mA, 80 V, 6 W (upper limits) for 1 h, and then 240 mA, 500 V, 78 W until the bromophenol blue dye front had run to ~1 cm above the bottom of the gels. Laemmli buffers (250mM Tris, 1.92M glycine, 1% SDS) were used in the lower (1X) and upper (2X) chambers, respectively.

Gel Imaging and Analysis

Labelled gels were scanned using a Typhoon TRIO (GE Healthcare) and Cy2, Cy3 and Cy5 images acquired using 520BP40, 580BP30 and 670BP30 laser emission filters, respectively, at 500 PMT and 100 μm resolution. Images were cropped to remove extraneous areas prior to analysis (ImageQuantTL, GE Healthcare), and image analysis performed using DeCyder V7.0 (GE Healthcare). The estimated number of spots for each co-detection procedure was set at 10,000 and an exclusion filter was applied to remove spots with a volume lower than 30,000. Differential expression of protein spots was examined by two-way ANOVA at a significance level of 0.05. After verifying that significant spots were well matched across the gels, two pick lists were generated with a total of 22 and 45 spots for the diet and genotype factors, respectively (Figure 1; Additional files 3 and 4).

Spot Picking and Protein Identification by Peptide Fragment Fingerprinting

Four preparative gels were run under the conditions described above but with higher amounts of protein (600 and 1000 μg). They were stained with colloidal Coomassie and, whenever possible, spots were excised and sequenced in the Mass Spectrometry Laboratory ITQB-UNL (Oeiras, Portugal), where in-gel digestion and extraction of the proteins from the gel was performed, followed by micropurification, and peptides identified by mass spectrometry (using an Applied Biosystems (ABI) 4800 MALDI TOF/TOF Analyzer). The search engine MASCOT was then used to identify and confirm protein ID's from the peptide mass fingerprinting (PMF; MS) and peptide fragment fingerprinting (PFF; MS/MS) data.

DECLARATIONS

Acknowledgements

This study was funded by the EU FP6 IP "AQUAMAX" (Sustainable Aquafeeds to Maximise the Health Benefits of Farmed Fish for Consumers; 016249–2). SM was supported by a Marie Curie Intra European Fellowship (FP7-PEOPLE-2007-2-1-IEF, Proposal N219667) and by AQUAGENOME Mobility and Resource Exchange Grants in 2008. The authors acknowledge Landcatch Ltd for provision of the Lean and Fat smolts and thank staff of Marine Harvest Ltd, Ardnish FTU, for assistance with fish husbandry and sample collection. Technical assistance from Jacquie Ireland in microarray hybridizations is deeply appreciated. We also acknowledge Mahaut de Vareilles for performing some of the protein identifications by mass spectrometry, and discussions on the use of DeCyder V7.0.

Electronic supplementary material

Additional file 1: Intestine transcripts corresponding to the top 100 most significant features exhibiting a significant diet × family interaction. This table shows the annotated features (62% of all clones) arranged by categories of biological function and, within these, by decreasing significance (assessed by two-way ANOVA). Also indicated are the GenBank accession numbers for each clone (or, when not available, the probe number is given instead) and the expression ratios between fish fed VO and those fed FO, for each one of the families, and between Lean and Fat fish fed either FO or VO.

Accession no.	Gene	VO/FO		Lean/Fat		p-value
		Fat	Lean	FO	VO	
Metabolism (41%)						
Lipid metabolism (12%)						
DW589103	ATP citrate lyase	1.1	- 1.3	1.3	- 1.1	0.0002
AM397481	Delta-6 fatty acyl desaturase	1.3	- 1.1	1.2	- 1.2	0.0004

AY458652	Delta-6 fatty acyl desaturase	1.9	- 1.4	1.6	- 1.7	0.0005
DW588445	Delta-6 fatty acyl desaturase	1.9	- 1.3	1.5	- 1.6	0.0005
CK877956	Triacylglycerol lipase	1.4	- 1.2	1.4	- 1.2	0.0012
AJ425157	Phospholipase B-like 1	- 1.3	1.6	- 1.5	1.4	0.0013
CK897463	ATP citrate lyase	1.2	- 1.3	1.3	- 1.2	0.0015
AY458652	Delta-6 fatty acyl desaturase	2.2	- 1.4	1.6	- 1.9	0.0017
CK888091	Delta-6 fatty acyl desaturase	1.8	- 1.3	1.5	- 1.6	0.0027
CK885553	Peroxisomal proliferator-activated receptor gamma	- 1.0	1.4	- 1.3	1.1	0.0049
Energy metabolism/generation of precursor metabolites (12%)						
CK896189	Calcium-binding mitochondrial carrier protein SCaMC-2	1.1	- 1.7	1.1	- 1.6	0.0006
CK890974	Calcium-binding mitochondrial carrier protein SCaMC-2	1.0	- 1.5	- 1.1	- 1.6	0.0010
EG647452	Cytochrome c oxidase subunit 4 isoform 1	- 1.2	1.1	- 1.1	1.1	0.0012
CK886135	Creatine kinase mitochondrial 1	1.1	- 1.2	1.3	- 1.1	0.0013
GE624328	Pyruvate dehydrogenase kinase, isoenzyme 2	1.5	- 1.3	1.1	- 1.6	0.0035
Protein and amino acid metabolism (10%)						

AM042268	N-acylaminoacyl--peptide hydrolase	- 1.1	1.3	- 1.2	1.1	0.0011
EG648672	Aspartate amino-transferase	1.2	- 1.2	1.2	- 1.2	0.0012
DW590246	Betaine aldehyde dehydrogenase	- 1.0	- 1.4	1.2	- 1.2	0.0012
DW588095	Heat shock protein DnaJ (Hsp40) homolog	- 1.1	1.2	- 1.3	1.1	0.0029
Carbohydrate metabolism (5%)						
EG649263	Endosulfine alpha	- 1.0	- 1.3	1.1	- 1.2	0.0021
CK885545	Glucose transporter type 8	1.5	1.0	1.3	- 1.1	0.0023
Xenobiotic and oxidant metabolism (2%)						
CK884553	UDP glucuronosyl-transferase 2-like	1.1	- 1.3	1.1	- 1.3	0.0011
CK885392	Catalase	1.1	1.6	- 1.4	1.0	0.0020
Transport/ intracellular trafficking (7%)						
CK878806	AP1 gamma subunit-binding protein 1	- 1.2	1.2	- 1.1	1.3	0.0001
AJ425217	Rhesus-associated glycoprotein	- 1.1	1.3	- 1.2	1.2	0.0006
AM041759	Mitochondrial import inner mem-brane translocase subunit TIM13	- 1.2	1.3	- 1.3	1.2	0.0045
Regulation of transcription (12%)						
DW588482	General transcrip-tion factor 3C polypeptide 4	1.2	- 1.2	1.1	- 1.3	0.0004

CN181253	COMM domain-containing protein 7	- 1.2	1.3	- 1.3	1.3	0.0013
CK879147	High-mobility group box 1	- 1.1	1.3	- 1.1	1.3	0.0028
CK889611	DNA-directed RNA polymerase	- 1.2	1.2	- 1.1	1.3	0.0039
BM413765	Inhibitor of DNA binding/differentiation 2B	- 1.0	1.5	- 1.3	1.2	0.0041
Translation (5%)						
CN181210	40S ribosomal protein S27	1.1	- 1.2	1.3	- 1.0	0.0005
CK887633	Ribosome biogenesis protein NSA2	- 1.2	1.1	- 1.3	- 1.0	0.0036
Signalling/Signal transduction (18%)						
CO471793	Rho GDP-dissociation inhibitor 1	1.1	- 1.1	1.3	1.0	0.0007
CK885243	Integrin-linked protein kinase	1.3	- 1.3	1.3	- 1.3	0.0022
CK885881	Tetraspanin-9	- 1.0	1.2	- 1.0	1.1	0.0024
EG647377	Regulator of G-protein signaling 5	- 1.2	1.1	- 1.0	1.3	0.0024
EG647545	Protein tyrosine phosphatase receptor type D	- 1.1	1.1	1.0	1.2	0.0026
CK890952	G protein gamma subunit	1.2	- 1.1	1.1	- 1.2	0.0041
pit_cpi_C2F04	Phosducin-like protein 3	- 1.2	1.1	- 1.3	1.1	0.0042
CK885079	Sonic hedgehog-like protein	1.0	1.3	- 1.3	- 1.0	0.0048
Structural proteins (15%)						
CO472006	Cytokeratin-13	- 1.3	1.0	- 1.2	1.1	0.0002
CK885977	Cytokeratin-18	1.1	- 1.1	1.1	- 1.2	0.0004
CN181168	Type II keratin	1.1	- 1.1	1.1	- 1.0	0.0016

EG648864	Alpha-actinin 1	- 1.1	1.3	- 1.1	1.3	0.0021
EG647396	Tubulin beta-2 chain	- 1.1	1.4	- 1.1	1.3	0.0030
CK884318	Procollagen-lysine 2-oxoglutarate 5-dioxygenase 2	- 1.2	1.3	- 1.2	1.3	0.0032
CK876976	Beaded filament structural protein 2	1.1	- 1.1	1.1	- 1.1	0.0045
Immune response (2%)						
CK881225	Nonspecific cyto-toxic cell receptor protein-1	1.2	- 1.3	1.2	- 1.2	0.0012
Miscellaneous						
CK876833	Keratocan	- 1.2	1.4	- 1.0	1.6	0.0003
DW592240	Death-associated protein 1	- 1.2	1.0	- 1.2	1.1	0.0010
CK895964	Transposase	1.1	- 1.1	1.2	- 1.0	0.0013
BE518538	Transformer 2 alpha homolog	1.3	- 1.1	1.1	- 1.2	0.0015
CK882254	ADAM8 precursor (A disintegrin and metalloproteinase domain 8)	1.0	- 1.1	1.1	- 1.1	0.0015
CK882779	40 kDa peptidyl--prolyl cis-trans isomerase	1.2	- 1.2	1.2	- 1.2	0.0018
kid_cki_A1E04	S100-like calcium binding protein	1.2	- 1.1	1.1	- 1.3	0.0023
BI468158	Nuclear protein 1	1.4	- 1.1	- 1.0	- 1.6	0.0025
DW588616	Kainate receptor	1.6	- 1.4	1.4	- 1.6	0.0025
CK882262	Endonuclease domain-containing 1	1.2	- 1.3	1.4	- 1.1	0.0040
CK877382	Septin 7	- 1.1	1.1	- 1.0	1.2	0.0045
EG649429	Mitofusin-2	1.2	- 1.3	1.2	- 1.3	0.0046

CK891845	Transmembrane and coiled-coil domain-containing protein1	1.1	- 1.3	1.2	- 1.2	0.0049

Additional file 2: Primers used for RT-qPCR analyses.

Transcript	Primer sequence	Fragment	Ta	Efficiency	Accession No.	Source
Δ5fad	GTGAATGGGGATCCATAGCA	192 bp	56°C	0.990	AF478472 [1]	[84]
	AAACGAACGGACAACCAGA					
NADH1	CGCTTTCCTCACCTTACTCG	162 bp	60°C	0.988	TC161581 [2]	New design
	AATAGGAAGGGGGAGGAGGT					
UCP2	GTTCATATATCTCGTCTTCAC-CACATC	73 bp	60°C	0.994	BQ035332 [1]	[85]
	CTGCATTCGGTGCAGGATT					
MTHFD1	TTAGGCGCTTCTCTCCCATA	216 bp	60°C	0.997	NM_001165368 [1]	New design
	CTCGCACATGTCCTTTCTCA					
PA2G4	AGACCGAAGCTGCGAAATAA	209 bp	60°C	0.998	TC120403 [2]	New design
	TCACCCGGTACAGTCTGACA					
RP60S	AGATCAGGCTGCCAATGTCT	93 bp	60°C	1.000	TC154862 [2]	New design
	CTCTCATCAAGGCTGCCAAT					
MYO	GGCCAAGAAGGCTATCACTG	248 bp	60°C	0.997	TC129320 [2]	New design
	CCGCGTCTACACCTCTTCTC					
PCNA	CGTGGAGAGTATGGATTC-GTCCCACG	80 bp	60°C	0.996	GO059618 [1]	[22]
	CGCAGCGGTAAGAGTCGAAC					
CYP1A	CCAACTTACCTCTGCTGGAAGC	68 bp	62°C	0.999	AF361643 [1]	[86]
	GGTGAACGGCAGGAAGGA					
ABCA1	GGACGAACCCTGTGTCTGTT	203 bp	60°C	0.993	EG836783 [1]	[10]
	ATTTGCATTGCGTTTCAGTG					
PSMB8	CGTTTGCGGGATATCAGATT	204 bp	60°C	0.999	TC138982 [2]	New design
	GTGCCATGCTCCAGGTTAAT					
TAGLN	AGGCTGGCATGACTGGATAC	192 bp	60°C	0.999	TC146580 [2]	New design
	TTCTCCTTGGTCCAGTTTGG					
COL1A2	CATTGCACCTTTGGACATTG	219 bp	60°C	0.975	TC148395 [2]	New design
	TCAGTCCATTGGTCCTCTCC					
Δ6fad_a	CCCCAGACGTTTGTGTCAG	181 bp	56°C	0.998	AY458652 [1]	[84]
	CCTGGATTGTTGCTTTGGAT					
elovl5a	ACAAGACAGGAATCTCTTTCA-GATTAA	137 bp	60°C	0.983	AY170327 [1]	[34]
	TCTGGGGTTACTGTGCTATAGT-GTAC					
elovl5b	ACAAAAAGCCATGTTTATCT-GAAAGA	141 bp	60°C	0.999	DW546112 [1]	[34]

	CACAGCCCCAGAGACCCACTT					
elovl2	CGGGTACAAAATGTGCTGGT	145 bp	60°C	0.997	TC91192 [2]	[34]
	TCTGTTTGCCGATAGCCATT					
FAS	GTGCCCACTGAAATACCATCC	212 bp	60°C	0.999	CK876943 [1]	[9]
	ATGAACCATTAGGCGGACAG					
ACO	AAAGCCTTCACCACATGGAC	230 bp	60°C	0.986	TC49531 [2]	[32]
	TAGGACACGATGCCACTCAG					
CPT1	CCTGTACCGTGGAGACCTGT	212 bp	60°C	0.998	AM230810 [1]	[32]
	CAGCACCTCTTTGAGGAAGG					
PPARα	TCCTGGTGGCCTACGGATC	111 bp	60°C	0.980	DQ294237 [1]	[87]
	CGTTGAATTTCATGGCGAACT					
PPARβ	GAGACGGTCAGGGAGCTCAC	151 bp	60°C	0.999	AJ416953 [1]	[87]
	CCAGCAACCCGTCCTTGTT					
PPARγ	CATTGTCAGCCTGTCCAGAC	144 bp	60°C	0.991	AJ416951 [1]	[87]
	TTGCAGCCCTCACAGACATG					
GST	ATTTTGGGACGGGCTGACA	81 bp	60°C	0.996	GE619558 [1]	[22]
	CCTGGTGCTCTGCTCCAGTT					
MT-A	CCTTGTGAATGCTCCAAAACTG	101 bp	60°C	0.993	GO049284 [1]	[22]
	CAGTCGCAGCAACTTGCTTTC					
CAT	GGGCAACTGGGACCTTACTG	59 bp	60°C	0.993	GE781438 [1]	[22]
	GCATGGCGTCCCTGATAAA					
SOD	CCACGTCCATGCCTTTGG	140 bp	60°C	0.922	BG936553 [1]	[88]
	TCAGCTGCTGCAGTCACGTT					
CASP3B	TGCGATCAAAGTGTTCTC-GAGTTT	85 bp	62°C	0.997	GE794765 [1]	[89]
	GGAAAGCAGCTGTTGTATCT-GTTG					
CASP6A/B	AGTGCAGGCCTTTGATAATCT-GAAG	215 bp	62°C	0.994	DQ008068 [1]	[89]
	ACACGAAGCAGTCTGCATCTG					
Reference genes						
Elf-1α	CTGCCCCTCCAGGACGTT-TACAA	175 bp	60°C	0.995	AF321836 [1]	[34]
	CACCGGGCATAGCCGATTCC					
Cofilin-2	AGCCTATGACCAACCCACTG	224 bp	60°C	1.000	TC63899 [2]	[34]
	TGTTCACAGCTCGTTTACCG					

Additional file 3: 2D-gel pick list for the factor diet. Table indicates protein spots identified, by two-way ANOVA ($P<0.05$), as being differentially expressed in salmon fed VO and FO diets, independently of the genotype.

Spot No.	Average ratio (VO/FO)		Two-Way ANOVA P-value		
	Lean	Fat	Diet factor	Family factor	Interaction
1151	1.38	1.54	0.0001	0.0120	0.5500
1148	1.32	1.56	0.0002	0.0270	0.3200
1135	1.24	1.27	0.0012	0.0200	0.8700
2114	-1.17	-1.58	0.0022	0.2900	0.1300
2275	1.58	1.11	0.0100	0.0990	0.0580
2687	-1.2	-1.64	0.0130	0.8000	0.1700
2195	1.58	1.12	0.0150	0.1900	0.0680
3014	-1.39	-1.24	0.0160	0.1700	0.5600
1119	-1.25	-1.17	0.0170	0.1100	0.6400
3168	-1.46	-1.18	0.0170	0.5500	0.2900
3445	1.17	1.15	0.0170	0.9100	0.8300
1136	1.23	1.4	0.0190	0.0660	0.4100
3108	-1.43	-1.19	0.0210	0.6800	0.3400
3334	-1.48	-1.18	0.0210	0.9500	0.3000
2266	1.09	1.11	0.0230	0.0800	0.8000
2766	-1.66	-1.03	0.0240	0.0620	0.0390
3331	-1.58	-1.18	0.0250	0.8600	0.2300
2932	-2.08	-1.22	0.0270	0.0024	0.0920
2683	1.14	1.09	0.0310	0.9500	0.7300
3253	-1.44	-1.22	0.0330	0.7000	0.5800
3109	-1.34	-1.25	0.0450	0.4100	0.8700
3243	-1.23	-1.11	0.0490	0.1600	0.5000

Additional file 4: 2D-gel pick list for the factor genotype. Table indicates protein spots identified, by two-way ANOVA ($P<0.05$), as being differentially expressed in Fat and Lean salmon family groups, independently of diet.

Spot No.	Average ratio (Fat/Lean)		Two-Way ANOVA P-value		
	FO	VO	Diet factor	Family factor	Interaction
2970	1.51	1.44	0.1400	0.0000	0.7000
2488	1.51	1.29	0.4500	0.0002	0.2300
2718	-1.24	-1.27	0.2600	0.0004	0.8600
2109	-1.55	-1.29	0.5400	0.0008	0.3700
2522	-1.26	-1.16	0.8900	0.0009	0.4200

2941	1.68	1.82	0.3200	0.0011	0.7200
2499	1.37	1.27	0.7700	0.0021	0.7100
2932	1.55	2.66	0.0270	0.0024	0.0920
1389	1.28	1.33	0.5600	0.0026	0.7800
2998	1.62	3.34	0.0640	0.0040	0.0850
2487	1.28	1.31	0.6000	0.0050	0.7600
3341	-1.03	-1.31	0.2900	0.0052	0.0220
1641	1.22	1.26	0.7200	0.0065	0.9400
1441	-1.28	-1.20	0.4700	0.0066	0.6600
2903	1.62	1.99	0.1800	0.0069	0.4100
2343	1.35	1.22	0.4100	0.0071	0.6200
2527	-1.12	-1.11	0.9500	0.0074	0.9000
1461	1.46	1.28	0.7200	0.0090	0.7100
2258	1.2	1.09	0.9500	0.0120	0.3500
2215	1.14	1.31	0.5900	0.0130	0.3300
2717	1.29	1.12	0.3100	0.0140	0.3300
2179	1.32	1.15	0.8500	0.0140	0.3500
2882	-1.07	-1.34	0.3100	0.0160	0.1600
2533	1.26	1.06	0.2500	0.0160	0.1300
2282	1.24	1.19	0.9400	0.0160	0.9300
311	-1.26	-1.15	0.4700	0.0170	0.4400
1135	1.14	1.17	0.0012	0.0200	0.8700
2746	1.37	1.23	0.4200	0.0230	0.6700
2157	1.34	1.47	0.3900	0.0240	0.7300
878	-1.59	-1.22	0.0740	0.0250	0.4500
2455	-1.11	-1.44	0.7700	0.0260	0.3100
1714	-1.16	-1.77	0.1700	0.0260	0.2900
1148	1.09	1.29	0.0002	0.0270	0.3200
2116	1.33	1.18	0.0870	0.0280	0.4200
2458	1.14	1.11	0.8000	0.0300	0.8400
2213	1.15	1.24	0.7400	0.0300	0.5300
3457	-1.03	-2.09	0.4100	0.0320	0.0560
3345	1.32	1.10	0.8900	0.0330	0.3300
2677	1.61	1.35	0.9100	0.0350	0.5900
2154	1.34	1.51	0.5500	0.0350	0.8300
2494	1.18	1.12	0.7200	0.0370	0.5800

2743	1.25	1.29	0.2600	0.0460	0.8300
2345	-1.29	-1.20	0.6700	0.0460	0.5800
2138	1.13	1.22	0.8700	0.0460	0.5800
2134	1.19	1.09	0.7100	0.0460	0.5100

Competing interests

The authors declare that they have no competing interests.

Authors' contributions

SM performed laboratory work and data analysis, with assistance from TS and OC in proteomics; DRG was responsible for family selection; JBT and JEB supported the microarray analysis; PR supported the proteomic analysis; SM wrote the first draft of the manuscript, followed by contributions from remaining authors; SM, JGB and DRT planned and coordinated the research; JGB and DRT were project leaders. All authors read and approved the final manuscript.

REFERENCES

1. Calder PC, Yaqoob P: Omega-3 polyunsaturated fatty acids and human health outcomes. Biofactors. 2009, 35: 266-272. 10.1002/biof.42.
2. FAO: The State of World Fisheries and Aquaculture (SOFIA) 2010. 2010, Food and Agriculture Organization of the United Nations, Rome
3. Tacon AGJ, Metian M: Global overview on the use of fish meal and fish oil in industrially compounded aquafeeds: Trends and future prospects. Aquaculture. 2008, 285: 146-158. 10.1016/j.aquaculture.2008.08.015.
4. Torstensen BE, Espe M, Sanden M, Stubhaug I, Waagbø R, Hemre G-I, Fontanillas R, Nordgarden U, Hevrøy EM, Olsvik P, Berntssen MHG: Novel production of Atlantic salmon (Salmo salar) protein based on combined replacement of fish meal and fish oil with plant meal and vegetable oil blends. Aquaculture. 2008, 285: 193-200. 10.1016/j.aquaculture.2008.08.025.
5. Bell JG, Pratoomyot J, Strachan F, Henderson RJ, Fontanillas R,

Hebard A, Guy DR, Hunter D, Tocher DR: Growth, flesh adiposity and fatty acid composition of Atlantic salmon (Salmo salar) families with contrasting flesh adiposity: Effects of replacement of dietary fish oil with vegetable oils. Aquaculture. 2010, 306: 225-232. 10.1016/j.aquaculture.2010.05.021.

6. Jordal AE, Torstensen BE, Tsoi S, Tocher DR, Lall SP, Douglas SE: Dietary rapeseed oil affects the expression of genes involved in hepatic lipid metabolism in Atlantic salmon (Salmo salar L.). J Nutr. 2005, 135: 2355-2361.
7. Panserat S, Kolditz C, Richard N, Plagnes-Juan E, Piumi F, Esquerré D, Médale F, Corraze G, Kaushik S: Hepatic gene expression profiles in juvenile rainbow trout (Oncorhynchus mykiss) fed fishmeal or fish oil-free diets. Br J Nutr. 2008, 100: 953-967. 10.1017/S0007114508981411.
8. Panserat S, Hortopan GA, Plagnes-Juan E, Kolditz C, Lansard M, Skiba-Cassy S, Esquerre D, Geurden I, Médale F, Kaushik S, Corraze G: Differential gene expression after total replacement of dietary fish meal and fish oil by plant products in rainbow trout (Oncorhynchus mykiss) liver. Aquaculture. 2009, 294: 123-131. 10.1016/j.aquaculture.2009.05.013.
9. Morais S, Pratoomyot J, Taggart JB, Bron JE, Guy DR, Bell JG, Tocher DR: Genotype-specific responses in Atlantic salmon (Salmo salar) subject to dietary fish oil replacement by vegetable oil: a liver transcriptomic analysis. BMC Genomics. 2011, 12: 255-10.1186/1471-2164-12-255.
10. Morais S, Pratoomyot J, Torstensen BE, Taggart JB, Guy DR, Bell JG, Tocher DR: Diet x genotype interactions in hepatic cholesterol and lipoprotein metabolism in Atlantic salmon (Salmo salar) in response to replacement of dietary fish oil with vegetable oil. Br J Nutr. 2011, 106: 1457-1469. 10.1017/S0007114511001954.
11. Geay F, Ferraresso S, Zambonino-Infante JL, Bargelloni L, Quentel C, Vandeputte M, Kaushik S, Cahu CL, Mazurais D: Effects of the total replacement of fish-based diet with plant-based diet on the hepatic transcriptome of two European sea bass (Dicentrarchus labrax) half-sibfamilies showing different growth rates with the plant-based diet. BMC Genomics. 2011, 12: 522-10.1186/1471-

2164-12-522.

12. Murray HM, Lall SP, Rajaselvam R, Boutilier LA, Blanchard B, Flight RM, Colombo S, Mohindra V, Douglas SE: A nutrigenomic analysis of intestinal response to partial soybean meal replacement in diets for juvenile Atlantic halibut, Hippoglossus hipploglossus, L. Aquaculture. 2010, 298: 282-293. 10.1016/j.aquaculture.2009.11.001.
13. Skugor S, Grisdale-Helland B, Refstie S, Afanasyev S, Vielma J, Krasnov A: Gene expression responses to restricted feeding and extracted soybean meal in Atlantic salmon (Salmo salar L.). Aquacult Nutr. 2011
14. Bell MV, Dick JR, Porter AEA: Pyloric ceca are a major site of 22:6n-3 synthesis in rainbow trout (Oncorhynchus mykiss). Lipids. 2003, 39: 39-44.
15. Fonseca-Madrigal J, Bell JG, Tocher DR: Nutritional and environmental regulation of the synthesis of highly unsaturated fatty acids and of fatty-acid oxidation in Atlantic salmon (Salmo salar L.) enterocytes and hepatocytes. Fish Physiol Biochem. 2006, 32: 317-328. 10.1007/s10695-006-9109-2.
16. Caballero MJ, Obach A, Rosenlund G, Montero D, Gisvold M, Izquierdo MS: Impact of different dietary lipid sources on growth, lipid digestibility, tissue fatty acid composition and histology of rainbow trout, Oncorhynchus mykiss. Aquaculture. 2002, 214: 253-271. 10.1016/S0044-8486(01)00852-3.
17. Caballero MJ, Izquierdo MS, Kjorsvik E, Montero D, Socorro J, Fernandez AJ, Rosenlund G: Morphological aspects of intestinal cells from gilthead seabream (Sparus aurata) fed diets containing different lipid sources. Aquaculture. 2003, 225: 325-340. 10.1016/S0044-8486(03)00299-0.
18. Olsen RE, Dragsnes BT, Myklebust R, Ringø R: Effect of soybean oil and soybean lecithin on intestinal lipid composition and lipid droplet accumulation of rainbow trout, Oncorhynchus mykiss Walbaum. Fish Physiol Biochem. 2003, 29: 181-192.
19. Olsen RE, Myklebust R, Kaino T, Ringø E: Lipid digestibility and ultrastructural changes in the enterocytes of Arctic char (Salvelinus alpinus L.) fed linseed oil and soybean lecithin. Fish Physiol

Biochem. 1999, 21: 35-44. 10.1023/A:1007726615889.

20. Olsen RE, Myklebust R, Ringø E, Mayhew TM: The influences of dietary linseed oil and saturated fatty acids on caecal enterocytes in Arctic char (Salvelinus alpinus L.): a quantitative ultrastructural study. Fish Physiol Biochem. 2000, 22: 207-216. 10.1023/A:1007879127182.
21. Morais S, Edvardsen RB, Tocher DR, Bell JG: Transcriptomic analyses of intestinal gene expression of juvenile Atlantic cod (Gadus morhua) fed diets with Camelina oil as replacement for fish oil. Comp Biochem Physiol B. -
22. Olsvik PA, Torstensen BE, Berntssen MHG: Effects of complete replacement of fish oil with plant oil on gastrointestinal cell death, proliferation and transcription of eight genes' encoding proteins responding to cellular stress in Atlantic salmon Salmo salar L. J Fish Biol. 2007, 71: 550-568. 10.1111/j.1095-8649.2007.01521.x.
23. Teitelbaum JE, Allan Walker W: Review: the role of omega 3 fatty acids in intestinal inflammation. J Nutr Biochem. 2001, 12: 21-32. 10.1016/S0955-2863(00)00141-8.
24. Leaver MJ, Taggart JB, Villeneuve L, Bron JE, Guy DR, Bishop SC, Houston RD, Matika O, Tocher DR: Heritability and mechanisms of n-3 long chain polyunsaturated fatty acid deposition in the flesh of Atlantic salmon. Comp Biochem Physiol Part D Genomics Proteomics. 2011, 6: 62-69. 10.1016/j.cbd.2010.04.002.
25. Dale OB, Tørud B, Kvellestad A, Koppang HS, Koppang EO: From chronic feed-induced intestinal inflammation to adenocarcinoma with metastases in salmonid fish. Cancer Res. 2009, 69: 4355-4362. 10.1158/0008-5472.CAN-08-4877.
26. Burrells C, Williams PD, Southgate PJ, Crampton VO: Immunological, physiological and pathological responses of rainbow trout (Oncorhynchus mykiss) to increasing dietary concentrations of soybean proteins. Vet Immunol Immunopathol. 1999, 72: 277-288. 10.1016/S0165-2427(99)00143-9.
27. Kolditz CI, Paboeuf G, Borthaire M, Esquerré D, SanCristobal M, Lefèvre F, Médale F: Changes induced by dietary energy intake and divergent selection for muscle fat content in rainbow trout (Oncorhynchus mykiss), assessed by transcriptome and proteome

analysis of the liver. BMC Genomics. 2008, 9: 506-10.1186/1471-2164-9-506.

28. Allendorf FW, Thorgaard GH: Tetraploidy and the evolution of salmonid fishes. Evolutionary Genetics of Fishes. Edited by: Turner BJ. 1984, Plenum Press, New York, 55-93.
29. Jump DB, Clarke SD, Thelen A, Liimatta M, Ren B, Badin M: Dietary polyunsaturated fatty acid regulation of gene transcription. Prog Lipid Res. 1996, 35: 227-241. 10.1016/S0163-7827(96)00007-0.
30. Nakamura MT, Nara TY: Gene regulation of mammalian desaturases. Biochem Soc Trans. 2002, 30: 1076-1079.
31. Wang Y, Botolin D, Christian B, Busik J, Xu J, Jump DB: Tissue-specific, nutritional, and developmental regulation of rat fatty acid elongases. J Lipid Res. 2005, 46: 706-715. 10.1194/jlr.M400335-JLR200.
32. Leaver MJ, Villeneuve LA, Obach A, Jensen L, Bron JE, Tocher DR, Taggart JB: Functional genomics reveals increases in cholesterol biosynthetic genes and highly unsaturated fatty acid biosynthesis after dietary substitution of fish oil with vegetable oils in Atlantic salmon (Salmo salar). BMC Genomics. 2008, 9: 299-10.1186/1471-2164-9-299.
33. Taggart JB, Bron JE, Martin SAM, Seear PJ, Høyheim B, Talbot R, Carmichael SN, Villeneuve LAN, Sweeney GE, Houlihan DF, Secombes CJ, Tocher DR, Teale AJ: A description of the origins, designs and performance of the TRAITS-SGP Atlantic salmon Salmo salar L. cDNA microarray. J Fish Biol. 2008, 72: 2071-2094. 10.1111/j.1095-8649.2008.01876.x.
34. Morais S, Monroig O, Zheng X, Leaver MJ, Tocher DR: Highly unsaturated fatty acid synthesis in Atlantic salmon: characterization of ELOVL5- and ELOVL2-like elongases. Mar Biotechnol. 2009, 1: 627-639.
35. van Schothorst EM, Flachs P, Franssen-van Hal NL, Kuda O, Bunschoten A, Molthoff J, Vink C, Hooiveld GJ, Kopecky J, Keijer J: Induction of lipid oxidation by polyunsaturated fatty acids of marine origin in small intestine of mice fed a high-fat diet. BMC Genomics. 2009, 10: 110-10.1186/1471-2164-10-110.

36. Leaver MJ, Boukouvala E, Antonopoulou E, Diez A, Favre-Krey L, Ezaz MT, Bautista JM, Tocher DR, Krey G: Three peroxisome proliferator-activated receptor isotypes from each of two species of marine fish. Endocrinology. 2005, 146: 3150-3162. 10.1210/en.2004-1638.

37. Rosen ED, Walkey CJ, Puigserver P, Spiegelman BM: Transcriptional regulation of adipogenesis. Genes Dev. 2000, 14: 1293-1307.

38. Schmid B, Rippmann JF, Tadayyon M, Hamilton BS: Inhibition of fatty acid synthase prevents preadipocyte differentiation. Biochem Biophys Res Commun. 2005, 328: 1073-1082. 10.1016/j.bbrc.2005.01.067.

39. Davidson MH: Mechanisms for the hypotriglyceridemic effect of marine omega-3 fatty acids. Am J Cardiol. 2006, 98: 27i-33i.

40. Mori T, Kondo H, Hase T, Tokimitsu I, Murase T: Dietary fish oil upregulates intestinal lipid metabolism and reduces body weight gain in C57BL/6J mice. J Nutr. 2007, 137: 2629-2634.

41. Madsen L, Frøyland L, Kryvi H, Auwerx J, Staels B, Berge RK: The mitochondrion is the principal target for nutritional and pharmacological control of plasma triglyceride. Lipids. 1999, 34 (Suppl): S167-

42. Kjær MA, Vegusdal A, Gjøen T, Rustan AC, Todorcević M, Ruyter B: Effect of rapeseed oil and dietary n-3 fatty acids on triacylglycerol synthesis and secretion in Atlantic salmon hepatocytes. Biochim Biophys Acta. 2008, 178: 112-122.

43. Jacobs MN, Covaci A, Schepens P: Investigation of selected persistent organic pollutants in farmed Atlantic salmon (Salmo salar), salmon aquaculture feed, and fish oil components of the feed. Environ Sci Technol. 2002, 36: 2797-2805. 10.1021/es011287i.

44. Bell JG, McGhee F, Dick JR, Tocher DR: Dioxin and dioxin-like polychlorinated biphenyls (PCBs) in Scottish farmed salmon (Salmo salar): effects of replacement of dietary marine fish oil with vegetable oils. Aquaculture. 2005, 243: 305-314. 10.1016/j.aquaculture.2004.10.016.

45. Takahashi M, Tsuboyama-Kasaoka N, Nakatani T, Ishii M, Tsutsumi S, Aburatani H, Ezaki O: Fish oil feeding alters liver gene expressions to defend against PPARα activation and ROS production. Am J

Physiol Gastrointest Liver Physiol. 2002, 282: G338-G348.

46. Saera-Vila A, Benedito-Palos L, Sitjà-Bobadilla A, Nácher-Mestre J, Serrano R, Kaushik S, Pérez-Sánchez J: Assessment of the health and antioxidant trade-off in gilthead sea bream (Sparus aurata L.) fed alternative diets with low levels of contaminants. Aquaculture. 2009, 296: 87-95. 10.1016/j.aquaculture.2009.07.028.
47. Moore MJ, Mitrofanov IV, Valentini SS, Volkov VV, Kurbskiy AV, Zhimbey EN, Eglinton LB, Stegeman JJ: Cytochrome P4501A expression, chemical contaminants and histopathology in roach, goby and sturgeon and chemical contaminants in sediments from the Caspian Sea, Lake Balkhash and the Ily River Delta, Kazakhstan. Mar Pollut Bull. 2003, 46: 107-119. 10.1016/S0025-326X(02)00325-9.
48. Lee RF, Anderson JW: Significance of cytochrome P450 system responses and levels of bile fluorescent aromatic compounds in marine wildlife following oil spills. Mar Pollut Bull. 2005, 50: 705-723. 10.1016/j.marpolbul.2005.04.036.
49. Bacanskas LR, Whitaker J, Di Giulio RT: Oxidative stress in two populations of killifish (Fundulus heteroclitus) with differing contaminant exposure histories. Mar Environ Res. 2004, 58: 597-601. 10.1016/j.marenvres.2004.03.048.
50. Galar-Martínez M, Gomez-Olivan LM, Amaya-Chavez A, Razo-Estrada C, Garcia-Medina S: Oxidative stress induced on Cyprinus carpio by contaminants present in the water and sediment of Madín Reservoir. J Environ Sci Health, Part A: Tox Hazard Subst Environ Eng. 2010, 45: 155-160. 10.1080/10934520903425780.
51. Wood ZA, Schröder E, Robin Harris J, Poole LB: Structure, mechanism and regulation of peroxiredoxins. Trends Biochem Sci. 2003, 28: 32-40. 10.1016/S0968-0004(02)00003-8.
52. Tolosano E, Altruda F: Hemopexin: structure, function, and regulation. DNA Cell Biol. 2002, 21: 297-306. 10.1089/104454902753759717.
53. Burk RF, Hill KE, Motley AK: Selenoprotein metabolism and function: evidence for more than one function for selenoprotein P. J Nutr. 2003, 133 (5 Suppl 1): 1517S-1520S.
54. Kang YJ: Metallothionein redox cycle and function. Exp Biol Med. 2006, 231: 1459-1467.

55. Berntssen MH, Julshamn K, Lundebye AK: Chemical contaminants in aquafeeds and Atlantic salmon (Salmo salar) following the use of traditional- versus alternative feed ingredients. Chemosphere. 2010, 78: 637-646. 10.1016/j.chemosphere.2009.12.021.

56. Moret S, Conte LS: Polycyclic aromatic hydrocarbons in edible fats and oils: occurrence and analytical methods. J Chromatogr A. 2000, 882: 245-253. 10.1016/S0021-9673(00)00079-0.

57. Shimada T: Xenobiotic-metabolizing enzymes involved in activation and detoxification of carcinogenic polycyclic aromatic hydrocarbons. Drug Metab Pharmacokinet. 2006, 21: 257-276. 10.2133/dmpk.21.257.

58. Hellou J, Leonard J, Anstey C: Dietary exposure of finfish to aromatic contaminants and tissue distribution. Arch Environ Contam Toxicol. 2002, 42: 470-476. 10.1007/s00244-001-0038-x.

59. Yuen BB, Wong CK, Woo NY, Au DW: Induction and recovery of morphofunctional changes in the intestine of juvenile carnivorous fish (Epinephelus coioides) upon exposure to foodborne benzo[a]pyrene. Aquat Toxicol. 2007, 82: 181-194. 10.1016/j.aquatox.2007.02.010.

60. Zhang Y, Lu Y, Zhou H, Lee M, Liu Z, Hassel BA, Hamburger AW: Alterations in cell growth and signaling in ErbB3 binding protein-1 (Ebp1) deficient mice. BMC Cell Biol. 2008, 9: 69-10.1186/1471-2121-9-69.

61. Celis JE, Madsen P, Celis A, Nielsen HV, Gesser B: Cyclin (PCNA, auxiliary protein of DNA polymerase delta) is a central component of the pathway(s) leading to DNA replication and cell division. FEBS Lett. 1987, 220: 1-7. 10.1016/0014-5793(87)80865-7.

62. Liu Z, Ahn JY, Liu X, Ye K: Ebp1 isoforms distinctively regulate cell survival and differentiation. Proc Natl Acad Sci USA. 2006, 103: 10917-10922. 10.1073/pnas.0602923103.

63. Sweet LI, Passino-Reader DR, Meier PG, Omann GM: Xenobiotic-induced apoptosis: significance and potential application as a general biomarker of response. Biomarkers. 1999, 4: 237-253. 10.1080/135475099230778.

64. Ramachandran A, Madesh M, Balasubramanian KA: Apoptosis in the intestinal epithelium: its relevance in normal and

pathophysiological conditions. J Gastroenterol Hepatol. 2000, 15: 109-120. 10.1046/j.1440-1746.2000.02059.x.

65. Paclik D, Lohse K, Wiedenmann B, Dignass AU, Sturm A: Galectin-2 and −4, but not galectin-1, promote intestinal epithelial wound healing in vitro through a TGF-beta-independent mechanism. Inflamm Bowel Dis. 2008, 14: 1366-1372. 10.1002/ibd.20499.

66. Loser K, Sturm A, Voskort M, Kupas V, Balkow S, Auriemma M, Sternemann C, Dignass AU, Luger TA, Beissert S: Galectin-2 suppresses contact allergy by inducing apoptosis in activated CD8+ T cells. J Immunol. 2009, 182: 5419-5429. 10.4049/jimmunol.0802308.

67. Tsukamoto K, Miura F, Fujito NT, Yoshizaki G, Nonaka M: Long-Lived Dichotomous Lineages of the Proteasome Subunit Beta Type 8 (PSMB8) Gene Surviving More than 500 Million Years as Alleles or Paralogs. Mol Biol Evol. 2012, May 2. [Epub ahead of print]

68. Nandi D, Tahiliani P, Kumar A, Chandu D: The ubiquitin-proteasome system. J Biosci. 2006, 31: 137-155. 10.1007/BF02705243.

69. Widmer CC, Pereira CP, Gehrig P, Vallelian F, Schoedon G, Buehler PW, Schaer DJ: Hemoglobin can attenuate hydrogen peroxide-induced oxidative stress by acting as an antioxidative peroxidase. Antioxid Redox Signal. 2010, 12: 185-198. 10.1089/ars.2009.2826.

70. Shimizu S, Matsuoka Y, Shinohara Y, Yoneda Y, Tsujimoto Y: Essential role of voltage-dependent anion channel in various forms of apoptosis in mammalian cells. J Cell Biol. 2001, 152: 237-250. 10.1083/jcb.152.2.237.

71. Jeon YJ, Kim DH, Jung H, Chung SJ, Chi SW, Cho S, Lee SC, Park BC, Park SG, Bae KH: Annexin A4 interacts with the NF-kappaB p50 subunit and modulates NF-kappaB transcriptional activity in a Ca2+−dependent manner. Cell Mol Life Sci. 2010, 67: 2271-2281. 10.1007/s00018-010-0331-9.

72. Parag HA, Raboy B, Kulka RG: Effect of heat shock on protein degradation in mammalian cells: involvement of the ubiquitin system. EMBO J. 1987, 6: 55-61.

73. Raboy B, Sharon G, Parag HA, Shochat Y, Kulka RG: Effect of stress on protein degradation: role of the ubiquitin system. Acta Biol Hung. 1991, 42: 3-20.

74. Pirkkala L, Alastalo TP, Zuo X, Benjamin IJ, Sistonen L: Disruption of heat shock factor 1 reveals an essential role in the ubiquitin proteolytic pathway. Mol Cell Biol. 2000, 20: 2670-2675. 10.1128/MCB.20.8.2670-2675.2000.

75. Di Lullo GA, Sweeney SM, Korkko J, Ala-Kokko L, San Antonio JD: Mapping the ligand-binding sites and disease-associated mutations on the most abundant protein in the human, type I collagen. J Biol Chem. 2002, 277: 4223-4231. 10.1074/jbc.M110709200.

76. Lussier C, Sodek J, Beaulieu JF: Expression of SPARC/osteonectin/BM4O in the human gut: predominance in the stroma of the remodeling distal intestine. J Cell Biochem. 2001, 81: 463-476. 10.1002/1097-4644(20010601)81:3<463::AID-JCB1060>3.0.CO;2-L.

77. Layland J, Solaro RJ, Shah AM: Regulation of cardiac contractile function by troponin I phosphorylation. Cardiovasc Res. 2005, 66: 12-21. 10.1016/j.cardiores.2004.12.022.

78. Olivares-Hernández R, Usaite R, Nielsen J: Integrative analysis using proteome and transcriptome data from yeast to unravel regulatory patterns at post-transcriptional level. Biotechnol Bioeng. 2010, 107: 865-875. 10.1002/bit.22868.

79. National Research Council (NRC): Nutrient requirements of fish and shrimp. 2011, National Academy Press, Washington

80. Folch J, Lees M, Sloane-Stanley GH: A simple method for the isolation and purification of total lipides from animal tissues. J Biol Chem. 1957, 226: 497-509.

81. Christie WW: Lipid analysis. 2003, The Oily Press, Bridgewater, UK, 3

82. Brazma A, Hingamp P, Quackenbush J, Sherlock G, Spellman P, Stoeckert C, Aach J, Ansorge W, Ball CA, Causton HC, Gaasterland T, Glenisson P, Holstege FC, Kim IF, Markowitz V, Matese JC, Parkinson H, Robinson A, Sarkans U, Schulze-Kremer S, Stewart J, Taylor R, Vilo J, Vingron M: Minimum information about a microarray experiment (MIAME)-toward standards for microarray data. Nat Genet. 2001, 29: 365-371. 10.1038/ng1201-365.

83. Pfaffl MW, Horgan GW, Dempfle L: Relative expression software tool (REST) for group-wise comparison and statistical analysis of

relative expression results in real-time PCR. Nucleic Acids Res. 2002, 30: e36-10.1093/nar/30.9.e36.

84. Zheng X, Tocher DR, Dickson CA, Bell JG, Teale AJ: Highly unsaturated fatty acid synthesis in vertebrates: new insights with the cloning and characterization of a Δ6 desaturase of Atlantic salmon. Lipids. 2005, 40: 13-24. 10.1007/s11745-005-1355-7.

85. Torstensen BE, Nanton DA, Olsvik PA, Sundvold H, Stubhaug I: Gene expression of fatty acid-binding proteins, fatty acid transport proteins (cd36 and FATP) and β-oxidation-related genes in Atlantic salmon (Salmo salar L.) fed fish oil or vegetable oil. Aquacult Nutr. 2009, 15: 440-451. 10.1111/j.1365-2095.2008.00609.x.

86. Rees CB, Li W: Development and application of a real-time quantitative PCR assay for determining CYP1A transcripts in three genera of salmonids. Aquat Toxicol. 2004, 66: 357-368. 10.1016/j.aquatox.2003.10.004.

87. Kleveland EJ, Ruyter B, Vegusdal A, Sundvold H, Berge RK, Gjøen T: Effects of 3-thia fatty acids on expression of some lipid related genes in Atlantic salmon (Salmo salar L.). Comp Biochem Physiol B Biochem Mol Biol. 2006, 145: 239-248. 10.1016/j.cbpb.2006.07.012.

88. Olsvik PA, Kristensen T, Waagbø R, Rosseland BO, Tollefsen KE, Baeverfjord G, Berntssen MHG: mRNA expression of antioxidant enzymes (SOD, CAT and GSH-Px) and lipid peroxidative stress in liver of Atlantic salmon (Salmo salar) exposed to hyperoxic water during smoltification. Comp Biochem Physiol C Toxicol Pharmacol. 2005, 141: 314-323. 10.1016/j.cbpc.2005.07.009.

89. Takle H, McLeod A, Andersen O: Cloning and characterization of the executioner caspases 3, 6, 7 and Hsp70 in hyperthermic Atlantic salmon (Salmo salar) embryos. Comp Biochem Physiol B Biochem Mol Biol. 2006, 144: 188-198. 10.1016/j.cbpb.2006.02.006.

CHAPTER

14

MICROARRAY GENE EXPRESSION PROFILES FROM MATURE GONAD TISSUES OF ATLANTIC BLUEFIN TUNA, THUNNUS THYNNUS IN THE GULF OF MEXICO

Luke D Gardner[1], Nishad Jayasundara[1], Pedro C Castilho[1,2] and Barbara Block[1]

[1]Biology Department, Hopkins Marine Station, Pacific Grove, Stanford University

[2]Universidade Federal Rural de Pernambuco (UFRPE/UAST)

ABSTRACT

Background

Bluefin tunas are highly prized pelagic fish species representing a significant economic resource to fisheries throughout the world. Atlantic bluefin tuna (*Thunnus thynnus*) populations have significantly declined due to overexploitation. As a consequence of their value and population decline, *T. thynnus* has been the focus of considerable research effort concerning many aspects of their life history. However, in-depth understanding of *T. thynnus* reproductive biology is still lacking. Knowledge of reproductive physiology is a very important tool for determining effective fisheries and aquaculture management. Transcriptome techniques are proving powerful and provide novel insights into physiological processes. Construction of a microarray from *T. thynnus* ESTs sourced from reproductive tissues has provided an ideal platform to study the reproductive physiology of bluefin tunas. The aim of this investigation was to compare transcription profiles from the ovaries and testes of mature *T. thynnus* to establish sex specific variations underlying their reproductive physiology.

Results

Male and females *T. thynnus* gonad tissues were collected from the wild and histologically staged. Sub-samples of sexually mature tissues were also measured for their mRNA differential expression among the sexes using the custom microarray design BFT 4X44K. A total of 7068 ESTs were assessed for differential expression of which 1273 ESTs were significantly different ($p<0.05$) with >2 fold change in expression according to sex. Differential expression for 13 of these ESTs was validated with quantitative PCR. These include genes involved in egg envelope formation, hydration, and lipid transport/accumulation more highly expressed in ovaries compared with testis, while genes involved in meiosis, sperm motility and lipid metabolism were more highly expressed in testis compared with ovaries.

Conclusions

This investigation has furthered our knowledge of bluefin tunas reproductive biology by using a contemporary transcriptome approach. Gene expression profiles in *T. thynnus* sexually mature testes and ovaries were characterized with reference to gametogenesis and potential alternative functions. This report is the first application of microarray technology for bluefin tunas and demonstrates the efficacy by which this technique may be used for further characterization of specific biological aspects for this valuable teleost fish.

BACKGROUND

Bluefin tunas are highly migratory species that represent a significant economic resource to fisheries globally[1]. Three species have been identified in the Atlantic Ocean (*Thunnus thynnus*), Pacific Ocean (*Thunnus orientalis*) and Southern Ocean (*Thunnus maccoyii*). The

Southern and Atlantic bluefin tuna populations have been in significant decline due to overexploitation[2]; the status of the Pacific bluefin is less well known. All three bluefin tuna species have been the subject of recent research efforts to better understand and manage their populations. In recent years, rapid advances in biological techniques for studying highly migratory species has enabled a better understanding of bluefin tuna population structure, oceanic migrations, habitat utilization and genetics[3, 4]. Despite these advances, significant questions remain regarding their reproductive dynamics. Aquacultural development of Pacific and Southern bluefin tuna has improved our knowledge of the reproductive biology for these species through the observation of these species in captivity[5].

Atlantic bluefin tuna has been the most challenging of the three bluefin species to understand, primarily due to their complex population structure, the late ages to reproduction in the western population, and logistical challenges of handling the largest of the three bluefin tuna species in captive environments. Currently, at least three populations of bluefin tuna are recognized in the Atlantic and Mediterranean basins[6, 7, 8, 9]. Knowledge of the reproductive biology of Atlantic bluefin tuna is still lacking[4] within these discrete populations and is complicated by the extensive mixing population evident from electronic tagging[10]. Evidence suggests that discrete ages to maturity exist among bluefin tuna populations in the Atlantic basin, with Gulf of Mexico spawning populations maturing later than eastern populations, with a year class range showing maturity as early as 8 years of age and a mean closer to 12 years of age for tuna that are spawning in the Gulf of Mexico[11]. In contrast, bluefin tuna in the eastern Mediterranean Sea have been observed to be mature as early as 4 years of age[12], however the mean age to maturity remains unclear due to the presence of western and eastern Mediterranean populations adding complexity to the population biology and fisheries assessments[13].

Knowledge of reproductive biology is a very important tool for effective fisheries and aquaculture management. To date such research has been largely limited to large-scale manifestations of field biology including, spawning season, sex ratio, batch fecundity, gonadosomatic index and gonad histological analysis[14]. Since knowledge of reproductive biology is a critical component of effective fisheries and aquaculture management, further research into the physiological mechanisms underlying Atlantic bluefin tuna reproductive biology is necessary.

Recently, researchers have begun to address the paucity of information in Atlantic bluefin tuna reproductive physiology. Endocrinology studies have measured circulating concentrations of hormones in reference to sex and maturity as well as the effects of hormonal administration on gonads[14, 15, 16, 17]. Similarly, Atlantic bluefin tuna gonad reproductive physiology has also been qualitatively investigated at the transcriptome level. Chini et al.[18] sequenced and partially annotated expressed sequence tag (EST) libraries containing 10163 EST from the testis, ovary and liver of mature *Thunnus thynnus*. Similar transcriptome based studies are becoming increasingly prevalent amongst researchers to investigate physiological processes in a number of fish, especially in model organisms or fish with commercial relevance[19, 20, 21]. These transcriptome techniques are becoming an established technology for providing novel insights into physiological processes. The availability of 10163 *T. thynnus* ESTs forms a foundation from which the reproductive physiology of bluefin tunas may be further elucidated.

The aim of this investigation was to identify and compare the transcriptomes of the ovaries and testes of mature Atlantic bluefin tuna to establish sex specific variations underlying their reproductive physiology. For this purpose we generated a novel microarray capable of measuring the differential transcriptional expression of 7068 ESTs from *T. thynnus*. The development of the microarray platform in conjunction with examination of gonadal histology and quantitative PCR (QPCR) identified a number of differentially expressed transcripts with particular relevance to the underlying processes of Atlantic bluefin tuna reproductive physiology.

RESULTS

Histology

Ovarian tissue from four female Atlantic bluefin tuna from the Gulf of Mexico were examined using histology. These fish ranged in curved fork length from 225 to 266 cm. Representative histological samples of the ovaries are shown in Figure1. Ovarian tissue from fish 1, 2 and 4 of the four Atlantic bluefin tuna were classed at stage 6 maturity[22] indicated by the presence of fully yolked oocytes, some of which were displaying coalescence of lipid droplets and early stage nucleus migration. Less than 50% of the fully yolked oocytes were undergoing atresia and post ovulatory follicles were also present. According to Itano et al.[22] this is typical of an actively reproductive and spawning fish. Histological observations of the ovarian tissue from fish 3 are more consistent with stage 4 of Itano's et al.[22] maturity index noting a high incidence of unyolked and partially yolked oocytes. Atresia of fully yolked oocytes was also noted. According to its designation, fish 3 is considered to have reached a fully yolked and potentially reproductive state but has regressed to a reproductively inactive state.

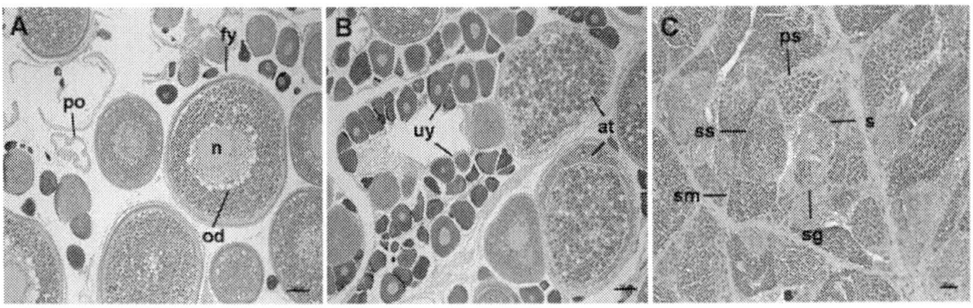

Figure 1: Histological analysis of *T. thynnus* gonad tissue. Transverse sections of mature *T. thynnus* gonads at different stages as per Itano et al.[22] and Schaefer[23] designations. (**A**) Stage 6 mature ovaries displaying fully yolked oocytes (fy), oil droplets (od), nucleus (n) and post ovulatory follicles (po), scale bar = 100 µm. (**B**) Stage 4 mature ovaries exhibiting a high proportion of unyolked oocytes (uy) and fully yolked oocytes undergoing atresia (at), scale bar = 100 µm. (**C**) Mature unrestricted spermatogonial testis displaying spermatogonia (sg), spermatocytes (ss), spermatids (sm) and spermatozoa (s), scale bar = 20 µm.

Testis tissue used in this investigation were obtained from four males ranging in curved fork length from 213 to 269 cm. Histological slide preparations of testis tissue taken from these four males all showed a mature unrestricted spermatogonial type, demonstrated by the presence of all stages of spermatogenesis occurring throughout the seminiferous tubules including spermatogonia, spermatocytes, spermatids and spermatozoa[23].

Microarray Findings

Gonad tissue from eight wild *T. thynnus* adults caught in the Gulf of Mexico were analysed for differential gene expression on the BFT 4X44K microarray in reference to gender (four males vs. four females). A total of 1820 microarray features were defined as significantly different ($p<0.05$) with a $\log_2$ transformed fold change greater than one between the conditions. A hierarchical clustered heat map of these 1820 features across all of the male and female gonads (4 per condition) shows their relative gene expression arranged according to similar feature expression profiles (Figure2). The upper limits of these features' $\log_2$ transformed fold change range from 4.5 in the female gonad specific features to 5.4 in the male gonad specific features. These $\log_2$ fold change values equate to an absolute fold change of 23 and 43 respectively. Within the 1820 microarray features, 1273 ESTs are represented, 737 of which are reported as having transcript abundances greater than 2-fold higher in the female gonad tissue in comparison to the male gonad tissue (Additional file1). Likewise, the remaining 536 ESTs were found to be expressed as greater than 2-fold higher in the male gonad tissue in reference to the female gonad tissue (Additional file2).

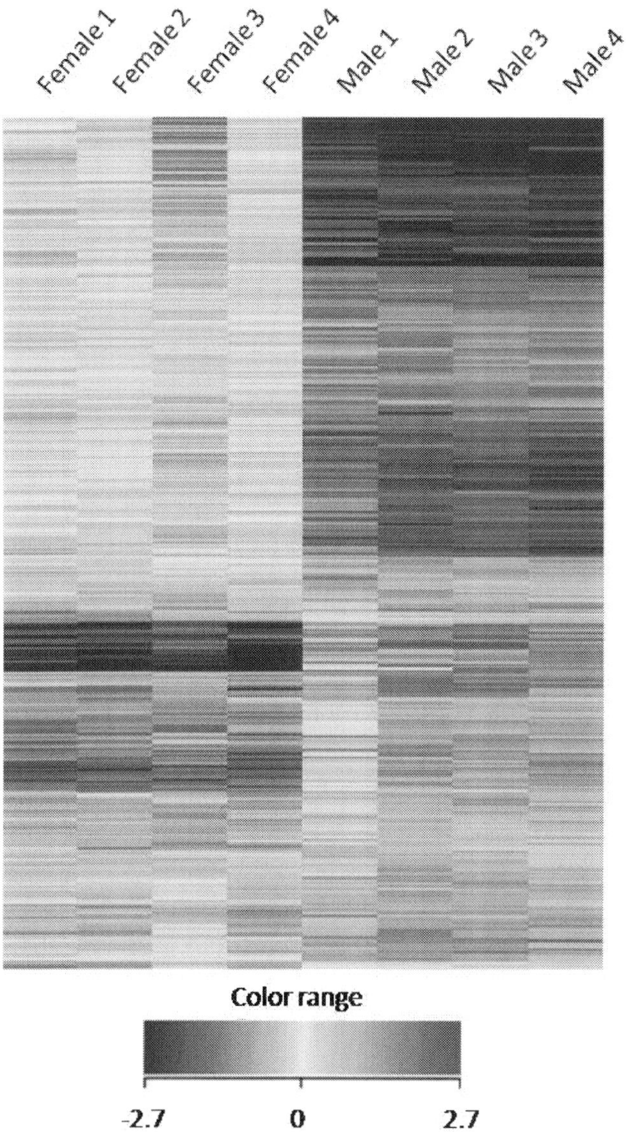

Figure 2:

Hierarchical heat map of *T. thynnus* ESTs differentially expressed among male and female gonad tissue. A hierarchical clustered heat map showing the $\log_2$ transformed expression values for microarray features on the BFT 4X44K Array following hybridization of mRNA preparations from *T. thynnus* male and female gonad tissue. The individual features are not labelled but are pictured horizontally showing their relative expression values across all replicates of the male and female gonad tissues. The intensity of the color scheme is calibrated to the log_2expression values such that red refers to higher transcript abundance and blue refers to lower transcript abundance. The features displayed are those which are significantly different ($p<0.05$) between the male and female conditions and have a log_2 fold change value greater than one.

Overall, gene expression in males shows a clear concordance among the individuals. The female gene expression profiles also show a high reproducibility among the individuals with the exception of Female 3 (Figure2). The expression profile of this female gonad sample appears to be partially similar in profile to both the male and female condition. Additional statistical analyses were performed excluding Female 3, however little variation in the list of microarray features deemed statistically significant was noted.

Annotation of the entire BFT 4X44K Array ESTs is relatively low compared to model organisms (zebrafish, killifish, mouse, rat), at 18% of the 7068 ESTs. A further comparison of the annotation percentage between the male and female specific EST lists showed a similarly low level of annotation at 16% for each condition. Considering this level of annotation, automated Gene Ontology annotation categorization techniques were not used to interpret biological meaning for the significant differentially expressed transcripts identified. Instead differentially expressed transcripts were considered individually for biological relevance and grouped accordingly. Gene categories identified from the microarray analysis as pertinent to this study include those concerned with egg envelope formation, yolk proteolysis, oocyte hydration, and lipid accumulation in the case of higher ovarian expression. Relevant categories identified as being expressed in higher concentrations in the testes include meiosis, sperm motility and lipid metabolism.

Quantitative PCR

T. thynnus transcripts shown by microarray analysis to be significantly differentially expressed between male and female gonad tissue were selected for QPCR relative abundance analyses, primarily to validate the microarray differential expression findings. The ESTs selected for QPCR analysis were chosen based on a process favouring ESTs with BLAST sequence similarities to genes that had established annotations and which also had a perceived relevance to reproduction in vertebrates. In the cases where several ESTs showed similar BLAST sequence similarities, usually only one EST was selected for QPCR analysis meant to serve as a proxy validation for the larger EST group (Table1). The relative differential expression between male and female *T. thynnus* gonads was assessed for a total of 13 transcripts. Nine transcripts were chosen to represent gene categories of interest with significant higher expression in the ovaries including: ZPC1-TTC00305, vitelline envelope protein gamma-TTC00056, alveolin-TTC00935, choriogenin L-TTC04136 (egg envelope formation); Cathepsin S-TTC00230 (yolk proteolysis); Aquaporin 1-TTC00180, Tmc6 related protein 1-TTC04625 (oocyte hydration); Fatty acid-binding protein, adipocyte-TTC00964, Epididymal secretory protein E1 precursor-TTC00209 (lipid accumulation) (Figure3). Four transcripts were chosen to represent the gene categories of interest with significantly higher expression in the testis including: Synaptonemal complex protein 3-TTC02745 (meiosis); T-complex-associated testis-expressed protein 1-TTC02749 (sperm motility); Brain-type fatty acid binding protein-TTC05153, Intestinal fatty acid-binding protein-TTC05128 (lipid metabolism) (Figure3).

Table 1:A subset of significant differentially expressed ESTs from *T. thynnus* female and male gonad tissue

T. thynnus-EST identifier	T. thynnus EST accession number	Relative fold change	Blast2GO sequence similarity description	Reference sequence	E-value*
Significantly differentially expressed at greater abundances in T. thynnus ovarian tissue relative to testes tissue					
TTC00054	EC091690	6.1	ZPC1 [Oryzias latipes]	AAN31188.1	1e-26
TTC00677	EC092378	5.2	ZPC1 [Cynoglossus semilaevis]	ABY81291.1	7e-21
TTC00305	EC091965	4.5	ZPC1 [Cynoglossus semilaevis]	ABY81291.1	1e-22
TTC00023	EC091658	4.2	ZPC1 [Cynoglossus semilaevis]	ABY81291.1	2e-73
TTC05527	EH000098	2.6	ZPC1 [Oryzias latipes]	AAN31188.1	2e-24
TTC00911	EC092645	4.1	zona pellucida sperm-binding protein 4 precursor [Felis catus]	NP_001009260.1	8e-12
TTC00906	EC092640	4.7	ZPB domain containing protein [Oryzias latipes]	NP_001098217.1	8e-51
TTC00944	EC092682	4.2	ZP1 precursor [Mus musculus]	AAC48480.1	1e-12
TTC03982	EG630106	5.8	ZPB [Oryzias latipes]	AAN31187.1	8e-61
TTC00652	EC092350	3.2	ZP2 [Cyprinus carpio]	CAA96572.1	2e-08
TTC04154	EG630316	8.6	ZPC1 [Cynoglossus semilaevis]	ABY81291.1	7e-52
TTC04773	EG631058	3.3	ZPC2 [Oryzias latipes]	AAN31189.1	1e-28
TTC00418	EC092092	3.1	ZPC1 [Cynoglossus semilaevis]	ABY81291.1	9e-15
TTC04501	EG630741	2.8	ZPC1 [Cynoglossus semilaevis]	ABY81291.1	2e-17
TTC00056	EC091692	6.4	vitelline envelope protein gamma [Oncorhynchus mykiss]	NP_001117746.1	3e-09
TTC00654	EC092352	4.2	ZP3 [Carassius auratus]	CAA88838.1	1e-14
TTC00493	EC092174	7.6	egg envelope glycoprotein [Xenopus laevis]	AAY22122.1	4e-05

TTC01165	EC092929	3.5	egg envelope glycoprotein [Xenopus laevis]	AAY22123.1	1e-07
TTC04306	EG630505	3.0	egg envelope glycoprotein [Xenopus laevis]	AAY22122.1	3e-10
TTC00935	EC092671	4.2	alveolin [Oryzias latipes]	NP_001098139.1	2e-12
TTC04136	EG630294	2.9	choriogenin L [Fundulus heteroclitus]	BAJ07538.1	1e-04
TTC00085	EC091723	10.3	Cathepsin Z precursor [Osmerus mordax]	ACO09238.1	8e-51
TTC01243	EC093018	3.9	cathepsin Z-like protein [Lutjanus argentimaculatus]	ACO82387.1	4e-24
TTC00340	EC092003	2.4	Cathepsin Z precursor [Osmerus mordax]	ACO09238.1	6e-85
TTC00230	EC091881	5.5	cathepsin S [Oryzias latipes]	NP_001098157.1	2e-97
TTC04658	EG630923	5.2	cathepsin S [Oryzias latipes]	NP_001098157.1	5e-42
TTC00689	EC092390	4.9	cathepsin S [Ictalurus punctatus]	ABD65539.1	3e-18
TTC00766	EC092478	4.8	cathepsin S [Oryzias latipes]	NP_001098157.1	7e-32
TTC04399	EG630616	4.7	cathepsin S [Ictalurus furcatus]	ADO27765.1	9e-13
TTC00605	EC092298	4.5	cathepsin K [Fundulus heteroclitus]	AAO64475.1	2e-13
TTC04464	EG630693	4.2	cathepsin S [Oryzias latipes]	NP_001098157.1	5e-43
TTC03943	EG630062	2.0	cathepsin S precursor [Fundulus heteroclitus]	AAO64477.1	1e-06
TTC00346	EC092009	4.8	cathepsin S [Oryzias latipes]	NP_001098157.1	3e-28
TTC00235	EC091886	3.8	cathepsin [Paralabidochromis chilotes]	AAQ01147.1	6e-15
TTC06758	EH667736	3.0	cathepsin L-like protein [Lutjanus argentimaculatus]	ACO82386.1	4e-50
TTC04596	EG630853	3.8	cathepsin K precursor	NP_031828.2	3e-28

TTC04425	EG630647	5.7	aquaporin [Solea senegalensis]	AAV34612.1	1e-21
TTC04368	EG630581	5.2	aquaporin 1o [Cynoglossus semilaevis]	ADG21867.1	2e-40
TTC00180	EC091829	4.5	aquaporin 1o [Cynoglossus semilaevis]	ADG21867.1	2e-34
TTC00584	EC092275	5.0	aquaporin [Solea senegalensis]	AAV34612.1	3e-11
TTC02704	EC918066	4.5	aquaporin 1o [Cynoglossus semilaevis]	ADG21867.1	2e-19
TTC04625	EG630885	23.1	Tmc6-related protein 1 [Takifugu rubripes]	AAP78785.1	4e-07
TTC01498	EC421545	9.6	fatty acid binding protein 11b [Danio rerio]	NP_001018394.1	3e-20
TTC00750	EC092461	4.7	fatty acid-binding protein liver-type [Ictalurus punctatus]	ADO29352.1	6e-34
TTC04564	EG630815	4.7	fatty acid-binding protein liver-type [Ictalurus punctatus]	ADO29352.1	2e-27
TTC07800	EL610584	4.4	Fatty acid-binding protein, heart [Anoplopoma fimbria]	ACQ57957.1	5e-43
TTC04356	EG630567	4.0	fatty acid-binding protein liver-type [Ictalurus punctatus]	ADO29352.1	1e-26
TTC03876	EG629978	3.5	fatty acid binding protein 1 [Mesocricetus auratus]	AAV33399.1	4e-07
TTC00964	EC092703	2.8	Fatty acid-binding protein, adipocyte [Salmo salar]	NP_001134675.1	5e-54
TTC00209	EC091860	7.6	Epididymal secretory protein E1 precursor [A. fimbria]	ACQ58497.1	2e-32
TTC04551	EG630798	7.2	Epididymal secretory protein E1 precursor [Osmerus mordax]	ACO09051.1	2e-38

TTC00127	EC091768	6.2	Epididymal secretory protein E1 precursor [A. fimbria]	ACQ58497.1	5e-38
TTC01068	EC092821	3.7	Epididymal secretory protein E1 precursor [A. fimbria]	ACQ58497.1	2e-13
TTC00421	EC092095	2.8	Epididymal secretory protein E1 precursor [A. fimbria]	ACQ58497.1	2e-30
Significantly differentially expressed at greater abundances in T. thynnus testes tissue relative to ovarian tissue					
TTC05519	EH000090	18.2	tctex1 domain-containing protein 1-A [Xenopus laevis]	NP_001090117.1	1e-28
TTC05755	EH000358	11.6	tctex1 domain-containing protein 1-B [Xenopus laevis]	NP_001106901.1	2e-25
TTC02749	EC918118	12.6	T-complex-associated testis-expressed protein 1 [R.norvegicus]	NP_001101676.1	2e-42
TTC02745	EC918114	12.4	synaptonemal complex protein 3 [Oncorhynchus mykiss]	NP_001117979.1	4e-39
TTC05128	EG999641	6.6	intestinal fatty acid-binding protein [Paralichthys olivaceus]	ABV91589.1	5e-12
TTC05153	EG999669	33.9	brain-type fatty acid binding protein [Oryzias latipes]	NP_001116389.1	8e-11

* Expect value is the likelihood that the sequence similarity match is a random occurrence on the given database.

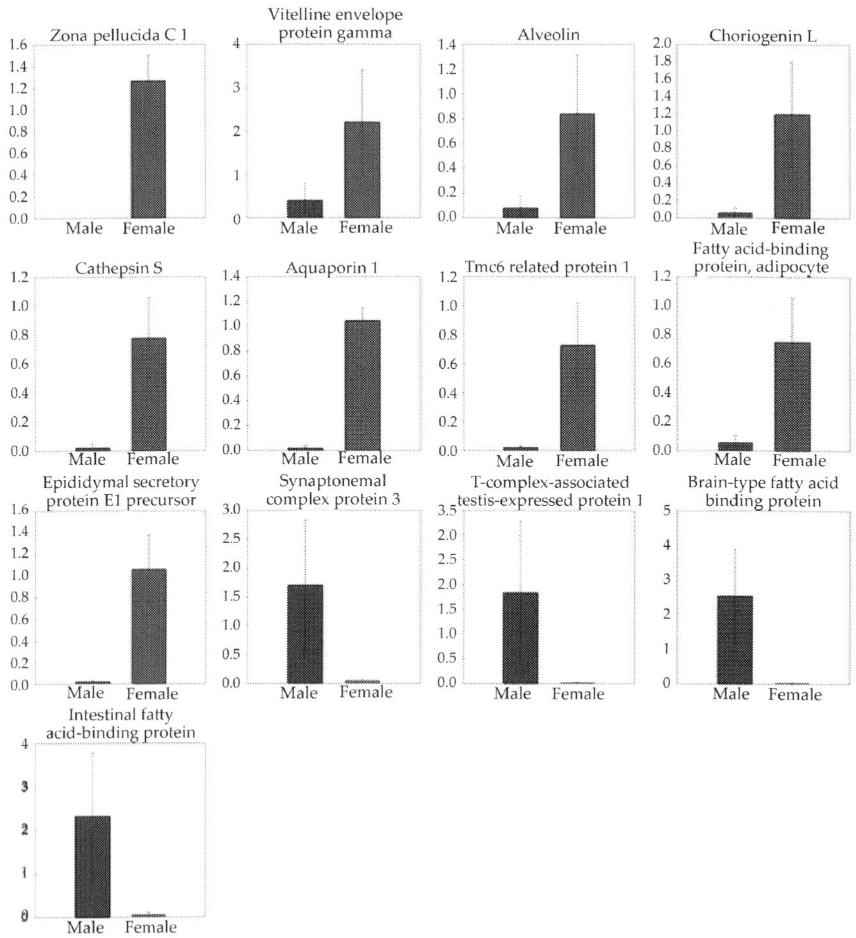

Figure 3: Mean relative EST abundance between *T. thynnus* male and female gonad tissues as determined by QPCR analysis. All ESTs graphed are significantly different ($p<0.05$) with error bars representing standard deviation. Putative EST sequence annotation identities are as follows: Zona pellucida C 1 – TTC00305; Vitelline envelope protein gamma –TTC00056; Alveolin – TTC00935; Choriogenin L – TTC04136; Cathepsin S – TTC00230; Aquaporin 1 – TTC00180; Tmc6 related protein 1 – TTC04625; Fatty acid-binding protein, adipocyte – TTC00964; Epididymal secretory protein E1 precursor – TTC00209;

Synaptonemal complex protein 3 – TTC02745; T-complex-associated testis-expressed protein 1 – TTC02749; Brain-type fatty acid binding protein – TTC05153; Intestinal fatty acid-binding protein – TTC05128. Refer to table1 for accession numbers associated with these ESTs and putative annotations. The colors blue and red refer to male and female gonad tissues respectively. The Y axis represents relative abundance of transcripts between male and female gonads.

DISCUSSION

Bluefin tunas are apex oceanic predators found in the Atlantic, Southern and Pacific oceans. They are an important commercial resource that are in decline primarily due to over-fishing. Effective management of the wild resources as well as the success of bluefin tuna aquaculture will rely heavily on a thorough understanding of the reproductive biology of these fishes. This investigation was initiated to further our knowledge of bluefin tuna reproductive biology by using a contemporary transcriptome approach novel to bluefin tuna research. The transcriptional expression profiles from *T. Thynnus* male and female mature gonad tissues and their transcript constituents are discussed here with relation to their importance to the reproductive processes of bluefin tuna. ESTs of interest that were over expressed in the *T. thynnus* ovarian tissue are discussed first followed by those that were significantly more highly expressed in the testis tissue. The results from this investigation are likely to be applicable for the *Thunnus* genus as a whole due to the relatively low genomic variation among its members. This is because the Genus is of relatively recent origin as evidenced by comparatively low levels of inter-specific nucleotide variation among member species, ranging from 0.01 for the mitochondrial CO1 gene to 0.03 for nuclear non-coding ITS-1 sequences[24]. Furthermore, because nuclear genomes mutate more slowly than mitochondrial genomes, and because nonsynonymous sites are much less free to vary than non-coding regions[25], we believe our transcriptional methods for Atlantic bluefin will be applicable to other Thunnus species.

EST Differential Expression in Ovarian Tissue

Reproductive strategies in fish vary greatly including aspects such as attraction, gonochorism and sex change, synchronous and asynchronous ovarian development, spawning temporal and spatial patterns and parental care[26, 27]. However among teleosts the fertilization strategy is dominated by an oviparous regime in which oocyte development leading up to external fertilization in teleosts appears to be a uniform process. This coordinated assembly of the fish egg is classified into six main sequential phases: oogenesis, primary oocyte growth, cortical alveolus stage, vitellogenesis, maturation and ovulation[28, 29].

The histological examination of the *T. thynnus* ovarian tissues used for this investigation showed that oocyte development had reached a mature stage when the fish were captured and sampled. This observation is in concordance with previous findings whereby large bluefin tuna are considered to be present in the Gulf of Mexico predominately from March to June for spawning[30]. The microarray and QPCR results generated from this study using the same *T. thynnus* ovarian samples provides a transcriptome profile of these mature individuals. This permits examining the portion of the transcriptome that may be utilized to generate a gender specific maturation profile. In this regard we detected differential expression of a number *T. thynnus* ESTs present on the BFT 4X44K array that were homologous with annotated genes consistent with mature oocyte presence. Specifically, *T. thynnus* ovarian differentially expressed transcripts pertaining to oogenesis related gene categories to be discussed herein will include those concerned with egg envelope formation, yolk proteolysis, oocyte hydration, and lipid accumulation.

The egg envelope often termed the vitelline envelope (VE) in teleosts is involved in processes including fertilization and protection of the egg and embryo[31]. The VE is formed during oocyte development between the follicle cells and plasma membrane of the oocyte[32] and is composed of a relatively thick proteinaceous extracellular matrix usually of a few major glycoproteins[33]. The terminology for VE proteins and genes is broad due in part to different names being ascribed to various vertebrate groups, including zona pellucida, vitelline membrane, chorion, egg shell protein, zona radiata and vitelline envelope. However, amino acid sequence homologies among these proteins aid in characterizing them via the typical presence of a specific zona pellucida like domain (ZP) [32].

Microarray analysis revealed a total of 21 *T. thynnus* ESTs with significant sequence similarities to genes encoding VE proteins (Table1). These transcripts were detected at greater abundance in the female gonad of *T. thynnus*. QPCR analyses of a subset of these VE gene homologs (ZPC1-TTC00305; vitelline envelope protein gamma-TTC00056; alveolin-TTC00935; choriogenin L-TTC04136) validated the differential expression shown by microarray analysis (Figure3). While these ESTs' identity and function as VE protein genes is inferred from their sequence similarities it should be noted that previous studies identifying the VE proteins of teleosts have routinely identified only a few major proteins and genes associated with the VE[33, 34, 35, 36, 37, 38]. This trend is perhaps an outcome from the 'protein-down' approach often employed in these studies including techniques whereby VE gene sequences were searched for using degenerative PCR primers designed from the amino acid sequences of the purified VE proteins. More recently, with the proliferation of genome sequencing among organisms, specifically teleost species zebrafish and medaka, has been established that both these fish have multiple gene isoforms containing ZP domains[37, 39]. These findings help to explain the similar detection of multiple ESTs containing ZP domains preferentially expressed in the mature ovaries of *T. thynnus*(Table1).

Hydration of the oocyte is an integrated process, essential to the maturation of pelagophil eggs[40]. During the final stages of oocyte maturation, an osmotic gradient is created between the oocyte and the interstitial fluid or oviduct leading to a rapid swelling of the oocyte. Hyperosmolality of the yolk drives the oocyte hydration due in part to the proteolytic cleavage of oocyte yolk proteins sequestered during vitellogenesis. This protein degradation results in a rapid increase in concentration of free amino acids (FAA) and small peptides. However, yolk hydrolysis is not the only mechanism driving oocyte hydration. The accumulation of inorganic ions, particularly Cl^- in the oocyte is also contributing to the hydration[41]. The rise in these molecules (FAA and inorganic ions) helps to build an osmotic potential between the oocyte and the surrounding environment. This gradient, in conjunction with the water channel protein, aquaporin, drive the hydration of the oocyte which is essential for the production of pelagophil buoyant eggs. Differential expression of transcripts in maturing ovarian tissue of *T. thynnus* with sequence similarities to genes putatively involved in these processes are discussed here together. Namely, protease-encoding genes from the cathepsin family as well as the transmembrane channel-like protein 6 (TMC6) and aquaporin genes are considered.

Cathepsin proteases have been reported to be responsible for yolk protein hydrolysis in teleosts[42, 43, 44, 45]. Significant sequence similarities with cathepsin coding genes of 15 *T. thynnus* transcripts that were differentially expressed in the maturing ovarian tissue of *T. thynnus* in this investigation indicate a similar putative function for these ESTs (Table1). However which members of the cathepsin family are responsible for the final stages of yolk protein hydrolysis is unclear. Cathepsin L[43] and Cathepsin B[44, 46], have both been purported separately as the main protease responsible for the final hydrolysis of the yolk proteins in different teleost species. Furthermore, cathepsin D has been reported as the initial protease responsible for cleaving the yolk precursor protein, vitellogenin, into the yolk proteins[43, 47]. Interestingly, none of these cathepsins seem likely to be involved in yolk proteolysis in *T. thynnus* maturing oocytes. In this investigation only one of the 15 *T. thynnus* transcripts preferentially expressed in maturing ovarian tissue showed a sequence similarity to cathepsin L while the remaining EST sequences aligned with cathepsin Z precursor (3) and cathepsin S (14) (Table1). The potential that the cathepsins putatively involved in yolk proteolysis were not detected in this investigation due to the *T. thynnus* ovarian tissues used being at a different oocyte developmental stage is unlikely. Raldua et al.[44] showed that temporal expression of these yolk proteolysis enzymes is detectable from early vitellogenesis and are stored in the egg as latent acid-activatable proenzymes. Histological examination of the ovarian tissues used in this study indicates female specimens (3 of 4) are in late vitellogenesis during which yolk proteolysis enzymes should be expressed. Therefore it seems likely that *T. thynnus* employs different cathepsin proteases for the initial cleavage of the vitellogenin as well as the final yolk proteolysis in comparison to the current reported enzymes for teleost species.

Although yolk proteolysis has been accounted for an increase in oocyte osmolality, the accumulation of inorganic ions is thought to provide approximately half of the osmolytes[41]. The mechanism by which these inorganic ions accumulate in pelagophil oocytes is largely unknown. *T. thynnus* – trans-membrane channel protein 6 (TMC6) may function to transiently accumulate inorganic ions, specifically Cl^- ions into the oocyte thus raising the osmolality required for full oocyte hydration. This transcript was expressed 23 times higher in the ovaries compared to the testes of *T. thynnus* and had the greatest differential expression profile of all 1273 ESTs highlighted in the microarray analysis (Table1). Although currently no annotation references for TMC6 to egg hydration are available, we note that TMC6 is a part of a larger transmembrane channel family. While the discovery of this gene family is relatively novel, a consensus is building in the literature that these proteins may function as ion channels, pumps or transporters[48, 49, 50]. Specifically, Hahn et al.[48] concluded that TMC6 is likely to have Cl^- channel activity based on sequence homology. Considering the ovarian specific expression of *T. thynnus Tmc6* and its homology-based annotation linking it to Cl^- channelling, we propose that TMC6 in the *T. thynnus* oocyte is involved in the oocyte hydration via the accumulation of Cl^-in the oocyte thus raising the internal osmolality and driving the influx of water through aquaporin channels.

Following yolk proteolysis and accumulation of inorganic ions, teleost ooycte hydration is typically achieved by an influx of water across an osmotic gradient mediated by the water channel protein aquaporin[41]. Five transcripts differentially expressed in *T. thynnus* ovarian tissue showed sequence similarities to aquaporin suggesting a similar mechanism may be employed in *T. thynnus* (Table1). Although oocyte hydration occurs just prior to spawning, the expression of the aquaporin protein begins during early vitellogenesis in a similar manner to yolk proteases and is stored within the egg until they are activated by an unknown mechanism[40].

Teleost ovarian development is a nutrient demanding process of which lipids are a substantial requirement for oocyte development. The lipids necessary for developing oocytes are mobilized from reserves within the animals including the liver, muscle and other tissues. Considering the importance of lipid transport and accumulation, differential expression of transcripts with sequence similarities to genes encoding lipid transport proteins in *T. thynnus* ovarian tissue reported in this study are discussed.

A vital component facilitating cellular lipid transport among others is the family of fatty acid binding proteins (FABP). FABPs are cytoplasmic proteins whose primary role is to regulate fatty acid uptake and intracellular transport[51]. Seven differentially expressed transcripts (TTC01498; TTC00750; TTC04564; TTC07800; TTC04356; TTC03876; TTC00964) in ovarian tissue of *T. thynnus* show sequence similarities with the *Fabp* gene family, particularly *Fabp1* and *Fabp4* (Table1). Differential expression for TTC00964 was confirmed with QPCR analyses (Figure3). Considering the intracellular nature of FABP we propose that the ovarian specific expression of these seven transcripts in *T. thynnus* maturing ovaries is likely involved in the membrane trafficking and sequestration of lipids in the oocytes which is a requirement for normal embryo development. This assertion is supported by observations that expression of both *Fabp1* and *Fabp4* are well documented in adipose tissue and liver, both of which are heavily involved in lipid sequestration[52, 53]. An additional function for *Fabp* homologues expressed in *T. thynnus* ovarian tissue beyond oocyte lipid accumulation is indicated by a significant sequence similarity between transcripts TTC01498 and *FABP11*. Agulleiro et al.[54] observed in a teleost fish (*Solea senegalensis*) that *Fabp11* (restricted to fishes) was expressed in ovarian follicle cells positively correlated with ovarian atresia (reabsorption of the oocyte), particularly postovulatory regression. While our histological examination of subsamples from the *T. thynnus* ovarian tissue used for microarray analysis was not consistent with postovulatory regression some minor oocyte atresia was observed. Minor oocyte atresia is known to occur normally during ovarian development. The differential expression of TTC01498 in ovarian tissue undergoing minor atresia used for this investigation and the observations of Agulleiro et al.[54] support a putative role for FABP11 in fatty acid trafficking specifically related to oocyte atresia.

Another lipid transport gene of interest to this study is Epididymal secretory protein E1 gene. Despite its title being suggestive of a testicular function, five *T. thynnus* ESTs bearing a significant sequence resemblance to this gene were observed to be preferentially expressed in the ovary in comparison to the testes of *T. thynnus*(Table1). The relative expression of one of the transcripts (TTC00209) was further examined with QPCR confirming the microarray analysis for this EST and serves as a proxy for the remaining four ESTs (Figure3). The alternative name for Epididymal secretory protein E1 is and Niemann-Pick disease type C2 (*Npc2*). When considering this alternative gene name and protein functions, its expression in *T. thynnus* ovaries seems more plausible. This protein is known to be involved in cholesterol homeostasis, specifically intracellular cholesterol trafficking[55]. It functions such that after lipoproteins carrying cholesterol are endocytosed and hydrolyzed in lysosomes, the NPC2 is responsible for their exit from the lysosome to where required[56]. This cholesterol homeostatic function likely explains the higher ovarian transcript abundance for the seven *Npc2*-like *T. thynnus* ESTs, given the high lipid requirement for fish oocyte maturation, of which cholesterol is a significant component. Greater concentration of these *Npc2*-like transcripts expressed in the maturing ovary in comparison to the testes of *T. thynnus* would be necessary for the NPC2 to process the influx of cholesterol via intercellular lipid transport proteins like low density lipoprotein (LDL)[57].

EST Differential Expression in the Testis Tissue

Although the development of eggs and sperm share common principles, many aspects of gametogenesis differ between the sexes. Knowledge on spermatogenesis in fish is limited to a few species used in basic research and/or aquaculture biotechnology[58]. The microarray and QPCR results generated from this study using mature *T. thynnus* testes tissue has helped to identify a number of these differences on a molecular level. Specifically, *T. thynnus* testes differentially expressed transcripts related to spermatogenesis is discussed including transcripts potentially involved in meiosis, sperm motility and lipid metabolism.

T. thynnus testes are described as having unrestricted spermatogonial distribution whereby gametes are synchronously produced in cysts spread throughout the germinal compartment[59]. Histological classification of the *T. thynnus* testes tissues used in this study (Figure1) was adapted from[23] classification system for yellowfin tuna. These cysts are present at all spermatogenesis stages during testicular maturation. Final sexual maturation in *T. thynnus* involves a significant enlargement of the testes, but unlike females no apparent noteworthy histological changes are present, with the exception of a marginally higher frequency of the most advanced stages of spermatogenesis in fully mature bluefin tuna[59].

Histological analysis of subsamples from the testes tissue used in this investigation shows spermatozoa are present which is the final product in mature tuna.

The synaptonemal complex (SC) is a meiosis specific structure formed during the first meiotic prophase of germ cells within sexually reproducing organisms. The SC is made up of three proteins - synaptonemal complex protein 1, 2 and 3 (SYCP), involved in chromosome pairing and recombination[60]. *T. thynnus* EST TTC02745 exhibits a significant sequence similarity to *Sycp3* and is shown to be highly differentially expressed in the mature testis of the *T. thynnus* in comparison to ovaries (Table1, Figure3). SYCP3 is the best characterized of all the SC proteins albeit studied predominately in mammals. However, recent investigations characterizing this protein in fish are highlighting some divergences with the mammalian SC protein[61, 62]. Based on previous SCYP3 annotations this protein is not considered to be sexually dimorphic, yet our expression analysis indicated the contrary for *Sycp3*-like EST TTC02745. This is an interesting observation when considering the spawning modes of the species. The testis development of *T. thynnus* is reported as that of the unrestricted spermatogonial testicular type whereby continuous spermatogenesis occurs in the testes tubules[59]. Thus the testes of *T. thynnus* contain germ cells at all stages of development including the first stages of meiosis during which SYCP3 is known to be expressed.

T. thynnus females have been described as serial spawners characterized by asynchronous ovarian development in which all stages of oogenesis are continuously present during the spawning season[63]. Despite this apparent continuous gametogenesis similarity, *Sycp3* in ovarian tissue is expressed at a reduced level in comparison to testes tissue. Mammalian studies explain this disparity in that the first stages of meiosis in female germ cells occur during embryonic development after which they go into meiotic arrest until meiosis resumption at puberty[64]. Males in contrast do not begin meiosis until puberty. This meiotic arrest explanation relies on an assumption that female fish like mammals have a finite number of germ cells that cannot be replenished or regenerated. However, this theory has not been established outside mammals and conversely there is some evidence that highly fecund lower vertebrates may produce new oocytes from mitotic oogonia[65]. Therefore considering that *T. thynnus* is a highly fecund species in conjunction with potentially continuously replenishing oocytes, the relatively low meiosis specific *Sycp3*-like EST ovarian expression in comparison to males may indicate some additional mechanism of SYCP3 sexual dimorphism beyond the established mammalian temporal patterns. Furthermore, potential indications for a sex specific role for SYCP3 have been reported in mouse knock-out studies. Male mice lacking *Sycp3* expression were rendered completely sterile while females were only marginally affected remaining largely fertile[66]. This disparity in fertility mediated by SYCP3 indicates that this protein may function differently between sexes with a particular importance for male fertility.

An important molecular component of spermatogenesis is the t-complex, a chromosomal region containing genes known to specifically influence male fertility[67]. *T. thynnus* ESTs possessing sequence similarities to two genes known to map to this t-complex, *Tctex1* (TTC05519; TTC05755) and *Tcte1* (TTC02749) genes were found to be highly differential expressed in the mature testis of the *T. thynnus* (Table1). Differential expression was confirmed with QPCR for EST TTC02749 (Figure3). TCTE1 is considered to be involved in the species specific molecular interaction between the sperm and egg zona pellucida permitting the penetration of the sperm and fertilization[68]. Species specific recognition of gametes is particularly relevant for *T. thynnus* along with other marine spawning fish as they release their gametes in a highly dynamic environment whereby cross-fertilization is undoubtedly an issue. The molecular mechanism by which this sperm-egg recognition is achieved is largely unknown for marine fish. However the relative differential expression of *Tcte1*-like EST TTC02749 in the mature testis of *T. thynnus* suggests a role in facilitating such species specific gamete fertilization. Also part of the t-complex, TCTEX1 has also been assessed as essential for male fertility in the animals studied. However unlike *Tcte1*, null mutations of *Tctex1* affect the phenotypic function of sperm. TCTEX1 has been characterized as a cytoplasmic dynein light chain subunit involved in ubiquitous intracellular transport processes and the proper attachment between the sperm nucleus and flagellar basal body[69]. Surprisingly, Li et al.[69] found that TCTEX1 contributions to the essential cytoplasmic dynein functions are dispensable while male fertility functions were not.

This was demonstrated in drosophila whereby *Tctex1* null mutants were deemed largely viable other than for complete male sterility. This further exemplifies the male specific influence that genes in the t-complex have on spermatid production. Considering ESTs TTC05519 and TTC05755 sequence similarities with *Tctex1* and their differential testis expression in *T. thynnus*, we propose these transcripts are likely to perform a similar function to that of TCTEX1 and as such are a notable aspect of the differentiation between the testis and ovarian transcriptomes of *T. thynnus*.

As previously described in the ovarian component of this discussion, fatty acid binding proteins are part of a multigene family responsible for a diverse array of functions centered on cytoplasmic fatty acid binding. The testis differential expression of transcripts TTC05153 and TTC05128 exhibiting sequence similarities to that of the brain type (*Fabp7*) and intestinal type (*Fabp2*) fatty acid binding proteins respectively is further evidence of the diversity of this gene family (Table1). Differential expression of these FABP-like ESTs is discussed below with reference to *T. thynnus* gonads and their possible function.

Much of FABP7 characterization thus far has largely focused on its expression in the brain of vertebrates, highlighting its association with essential highly polyunsaturated long-chain fatty acids present there, particularly docosahexaenoic acid (DHA)[70]. FABP7 has been shown to have the highest affinity for DHA among all the FABP[71]. Although little is known regarding FABP7 function in the testis, similarities in the fatty acid profiles between the two tissues suggests an explanation for the expression of *Fabp7*-like transcript TTC05153 in the testis of *T. thynnus*. Like the brain, DHA is also present at high concentrations in the retinitis and in mature testis (sperm tail) of vertebrates[72]. Apart from the presence of DHA an additional similarity between these tissues is the presence of axonemes (organelles composed of microtubules). DHA is theorized to contribute to membrane fluidity necessary for the motility of the axoneme[73]. This link is further supported in that DHA deficiencies have been noted to cause retina pigmentosa as well as sperm abnormalities[72]. Taken together we propose that TTC05153 functions similarly to *Fabp7* and its expression in the mature testis of *T. thynnus* is involved in DHA intracellular transport necessary for sperm motility. However it should be noted that this hypothesis does not adequately explain how DHA is incorporated into oocytes. DHA is a well known essential fatty acid present in marine fish eggs, required during embryogenesis particularly for eye and brain development[28]. It may be that the large difference in relative expression of *Fabp7*-like EST TTC05153 between the mature testes and ovaries from *T. thynnus* is simply due to a sexually dimorphic requirement for DHA; oocyte requirements for DHA while significant may be met with considerably less than that of sperm.

Similar to FABP7, much of the previous characterization efforts related to FABP2 function have involved tissues other than testes. Specifically, functional investigations for FABP2 have focused on the intestinal tissues, noting specific polymorphisms in this genes sequence are correlated with obesity and insulin resistance in vertebrates. This has lead to the hypothesis that FABP2 is involved in the transmembrane uptake of dietary fatty acids[74, 75, 76]. However, gene knock-out studies in mice showed that FABP2 is not essential to dietary fat absorption but may instead function as a lipid-sensing component of energy homeostasis that alters energy balance and thus body mass in a gender-specific fashion[77]. While this hypothesis is derived from the intestinal studies, a gender-specific role may explain the differential expression of *Fabp*-like EST TTC05123 in the testis of mature *T. thynnus* (Figure3). Sex-specific energy budgets are a well established concept owing to different reproductive requirements, especially in highly migratory pelagic fish[78, 79, 80]. Therefore based on *Fabp2*-like EST TTC05123 observed sexual dimorphism in expression and the established gender-specific energy requirements for fish, we propose *Fabp2*-like EST TTC05123 is involved in energy homeostasis. Furthermore we suggest that the greater expression of *Fabp2*-like EST TTC05123 in testis tissue compared to ovarian tissue may indicate that male *T. thynnus* are using and/or mobilizing a greater proportion of lipids for energy homeostasis than females who are presumed while present in spawning locations such as the Gulf of Mexico to be sequestering their lipid reserves for oocyte development[81].

CONCLUSIONS

In summary, the transcriptomic approach is a useful tool for examining differential gene expression profiles for the gonadal tissue of *T. thynnus*. In this study, 7068 transcripts were assessed for their differential gene expression in testis and ovarian tissues of sexually mature *T. thynnus*. ESTs bearing sequence similarities to annotated genes and that were significantly differentially over-expressed between the two tissues were considered with respect to their organ's primary gametogenic functions. A number of important components of oogenesis were discussed in relation to identified ESTs including egg envelope formation, yolk proteolysis, oocyte hydration, and lipid metabolism. Similarly, ESTs potentially related to components significant to spermatogenesis were also discussed, including meiosis, sperm motility and lipid metabolism. These ESTs were characterized with reference to their potential role in conventional gametogenesis and where appropriate, alternative functions were proposed. Additional research is required to corroborate the hypothesized functions of the ESTs identified herein in *T. thynnus* gametogenesis. Although Atlantic bluefin reproductive research is a challenging undertaking with tissues difficult to obtain, further transcriptome styled investigation of other tissues taken in a non-destructive method may provide valuable insight and proxy indications for bluefin reproductive condition. The specific ESTs identified herein are important as potential biomarkers for reproductive development, gender distinction and maturation. Such information and non-destructive tools would be highly desirable for both fisheries management and aquaculture development of the Atlantic bluefin by allowing managers to make informed decisions on sexual maturity with little impact to these already over-exploited and highly prized animals. This investigation is the first application of microarray technology for bluefin tunas and has demonstrated the efficacy by which this technique may be used for further characterization of the many unknown biological aspects for this valuable fish.

METHODS

Animal and Tissue Collection

Gonad tissue samples were collected from mature *T. thynnus* by observers as part of the Pelagic Longline Observer Program coordinated by the National Oceanic and Atmospheric Administration (NOAA). Observers stationed aboard commercial longline vessels in the Gulf of Mexico during May of 2009 sampled tissues from *T. thynnus* taken as bycatch on pelagic longlines within 30 minutes of death. Where possible, curved fork length of captured *T. thynnus* were recorded. Gender was determined by gross dissection and review of gonad histology. Gonad tissue was sampled from the fish in a sterile and RNAse free manner and fixed immediately in RNA*later*® (Applied Biosystems, Foster City, CA, USA) for gene expression analyses and 10% phosphate buffered formalin for subsequent histological processing. A 1:10 sample to fixative volume ratio was used for both fixatives. Tissues for RNA analysis were immediately stored at 4°C overnight and then transferred to −20°C until processing.

Histology

Histological sections were sliced from formalin preserved and paraffin embedded gonad samples by IDEXX Laboratories (Sacramento, CA, USA) and stained with with haematoxylin and eosin. Sexual maturity and spawning status were determined histologically according to the methods used by Schaefer[23] and Itano et al.[22].

RNA isolation

Tissue samples were homogenized with a TissueLyser II and stainless steel beads (Qiagen, Valencia, CA, USA). Total RNA was purified from gonad samples fixed in RNA*later*® using TRIZOL® reagent as recommended by the manufacturer (Invitrogen Life Technologies, Carlsbad, CA, USA). Concentration and purity of the RNA were determined using a spectrophotometer (NanoDrop® ND-1000, NanoDrop Technologies Inc., Wilmington, DE, USA) with 230, 260 and 280 nm readings. RNA quality was assessed for all samples by visualization on a denaturing formaldehyde RNA gel (protocol recommended by Qiagen, Valencia, CA, USA) and ethidium bromide staining.

Microarray

Microarray Platform Description

The microarray platform used during this investigation for the purposes of gonad transcriptome expression profiling is entitled BFT 4X44K Array. The platform is a custom commercial Agilent Technologies 4 X 44K format (Agilent Technologies, Santa Clara, CA, USA), capable of the independent hybridization of four separate microarrays per microarray chip. Each independent microarray contains 44 000 *in situ* synthesized oligonucleotide probes with an average length of 50 nucleotides. The probes represent published ESTs derived from Atlantic bluefin tuna liver, ovaries and testis[18]. A total of 10163 ESTs were downloaded from GenBank (accession numbers: EC091633-EC093160, EG629962-EG631176, EC917676-EC919417, EG999340-EG999999, EH000001-EH000505, EH667253-EH668984, and EL610526-EL611807) using the Trace2bdEST component of the PartiGene EST-software pipeline[82]. These ESTs were clustered using TIGR Gene Indices Clustering software to create a non-redundant set of ESTs[83]. A total of 7068 EST clusters were generated and annotated with the Blastx algorithm[84] and Blast2GO software suite[85]. Array Designer software (Premier Biosoft International, Palo Alto, CA, USA) was used to select from the 7068 ESTs approximately three sense strand probes per EST each being a unique complimentary sequence for that EST and approximately 50 oligonucleotides in length. These probes were all duplicated on the microarray to a combined total of 44 000 probes per array. The BFT 4X44K Array platform design is publically accessible at the Gene Expression Omnibus (GEO) repository (http://www.ncbi.nlm.nih.gov/geo/), accession number GPL10007.

Microarray Experimental Design, Target Preparation and Hybridization

A total of eight *T. thynnus* gonad specimens were utilized for the microarray hybridizations, four females and four males representing biological replicates for both sexes. Total RNA was extracted individually from each of these samples as previously described. Aliquots of the extracted total RNA samples were taken and pooled equally among the eight individuals according to RNA concentration. This pooled RNA extraction was used as a common reference to compare variations in mRNA expression among the eight gonad samples. Each gonad sample was individually hybridized against the reference sample in a two-color design, totalling eight microarrays that were hybridized for the investigation in this manner. Target preparation including reverse transcription of total RNA, RNA amplification by *in vitro* transcription and aRNA labelling, was conducted using the Amino Allyl MessageAmp™ II aRNA Amplification Kit (Applied Biosystems, Foster City, CA, USA) according to manufacturer's instruction. Briefly, reverse transcription reactions were performed using T7 Oligo(dT) primer with Arrayscript reverse transcriptase. Second strand synthesis was then performed and cDNA purified using cDNA filter cartridges and used as template for aRNA synthesis in *in vitro* transcription reactions. aRNA was then purified using aRNA filter cartridges and yield and quality of aRNA was assessed by spectrophotometer as previously described. Up to 20 μg per sample of amino allyl-modified aRNA was labeled with Cy3 or Cy5 NHS ester dyes and subsequently column purified.

Hybridization of the microarrays followed the Two-Color Microarray-Based Gene Expression Analysis protocol with reagents from the Gene Expression Hybridization Kit (Agilent Technologies, Santa Clara, CA, USA). Briefly, 825 ng each of Cy3 and Cy5 labelled aRNA was fragmented to nucleotide lengths of 60 – 200 in a fragmentation mix and then added to the hybridization buffer. The hybridization mixes were then applied separately to all four microarrays present on the BFT 4X44K Array and individually sealed with gaskets dividers within a Hybridization chamber (Agilent Technologies, Santa Clara, CA, USA). Microarrays were hybridized at 65°C for 17 hours in a rotating hybridization oven. Following hybridization, microarrays were washed according to the Two-Color Microarray-Based Gene Expression Analysis protocol (Agilent Technologies, Santa Clara, CA, USA). Briefly, the hybridisation chamber is disassembled and the microarray slide is washed for 1 minute in GE wash buffer 1 at room temperature and then transferred to GE wash buffer 2 at 37°C for an additional 1 minute. Following the second wash the slide is slowly removed from the buffer minimising droplet adherence to the slide and scanned immediately. Scanning was performed using the Axon GenePix® 4000B microarray scanner (Molecular Devices, Sunnyvale, CA, USA) at 5 μm pixel resolution with automated photomultiplier balancing used to determine signal intensity and channel balancing.

Analysis

Each microarray on the BFT 4X44K Array slide was scanned separately with the resulting images saved as .tiff files. Feature Extraction 4.0 image analysis software (Agilent Technologies, Santa Clara, CA, USA) was used to extract and process these raw microarray images. Briefly, the Feature Extraction processing pipeline begins by placing a grid specific to the BFT 4X44K Array on the scanned image and identifying each microarray feature. Non-uniform outliers are excluded followed correction of the raw mean signal intensity values by computing background, bias and error values. Global Lowess normalization was used to correct for dye biases, before calculating the log ratios of dye-normalized signals for each feature. The microarray data is publically accessible at the Gene Expression Omnibus (GEO) repository (http://www.ncbi.nlm.nih.gov/geo/) accession number GSE34084.

The Feature Extraction output files generated were imported into GeneSpring GX 11.0 (Agilent Technologies, Santa Clara, CA, USA) software for microarray analysis. Statistical significance for difference in gene expression between the male and female gonad samples by using an unpaired Student's *t*-test on the processed expression ratio values. A p-value cut-off criteria of <0.05 was used in conjunction with the Benjamini-Hochberg multiple testing correction. Fold change analysis was further applied to those features that were deemed statistically significant with a cut-off of two fold difference in expression. The final list of microarray features meeting the above criteria were hierarchically clustered with the Euclidean distance metric and visualized through a heat map plot.

Quantitative Real-time PCR Analysis

Quantitative real-time PCR (QPCR) was used to confirm the relative expression profiles of 13 transcripts identified as significantly differentially expressed by microarray analysis. QPCR primers were designed based on the EST sequence from which the microarray oligonucleotide probe was derived (Table2). In preparation for QPCR cDNA generation, total RNA was treated to remove any contaminating DNA with the DNA-free™ kit as per the manufacturer's protocol (Applied Biosystems, Foster City, CA, USA). Following DNAse treatment, 5 μg of total RNA from each gonad sample was used to generate cDNA with the Superscript™ III Reverse Transcriptase enzyme and random primers following the manufacturer's protocol (Invitrogen Life Technologies, Carlsbad, CA, USA). QPCR amplification reactions were performed in triplicate for each gonad cDNA sample using a Bio-Rad, iCycler iQ, Real-Time

PCR Detection System using iQ™ SYBR® Green Supermix (Bio-Rad, Hercules, CA, USA) in 25 μl reaction volumes as per the manufacturer's protocol. The thermal profile for the qPCRs consisted of an activation step at 95°C for 15 min and 40 cycles of denaturing at 94°C for 40 s, annealing at 55°C for 40 s and elongation at 72°C for 1 min. After the last amplification cycle, a melt curve procedure was performed by which the temperature was cycled from 95°C for 1 min to 55°C for 10 s and then repeated another 81 times, increasing by 0.5°C per cycle for the latter temperature, confirming the presence or absence of non-specific PCR products and primer dimer. β-actin was evaluated for its suitability as a stable transcript to be used for QPCR normalization[86]. Specifically, β-actin transcript variation between all samples ranged between 18 and 20 amplification threshold cycles. Both QPCR and microarray results showed that β-actin expression levels were independent of the sex in relation to gonads thereby validating the use of β-actin for normalization of QPCR results for this investigation.

Table 2: Quantitative PCR specifics

Putative Gene Annotation	Microarray identifier	EST Accession No.	Sequence (5′-3′)[#]	Amplicon length (bp)	Amplification Efficiency (%)
ZPC1	TTC00305	EC091965	F – CATACCACCCTTCACCCATC	212	102
			R – GCTCCACACTAGCCCATGAT		
Vitelline egg envelope gamma	TTC00056	EC091692	F – GCTTGCATGTGTCAGGCTTA	198	94
			R – GGAGAATGGCTTGACTGCTC		
Aveolin	TTC00935	EC092671	F – GCTTCTGCTGCTCTTCGTCT	198	90
			R – AACAACACCTGAGGCAGGAC		
Choriogenin L	TTC04136	EG630294	F – GAGCAGTCAAGCCATTCTCC	230	94
			R – CGTCAATCTCAGTGGCTGAA		
Cathepsin S	TTC00230	EC091881	F – GGATCGACACTGGGAACTGT	297	92
			R – CCCTCCAGTCCAGTGATTGT		

Aquaporin 1	TTC00180	EC091829	F – CCTGTTTCG-CAGTCTTGGAT	191	96
			R – GGTCGGGGTAG-GAATCATTT		
Tmc6-related protein 1	TTC04625	EG630885	F – CCGGTTTCTCCT-CACCAATA	135	91
			R – TTGTGCGT-GACATTCCTGAT		
Fatty acid-binding protein, adipocyte	TTC00964	EC092703	F – ACTGCAATGACC-GAAAGACC	175	97
			R – CCTCCTTTCCGTAG-GTCCTC		
Epididymal secretory protein E1 precursor	TTC00209	EC091860	F – GCTT-GATGGGATTCACCTGT	215	94
			R – CGATTATTCCCATG-GACCAC		
Synaptonemal complex protein 3	TTC02745	EC918114	F – AAGAGCTGAGCG-GTTCAGAG	264	91
			R – TGACCGTGG-TAGTTGTTCCA		
T-complex-associated testis-expressed protein 1	TTC02749	EC918118	F – TGCTGGTGAAA-CACCTCTTG	208	92
			R – AGGGA-CAAAAGGGTGGAGTT		
Brain-type fatty acid binding protein	TTC05153	EG999669	F – CCTACACCTGAT-GACCGACA	212	98
			R – GCTGGGATGATTT-GCTCATT		
Intestinal fatty acid-binding protein	TTC05128	EG999641	F – CGCAGCGAGAAT-TATGACAA	244	95
			R – AGCATGTCACCCTC-CATCTC		

[#]F refers to the forward primer while R indicates the reverse primer.

Relative mRNA transcription abundances were calculated using the delta-delta Ct method[87]. Standard curves were generated from a five point serial dilution of cDNA synthesized from total RNA of a selected gonad sample. A reference cDNA preparation from the gonads of the female samples and another from the males were selected and used as the basis for the standard dilution series. The cDNA template differed depending on whether the target is predominantly considered expressed in male gonads or females. The dilution series forms a standard curve from which a linear relationship between threshold cycle (C_t) and $\log_{10}$of template concentration is determined. This standard curve was used as the basis for calculating/calibrating relative abundance values in the remaining samples. Amplification efficiency of the reactions was also determined from the standard curves of each primer pair (Table2). All relative abundance values were normalized with respect to the relative abundance of β-actin pertaining to that specific gonad cDNA preparation. A Student's t-test was used to determine whether differences in relative expression of the QPCR targeted ESTs were statistically significant among the male and female gonads of *T. thynnus* used in this investigation (Table2).

DECLARATIONS

Acknowledgements

We thank the Southeast Fisheries Center and the NOAA observer program for bluefin tuna in obtaining samples from the Gulf of Mexico spawning ground. We also thank Carol Reeb from Stanford University for her advice regarding the evolutionary genetics of tuna. The authors are grateful for the peer reviewers who contributed their time and valuable comments to the manuscript. This research was supported by funding from NOAA National Sea Grant Office, the California Sea Grant, the Monterey Bay Aquarium Foundation and the TAG-A-Giant Foundation.

Competing Interests

The authors declare they have no competing interests in the manuscript.

Authors' Contributions

LG performed RNA extractions, microarray hybridizations, bioinformatics analysis, QPCR, histological analysis, interpretation of the data, drafting the manuscript and contributed to the overall experimental design and conception of the project. NJ contributed to the microarray design, microarray hybridization, QPCR and drafting of the manuscript. PC contributed to drafting of the manuscript and overall experimental design and conception. BB contributed to histological analysis, drafting of the manuscript and overall experimental design and conception. All authors read and approved the final manuscript.

REFERENCES

1. Chen K-S, Crone P, Hsu C-C: Reproductive biology of female Pacific bluefin tuna Thunnus orientalis from south-western North Pacific Ocean. Fisheries Sci. 2006, 72: 985-994. 10.1111/j.1444-2906.2006.01247.x.Taylor NG, McAllister MK, Lawson GL, Carruthers T, Block BA: Atlantic Bluefin Tuna: A Novel Multistock Spatial Model for Assessing Population Biomass. PLoS One. 2011, 6: e27693-10.1371/journal.pone.0027693.Block BA, Jonsen ID, Jorgensen SJ, Winship AJ, Shaffer SA, Bograd SJ, Hazen EL, Foley DG, Breed GA, Harrison A-L, Ganong JE, Swithenbank A, Castleton M, Dewar H, Mate BR, Shillinger GL, Schaefer KM, Benson SR, Weise MJ, Henry RW, Costa DP: Tracking apex marine predator movements in a dynamic ocean. Nature. 2011, 475: 86-90. 10.1038/nature10082.Fromentin J-M, Powers JE: Atlantic bluefin tuna: population dynamics, ecology, fisheries and management. Fish Fisheries. 2005, 6: 281-306. 10.1111/j.1467-2979.2005.00197.x.Sawada Y, Okada T, Miyashita S, Murata O, Kumai H: Completion of the Pacific bluefin tuna Thunnus orientalis (Temminck et Schlegel) life cycle. Aquac Res. 2005, 36: 413-421. 10.1111/j.1365-2109.2005.01222.x.Reeb CA: Genetic discontinuity of big fish in a small sea. P Natl Acad Sci USA. 2010, 107: 2377-2378. 10.1073/pnas.0914639107.Carlsson J, McDowell JR, Carlsson JEL, Graves JE: Genetic Identity of YOY Bluefin Tuna from the Eastern and Western Atlantic Spawning Areas. J Hered. 2007, 98: 23-28.Boustany AM, Reeb CA, Block BA: Mitochondrial DNA and electronic tracking reveal population structure of Atlantic bluefin tuna (Thunnus thynnus). Mar Biol. 2008, 156: 13-24. 10.1007/s00227-008-1058-0.Riccioni G, Landi M, Ferrara G, Milano I, Cariani A, Zane L, Sella M, Barbujani G, Tinti F: Spatio-temporal population structuring and genetic diversity retention in depleted Atlantic Bluefin tuna of the Mediterranean Sea. P Natl Acad Sci USA. 2010, 107: 2102-2107. 10.1073/pnas.0908281107. Block BA, Teo SLH, Walli A, Boustany A, Stokesbury MJW, Farwell CJ, Weng KC, Dewar H, Williams TD: Electronic tagging and population structure of Atlantic bluefin tuna. Nature. 2005, 434: 1121-1127. 10.1038/nature03463.Restrepo VR, Diaz GA, Walter JF, Neilson JD, Campana SE, Secor D, Wingate RL: Updated

Estimate of the Growth Curve of Western Atlantic Bluefin Tuna. Aquat Living Resour. 2010, 23: 335-342. 10.1051/alr/2011004. Corriero A, Karakulak S, Santamaria N, Deflorio M, Spedicato D, Addis P, Desantis S, Cirillo F, Fenech-Farrugia A, Vassallo-Agius R, Serna JM, Oray Y, Cau A, Megalofonou P, Metrio G: Size and age at sexual maturity of female bluefin tuna (Thunnus thynnus L. 1758) from the Mediterranean Sea. J Appl Ichthyol. 2005, 21: 483-486. 10.1111/j.1439-0426.2005.00700.x.Schirripa MJ: A literature review of Atlantic bluefin tuna age at maturity. Collect Vol Sci Pap ICCAT. 2011, 66: 898-914.Susca V, Corriero A, Deflorio M, Bridges CR, De Metrio G: New results on the reproductive biology of the bluefin tuna (Thunnus thynnus) in the Mediterranean. ICCAT. 2001, 52: 745-751

2. Corriero A, Medina A, Mylonas CC, Abascal FJ, Deflorio M, Aragón L, Bridges CR, Santamaria N, Heinisch G, Vassallo-Agius R, Belmonte A, Fauvel C, Garcia A, Gordin H, De Metrio G: Histological study of the effects of treatment with gonadotropin-releasing hormone agonist (GnRHa) on the reproductive maturation of captive-reared Atlantic bluefin tuna (Thunnus thynnus L.). Aquaculture. 2007, 272: 675-686. 10.1016/j.aquaculture.2007.08.030

3. Corriero A, Medina A, Mylonas CC, Bridges CR, Santamaria N, Deflorio M, Losurdo M, Zupa R, Gordin H, de la Gandara F, Belmonte Rìos A, Pousis C, De Metrio G: Proliferation and apoptosis of male germ cells in captive Atlantic bluefin tuna (Thunnus thynnus L.) treated with gonadotropin-releasing hormone agonist (GnRHa). Anim Reprod Sci. 2009, 116: 346-357. 10.1016/j.anireprosci.2009.02.013

4. De Metrio G, Bridges CR, Mylonas CC, Caggiano M, Deflorio M, Santamaria N, Zupa R, Pousis C, Vassallo-Agius R, Gordin H, Corriero A: Spawning induction and large-scale collection of fertilized eggs in captive Atlantic bluefin tuna (Thunnus thynnus L.) and the first larval rearing efforts. J Appl Ichthyol. 2010, 26: 596-10.1111/j.1439-0426.2010.01475.x

5. Chini V, Cattaneo AG, Rossi F, Bernardini G, Terova G, Saroglia M, Gornati R: Genes expressed in Blue Fin Tuna (Thunnus thynnus) liver and gonads. Gene. 2008, 410: 207-213. 10.1016/j.gene.2007.12.012

6. von Schalburg K, Rise M, Cooper G, Brown G, Gibbs AR, Nelson C, Davidson W, Koop B: Fish and chips: Various methodologies demonstrate utility of a 16,006-gene salmonid microarray. BMC Genomics. 2005, 6: 126-10.1186/1471-2164-6-126
7. Mazurais D, Montfort J, Delalande C, Gac FL: Transcriptional analysis of testis maturation using trout cDNA macroarrays. Gen Comp Endocr. 2005, 142: 143-154. 10.1016/j.ygcen.2005.02.018
8. Santos EM, Paull GC, Van Look KJW, Workman VL, Holt WV, van Aerle R, Kille P, Tyler CR: Gonadal transcriptome responses and physiological consequences of exposure to oestrogen in breeding zebrafish (Danio rerio). Aquat Toxicol. 2007, 83: 134-142. 10.1016/j.aquatox.2007.03.019
9. Itano DG, Marine U of H (Honolulu). JI for, Research A, Program PFR: The reproductive biology of yellowfin tuna (Thunnus albacares) in Hawaiian waters and the western tropical Pacific Ocean: project summary. 2000, Joint Institute for Marine and Atmospheric Research, University of Hawar
10. Schaefer K: Reproductive biology of yellowfin tuna (Thunnus albacares) in the eastern pacific ocean. Bulletin. 1998, 21 (5): 205-27r
11. Viñas J, Tudela S: A Validated Methodology for Genetic Identification of Tuna Species (Genus Thunnus). PLoS One. 2009, 4: e7606-10.1371/journal.pone.000760r
12. Lynch M, Koskella B, Schaack S: Mutation Pressure and the Evolution of Organelle Genomic Architecture. Science. 2006, 311: 1727-1730. 10.1126/science.111888r
13. Wootton RJ: Ecology of teleost fishes. 1998, Kluwer Academic Publishers, Boston, 396r
14. Jalabert B: Particularities of reproduction and oogenesis in teleost fish compared to mammals. Reprod Nutr Dev. 2005, 45: 1r
15. Brooks S, Tyler CR, Sumpter JP: Egg quality in fish: what makes a good egg?. Rev Fish Biol Fisher. 1997, 7: 387-416. 10.1023/A:101840013069r
16. Coward K, Ponting CP, Chang H-Y, Hibbitt O, Savolainen P, Jones KT, Parrington J: Phospholipase C{zeta}, the trigger of egg activation in mammals, is present in a non-mammalian species.

Reproduction. 2005, 130: 157-163. 10.1530/rep.1.0070r

17. Nemerson D, Berkeley S, Safina C: Spawning site fidelity in Atlantic bluefin tuna, Thunnus thynnus: the use of size-frequency analysis to test for the presence of migrant east Atlantic bluefin tuna on Gulf of Mexico spawning grounds. Fish B-NOAA. 2000, 98: 118-12r
18. Mold DE, Dinitz AE, Sambandan DR: Regulation of Zebrafish Zona Pellucida Gene Activity in Developing Oocytes. Biol Reprod. 2009, 81: 101-110. 10.1095/biolreprod.108.07172r
19. Sugiyama H, Yasumasu S, Murata K, Iuchi I, Yamagami K: The third egg envelope subunit in fish: cDNA cloning and analysis, and gene expression. Dev Growth Differ. 1998, 40: 35-45. 10.1046/j.1440-169X.1998.t01-5-00005.r
20. Hyllner SJ, Westerlund L, Olsson P-E, Schopen A: Cloning of Rainbow Trout Egg Envelope Proteins: Members of a Unique Group of Structural Proteins. Biol Reprod. 2001, 64: 805-811. 10.1095/biolreprod64.3.80r
21. Chang YS, Wang SC, Tsao CC, Huang FL: Molecular cloning, structural analysis, and expression of carp ZP3 gene. Mol Reprod Dev. 1996, 44: 295-304. 10.1002/(SICI)1098-2795(199607)44:3<295::AID-MRD3>3.0.CO;2-r
22. Del Giacco L, Diani S, Cotelli F: Identification and spatial distribution of the mRNA encoding an egg envelope component of the Cyprinid zebrafish, Danio rerio, homologous to the mammalian ZP3 (ZPC). Dev Gene Evol. 2000, 210: 41-46. 10.1007/PL0000818r
23. Lyons CE, Payette KL, Price JL, Huang RC: Expression and structural analysis of a teleost homolog of a mammalian zona pellucida gene. J Biol Chem. 1993, 268: 21351-2135r
24. Mold DE, Kim IF, Tsai C-M, Lee D, Chang C-Y, Huang RCC: Cluster of genes encoding the major egg envelope protein of zebrafish. Mol Reprod Dev. 2001, 58: 4-14. 10.1002/1098-2795(200101)58:1<4::AID-MRD2>3.0.CO;2-r
25. Murata K, Sasaki T, Yasumasu S, Iuchi I, Enami J, Yasumasu I, Yamagami K: Cloning of cDNAs for the Precursor Protein of a Low-Molecular-Weight Subunit of the Inner Layer of the Egg Envelope (Chorion) of the Fish Oryzias latipes. Dev Biol. 1995, 167: 9-17. 10.1006/dbio.1995.100r

26. Kanamori A, Naruse K, Mitani H, Shima A, Hori H: Genomic organization of ZP domain containing egg envelope genes in medaka (Oryzias latipes). Gene. 2003, 305: 35-45. 10.1016/S0378-1119(02)01211-r
27. Fabra M, Raldúa D, Bozzo MG, Deen PMT, Lubzens E, Cerdà J: Yolk proteolysis and aquaporin-1o play essential roles to regulate fish oocyte hydration during meiosis resumption. Dev Biol. 2006, 295: 250-262. 10.1016/j.ydbio.2006.03.03r
28. Finn RN, Østby GC, Norberg B, Fyhn HJ: In vivo oocyte hydration in Atlantic halibut (Hippoglossus hippoglossus); proteolytic liberation of free amino acids, and ion transport, are driving forces for osmotic water influx. J Exp Biol. 2002, 205: 211-22r
29. Kestemont P, Cooremans J, Abi-Ayad A, Mélard C: Cathepsin L in eggs and larvae of perch Perca fluviatilis: variations with developmental stage and spawning period. Fish Physiol Biochem. 1999, 21: 59-64. 10.1023/A:100771431407r
30. Carnevali O, Carletta R, Cambi A, Vita A, Bromage N: Yolk Formation and Degradation during Oocyte Maturation in Seabream Sparus aurata: Involvement of Two Lysosomal Proteinases. Biol Reprod. 1999, 60: 140-146. 10.1095/biolreprod60.1.14r
31. Raldua D, Fabra M, Bozzo MG, Weber E, Cerda J: Cathepsin B-mediated yolk protein degradation during killifish oocyte maturation is blocked by an H+−ATPase inhibitor: effects on the hydration mechanism. Am J Physiol Regul Integr Comp Physiol. 2006, 290: R456-R46r
32. Wood AW, Kraak GVD: Yolk proteolysis in rainbow trout oocytes after serum-free culture: Evidence for a novel biochemical mechanism of atresia in oviparous vertebrates. Mol Reprod Dev. 2003, 65: 219-227. 10.1002/mrd.1027r
33. Carnevali O, Mosconi G, Cambi A, Ridolfi S, Zanuy S, Polzonetti-Magni AM: Changes of lysosomal enzyme activities in sea bass (Dicentrarchus labrax) eggs and developing embryos. Aquaculture. 2001, 202: 249-256. 10.1016/S0044-8486(01)00775-r
34. Hiramatsu N, Ichikawa N, Fukada H, Fujita T, Sullivan CV, Hara A: Identification and characterization of proteases involved in specific proteolysis of vitellogenin and yolk proteins in salmonids. J Exp

Zool. 2002, 292: 11-25. 10.1002/jez.113r

35. Hahn Y, Kim DS, Pastan IH, Lee B: Anoctamin and transmembrane channel-like proteins are evolutionarily related. Int J Mol Med. 2009, 24: 51-5r
36. Keresztes G, Mutai H, Heller S: TMC and EVER genes belong to a larger novel family, the TMC gene family encoding transmembrane proteins. BMC Genomics. 2003, 4: 24-10.1186/1471-2164-4-2r
37. Kurima K, Yang Y, Sorber K, Griffith AJ: Characterization of the transmembrane channel-like (TMC) gene family: functional clues from hearing loss and epidermodysplasia verruciformis. Genomics. 2003, 82: 300-308. 10.1016/S0888-7543(03)00154-r
38. Storch J, Thumser AEA: The fatty acid transport function of fatty acid-binding proteins. BBA-Mol Cell Biol L. 2000, 1486: 28-4r
39. Wu Q, Ortegon AM, Tsang B, Doege H, Feingold KR, Stahl A: FATP1 Is an Insulin-Sensitive Fatty Acid Transporter Involved in Diet-Induced Obesity. Mol Cell Biol. 2006, 26: 3455-3467. 10.1128/MCB.26.9.3455-3467.200r
40. Weaver RE, Donnelly D, Wabitsch M, Grant PJ, Balmforth AJ: Functional expression of glucose-dependent insulinotropic polypeptide receptors is coupled to differentiation in a human adipocyte model. Int J Obes. 2008, 32: 1705-1711. 10.1038/ijo.2008.14r
41. Agulleiro MJ, André M, Morais S, Cerdà J, Babin PJ: High Transcript Level of Fatty Acid-Binding Protein 11 but Not of Very Low-Density Lipoprotein Receptor Is Correlated to Ovarian Follicle Atresia in a Teleost Fish (Solea senegalensis). Biol Reprod. 2007, 77: 504-516. 10.1095/biolreprod.107.06159r
42. Carstea ED, Morris JA, Coleman KG, Loftus SK, Zhang D, Cummings C, Gu J, Rosenfeld MA, Pavan WJ, Krizman DB, Nagle J, Polymeropoulos MH, Sturley SL, Ioannou YA, Higgins ME, Comly M, Cooney A, Brown A, Kaneski CR, Blanchette-Mackie EJ, Dwyer NK, Neufeld EB, Chang TY, Liscum L, Strauss JF, Ohno K, Zeigler M, Carmi R, Sokol J, Markie D, O'Neill RR, van Diggelen OP, Elleder M, Patterson MC, Brady RO, Vanier MT, Pentchev PG, Tagle DA: Niemann-Pick C1 disease gene: homology to mediators of cholesterol homeostasis. Science. 1997, 277: 228-231. 10.1126/

science.277.5323.22r

43. Higgins ME, Davies JP, Chen FW, Ioannou YA: Niemann-Pick C1 is a late endosome-resident protein that transiently associates with lysosomes and the trans-Golgi network. Mol Genet Metab. 1999, 68: 1-13. 10.1006/mgme.1999.288r
44. Babin PJ, Cerda J, Lubzens E: The Fish Oocyte. 2007, Springer Netherlands, Dordrecr
45. Schulz RW, de França LR, Lareyre J-J, LeGac F, Chiarini-Garcia H, Nobrega RH, Miura T: Spermatogenesis in fish. Gen Comp Endocr. 2010, 165: 390-411. 10.1016/j.ygcen.2009.02.01r
46. Abascal FJ, Megina C, Medina A: Testicular development in migrant and spawning bluefin tuna (Thunnus thynnus (L.)) from the eastern Atlantic and Mediterranean. Fish B-NOAA. 2004, 102: 407-41r
47. Di Carlo A, Travia G, De Felici M: The meiotic specific synaptonemal complex protein SCP3 is expressed by female and male primordial germ cells of the mouse embryo. Int J Dev Biol. 2000, 44: 241-24r
48. Baier A, Alsheimer M, Benavente R: Synaptonemal complex protein SYCP3: Conserved polymerization properties among vertebrates. BBA-Proteins Proteom. 2007, 1774: 595-602. 10.1016/j.bbapap.2007.03.00r
49. Iwai T, Yoshii A, Yokota T, Sakai C, Hori H, Kanamori A, Yamashita M: Structural components of the synaptonemal complex, SYCP1 and SYCP3, in the medaka fish Oryzias latipes. Exp Cell Res. 2006, 312: 2528-2537. 10.1016/j.yexcr.2006.04.01r
50. Baglin RE: Reproductive biology of western Atlantic bluefin tuna. Fish Bull. 1982, 80: 121-13r
51. Masui Y, Clarke HJ: Oocyte Maturation. In Academic Press;. 1979, 57: 185-28r
52. Nakamura S, Kobayashi K, Nishimura T, Higashijima S, Tanaka M: Identification of germline stem cells in the ovary of the teleost medaka. Science. 2010, 328: 1561-1563. 10.1126/science.118547r
53. Hunt PA, Hassold TJ: Sex matters in meiosis. Science. 2002, 296: 2181-2183. 10.1126/science.107190r
54. Fraser LR, Dudley K: New insights into the t-complex and control of sperm function. BioEssays. 1999, 21: 304-312. 10.1002/

(SICI)1521-1878(199904)21:4<304::AID-BIES6>3.0.CO;2-r

55. Juneja R, Agulnik SI, Silver LM: Sequence divergence within the sperm-specific polypeptide TCTE1 is correlated with species-specific differences in sperm binding to zona-intact eggs. J Androl. 1998, 19: 183-18r

56. Li M, Serr M, Newman EA, Hays TS: The Drosophila tctex-1 Light Chain Is Dispensable for Essential Cytoplasmic Dynein Functions but Is Required during Spermatid Differentiation. Mol Biol Cell. 2004, 15: 3005-3014. 10.1091/mbc.E04-01-001r

57. Zimmerman AW, Veerkamp JH: New insights into the structure and function of fatty acid-binding proteins. Cell Mol Life Sci (CMLS). 2002, 59: 1096-1116. 10.1007/s00018-002-8490-r

58. Zhong Xu L, Sánchez R, Sali A, Heintz N: Ligand Specificity of Brain Lipid-binding Protein. J Biol Chem. 1996, 271: 24711-24719. 10.1074/jbc.271.40.2471r

59. Connor WE, Weleber RG, DeFrancesco C, Lin DS, Wolf DP: Sperm abnormalities in retinitis pigmentosa. Invest Ophthalmol Vis. Sci. 1997, 38: 2619-262r

60. Connor WE, Lin DS, Wolf DP, Alexander M: Uneven distribution of desmosterol and docosahexaenoic acid in the heads and tails of monkey sperm. J Lipid Res. 1998, 39: 1404-141r

61. André M, Ando S, Ballagny C, Durliat M, Poupard G, Briançon C, Babin PJ: Intestinal fatty acid binding protein gene expression reveals the cephalocaudal patterning during zebrafish gut morphogenesis. Int J Dev Biol. 2000, 44: 249-25r

62. Levy E, Ménard D, Delvin E, Stan S, Mitchell G, Lambert M, Ziv E, Feoli-Fonseca JC, Seidman E: The Polymorphism at Codon 54 of the FABP2 Gene Increases Fat Absorption in Human Intestinal Explants. J Biol Chem. 2001, 276: 39679-39684. 10.1074/jbc.M10571320r

63. Albala C, Santos JL, Cifuentes M, Villarroel AC, Lera L, Liberman C, Angel B, Perez-Bravo F: Intestinal FABP2 A54T Polymorphism: Association with Insulin Resistance and Obesity in Women. Obesity. 2004, 12: 340-345. 10.1038/oby.2004.4r

64. Vassileva G, Huwyler L, Poirier K, Agellon LB, Toth MJ: The intestinal fatty acid binding protein is not essential for dietary fat

absorption in mice. FASEB J. 2000, 14: 2040-2046. 10.1096/fj.99-0959cor

65. Ballantyne JS, Mercure F, Gerrits MF, Van Der Kraak G, McKinley S, Martens DW, Hinch SG, Diewert RE: Plasma nonesterified fatty acid profiles in male and female sockeye salmon,(Oncorhynchus nerka), during the spawning migration. Can J Fish Aquat Sci. 1996, 53: 1418-1426. 10.1139/f96-05r

66. Hutchings JA, Pickle A, McGregor-Shaw CR, Poirier L: Influence of sex, body size, and reproduction on overwinter lipid depletion in brook trout. J Fish Biol. 1999, 55: 1020-1028. 10.1111/j.1095-8649.1999.tb00737.r

67. Pérez MJ, Rodríguez C, Cejas JR, Martín MV, Jerez S, Lorenzo A: Lipid and fatty acid content in wild white seabream (Diplodus sargus) broodstock at different stages of the reproductive cycle. Comp Bioche. Physiol, Part B Biochem Mol Biol. 2007, 146: 187-196. 10.1016/j.cbpb.2006.10.09r

68. Mourente G, Megina C, Díaz-Salvago E: Lipids in female northern bluefin tuna (Thunnus thynnus thynnus L.) during sexual maturation. Fish Physiol Biochem. 2001, 24: 351-363. 10.1023/A:101501160901r

69. Parkinson J, Anthony A, Wasmuth J, Schmid R, Hedley A, Blaxter M: PartiGene–constructing partial genomes. Bioinformatics. 2004, 20: 1398-1404. 10.1093/bioinformatics/bth10r

70. Pertea G, Huang X, Liang F, Antonescu V, Sultana R, Karamycheva S, Lee Y, White J, Cheung F, Parvizi B, Tsai J, Quackenbush J: TIGR Gene Indices clustering tools (TGICL): a software system for fast clustering of large EST datasets. Bioinformatics. 2003, 19: 651-652. 10.1093/bioinformatics/btg03r

71. Altschul SF, Madden TL, Schäffer AA, Zhang J, Zhang Z, Miller W, Lipman DJ: Gapped BLAST and PSI-BLAST: a new generation of protein database search programs. Nucleic Acids Res. 1997, 25: 3389-3402. 10.1093/nar/25.17.338r

72. Götz S, García-Gómez JM, Terol J, Williams TD, Nagaraj SH, Nueda MJ, Robles M, Talón M, Dopazo J, Conesa A: High-throughput functional annotation and data mining with the Blast2GO suite. Nucleic Acids Res. 2008, 36: 3420-3435. 10.1093/nar/gkn17r

73. Mladineo I, Block BA: Expression of Hsp70, Na+/K+ ATP-ase, HIF-1α, IL-1β and TNF-α in captive Pacific bluefin tuna (Thunnus orientalis) after chronic warm and cold exposure. J Exp Mar Biol Ecol. 2009, 374: 51-57. 10.1016/j.jembe.2009.04.00r

74. Livak KJ, Schmittgen TD: Analysis of relative gene expression data using real-time quantitative PCR and the 2(−Delta Delta C(T)) Method. Methods. 2001, 25: 402-408. 10.1006/meth.2001.1262.

CHAPTER

15

THE CULTURABLE INTESTINAL MICROBIOTA OF TRIPLOID AND DIPLOID JUVENILE ATLANTIC SALMON (SALMO SALAR) - A COMPARISON OF COMPOSITION AND DRUG RESISTANCE

Leon Cantas[1,2], Thomas WK Fraser[3], Per Gunnar Fjelldal[4] , Ian Mayer[3] and Henning Sørum[1]

[1]Department of Food Safety and Infection Biology, Norwegian School of Veterinary Science

[2]Norwegian Food Safety Authority, District Office for Romerike

[3]Department of Production Animal Clinical Sciences, Norwegian School of Veterinary Science

[4]Institute of Marine Research (IMR), Matre Research Station

ABSTRACT

Background

With the increased use of ploidy manipulation in aquaculture and fisheries management this investigation aimed to determine whether triploidy influences culturable intestinal microbiota composition and bacterial drug resistance in Atlantic salmon (*Salmo salar*). The results could provide answers to some of the physiological differences observed between triploid and diploid fish, especially in terms of fish health.

Results

No ploidy effect was observed in the bacterial species isolated, however, triploids were found to contain a significant increase in total gut microbiota levels, with increases in *Pseudomonas* spp., *Pectobacterium carotovorum*, *Psychrobacter* spp., *Bacillus* spp., and *Vibrio* spp., (12, 42, 9, 10, and 11% more bacteria in triploids than diploids, respectively), whereas a decrease in *Carnobacterium* spp., within triploids compared to diploids was close to significant (8% more bacteria in diploids). With the exception of gentamicin, where no bacterial resistance was observed, bacterial isolates originating from triploid hosts displayed increased resistance to antibacterials, three of which were significant (tetracycline, trimethoprim, and sulphonamide).

Conclusion

Results indicate that triploidy influences both the community and drug resistance of culturable intestinal microbiota in juvenile salmon. These results demonstrate differences that are likely to contribute to the health of triploid fish and have important ramifications on the use of antibacterial drugs within aquaculture.

BACKGROUND

Artificial triploid fish, that possess three complete sets of chromosomes compared to the more natural state of diploidy that possess two complete chromosome sets, are sterile due to reduced gonadal development and non-functional gametes [1]. Consequently, triploids are appealing to the aquaculture industry as a method for eliminating genetic interactions between wild and cultured stocks and reducing the financial costs associated with early maturation [2, 3]. Today, triploids (mainly salmonids) are commercially produced in several countries around the globe including France and Australia [1, 4].

Previously, is has been commented that the physiology of triploids is sufficiently different to diploids that they should be treated as a separate species [5]. Of most interest, triploids are composed of cells that are typically 50% larger than diploids due to the accommodation of the extra genetic material within the cell nucleus [6]. However, no size increase of the fish body is achieved through triploidy as the majority of tissues are composed of fewer cells, resulting in similar organ and whole body sizes between the ploidies [6, 7, 8]. The reduction in cell number may reduce oxygen delivery [9] and there is also evidence to suggest triploids have a lower optimum metabolic temperature compared to diploids [10], altered nutritional requirements [11], and differences in ontogenetic growth [1]. It is also widely reported that triploids are more susceptible to stress and disease [12, 13, 14, 15], although this has yet to be confirmed in controlled experiments [16, 17, 18, 19]. In addition, in contrast to male triploids, triploid females typically show no morphological or physiological signs of sexual maturation [8].

Evidence from both mammalian and fish studies suggests the gut microbiota is both influenced by, and has an influence on, host physiology [20, 21, 22, 23]. For instance, whilst the gastrointestinal (GI) tract provides an ecological niche for bacterial growth [20], the microbiota present within the GI tract is dependent on host species, strain, ontogenetic growth, stress levels, and gut structure, whereas nutrient processing and absorption, development of the mucosal immune system, angiogenesis, and epithelial renewal are influenced by the gut microbiota [20, 21, 22, 23, 24, 25]. Therefore, the gut microbiota plays a significant role in an individual's physiological status. As several of the above listed host factors are either known or suspected to be affected by triploidy (ontogenetic growth, stress response, immunological functioning, and gut structure/cell size and number), it is likely that gut microbiota is also affected by triploidy. Significantly, a changed microbiota may ease the growth and survival of potential fish pathogenic bacteria and also result in a different maturation of the mucosal immunity in young fish.

Previous studies have focused on characterizing the microbiota of different fish species and the biological significance of those bacteria present. For instance, the GI microbiota of several commercially important species has been studied during various life stages, such as in the salmonids [20]. Subsequently, bacteria have been classified as either allochthonous (transient) or autochthonous (permanent), with the allochthonous bacteria supplying the stock from which the autochthonous bacteria may become established. In addition, some bacteria have been identified as pathogens [i.e. *Vibrio*; [26]], involved in enteritis [27], or inhibit the growth of other bacteria [*Carnobacterium* spp.; [28, 29]].

These findings have led to attempts to manipulate the gut microflora of fish through pre- and probiotics to improve fish health and growth [30].

A current area of concern within the field of environmental microbiology is the increase in the occurrence of antibiotic resistance among bacteria related to animal production. In this context it is considered that drug resistant bacteria in the environment may potentially transfer their resistance features to not only animal but also human pathogens [31]. This assumption has led to intensified efforts to monitor bacterial resistance, and is extensively done within bacteria of production animals and their environments [[32, 33], monitoring schemes, Norway NORM/NORM-VET, Denmark DANMAP, Sweden SVARM]. Significantly, bacteria are capable of transferring antibiotic resistance both within and between species through processes such as conjugation of transferable R-plasmids [34, 35]. Traditionally, the use of antibiotics was assumed the main selection factor for increasing antibiotic resistance in aquaculture and the surrounding environment [36, 37]; however, recent evidence would suggest that feed and just the presence of an aquaculture facility may also influence the antibiotic resistance levels of bacteria [38, 39].

Differences in physiology related to ploidy may influence the gut microbial community and therefore, through a changed microbiota, secondarily affect the growth and health of triploid fish; two key elements in successful aquaculture, however no information is currently available. The objective of this experiment was to compare the culturable intestinal microbiota of cultured triploid and diploid juvenile Atlantic salmon (*Salmo salar*) for differences in the microbial community and the antibacterial resistance profiles of the bacteria based on primary traditional bacterial culturing.

METHODS

The present experiment was approved by the Norwegian Animal Research Authority and performed according to prevailing animal welfare regulations.

Fish Stock and Rearing Conditions

Fish eggs were provided by Aquagen (Trondheim, Norway) and reared at Matre Research Station, Norway. On Nov 03 2009, eggs (12,000-22,500/

female, n = 209,700) from twelve female Atlantic salmon were fertilized by three different males, each male crossed with four different females (male 1 crossed with females 1-4, male 2 with females 5-8 etc), leaving four half-sibling families per male and three groups of independent families. Hydrostatic pressure, which interferes with the release of the second polar body, was used to create triploids [1, 40]. Thirty-seven minutes and 30 seconds after fertilization at 8°C, half of the eggs from each female were subjected to a hydrostatic pressure of 655 bar for 6 min and 15 s (TRC-APV, Aqua Pressure Vessel, TRC Hydraulics inc., Dieppe, Canada), giving a total of twenty four groups (12/ploidy). Thereafter, each group was incubated in single incubation trays [Sterner 2003, Sterner Fish Tech AS] in an isolated (UV treatment) flow-through system (water was buffered with seawater to 0.7 ppt salinity; oxygen saturation 95%; pH 6.9) under darkness. The eggs were mechanically agitated to allow dead eggs to be sorted from live eggs at the eye-egg stage on December 21. Eggs from one of the females showed very high mortality at this stage, both as diploid and triploid, and were excluded from the study. After removing the dead eggs (mean 6.7% per female, min 0.9%, max 21%) in each of the remaining twenty two trays, all the family groups within each ploidy were mixed (1 L per female) in a bucket and re-distributed back into six trays (1 L eggs/tray; n = 3 trays/ploidy) leaving three trays/ploidy that each contained a mixture of all three independent families. Hatching started on February 05 2010, and was completed by February 18. On April 26 2010, the yolk sack larvae of each incubation tray were put into single, square, grey, covered, fiberglass tanks (dimensions 1.5 × 1.5 × 0.5 m; water depth 30 cm; water flow 4 L/min/kg fish; n = 3 tanks/ploidy; approx. 5400 larvae/tank) under continuous light (2 × 18 W fluorescent daylight tubes, OSRAM L 18W/840 LUMILUX OSRAM GmbH, Ausburg, Germany) and continuous feeding using automatic disc feeders and a commercial start feed (NUTRA ST 0.5, Skretting AS, Fontaine-les-Vervins, France) that increased in size (up to 1.5 mm pellets, Skretting AS, Norway). On July 22 2010, the fish from each tank were transferred to single, green, fiber glass, 10.7 m^3 tanks (Namdal Plast AS, 3 m Ø; water depth 1 m; n = 3 tanks/ploidy; n = 6 tanks). The fish were reared under continuous light (light source as above) from start-feeding onwards. The mean temperature in the period fertilization to start-feeding was 4.9°C (min 3.2°C, max 6.8°C), and in the period start-feeding to August 2010 it was 13.2°C (min 10.7°C, max 15.6°C). The oxygen saturation in the outlet water was always above 80%.

To determine ploidy status, the diameter of the erythrocytes was measured. This method has previously been validated for identification of triploid salmonids, whose erythrocytes are consistently larger than those in diploids [41]. Blood smears were taken from a total of eighty fish per ploidy (n = 160). The diameters of 10 random red blood cells were measured (Image-Pro Plus, version 4.0, Media Cybernetics Silver Spring, MD, USA) on each blood smear. There was no overlap in mean red blood cell diameters between diploid (mean 16.8 μm, max 18.1 μm) and triploid (mean 20.8 μm, min 19.8 μm) individuals, suggesting 100% efficiency of ploidy manipulation.

Sampling Procedure

On Aug 24 2010, twelve juvenile fish (n = 6/ploidy) of unknown sex (mean weight: diploid 25.5 g, triploid 28.0 g; mean length: diploid 112 mm, triploid 118 mm) from six tanks (three tanks/ploidy; two fish/tank) were euthanized by a sharp blow to the head followed by decapitation and sampled for gut microbiology. Sterile latex gloves or latex gloves wiped with 70% ethanol were used throughout the dissection. The underside of the fish was sterilized using 70% ethanol and the ventral belly surface was opened with a sterile surgical blade and forceps. The spleen, gall bladder, liver, and fat deposits surrounding the GI tract were gently removed to expose the peritoneal cavity without disturbing the intestine. The intestine was divided into foregut, midgut, and hindgut as previously described [42]. Thereafter, the intestinal content from within each gut section was squeezed into separate sterile tubes and the appropriate 10-fold dilutions were prepared in physiological saline. Then 100 μl aliquots were pipetted onto the surface of 5% cattle blood agar (blood agar base no 2, Difco) and Brocalin agar (Merck, Darmstadt, Germany) both with 1% NaCl. Dry feed samples and water samples were also appropriately diluted and streaked onto blood agar and Brocalin agar. The plates were incubated at 12°C (± 1) and inspected regularly for up to 4 weeks.

Bacterial Isolation, Genus Identification, and Quantification

After incubation, each morphologically distinct colony (form, size, surface, color, texture, elevation, margin, and hemolysis) was coded. To obtain pure growth, two randomly selected representatives of all the coded bacterial colonies from each section of the intestine, were subcultured on blood agar. Afterwards, pure cultured colonies were checked for motility under a light microscope (Leitz, 301-314.001, Germany) at 1000 × magnification using a glass slide with suspended and Gram stained exponentially grown bacteria in physiological salt water [43]. Catalase and oxidase activity were tested as well as the ability to reduce nitrate. The IMViC reactions; indol production, methyl red test (fermentation of glucose), Voges-Proskauer reaction (diacetyl-production) and the ability to use citrate as the sole carbon source (Simmons citrate medium) were tested. H_2S production, gelatinase, urease and β-galactosidase (ONPG) production and ability to degrade glucose, trehalose, lactose, maltose, mannitol and esculin were tested [Table 1].

Table 1: Phenotypical properties of bacterial isolates.

Gram stain and Colony descriptions	Hae-molysis	Motility	Cata-lase	Oxidase	Nitrate reductase	Indole	Meth-yl-Red	Voges-Pros-kauer	Citrate Utiliza-tion	H_2S Produc-tion	Gelatin-ase	Urea hydro-lysis	O/129 disk	ONPG	Glu-cose	Treha-lose	Lac-tose	Malt-ose	Man-nitol	Esculin	Organisms and No of strains
GN rods, short, plump, in coccoid pairs or chains of variable length (Translucent to opaque, convex, entire colonies 1.5-2 mm)	-	-	+	-	-	-	-	-	V	-	-	-	-	-	V	-	V	-	-	-	*Acineto-bacter*spp. n: 24
GN slightly curved rods, (Large, smooth, yellowish-white mucoid 2-4 mm colonies)	V	V	+	+	V	-	-	-	+	-	V	V	-	V	-	-	-	V	V	-	*Pseudo-monas*spp. n: 72
GN rods (Flat, green-ish, 1-2 mm colonies)	V	-	+	+	V	-	-	-	+	-	+	V	-	-	-	V	-	V	V	-	*P. fluores-cens*n: 72
GN coccoba-cilli (Slightly, convex, cream colored 2 mm colonies)	-	-	-	+	V	-	-	-	-	-	-	V	-	-	-	-	-	-	-	-	*Psychro-bacter*spp. n: 48
GN rods (Opaque, circular, convex, white to pink entire colonies 2-3 mm)	-	+	-	-	V	-	V	+	+	-	+	-	-	+	+	+	-	+	+	V	*Serratia* spp. n: 24

GN curved or comma shaped rod (smooth, moist, opaque-white\yellow green, round 2-3 mm colonies)	V	+	+	+	+	+	V	-	V	-	V	V	+	V	+	+	V	V	V	V	*Vibrio* spp. n; 72
GN regular rods (Shiny, orange-yellow, circular, smooth, concave, 2 mm colonies)	-	-	+	-	+	ND	ND	ND	+	ND	+	+	ND	ND	+	+	+	ND	ND	-	*Pectobacterium carotovorum* n; 48
GP irregular rods with jointed rods 'V-shape' (Yellow-Greenish-metallic center,1-2 mm colonies)	-	-	-	+	V	-	-	-	-	-	V	V	-	-	-	V	V	V	V	-	*Arthrobacter*spp. n; 48
GP straight long-narrow rods in chain, contain oval endospore (Gray-white, dry appearance 3-4 mm)	V	+	+	V	V	V	V	V	V	-	V	V	V	V	+	+	V	V	V	V	*Bacillus* spp. n; 48
GP slightly slender singly rods (Whitish, round, convex, 1-2 mm colonies)	-	-	-	-	-	-	-	+	+	-	-	-	-	-	V	+	+	V	+	+	*Carnobacterium*spp. n; 48

GP cocci singly, in pairs (Circular, convex to slightly peaked, smooth, glossy, opaque 1-5 mm colonies)	-	ND	+	-	-	ND	ND	ND	ND	ND	ND	+	ND	-	+	+	V	+	V	+	*Staphylococcus* spp. n: 48
Negative Control (Sterilized 0.9% NaCl)	-	-	-	-	-	-	-	-	-	-		-	-	-	-	-	-	-	-	-	-

Key: GP = Gram Positive; GN = Gram Negative; ONPG: *ortho* -nitrophenyl-, 3-D-galactopyranoside. ND = Not done; + = 90% or more strains are positive; - = 90% or more strains are negative; V = Isolate variability (% 11.3-87.6).

From both fish lines a total of 552 different isolates were identified to the genus level based on morphological and biochemical characteristics and compared with Bergey›s Manual of Determinative Bacteriology, Ninth Edition (2000). The total count of named viable bacteria was indicated in terms of colony forming units per mg (cfu/mg).

Antibiotic Susceptibility Testing of Isolated Strains

Two identified colonies of *acinetobacter* spp., *Pseudomonas* spp., *Pseudomonas fluorescens*, *Serratia* spp., *Vibrio* spp. and *Psychrobacter* spp. were randomly selected from each individual fish. All coded bacterial colonies were identified from the hindgut, with the exception of *Acinetobacter* spp. that was only sampled from the foregut. Antimicrobial susceptibility of a total of 144 isolates was tested against tetracycline (Tet 80 µg), sulphonamide (Sul 240 µg), trimethoprim (Trim 5.2 µg), streptomycin (Str 100 µg), gentamicin (Gt 240 µg), nalidixic acid (Nal 130 µg) and chloramphenicol (Chl 60 µg) by the disc diffusion method [Neo-Sensitabs, Rosco, Taastrup, Denmark]. Inhibition zones were measured and resistance categorization was carried out according to Minimal Inhibitory Concentration (MIC) break points of the Norwegian AFA group (2005, 2006), described in ‹User›s guide Neo-Sensitabs®› [http://www.rosco.dk]. Intermediate zones were recorded as resistant.

DNA isolation, 16S rDNA Amplification

DNA extraction was performed from a loop-full of overnight growth, identified from 18 multidrug resistant (≥ 3 drugs), Gram negative isolates, including two representatives from a biochemically unclassified Gram negative bacteria group, using the DNAeasy Blood & Tissue Kit® [Qiagen S.A., France], according to the manufacturers guidelines. Following nucleic acid purification, the 5› part of the 16S rRNA gene (corresponding to *Escherichia coli* positions 10 to 806) was amplified using primers V1 [5›-AGA GTT TGA TCA TGG CTC AGA] and V3 [5›-GGT TAC CTT GTT ACG ACT TC]. Briefly, cycling parameters included an initial denaturizing step for 3 min at 94°C; 30 cycles of 30 s at 94°C, 30 s at 56°C, and 2 min at 72°C; and a final extension for 10 min at 72°C. Two µl of the DNA extract was used for amplification in a total volume of 25 µl containing 2.5 µl 10 × PCR Buffer, 1 µl 50 mM $MgCl_2$, 1 µl 10 mM dNTP mix, 5 pMol each of forward and reverse primers, 1 U of Taq DNA polymerase [Fermentas, Vilnius, Lithuania] and 16.4 µl sterile nuclease-free PCR grade water. DNA amplifications were performed in a GeneAmp 9700 PCR system thermocycler [Applied Biosystems, Foster City, USA]. Positive (*Escherichia coli* NVH 1067/03) and negative control (sterilized dH_2O) samples were included in all amplifications. The PCR products were analyzed on 1.5% agarose gel stained with SYBR Safe® DNA Gel Stain [Invitrogen]. After electrophoresis at 100 V for 60 min, DNA bands were visualized by a Gel Doc™ XR+ Imaging System [Bio-Rad, Hercules, CA].

Sequence Analysis

Amplicons were purified by using QIAquick PCR Purification Kit® [Qiagen S.A., France] and sequenced by GATC laboratories [Konstanz, Germany] with V1 and V3 primers. Afterwards 16S rRNA gene sequences were compared with those available in the GenBank, EMBL, and DDBJ databases using a two-step procedure. A first search was performed with the FASTA algorithm of the Wisconsin GCG program package [44]. All positions showing differences to the best-scoring reference sequence were visually inspected in the electropherogram, and the sequence was corrected manually if necessary as previously described [45]. Thereafter, a second search was done using BLASTN. Undetermined nucleotides (designated N) in either the determined sequence or the reference sequence were counted as matches. All derived sequences have been submitted to GenBank, with accession numbers listed in [Table 2].

Table 2: The GenBank accession numbers of partial 16s rRNA gene sequences of intestinal MDR Gram negative triploid and diploid Atlantic salmon isolates.

Fish	Source	Drug Resistance	Strain and Gene	GenBank accession number
Diploid	Foregut	Chl, Nal, Tet	*Acinetobacter* sp. Cantas1,16S ribosomal RNA gene, partial sequence	JN609530
Diploid	Foregut	Nal, Str, Tet, Trm	*Acinetobacter* sp. Cantas2 16S ribosomal RNA gene, partial sequence	JN609531
Diploid	Foregut	Nal, Tet, Trm	*Acinetobacter* sp. Cantas3 16S ribosomal RNA gene, partial sequence	JN609532
Triploid	Foregut	Nal, Str, Tet. Trm	*Acinetobacter* sp. Cantas4 16S ribosomal RNA gene, partial sequence	JN609533
Triploid	Foregut	Str, Sul, Tet	*Acinetobacter* sp. Cantas5 16S ribosomal RNA gene, partial sequence	JN609534
Triploid	Foregut	Nal, Sul, Tet	*Acinetobacter* sp. Cantas616S ribosomal RNA gene, partial sequence	JN609535
Triploid	Foregut	Str, Tet. Trm	*Acinetobacter* sp. Cantas7 16S ribosomal RNA gene, partial sequence	JN609536
Diploid	Hindgut	Sul, Tet, Trm	*Pseudomonas* sp. Cantas8 16S ribosomal RNA gene, partial sequence	JN609537
Diploid	Hindgut	Nal, Sul, Tet	*Pseudomonas* sp. Cantas9 16S ribosomal RNA gene, partial sequence	JN609538
Triploid	Hindgut	Sul, Tet, Trm	*Pseudomonas sp.* Cantas10 16S ribosomal RNA gene, partial sequence	JN609539
Triploid	Hindgut	Sul. Tet, Trm	*Pseudomonas* sp. Cantas11 16S ribosomal RNA gene, partial sequence	JN609540
Triploid	Hindgut	Nal, Sul Tet, Trm	*Pseudomonas* sp. Cantas12 16S ribosomal RNA gene, partial sequence	JN609541
Triploid	Hindgut	Nal, Tet, Trm	*Pseudomonas* sp. Cantas13 16S ribosomal RNA gene, partial strain	JN609542

Diploid	Hindgut	Sul, Tet, Tm	*Pseudomonas fluorescens* Cantas14 16S ribosomal RNA gene, partial sequence	JN609543
Triploid	Hindgut	Nal, Tet, Sul	*Pseudomonas fluorescens* Cantas15 16S ribosomal RNA gene, partial sequence	JN609544
Diploid	Hindgut	Tet, Str, Sul	*Psychrobacter* sp. Cantas16 16S ribosomal RNA gene, partial sequence	JN609545
Triploid	Hindgut	Tet, Trm, Sul	*Psychrobacter* sp. Cantas17 16S ribosomal RNA gene, partial sequence	JN609546
Triploid	Hindgut	Chl, Str, Tet, Trim	*Serratia* sp. Cantas18 16S ribosomal RNA gene, partial sequence	JN609547

Statistical Analysis

GI Microflora

The data was recorded in a Microsoft Office Excel 2003® spread sheet and then transferred to JMP® 8.0 Statistical Discovery Software from SAS for statistical analysis. After descriptive analyses of data using standard summary measures and graphical representation, two least-squares models were established, with total bacterial count and individual bacterial species count as the outcomes. First, the effect of ploidy (ordinal, diploid versus triploid) and gut section (ordinal, foregut, midgut and hindgut) on total bacteria count was tested. Secondly, a similar model was established for individual bacterial species. The model gave a standard analysis of variance table and the corresponding regression coefficients. Model fit was assessed using graphical techniques, plotting actual versus predicted values, and residuals were assessed using the normal quantile plot. To investigate for possible tank effects, parametric and non-parametric tests were used with i) the tank considered the unit of measure (n = 3/ploidy) and ii) the tank included within a nested model design (with individual fish the unit of measure, n = 6/ploidy). No tank effects were observed, with the results (the effect of ploidy on gut microbiota) remaining the same whether tank was included in the model or not (data not shown). Graphs were made in Microsoft Word Excel and JMP®. Results were considered significant at $P < 0.05$.

Antibacterial Resistance

Antibiotic sensitivity records for each isolate at ploidy level were coded in a Microsoft Excel 2003® spread sheet and the mean antibiotic resistance displayed as a histogram. For further statistical analysis, all data were transferred to Stata [Stata SE/10 for Windows, Stata Corp., College Station, TX]. For each antibiotic, an ordered linear logistic regression model analysis was built and odds ratio and 95% confidence intervals detected.

RESULTS

Fish growth was not affected by the triploid treatment (data not shown), and all fish appeared healthy with no gross deformities or lesions at the macroscopic level within the gut after visual inspection. Furthermore, by naked eye, there was no discernable difference in the amount of digesta within the gut of individual fish.

Fish Rearing Environment

Bacterial counts in the food source included 3.7×10^4 cfu/g *Staphylococcus* spp., 5.5×10^3 cfu/g *Bacillus* spp., 1.1×10^3 cfu/g *Pseudomonas* spp., and 4.8×10^2 cfu/g *Carnobacterium* spp. Bacterial counts from the system inflow water contained 8.0×10^5 cfu/ml *Pseudomonas* spp., 2.0×10^3 cfu/ml *P. fluorescens*, and 1.9×10^2cfu/ml *Psychrobacter* spp. In the tank water, the culturable bacteria consisted of a mixture of intestinal and feed bacterial flora, where *Bacillus* spp. and *Vibrio* spp. dominated, with differences between the tanks for each ploidy, triploids having 10^2 - 10^3 cfu/ml more *Bacillus* spp. and *Vibrio* spp. respectively, than diploids. For the other bacteria present, *Acinetobacter* spp., *Carnobacterium* spp., *Psychrobacter* spp., *Pectobacterium carotovorum*, *Arthrobacte*r spp., and *Pseudomonas* spp. ranging from 10^3 - 10^5 cfu/ml, there was no difference between the tanks of each ploidy.

GI Microflora

On the basis of physiological and biochemical identification 10 different bacterial genera were isolated and characterized within the GI tract of each ploidy and these corresponded to the genetic analysis [Tables 1 and 2]. There was no effect of ploidy on the bacterial diversity of the GI tract, or from which gut section the individual species were isolated [Figure 1]. No bacteria dominated the bacterial counts either as a percentage of the whole gut, or within specific gut sections, with *Pseudomonas* spp., *P. fluorescens*, *Pectobacterium carotovorum*, *Psychrobacter* spp., *Arthrobacter* spp., *Staphylococcus* spp., *Bacillus* spp., and *Carnobacterium*spp., each being between 8 and 13% of the total GI bacterial counts. There was a statistical significance in level of total bacterial counts from the foregut to the hindgut in both ploidy; however triploids had significantly greater total bacterial counts within each gut section compared to diploids [Figure 2]. Triploids had on average 7.3% more bacteria within the whole gut than diploids (max diploid and min triploid value 102.1 and 105.8 cfu/ml respectively), with the largest increase observed within the foregut (10%).

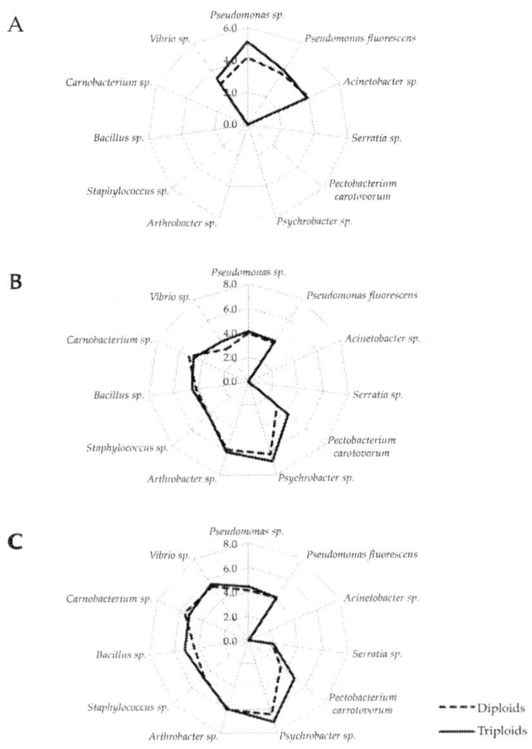

Figure 1: Spider graphs showing the mean bacterial levels (CFU/mg) within (A) foregut, (B) midgut, and (C) hindgut of juvenile triploid and diploid Atlantic salmon.

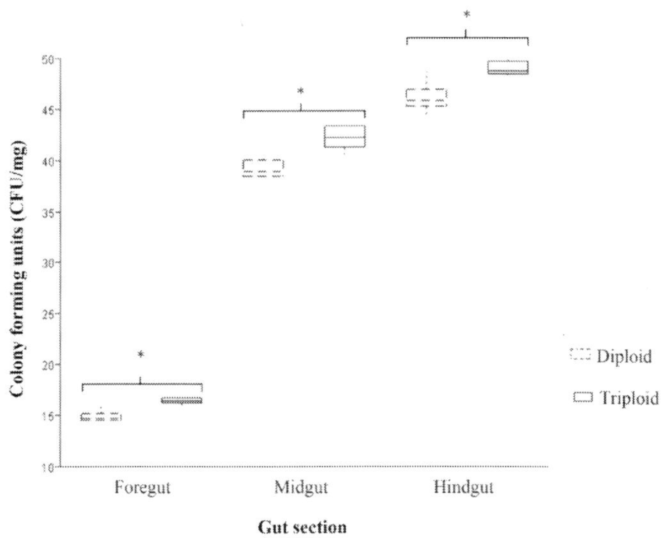

Figure 2: Box plot showing the total bacteria counts within the different gut sections of triploid and diploid juvenile Atlantic salmon. There was a significant difference ($p < 0.05$) in colony forming units (CFU) between all gut sections. Compared to diploids, triploid salmon had a statistically significantly ($p < 0.05$, indicated by an asterisk) greater number of bacteria within each gut section.

Triploidy had a significant effect on a number of the individual bacterial species isolated [Table 3]. Significantly increased levels of *Pseudomonas* sp., *Pectobacterium carotovorum*, *Psychrobacter* spp., *Bacillus* spp., and *Vibrio* spp., were observed in triploid fish compared to diploids (triploids having on average 12, 42, 9, 10, and 11% more bacteria than diploids, respectively), whereas a decrease in *Carnobacterium* spp., within triploids compared to diploids was close to significant ($P = 0.1$, diploids having on average 8% more bacteria than triploids). Where significant increases in bacteria were observed between the ploidies, the increase was not always universal along the whole gut. The increase in *Pseudomonas* spp., and *Vibrio* spp., over diploid counterparts was more pronounced in the forgut and midgut respectively, compared to *Pectobacterium carotovorum*, *Psychrobacter* spp., and *Bacillus* spp., where there was a similar increase in triploid bacterial populations along the entire gut length [Figure 1].

Table 3: *P* values for the effect of ploidy and gut section on bacterial counts within the gastrointestinal tract of juvenile Atlantic salmon.

Bacteria	Gut section	Ploidy
Pseudomonas spp.	0.0043	0.0013
Pseudomonas fluorescens	0.0246	0.1522
Acinetobacter spp.	0.0001	0.5552
Serratia spp.	0.0001	0.6727
Pectobacterium carotovorum	0.0001	0.0001
Psychrobacter spp.	0.0001	0.0150
Arthrobacter spp.	0.0001	0.4881
Staphylococcus spp.	0.0001	0.4250
Bacillus spp.	0.0001	0.0005
Carnobacterium spp.	0.0001	0.0989
Vibrio spp.	0.0001	0.0001

Antibacterial Resistance

No ploidy effect was observed on bacterial multiple drug resistance, or the numbers of antibiotics the bacteria were resistant to. For diploid and triploid isolated bacteria respectively, the frequency of resistant bacteria to two antibacterials was 16 and 20.8% with 9.7 and 13.2% of those resistant to more than two antibacterials. All strains were susceptible to gentamicin, whilst the least resistance was demonstrated against chloramphenicol which inhibited more than 95% of the strains.

For all the antibiotics where resistance was observed, there was a general trend of increased resistance in isolates from triploid fish compared to isolates from diploid fish [Figure 3]. Statistically, the presence of tetracycline (OR = 1.117; 95% CI = 0.52-2.37), trimethoprim (OR = 1.609; 95% CI = 0.71-3.64) and sulphonamide (OR = 1.46; 95% CI = 0.65-3.26) resistance was significantly higher in triploid fish intestinal Gram negative isolates compared to their diploid counterparts, whereas no significant differences related to ploidy were observed for streptomycin (OR = 0.496; 95% CI = 0.38-7.25), nalidixic acid (OR = 0.97; 95% CI = 0.29-3.16), or chloramphenicol (OR = 0.97; 95% CI = 0.23-7.09).

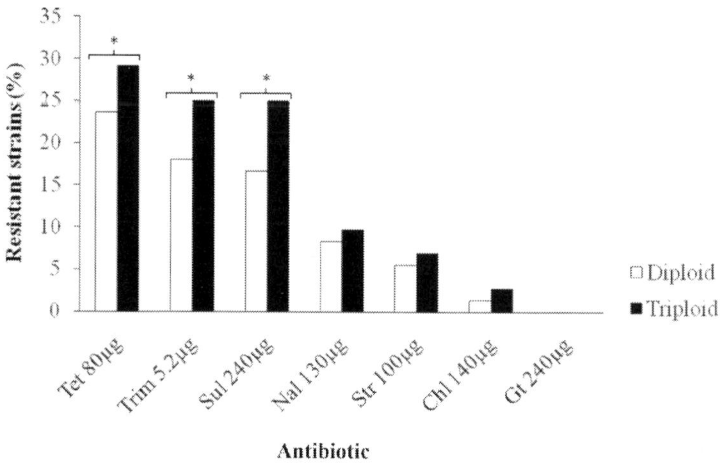

Figure 3: The average frequency distributions of antibiotic resistant gram negative isolates (N = 144) among triploid and diploid fish. There was a trend of increased antibacterial resistance from bacteria isolated from triploids compared to those bacteria isolated from diploids, this was statistically significant ($p < 0.05$, indicated by an asterisk) for some antibiotics.

DISCUSSION

To our knowledge, this is the first study into the intestinal microbiota of triploid Atlantic salmon, and the first into any species of triploid fish. The objective was to compare the GI microflora and the antibiotic resistance profiles of the intestinal microflora of Atlantic salmon after chromosome set manipulation (triploidy) to see if it resulted in an altered physiological profile. No differences were observed in the diversity of bacteria isolated from the gut of triploid or diploid fish, however, triploid fish were found to contain greater total bacterial counts within each gut section [Figure 2], although this was not due to a general increase in all the isolated bacteria [Table 3]. For specific bacteria, triploids were found to contain significantly more *Pseudomonas* sp., *Pectobacterium carotovorum*, *Psychrobacter* spp., *Bacillus* spp., and *Vibrio* spp., whereas *Carnobacterium*spp. was close to being significantly lower in triploids compared to diploids. These results suggest the physiological differences associated with triploidy have an influence on the gut microbiota.

Bacterial culturing and 16S rRNA PCR were used in this study to identify and quantify the bacteria within the fish gut. It is acknowledged that traditional culturing may only identify 11-50% of the present bacteria [46, 47]. Additionally, competitive differences between the bacteria on the culture medium compared to the fish gut may lead to misrepresentation of the microbiota community [42], and the media used in this study, blood agar and Brocalin agar, are not traditionally used in fish studies and may not have supported growth of *Lactobacillus* spp., previously found to be dominant in the GI tract of Atlantic salmon [48]. Metagenomic analysis may have led to a more realistic representation of the GI microflora in this study, however, this method is relatively underdeveloped in Atlantic salmon and not comparable to the existing literature that uses methods that are still routinely, and historically, used in the field of microbiology. As such, previously identified increases in the level of bacteria from the foregut to the hindgut were observed in this study [23, 42], along with the appearance of certain bacteria inhabiting the different gut sections within salmonids [20, 42, 49]. Therefore, the authors believe the results of this study are reliable for making an accurate comparison related to ploidy within this study.

Differences were observed in the culturable bacteria levels between the triploid and diploid tank water. Increases in *Bacillus* spp. and *Vibrio* spp. within triploid tanks were consistent compared to diploid tanks, however the levels of the other isolated bacteria were similar. All tanks were supplied with the same inflow water that contained neither *Vibrio* spp., nor *Bacillus* spp., however, feed did contain *Bacillus* spp. Therefore, increases in *Bacillus* spp. could be explained by an increase in uneaten feed in the triploid tanks; however, feed intake was not assessed and uneaten feed was removed by the self-cleaning design of the tank, and this would not explain the increase in *Vibrio* spp. Alternatively, the increases in *Vibrio* spp. and *Bacillus* spp., could be explained by the release of allochthonous bacteria that are only transient in the gut and are released over time in faeces. Therefore the increase in *Bacillus* spp., and *Vibrio* spp., in triploid compared to diploid tank water could be expected due to the increases in these two bacteria observed within the triploid gut. However, this was not the case for *Pseudomonas* spp., *Pectobacterium carotovorum*, and *Psychrobacter* spp. that were found in equal quantities in tank water from both ploidies, despite being found to have increased numbers within the triploid gut. The reasons for this are unknown, but may be related to the ability of different bacterial species to adhere to the intestinal mucosa as part of the autochthonous flora and therefore not be released in such great quantities in the faeces, or the ability of these bacteria to survive and compete within the tank water.

Limitations in our study include no control for the effect of pressure treatment independent of ploidy, and the effect of genetic variation within the study fish. Firstly, the current evidence would suggest that triploid induction procedures (when optimized) do not influence triploid performance [12, 13, 16], although Malison et al. [50] concluded otherwise, this study failed to have an adequate control. With regards to the genetic composition of the study population, it is recognized family effects are observed on triploid survival and physiological performance [51] and this may influence gut microbiota, however we did not test for this. Therefore, the study should be repeated with an emphasis on the family effect on gut microbiota in fish of both ploidy.

Previously, 10 of the 11 isolated bacteria have been recorded in Atlantic salmon [27, 42, 52, 53]. In addition, the levels of bacteria isolated in this study agree with those previously found in Atlantic salmon [42]. However, we could not find any previous report of *Pectobacterium carotovorum* subsp. *carotovorum*, previously isolated from Chinese cabbage [Zhang et al. 2008, unpublished data], in Atlantic salmon. Its source and biological significance in this study are unknown.

Our results suggest the triploid gut provides a different environment to that of the diploid gut, based on bacterial levels. Unfortunately, no literature exists on the triploid gut, or triploid physiology that may affect bacterial growth, therefore some of the major factors that influence gut microbiota cannot be discussed (i.e. gut mucous, physiochemical environment) [52, 54, 55]. Although no significant differences on weight, length, or condition factor were seen within this study, significant differences were seen in the growth (triploids were heavier and longer), condition factors (triploids had a lower value), and the hepatosomatic index (higher in triploids) in pre-smolts. Previously, differences in gut microbiota have been identified between fast and slow growing members of fish from the same population [25], and may explain our results. Alternatively, we also observed differences in the leucocyte composition (neutrophilic and B-cells proportions) in the same study fish three weeks after seawater transfer (Fraser et al, in preparation). It is well established that the gut microbiota influences the development of the immune system in mammals [56] and evidence suggests the same occurs in fish [22]. Therefore, it is possible that differences in the immune system related to ploidy (i.e. leucocyte proportions) are influencing the size and composition of the gut microbiota through interactions with mucosal immune cells, or the mechanisms by which the microbiota influences the immune development in fish may be altered by the ploidy status. Cortisol levels and female sex hormone levels have been found to be lower in triploids compared to diploids [18, 57, 58], and these may also influence the immune system in fish [59, 60, 61, 62, 63, 64, 65]. However, the majority of studies have observed no differences in cortisol [18, 62, 63, 64, 65] due to ploidy and, up until the time of maturing sex hormone levels are typically below detectable levels in salmon and therefore unlikely to influence the immune system in this study. A final explanation is that of differences in cell size and number relating to ploidy. It is well established that cell size increases with increasing numbers of chromosome sets [6]. This phenomenon is more apparent in cells where the nucleus occupies a greater proportion of the total cell volume, so may have little effect on some intestinal cells (i.e. enterocytes). However, triploids have organs of equal sizes to diploids due to reductions in cell number [6, 7], and it may be that triploids have fewer cells along the gastrointestinal tract, such as goblet cells, altering the conditions within the gut. Therefore, differences in the ontogenetic growth, the immune system, or gut environment in the triploids may explain our results.

A difference in the gut microbiota between the ploidies could have a number of implications on triploid aquaculture practices. For instance, *Vibrio* spp., are known to be opportunistic pathogens and causative agents of disease and mass mortalities [26], *Pseudomonas* spp., *Bacillus* spp, and *Carnobacterium* spp., are capable of inhibiting pathogenic bacteria and improving survival *in vivo* after challenge tests [47, 66], *Psychrobacter* spp., and *Arthrobacter agillis* have been linked to enteritis in Atlantic salmon [27], and *Bacillus*spp., and *Carnobacterium* spp., are commonly used in fish pro- and pre-biotics [23, 30]. These results suggest that the susceptibility of triploids to opportunistic pathogens (important in both fish and human health) or potential probiotic treatments may be increased in comparison to diploids. Indeed, several studies have reported triploids to be more prone to disease than diploids [12, 13, 14, 15].

Despite no known antibiotic exposure within our study population, we detected a variety of drug resistance patterns within bacterial isolates. Previous studies have reported antibacterial drug resistance in various aquatic environments [67, 68, 69] at levels higher than those found within our study, possibly as a result of antibiotic use that exerts an ecological pressure on bacteria [70]. However, increased levels of antibiotic resistant bacteria, over that of the local environment, have been reported in aquaculture facilities that are not utilizing antibiotics [39], and the reasons for this remain unclear.

Triploids were found to have consistently higher levels of antibacterial resistance in their bacterial isolates than in diploids; with 3 of the 6 effective antibacterials significantly affected by ploidy [Figure 3] but no ploidy effect was seen on the occurrence of multiple drug resistance. We suggest two possible explanations for the first observation i) possible physiological differences between the two ploidies (as discussed above) impacting upon the host-bacterium interaction, and ii) the significantly greater levels of bacteria found within the triploid compared to diploid intestine might provide better conditions for more efficient bacterial colonization, which is very critical for horizontal gene transfer [71]. Unfortunately, both of these hypotheses were outside the scope of this study.

CONCLUSIONS

This study would suggest the altered physiology of triploid fish affects the numbers of microbiota within GI tract and their drug resistance profiles. The observation of altered drug resistance profiles between triploid and diploid fish would also make for an interesting follow up study to determine the factors that cause such a phenomenon. It would also be of interest to see how the observed differences in bacterial communities are affected by fish age, genotype, and rearing environment, as these factors are known to alter the microbiota of the GI tract in diploids. It has also been observed that triploids can demonstrate altered growth patterns compared to diploids, such as delayed time of first feeding and slower growth in the juvenile phase, and the effects on/of the GI microbiota may be of interest.

NOTES

Leon Cantas, Thomas WK Fraser contributed equally to this work.

DECLARATIONS

Acknowledgements

The authors thank the Norwegian School of Veterinary Science (NSVS) for covering the running costs involved in the microbiology and the NSVS and UFE for providing travel money. Eystein Skjerve and Erik Ropstad are thanked for their help with statistical analysis. Daniel Merrifield is thanked for general discussions around the topic and comments on the manuscript. This study was funded by the EC project Salmotrip (FP7, 222115).

Authors' Contributions

LC contributed to the establishment of the collaborations, design, performed the sampling and all traditional bacterial culturing, molecular genetic studies, statistical analysis, data collection, drafting and writing of the manuscript. TF contributed to the establishment of the collaborations, design, data collection, data analysis, and drafting and writing the manuscript. P-GF contributed to funding, design, acquisition of the fish, and writing the manuscript. IM contributed to acquisition of funds, the supervision, and drafting of the manuscript. HS contributed to acquisition of funds, the supervision, and drafting and writing the manuscript. All authors read and approved the final manuscript.

REFERENCES

1. Piferrer F, Beaumont A, Falguiere JC, Flajshans M, Haffray P, Colombo L: Polyploid fish and shellfish: production, biology and applications to aquaculture for performance improvement and genetic containment. Aquaculture. 2009, 293: 125-156. 10.1016/j.aquaculture.2009.04.036.
2. McGinnity P, Prodohl P, Ferguson K, Hynes R, O'Maoileidigh N, Baker N, Cotter D, O'Hea B, Cooke D, Rogan G, Taggart J, Cross T: Fitness reduction and potential extinction of wild populations

of Atlantic salmon, *Salmo salar*, as a result of interactions with escaped farm salmon. Proceedings of the Royal Society of London Series B-Biological Sciences. 2003, 270: 2443-2450. 10.1098/rspb.2003.2520.

3. Taranger GL, Carillo M, Schulz RW, Fontaine P, Zanuy S, Felip A, Weltzien FA, Dufour S, Karlsen O, Norberg B, Andersson E, Hausen T: Control of pubert in farmed fish. General and Cpmperative Endoctinology. 2010, 165: 483-515. 10.1016/j.ygcen.2009.05.004.
4. Powell MD, Jones MA, Lijalad M: Effects of skeletal deformities on swimming performance and recovery from exhaustive exercise in triploid Atlantic salmon. Diseases of Aquatic Organisms. 2009, 85: 59-66.
5. Benfey TJ: Use of sterile triploid Atlantic salmon (*Salmo salar* L.) for aquaculture in New Brunswick, Canada. ICES Journal of Marine Science. 2001, 58: 525-529. 10.1006/jmsc.2000.1019.
6. Small SA, Benfey TJ: Cell-size in triploid salmon. Journal of Experimental Zoology. 1987, 241: 339-342. 10.1002/jez.1402410309.
7. Aliah RS, Yamaoka K, Inada Y, Taniguchi N: Effects of triploidy on tissue structure of some organs in ayu. Nippon Suisan Gakkaishi. 1990, 56: 569-575. 10.2331/suisan.56.569.
8. Benfey TJ, Dye HM, Solar II, Donaldson EM: The growth and reproductive endocrinology of adult triploid Pacific salmonids. Fish Physiology and Biochemistry. 1989, 6: 113-120. 10.1007/BF01875483.
9. Bernier NJ, Brauner CJ, Heath JW, Randall DJ: Oxygen and carbon dioxide transport during sustained exercise in diploid and triploid chinook salmon (*Oncorhynchus tshawytscha*). Canadian Journal of Fisheries and Aquatic Sciences. 2004, 61: 1797-1805. 10.1139/f04-110.
10. Atkins ME, Benfey TJ: Effect of acclimation temperature on routine metabolic rate in triploid salmonids. Comparative Biochemistry and Physiology A-Molecular & Integrative Physiology. 2008, 149: 157-161. 10.1016/j.cbpa.2007.11.004.
11. Fjelldal PG, Lock EJ, Hansen T, Wargelius A, El-Mowafi A, Ørnsrud R: Extra dietary phosphorous prevents vertebral deformities in

triploid Atlantic salmon (*Salmo salar* L.) smolts. Oral presentation at the Interdisciplinary Approach in Fish Skeletal Biology Second Meeting, 26-28 April. 2011, , Tavira, Portugal

12. Myers JM, Hershberger WK: Early growth and survival of heat-shocked and tetraploid-derived triploid rainbow-trout (*Oncorhynchus-mykiss*). Aquaculture. 1991, 96: 97-107. 10.1016/0044-8486(91)90142-T.
13. Blanc JM, Chourrout D, Krieg F: Evaluation of juvenile rainbow trout survival and growth in half-sib families from diploid and tetraploid sires. Aquaculture. 1987, 65: 215-220. 10.1016/0044-8486(87)90233-X.
14. Ojolick EJ, Cusack R, Benfey TJ, Kerr SR: Survival and growth of all-female diploid and triploid rainbow-trout (*Oncorhynchus-mykiss*) reared at chronic high-temperature. Aquaculture. 1995, 131: 177-187. 10.1016/0044-8486(94)00338-O.
15. Ozerov MY, Lumme J, Pakk P, Rintamaki P, Zietara MS, Barskaya Y, Lebedeva D, Saadre E, Gross R, Primmer CR, Vasemagi A: High *Gyrodactylus salaris* infection rate in triploid Atlantic salmon Salmo salar. Diseases of Aquatic Organisms. 2010, 91: 129-136.
16. Galbreath PF, Adams ND, Sherrill LW, Martin TH: Thermal tolerance of diploid versus triploid rainbow trout and brook trout assessed by time to chronic lethal maximum. Environmental Biology of Fishes. 2006, 75: 183-193. 10.1007/s10641-006-0008-2.
17. Benfey TJ, McCabe LE, Pepin P: Critical thermal maxima of diploid and triploid brook charr, *Salvelinus fontinalis*. Environmental Biology of Fishes. 1997, 49: 259-264. 10.1023/A:1007361231325.
18. Benfey TJ, Biron M: Acute stress response in triploid rainbow trout (*Oncorhynchus mykiss*) and brook trout (*Salvelinus fontinalis*). Aquaculture. 2000, 184: 167-176. 10.1016/S0044-8486(99)00314-2.
19. Budino B, Cal RM, Piazzon MC, Lamas J: The activity of several components of the innate immune system in diploid and triploid turbot. Comparative Biochemistry and Physiology A-Molecular & Integrative Physiology. 2006, 145: 108-113. 10.1016/j.cbpa.2006.05.007.
20. Ringø E, Strøm E, Tabachek JA: Intestinal microflora of salmonids:

a review. Aquaculture Research. 1995, 26: 773-789. 10.1111/j.1365-2109.1995.tb00870.x.

21. Sissons JW: Potential of probiotic organisms to prevent diarrhea and promote digestion in farm-animals - a review. Journal of the Science of Food and Agriculture. 1989, 49: 1-13. 10.1002/jsfa.2740490102.
22. Rawls JF, Samuel BS, Gordon JI: Gnotobiotic zebrafish reveal evolutionarily conserved responses to the gut microbiota. Proceedings of the National Academy of Sciences of the United States of America. 2004, 101: 4596-4601. 10.1073/pnas.0400706101.
23. Nayak SK: Role of gastrointestinal microbiota in fish. Aquaculture Research. 2010, 41: 1553-1573. 10.1111/j.1365-2109.2010.02546.x.
24. Olsen RE, Sundell K, Mayhew TM, Myklebust R, Ringo E: Acute stress alters intestinal function of rainbow trout, *Oncorhynchus mykiss* (Walbaum). Aquaculture. 2005, 250: 480-495. 10.1016/j.aquaculture.2005.03.014.
25. Sun Y, Yang H, Ling Z, Chang J, Ye J: Gut microbiota of fast and slow growing grouper *Epinephelus coioides*. African Journal of Microbiology Research. 2009, 3: 713-720.
26. Egidius E: Vibriosis - Pathogenicity and pathology - a review. Aquaculture. 1987, 67: 15-28. 10.1016/0044-8486(87)90004-4.
27. Bakke-McKellep AM, Penn MH, Salas PM, Refstie S, Sperstad S, Landsverk T, Ringo E, Krogdahl A: Effects of dietary soyabean meal, inulin and oxytetracycline on intestinal microbiota and epithelial cell stress, apoptosis and proliferation in the teleost Atlantic salmon (*Salmo salar* L.). British Journal of Nutrition. 2007, 97: 699-713. 10.1017/S0007114507381397.
28. Joborn A, Olsson JC, Westerdahl A, Conway PL, Kjelleberg S: Colonization in the fish intestinal tract and production of inhibitory substances in intestinal mucus and faecal extracts by *Carnobacterium* sp. strain K1. Journal of Fish Diseases. 1997, 20: 383-392. 10.1046/j.1365-2761.1997.00316.x.
29. Robertson PAW, O›Dowd C, Burrells C, Williams P, Austin B: Use of *Carnobacterium* sp as a probiotic for Atlantic salmon (*Salmo salar* L.) and rainbow trout (*Oncorhynchus mykiss*, Walbaum). Aquaculture. 2000, 185: 235-243. 10.1016/S0044-8486(99)00349-X.

30. Merrifield DL, Dimitroglou A, Foey A, Davies SJ, Baker RTM, Bogwald J, Castex M, Ringo E: The current status and future focus of probiotic and prebiotic applications for salmonids. Aquaculture. 2010, 302: 1-18. 10.1016/j.aquaculture.2010.02.007.
31. Cabello FC: Heavy use of prophylactic antibiotics in aquaculture: a growing problem for human and animal health and for the environment. Environmental Microbiology. 2006, 8: 1137-1144. 10.1111/j.1462-2920.2006.01054.x.
32. Barton MD: Antibiotic use in animal feed and its impact on human health. Nutrition Research Reviews. 2000, 13: 279-299. 10.1079/095442200108729106.
33. Sørum H: Antimicrobial drug resistance in fish pathogens. ASM Pres. 2006, 213-237.
34. Kruse H, Sørum H: Transfer of multiple-drug resistance plasmids between bacteria of diverse origins in natural microenvironments. Applied and Environmental Microbiology. 1994, 60: 4015-4021.
35. Rhodes G, Huys G, Swings J, McGann P, Hiney M, Smith P, Pickup RW: Distribution of oxytetracycline resistance plasmids between aeromonads in hospital and aquaculture environments: Implication of Tn1721 in dissemination of the tetracycline resistance determinant Tet A. Applied and Environmental Microbiology. 2000, 66: 3883-3890. 10.1128/AEM.66.9.3883-3890.2000.
36. Björklund H, Rabergh CMI, Bylund G: Residues of oxolinic acid and oxytetracycline in fish and sediments from fish farms. Aquaculture. 1991, 97: 85-96. 10.1016/0044-8486(91)90281-B.
37. McPhearson RM, Depaola A, Zywno SR, Motes ML, Guarino AM: Antibiotic-resistance in Gram-negative bacteria from cultured catfish and aquaculture ponds. Aquaculture. 1991, 99: 203-211. 10.1016/0044-8486(91)90241-X.
38. Miranda CD, Zemelman R: Antibiotic resistant bacteria in fish from the Concepcion Bay, Chile. Marine Pollution Bulletin. 2001, 42: 1096-1102. 10.1016/S0025-326X(01)00093-5.
39. Tamminen M, Karkman A, Lohmus A, Muziasari WI, Takasu H, Wada S, Suzuki S, Virta M: Tetracycline resistance genes persist at aquaculture farms in the absence of selection pressure. Environmental Sciene and Technology. 2011, 45: 386-391. 10.1021/es102725n.

40. Benfey TJ, Sutterlin AM: Triploidy induced by heat-shock and hydrostatic-pressure in landlocked Atlantic salmon (*Salmo-salar* L). Aquaculture. 1984, 36: 359-367. 10.1016/0044-8486(84)90328-4.

41. Benfey TJ, Sutterlin AM, Thompson RJ: Use of erythrocyte measurements to identify triploid salmonids. Canadian Journal of Fisheries and Aquatic Sciences. 1984, 41: 980-984. 10.1139/f84-112.

42. Hovda MB, Lunestad BT, Fontanillas R, Rosnes JT: Molecular characterisation of the intestinal microbiota of farmed Atlantic salmon (*Salmo salar* L.). Aquaculture. 2007, 272: 581-588. 10.1016/j.aquaculture.2007.08.045.

43. Hucker GJ, Conn HJ: Methods of Gram staining. NY State Agr Exp Sta Tech Bull. 1923, No. 93

44. Devereux J, Haeberli P, Smithies O: A comprehensive set of sequence analysis programs for the VAX. Nucleic Acids Res. 1984, 12: 387-395. 10.1093/nar/12.1Part1.387.

45. Bosshard PP, Zbinden R, Abels S, Böddinghaus B, Altwegg M, Böttger EC: 16S rRNA gene sequencing versus the API 20 NE system and the VITEK 2 ID-GNB card for identification of nonfermenting Gram-negative bacteria in the clinical laboratory. Journal of Clinical Microbiology. 2006, 44: 1359-1366. 10.1128/JCM.44.4.1359-1366.2006.

46. Huber I, Spanggaard B, Rossen KF, Nielsen T, Gram L: Phylogenetic analysis and in situ identification of the intestinal microbial community of rainbow trout (*Oncorhynchus mykiss*, Walbaum). Journal of Applied Microbiology. 2004, 96: 117-132. 10.1046/j.1365-2672.2003.02109.x.

47. Spanggaard B, Huber I, Nielsen J, Nielsen T, Appel KF, Gram L: The microflora of rainbow trout intestine: a comparison of traditional and molecular identification. Aquaculture. 2000, 182: 1-15. 10.1016/S0044-8486(99)00250-1.

48. Strøm E: Melkesyrebakterier i fisketarm. Isolasjon, karakterisering og egenskaper. M.Sci Thesis, The Norwegian College of Fishery Science 88. 1988

49. Ringø E: The effect of chromic oxide (Cr2O3) on aerobic bacterial-

populations associated with the intestinal epithelial mucosa of Arctic charr, *Salvelinus-alpinus* (L). Canadian Journal of Microbiology. 1993, 39: 1169-1173. 10.1139/m93-177.

50. Malison JA, Procarione LS, Held JA, Kayes TB, Amundson CH: The influence of triploidy and heat and hydrostatic-pressure shocks on the growth and reproductive development of juvenile yellow perch (*Perca-flavescens*). Aquaculture. 1993, 116: 121-133. 10.1016/0044-8486(93)90003-H.
51. Johnson RM, Shrimpton JM, Heath JW, Heath DD: Family, induction methodology and interaction effects on the performance of diploid and triploid chinook salmon (*Oncorhynchus tshawytscha*). Aquaculture. 2004, 234: 123-142. 10.1016/j.aquaculture.2004.01.024.
52. Navarrete P, Espejo RT, Romero J: Molecular analysis of microbiota along the digestive tract of juvenile Atlantic salmon (*Salmo salar* L.). Microbial Ecology. 2009, 57: 550-561. 10.1007/s00248-008-9448-x.
53. Ringø E, Sperstad S, Kraugerud OF, Krogdahl A: Use of 16S rRNA gene sequencing analysis to characterize culturable intestinal bacteria in Atlantic salmon (*Salmo salar*) fed diets with cellulose or non-starch polysaccharides from soy. Aquaculture Research. 2008, 39: 1087-1100. 10.1111/j.1365-2109.2008.01972.x.
54. Kim DH, Brunt J, Austin B: Microbial diversity of intestinal contents and mucus in rainbow trout (*Oncorhynchus mykiss*). J Appl Microbiol. 2007, 102: 1654-1664. 10.1111/j.1365-2672.2006.03185.x.
55. Nikolopoulou D, Moutou KA, Fountoulaki E, Venou B, Adamidou S, Alexis MN: Patterns of gastric evacuation, digesta chatacretistics and pH changes along the gastrointestinal tract of gilthead sea bream (*Sparus aurata* L.) and European sea bass (*Dicentrarchus labrax* L.): a comparative study. Comp Biochem Fisiol. 2010, 4: 406-414.
56. Round JL, Mazmanian SK: The gut microbiota shapes intestinal immune responses during health and disease. Nature Reviews Immunology. 2009, 9: 313-323. 10.1038/nri2515.
57. Garner SR, Madison BN, Bernier NJ, Neff BD: Juvenile growth and aggression in diploid and triploid chinook salmon *Oncorhynchus tshawytscha* (Walbaum). Journal of Fish Biology. 2008, 73: 169-185. 10.1111/j.1095-8649.2008.01923.x.

58. Peruzzi S, Varsamos S, Chatain B, Fauvel C, Menu B, Falguiere JC, Severe A, Flik G: Haematological and physiological characteristics of diploid and triploid sea bass, *Dicentrarchus labrax* L. Aquaculture. 2005, 244: 359-367. 10.1016/j.aquaculture.2004.11.028.

59. Aaltonen TM, Jokinen EI, Lappivaara J, Markkula SE, Salo HM, Leppanen H, Lammi R: Effects of primary- and secondary-treated bleached kraft mill effluents on the immune system and physiological parameters of roach. Aquatic Toxicology. 2000, 51: 55-67. 10.1016/S0166-445X(00)00101-6.

60. Pickering AD, Pottinger TG: Stress responses and disease resistance in salmonid fish - effects of chronic elevation of plasma-cortisol. Fish Physiology and Biochemistry. 1989, 7: 253-258. 10.1007/BF00004714.

61. Pickering AD, Pottinger TG, Carragher J, Sumpter JP: The effects of acute and chronic stress on the levels of reproductive hormones in the plasma of mature male brown trout, *Salmo-trutta*-l. General and Comparative Endocrinology. 1987, 68: 249-259. 10.1016/0016-6480(87)90036-0.

62. Biron M, Benfey TJ: Cortisol, glucose and hematocrit changes during acute stress, cohort sampling, and the diel cycle in diploid and triploid brook trout (*Salvelinus-fontinalis* Mitchill). Fish Physiology and Biochemistry. 1994, 13: 153-160. 10.1007/BF00004340.

63. Sadler J, Pankhurst NW, Pankhurst PM, King H: Physiological stress responses to confinement in diploid and triploid Atlantic salmon. Journal of Fish Biology. 2000, 56: 506-518. 10.1111/j.1095-8649.2000.tb00751.x.

64. Beyea MM, Benfey TJ, Kieffer JD: Hematology and stress physiology of juvenile diploid and triploid shortnose sturgeon (*Acipenser brevirostrum*). Fish Physiology and Biochemistry. 2005, 31: 303-313. 10.1007/s10695-005-1552-y.

65. Leggatt RA, Scheer KW, Afonso LOB, Iwama GK: Triploid and diploid rainbow trout do not differ in their stress response to transportation. North American Journal of Aquaculture. 2006, 68: 1-8. 10.1577/A05-035.1.

66. Moriarty DJW: Control of luminous Vibrio species in penaeid aquaculture ponds. Aquaculture. 1998, 164: 351-358. 10.1016/S0044-8486(98)00199-9.

67. Chelossi E, Vezzulli L, Milano A, Branzoni M, Fabiano M, Riccardi G, Banat IM: Antibiotic resistance of benthic bacteria in fish-farm and control sediments of the Western Mediterranean. Aquaculture. 2003, 219: 83-97. 10.1016/S0044-8486(03)00016-4.

68. Miranda CD, Zemelman R: Bacterial resistance to oxytetracycline in Chilean salmon farming. Aquaculture. 2002, 212: 31-47. 10.1016/S0044-8486(02)00124-2.

69. Schmidt AS, Bruun MS, Dalsgaard I, Pedersen K, Larsen JL: Occurrence of antimicrobial resistance in fish-pathogenic and environmental bacteria associated with four Danish rainbow trout farms. Applied and Environmental Microbiology. 2000, 66: 4908-4915. 10.1128/AEM.66.11.4908-4915.2000.

70. Hawkey PM, Jones AM: The changing epidemiology of resistance. Journal of Antimicrobial Chemotherapy. 2009, 64: 3-10. 10.1093/jac/dkp256.

71. Licht TR, Wilcks A: Conjugative gene transfer in the gastrointestinal environment. Adv Appl Microbiol. 2006, 58: 77-95.

CHAPTER

16

THE FREQUENCY OF SPONTANEOUS TRIPLOIDY IN FARMED ATLANTIC SALMON PRODUCED IN NORWAY DURING THE PERIOD 2007–2014

Kevin A Glover, Abdullah S Madhun, Geir Dahle, Anne G E Sørvik, Vidar Wennevik, Øystein Skaala, H Craig Morton, Tom J Hansen and Per G Fjelldal

Institute of Marine Research, Bergen, Norway

ABSTRACT

Background

Spontaneous triploidy has been reported in a number of fish species, and is often linked with *in vivo* or *in vitro* ageing of eggs post ovulation. Here, we provide the first investigation into the frequency of spontaneous triploidy in farmed Atlantic salmon by analysing more than 4000 fish from 55 farms, and approximately 1000 recaptured escapees, all sampled in the period 2007–2014. In addition, we compare microsatellite genotyping against flow cytometry and red blood cell diameter in a set of 45 putatively diploid and 45 putatively triploid Atlantic salmon.

Results

The three methods implemented for ploidy determination gave consistent results, thus validating the methods used here. Overall, 2.0% spontaneous triploids were observed in salmon sampled on farms. The frequency of spontaneous triploids varied greatly among sea cages (0-28%), but they were observed in similar frequencies among the three primary breeding companies (1.8-2.4%). Spontaneous triploids were observed in all farming regions in Norway, and in all years sampled. Spontaneous triploids were also observed among the escapees recaptured in both the marine environment and in rivers.

Conclusions

Spontaneous triploidy in commercially produced Atlantic salmon is likely to be a result of the practices employed by the industry. For logistical reasons, there is sometimes a pause of hours, and in some cases overnight, between killing the female broodfish, removal of her eggs, and fertilization. This gives the eggs time to age post ovulation, and increases the probability of duplication of the maternal chromosome set by inhibition of the second polar body release after normal meiosis II in the oocyte.

Keywords: Autopolyploidy, Autotriploid, Microsatellite, Breeding, Aquaculture, Escapees genetic

BACKGROUND

The Atlantic salmon (*Salmo salar*) farming industry was first initiated in Norway in the late 1960's, and has now grown to become an economically significant industry in several countries. Current global production exceeds 2 million tons, and primarily involves rearing domesticated strains that have been selected for a range of commercially important traits for up to 10 generations or more [1]. A number of breeding programs for Atlantic salmon exist, for example in Norway [1-3], Scotland [4,5], Atlantic Canada [6], British Colombia Canada [7] and Australia [8]. However, as Norway is the world's largest salmon producing nation, and genetic material from the three primary Norwegian breeders (Aqua Gen

AS, Marine Harvest and Salmobreed AS) has been distributed to fish farms in other regions of the globe (e.g., [9]), farmed salmon originating from Norwegian breeding programs dominate global production.

Each year, thousands or hundreds of thousands of farmed fish escape from their net pens into the wild. While many of these escapees disappear never to be seen again, some return to freshwater and can interbreed with wild salmon [10,11]. As a result of interbreeding, genetic changes in some wild salmon populations have been reported [12-16]. In response to requests from the Norwegian Directorate of Fisheries (NDF), who are responsible for production and implementation of aquaculture regulations in Norway, the Institute of Marine Research (IMR) developed a genetic method to identify the farm of origin for farmed escapees. The method, which was named the "stand-by method", was initially produced for the identification of salmon escapees back to their farm(s) of origin [17]. However, it has subsequently been adapted to identify the farm of origin for Atlantic cod (*Gadus morhua*) [18] and rainbow trout (*Oncorhynchus mykiss*) escapees also [19]. In short, the method works by taking samples of salmon from cages on farms in the vicinity of the recaptured escapees (termed the baseline samples=potential source of the escapees), and matches the multi locus genetic profile of each recaptured escapee to each of the baseline samples using a range of complimentary genetic assignment methods.

In the period 2007–2014, the "stand-by method" was implemented to identify the farm of origin for escaped salmon in >12 episodes [20-25], some of which have resulted in legal cases [22]. In some of these cases, the occasional triploid farmed salmon has been detected based upon its multi locus microsatellite genetic profile (i.e., displayed three alleles at multiple loci). While the Atlantic salmon aquaculture industry is currently conducting research into the potential use of triploid salmon for commercial production [26-29], during the period in which samples from this study were collected, there was either no or next to no commercial production of triploid salmon in Norway, and this only occurred on a very low number of farms. This has been confirmed by the three principal breeding companies in Norway (see acknowledgements). Thus, these observed triploid salmon arose as a result of a spontaneous event and not a deliberate protocol.

Spontaneous triploidy is a phenomena that has been observed in a number of fish species [30], for example rainbow trout [31], tench (*Tinca tinca*) [32-34], Japanese eel (*Anguilla japonica*) [35], coho salmon (*Oncorhynchus kisutch*) [36], European catfish (*Silurus glanis*) [37], and sterlet (*Acipenser ruthenus*) [38]. Within tetraploid fish, such as the Siberian sturgeon (*Acipenser baerii*), spontaneous hexaploidy, which involves the same mechanism as spontaneous triploidy, has also been observed [39]. While spontaneous triploidy has been observed in natural populations [30], it is primarily documented in cultured and farmed populations. Spontaneous triploidy has also been reported in cultured Atlantic salmon [40,41]. However, the extent and frequency of this phenomenon in commercial Atlantic salmon farming has yet to be studied. Here, based upon the re-analysis of an extensive database of genetic data for farmed salmon collected in association with the "stand-by method", we investigate the frequency of spontaneously occurring triploid salmon in commercial farms in Norway in the period 2007–2014.

METHODS

Samples

The present study is based upon the analysis of 5051 farmed salmon. Of these samples, 4089 were collected from 86 fish-cages located on 55 Norwegian commercial salmon farms spanning the entire Norwegian coastline (Figure 1), while the remaining 962 salmon were sampled as escapees in both fresh and saltwater (Figure 2). All of these samples were taken in association with the DNA stand-by method in the period 2007–2014. For legal reasons, only the approximate locations of the farms and the recapture sites of the escapees are revealed (Figures 1 and 2). As the primary objective of the study was to identify the frequency of spontaneous triploidy in farmed Norwegian salmon, precise farm information is not of importance.

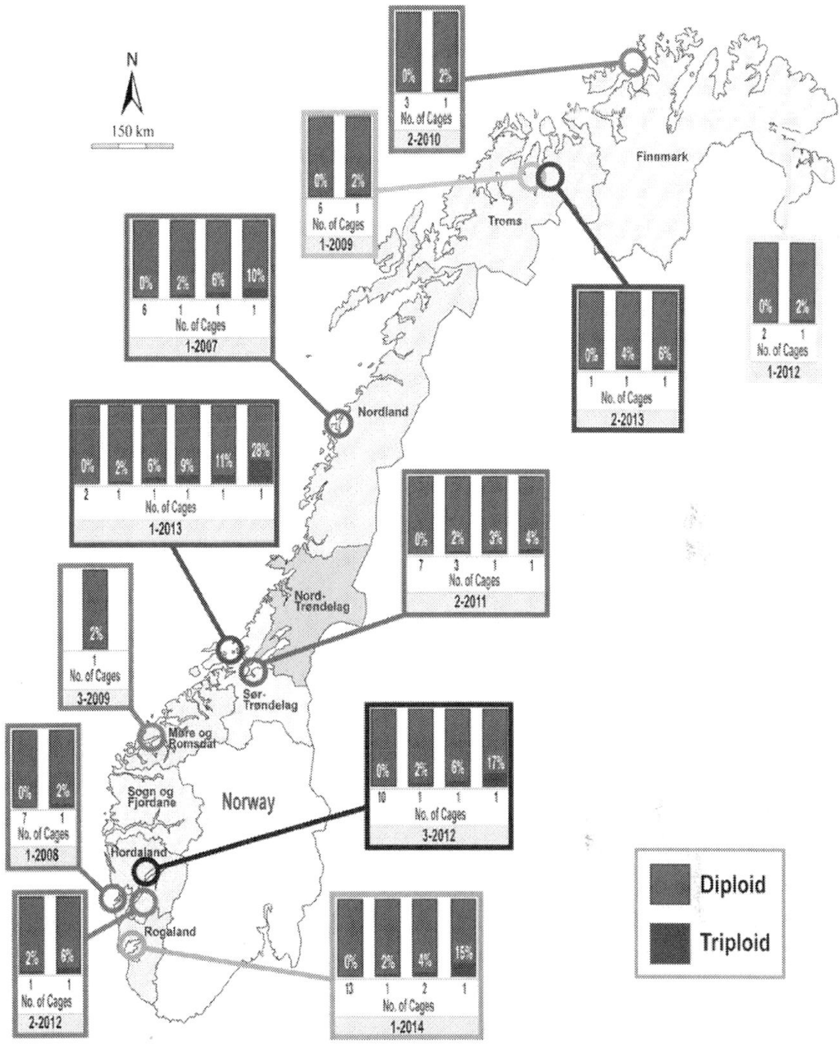

Figure 1: The observed frequencies of spontaneous tripoloid salmon sampled from 86 marine cages located on 55 marine farms in Norway in the period 2007–2014. Colour is only used for visual enhancement.

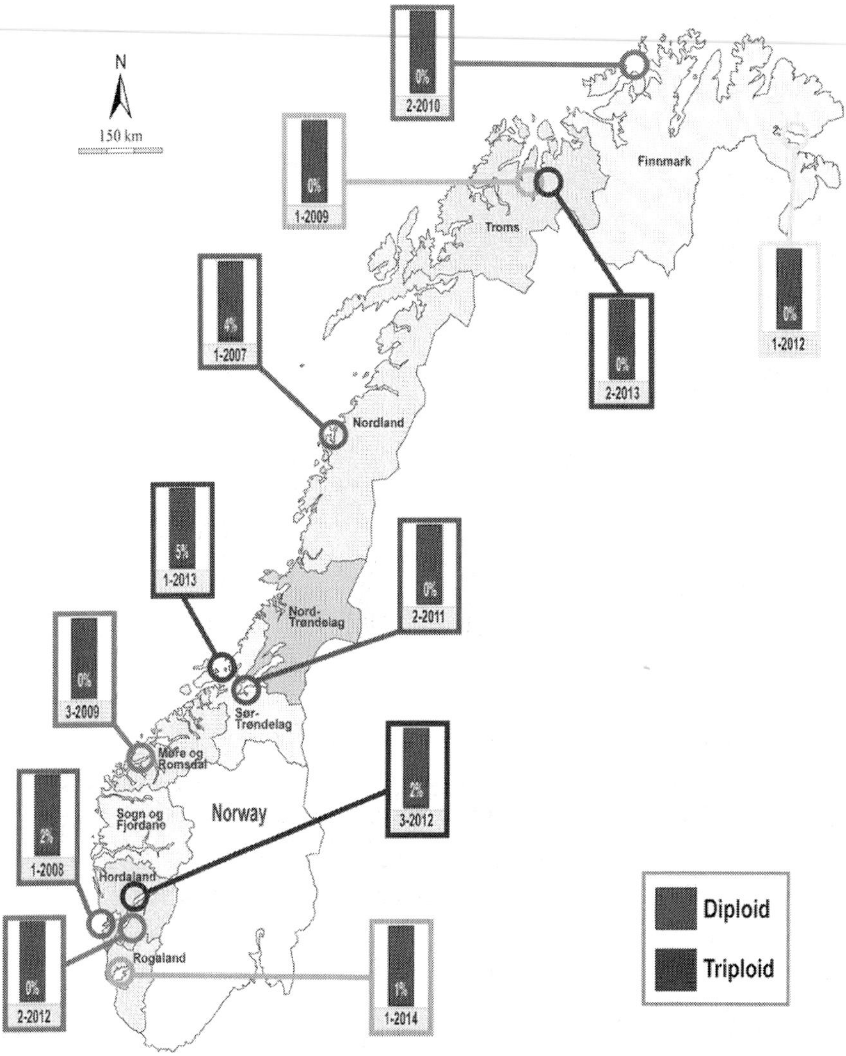

Figure 2: The observed frequencies of spontaneous triploid salmon farmed escapees recaptured in Norway in the period 2007–2014. Colour is only used for visual enhancement.

From each farm, adipose fin clips were randomly sampled from approximately 45–47 salmon (range 41–90) from a single cage using a wet net. The samples were thereafter placed into a single tube containing alcohol for preservation. No phenotypic nor rearing information was taken from these fish nor the farms from which they were sampled. These tissue samples were initially used to create a baseline genetic profile of the cage of fish for genetic assignment analysis [22]. Multiple cages were sampled on the same farm in cases where the farm had salmon originating from more than one juvenile or smolt producer, and therefore potentially contained fish of more than 1 genetic origin. In addition to the samples originating from fish farms, farmed escapees, originating from the same regions as the farms themselves (Figure 2) were collected. The numbers of farmed escapees varied between 16 and 343 per data set.

Tissue samples of salmon from commercial farms were taken by employees of the Norwegian Directorate of Fisheries together with the farmer owning the fish. Tissue samples of the farmed escapees were taken from fish that had been already captured and killed as a result of recreational angling or netting. The researchers of this research played no part in the sampling conducted. Consequently, specific permits for collection of these samples were not required by Norwegian or international law for the use of animals in research.

Genotyping

Prior to genetic analysis, all samples were organized into a database which allocates a unique number for all samples which includes its initial tissue tube and subsequent DNA isolation plate number. Blind genotyping control is routinely conducted in this laboratory [21,42,43]. Although the frequency of blind re-genotyping controls varied between the data sets included in this analysis, in all cases, a number of randomly selected fish had their DNA isolated twice in order to conduct blind genotyping and sample control. This equated to a minimum of 10% of the samples included in this study (>500 samples re-analysed). Genotyping error rates for parts of the data sets included in this study have been published previously [21], and for other data sets using the same markers in this laboratory [13]. In all cases, the genotyping error rates are very low, and are almost exclusively associated with homozygote and heterozygote errors. Thus far, no mis-identification between triploid nor diploid offspring have been detected, also when the pedigree of the parents and triploid offspring from single families has been double checked using parentage testing approaches [40].

DNA was isolated in 96 well format using the DNeasy blood and tissue kit from Qiagen. Each plate contained at least 2 blank cells as negative controls. As these analyses were conducted over an extended period of time, the exact protocols used for PCR amplification varied, however, the latest laboratory protocols for amplification of the microsatellite markers used here are available upon request. For each sample, between 15–18 microsatellite loci were amplified. These included the following loci organized into three multiplex reactions: *SSsp3016*(Genbank no. AY372820), *SSsp2210, SSspG7, SSsp2201, SSsp1605, SSsp2216* [44], *Ssa197, Ssa171, Ssa202* [45], *SsaD157, SsaD486, SsaD144* [46], *Ssa289, Ssa14* [47], *SsaF43* [48], *SsaOsl85* [49], *MHC I* [50] *MHC II* [51]. PCR products were analysed on an ABI 3730 Genetic Analyser and sized by a 500LIZ™ size-standard. Automatically binned alleles were manually checked by two researchers prior to exporting data for statistical analyses.

Protocol for Triploid Identification Using Microsatellites

Triploid organisms are readily identified through microsatellite DNA genotyping as they often display three clearly identifiable alleles per locus [52]. However, not all loci will display trisomy in a triploid individual. This depends upon the genotype of the mother and the father, and the distance of the given microsatellite locus from the centromere, which is in turn linked with the probability of crossing over.

In order to provide a conservative estimate of the frequency of triploidy in the present study, we only reported an individual as triploid in the case that it displayed three clear alleles at two or more of the 15–18 loci genotyped. Individual salmon displaying three alleles at only one locus were not reported as triploid (this only occurred for two individuals and thus does not influence the results of the present study). Thus, the approach we have implemented to identify the frequency of triploids is identical to a recent study documenting spontaneous triploids in cultured Atlantic salmon in a Baltic fish hatchery [41], and is similar to the approach used to identify triploids in a variety of other fish and insect studies of triploidy [52-55]. Finally, these genetic markers in this laboratory have been used to extensively conduct parentage testing for Atlantic salmon in common-garden experiments [40,56-58]. In one of these parentage-based studies, triploid salmon offspring were identified in the pedigree.

All of these individuals were re-genotyped and accurately re-identified a second time using these markers demonstrating the reliability of the approach to consequently differentiate between diploid and triploid individuals [40].

Validation of Microsatellite Genotyping for Triploid Identification

Microsatellite genotyping for triploid identification is an established technique (see above), and has also been validated against other triploid identification methods in species such as the *Carassius auratus* complex [59], and turbot (*Scophthalmus maximus*) [52]. Nevertheless, we validated microsatellite genotyping against a set of 45 putatively diploid and 45 putatively triploid farmed salmon samples in the present study using red blood cell (RBC) diameter measurements and flow cytometry which estimates DNA content. The putatively triploid salmon used for this validation had been produced by a deliberate pressure shock treatment. All of these fish had been killed by a sharp blow to the head prior to taking samples for analysis.

For each putatively diploid and putatively triploid salmon used in the ploidy method validation, fin clips for DNA analysis, and blood was collected. From each sample, 10 µl heparinised blood was fixed for 30–60 min with 4% PFA. The blood cells were washed three times in PBS, and re-suspended in 300 µl PBS containing 50 ug/ml propidium iodide (Life Technologies), 50 µg/ml Rnase A (Life Technologies), and 0.3% (v/v) Tween-20 (Sigma), and kept at room temperature overnight in the dark. Next day the cells were analyzed on a FACSCanto II flow cytometer (BD Biosciences). Cell debris and doublets were eliminated from the analysis by gating, and the data were analyzed using BD FACSDIVA software. The average diameter of the salmon RBCs were measured from blood smears (IMAGEPRO PLUS, version 4.0; Media Cybernetics, Silver Spring, MD). Ten erythrocytes per fish were measured. Microsatellites were genotyped according to the same protocols as described for the main data set above. Once data was produced using all three methods, ploidy results were compared.

RESULTS

Validation of Ploidy Determination Method

In total, 81 of the 90 samples used for method validation had their ploidy determined by all three methods, with the remaining 9 samples identified by two of the methods. There was complete agreement between the identification of ploidy using all three approaches. A single triploid salmon was identified in the putatively diploid group of salmon, while 11 diploid salmon were identified in the putatively triploid group of salmon. Thus, microsatellites can be used to reliably determine ploidy in Atlantic salmon where blood samples are not available, as in the main part of the current study.

General Trends in Samples from Cages and Escapees

Of the 5051 farmed salmon analysed in the period 2007–2014, a total of 91 (1.8%) spontaneously arising triploids were detected (Figure 1, Figure 2, Table 1). The number of loci displaying two or three alleles for each of the 91 spontaneous triploid salmon is presented in table form, as is a graphical representation of the profiles for a single diploid and triploid salmon example.

Table 1: The observed frequencies of diploid and spontaneous triploid salmon according to sampling source and genetic background

Data source	Diploid (n)	Triploid (n)	Total (n)	% Triploid
Samples from farms				
Aqua Gen AS	1635	37	1672	2.2
Salmobreed	783	19	802	2.4
Mowi	609	11	620	1.8
Unknown	1062	15	1077	1.4
Farms total	4089	82	4171	2.0
Samples of recaptured escapees				
Recaptured freshwater	376	4	380	1.1
Recaptured saltwater	586	5	591	0.8
Escapees total	962	9	971	0.9
All samples combined	5051	91	5142	1.8

Note, escapees are not assigned to genetic strain and therefore represent an unquantified mixture of fish from all strains. Fish of "unknown" strain sampled on commercial farms represent an undocumented mixture of all strains.

Samples collected in the period 2007–2014.

Triploids were observed in all regions in Norway, and across the entire time period. Triploids were observed among the fish sampled in cages, as well as among the farmed escapees recaptured in the wild.

Samples from Fish Cages

Looking specifically at the samples collected from the 86 cages located on 55 farms, the frequency of spontaneous triploidy varied greatly (Figure 1). In most cages sampled, no triploid salmon were observed. In a few cages, a low frequency, which typically meant 1–3 individuals from 45–47 fish typically sampled, were detected (i.e., 2-5%). In five of the cages sampled, the reported frequency of triploid salmon ranged from 10-28%. These five cages displaying notably higher frequencies of spontaneous triploidy, originated from separate case studies, and were therefore not connected to any single farm (Figure 1).

In many, but not all of the cages where samples were taken, the genetic breeding line of the salmon was unequivocally determined (based upon document information collected by the NDF when the samples were taken on each farm). When the triploid frequency data was split into the three main breeding companies producing salmon in Norway, these data revealed that triploids were observed in all three breeding programs, and notably, that the observed frequency of spontaneous triploidy was similar among strains (Table 1). Among the salmon of unidentified breeding origin collected from fish farms (i.e., the paperwork associated with the sample was not precise enough to unequivocally determine the genetic strain), the frequency of triploid salmon was slightly lower, but still close to the average numbers (Table 1).

Samples of Escapees

Overall, the frequency of triploid salmon among the escapees was 0.9% which is lower than the observed frequencies for the samples taken in cages (Table 1). Triploid salmon escapees were observed among the recaptured escapees in 6 of the 12 cases investigated, and in different regions (Figure 2). Most of the escapees were sampled in salt water (Table 1), however, the observed frequency of triploids recaptured in freshwater was similar to the observed frequency recaptured in salt water.

DISCUSSION

This study represents the first systematic investigation into the frequency of spontaneous triploidy in farmed Atlantic salmon. The genetic analysis of more than 5000 salmon collected in the period 2007–2014 gave the following main results: 1. The overall observed frequency of spontaneous triploidy among salmon collected from 55 farms was 2.0%, 2. Spontaneous triploidy occurred in farmed salmon originating from all three major breeding lines in Norway, and in similar frequencies, 3. Spontaneous triploidy was observed in farms located in all regions of Norway, and in all years sampled spanning from 2007 to 2014, 4. The frequency of spontaneous triploidy varied greatly among cages (ranging 0-28%), 5. Spontaneous triploids were observed amongst the escapees recaptured in both freshwater and the marine environment, 6. The validation tests implemented here demonstrated that microsatellite genotyping gives consistent results for ploidy determination as RBD diameter measurements and flow cytometry in this species.

Based upon personal communications with the three primary breeding companies operating in Norway, we were able to exclude the possibility that the triploid salmon observed in this study were the result of a deliberate pressure shock protocol to produce triploid fish as is used in Atlantic salmon experiments [26-28]. Thus, the reported triploids arose as a result of spontaneous event, as has been observed previously for Atlantic salmon [40,41], and a range of other fish species in culture [31-33,35,36], and in the wild [30]. While the Atlantic salmon farming industry has very recently started experimental production of triploid salmon in Norway [26-29], for the timescale in which the samples in this study were collected, very few triploid salmon were commercially produced. Furthermore, in all cases where triploid salmon were deliberately produced by these farming companies, triploids were sold to a very limited number of farms in specific locations. It has been verified by these three companies that none of these locations nor farms overlapped with the farms from which samples upon which the present study is based. While it is still theoretically possible that some of the recaptured triploid escapees could have arisen from a farm(s) outside of the sampling regions in the present analyses, and therefore from a farm that reared deliberately produced triploids, this remote possibility is highly unlikely given the fact that the vast majority of the escapees analysed in the present study had already been assigned to the farms sampled here based upon their genetic profiles [20-24]. Furthermore, the observed frequency in the escapees was lower than from the farms with a documented background, and has therefore not spuriously contributed to an inflated estimate of spontaneous triploid frequency within the industry as a whole.

Spontaneous triploidy originates from the duplication of the maternal chromosome set by inhibition of the second polar body release after normal meiosis II (crossing over) in the oocyte. As a result of observing the non-random distribution of this phenomena, it has been suggested that it may have an underlying genetic basis or predisposition [60]. This is supported by positive heritability estimates for propensity of this phenomena in common carp (*Cyprinus carpio*) [61], and evidence for a paternal contribution to autopolyploidy in white sturgeon (*Acipenser transmontanus*) [62]. However, in other studies, the frequency of spontaneous triploidy has been clearly linked with *in vitro* or *in vivo* post ovulatory aging of eggs prior to fertilisation [31,33,35], and is further enhanced by increased temperatures during aging [31,33]. Incidentally, the eggs used in the Atlantic salmon experiment by [40], where spontaneous triploids were observed in the resulting offspring, had been transported approximately 6–8 hours by car prior to fertilisation. It appears possible that the delayed fertilization combined with possible temperature increase in the car during transport could have been the trigger for the observed spontaneous triploidy in that case [40].

Turning attentions back to the observations in the present study, it is important to note that spontaneous triploidy was reported in all three breeding companies, and at similar frequencies (Table 1). This strongly suggests that the genetic material produced by all three of these companies, which account for most of the farmed salmon produced globally, display similar propensities for this phenomena. Furthermore, due to the logistics of the breeding practices on commercial farms, eggs are sometimes removed from female broodfish up to several hours after they have been killed. In addition, once the eggs have been removed from the female broodfish, they may be stored further for several hours or even until the next day before they are fertilised. Given the fact that storage and aging of eggs post ovulation has been documented to increase the frequency of spontaneous triploidy [31,33,35,36,63], it is concluded that stripping practices often implemented by the Atlantic salmon aquaculture industry, with pauses between killing the broodfish and fertilsation of eggs, is the most likely explanation for the observed frequency of 2.0% spontaneous triploidy in Norwegian farmed salmon in the period 2007–2014. This conclusion is consistent with the fact that spontaneous triploidy varied greatly among the cages and farms sampled here (Figure 1), which may in turn reflect differences in treatment of unfertilized eggs for the fish reared in those cages. Controlled experiments using aged eggs could help identify the underlying mechanism(s) driving this process specifically within the Atlantic salmon aquaculture industry.

Farmed escaped salmon have successfully interbred and caused genetic changes in a number of wild Atlantic salmon populations inhabiting rivers in Norway and Ireland [12-16,64]. Thus, the Atlantic salmon farming industry is currently investigating the potential to deliberately produce triploid salmon in order to mitigate potential genetic impacts of escapees on native wild populations. As a consequence of this, the numbers of deliberately produced triploid salmon in commercial fish farms is likely to expand within the near future in both Norway and other countries. Triploid salmon are sterile, and will therefore not be able to hybridise with wild populations. Nevertheless, in a recent experiment conducted in semi-natural spawning arenas, deliberately produced triploid farmed male salmon displayed wild salmon spawning behaviors, and importantly, managed to coax a wild female salmon to spawn [65]. Thus, the potential for an ecological interaction between triploid (sterile) escapees and wild salmon, through mate competition and non-productive spawning exists. However, in order for this to occur, triploid farmed escapees first need to migrate to freshwater where wild salmon spawn. Here, we demonstrate for the first time, that spontaneous triploid salmon, escaping from commercial fish farms, can enter freshwater. Furthermore, within this study, the frequency of triploids observed among the escapees recaptured in freshwater was similar to the frequency observed among the escapees recaptured in salt water (Table 1). Nevertheless, while these data demonstrate "proof of concept", we urge caution in interpreting the relative frequency of this occurrence between diploid and triploid salmon. First, only a small number of the escapees investigated in the present study were captured in freshwater, and second, freshwater sampling only occurred in three of the cases investigated. More extensive and representative sampling of escapees in a larger number of rivers is required in order to fully evaluate the relative frequency of this occurrence.

An experimental release study conducted with diploid and triploid salmon smolts in Ireland reported significantly lower freshwater return rates for triploid fish than their diploid counterparts [66]. The fish from that release experiment first had to survive in the marine environment, migrate to the ocean feeding grounds, and then return back to the coastline and ultimately freshwater. In the present study, almost all of the escapees were the result of larger fish (typically 1-5 kg) escaping from net pens and either being captured in the sea or a local river immediately or very shortly after escape. This is based upon the fact that the "stand-by method" is almost exclusively implemented in escape events when there is a distinct and sudden appearance of escapees in a local area [17,22]. Thus, the majority of the escapees analysed in the present study have not undergone an oceanic migration as in the smolt release experiment above, which could explain the difference in the results between these two studies. Immature escaped diploid salmon have been documented to occasionally enter freshwater soon after escape [25,67]. It is therefore possible that the triploid escapees that were reported in freshwater in the present study, have displayed a similar maladapted behavior of migrating to freshwater without any maturation as has been observed for immature diploids.

CONCLUSIONS

This study represents the first systematic investigation into the occurrence and frequency of spontaneous triploidy in farmed Atlantic salmon. Based upon the analysis of microsatellite DNA profiles of more than 4000 salmon collected from 55 fish farms, and a group of nearly 1000 farmed escapees recaptured in the wild, all sampled in the period 2007–2014, we were able to document that spontaneous triploidy occurred at 2.0% in the samples taken on farms, and 1.8% in the total material. We also documented for the first time, that triploid farmed fish escaping from commercial farms, may enter freshwater. We suggest that spontaneous triploidy occurs in farmed Atlantic salmon due to the occasional delay between stripping eggs from female broodfish and their fertilization. This has been documented in other fish species to increase the chance of duplication of the maternal chromosome set by inhibition of the second polar body release after normal meiosis II (crossing over) in the oocyte.

In the past decade, Norwegian Atlantic salmon aquaculture has produced close to or in excess of 1 million tons of farmed salmon annually. Taking an average slaughter weight of 5 kg, and an average spontaneous triploid frequency of 2%, our analyses could suggest that as many as 40 million spontaneous triploid salmon have been produced in Norwegian Atlantic salmon farming in the past decade.

NOTES

Acknowledgements

This study was financed using resources from the Norwegian Department of Industry and Fisheries. We gratefully acknowledge essential information regarding the deliberate production of triploid salmon from the three primary Atlantic salmon breeding companies in Norway: AquaGen AS (supplied by Maren Mommens and Svein Arild Korsvoll), Marine Harvest (supplied by Roy Hjelmeland, Reidar Våge and Olav Breck) and Salmobreed (supplied by Håvard Bakke). The information provided made it possible to unequivocally exclude the possibility that the triploid salmon identified in this study were the result of deliberate triploid production, which has been recently initiated at a pilot-scale level in Norway. Finally, we would like to acknowledge the constructive comments from the four anomynous referees that have helped improve the manuscript.

Competing Interests

The authors declare that they have no competing interests.

Authors' Contributions

All authors contributed to the concept and design of the study, and interpretation of data. AGES, VW, GD, ØS and KAG conducted genetic analyses and data quality checks, ASM created Figures, HCM, TJH, and PGF conducted validation analysis with red blood cell measurements and flow cytometry. KAG managed the project and wrote first draft of the manuscript. All authors read and approved the final version of the manuscript.

REFERENCES

1. Gjedrem T. Genetic improvement of cold-water fish species. Aquac Res. 2000;31(1):25–33.
2. Gjedrem T. The first family-based breeding program in aquaculture. Rev Aquac. 2010;2(1):2–15.
3. Skaala O, Hoyheim B, Glover K, Dahle G. Microsatellite analysis in domesticated and wild Atlantic salmon (*Salmo salar* L.): allelic diversity and identification of individuals. Aquaculture. 2004;240(1–4):131–43.
4. Youngson AF, Martin SAM, Jordan WC, Verspoor E. Genetic protein variation in Atlantic salmon in Scotland - comparisons of wild and farmed fish. Aquaculture. 1991;98(1–3):231–42.
5. Powell J, White I, Guy D, Brotherstone S. Genetic parameters of production traits in Atlantic salmon (Salmo salar). Aquaculture. 2008;274(2–4):225–31.
6. Quinton CD, McMillan I, Glebe BD. Development of an Atlantic salmon (Salmo salar) genetic improvement program: genetic parameters of harvest body weight and carcass quality traits estimated with animal models. Aquaculture. 2005;247(1–4):211–7.
7. Withler RE, Supernault KJ, Miller KM. Genetic variation within and among domesticated Atlantic salmon broodstocks in British Columbia, Canada. Anim Genet. 2005;36(1):43–50.
8. Taylor RS, Wynne JW, Kube PD, Elliott NG. Genetic variation of resistance to amoebic gill disease in Atlantic salmon (Salmo salar) assessed in a challenge system. Aquaculture. 2007;272:S94–9.
9. Norris AT, Bradley DG, Cunningham EP. Microsatellite genetic variation between and within farmed and wild Atlantic salmon (Salmo salar) populations. Aquaculture. 1999;180(3–4):247–64.

10. Saegrov H, Hindar K, Kalas S, Lura H. Escaped farmed Atlantic salmon replace the original salmon stock in the River Vosso, western Norway. Ices J Marine Sci. 1997;54(6):1166–72.

11. Fleming IA, Hindar K, Mjolnerod IB, Jonsson B, Balstad T, Lamberg A. Lifetime success and interactions of farm salmon invading a native population. Proc R Soc Lond Ser B-Biol Sci. 2000;267(1452):1517–23.

12. Skaala O, Wennevik V, Glover KA. Evidence of temporal genetic change in wild Atlantic salmon, *Salmo salar* L., populations affected by farm escapees. Ices J Marine Sci. 2006;63(7):1224–33.

13. Glover KA, Quintela M, Wennevik V, Besnier F, Sørvik AGE, Skaala O. Three decades of farmed escapees in the wild: a spatio-temporal analysis of population genetic structure throughout Norway. Plos One. 2012;7(8):e43129.

14. Glover KA, Pertoldi C, Besnier F, Wennevik V, Kent M, Skaala Ø. Atlantic salmon populations invaded by farmed escapees: quantifying genetic introgression with a Bayesian approach and SNPs. BMC Genet. 2013;14:4.

15. Clifford SL, McGinnity P, Ferguson A. Genetic changes in Atlantic salmon (*Salmo salar*) populations of northwest Irish rivers resulting from escapes of adult farm salmon. Can J Fish Aquat Sci. 1998;55(2):358–63.

16. Crozier WW. Evidence of genetic interaction between escaped farmed salmon and wild Atlantic salmon (*Salmo salar* L) in a Northern Irish river. Aquaculture. 1993;113(1–2):19–29.

17. Glover KA, Skilbrei OT, Skaala O. Genetic assignment identifies farm of origin for Atlantic salmon *Salmo salar* escapees in a Norwegian fjord. Ices J Marine Sci. 2008;65(6):912–20.

18. Glover KA, Dahle G, Westgaard JI, Johansen T, Knutsen H, Jorstad KE. Genetic diversity within and among Atlantic cod (Gadus morhua) farmed in marine cages: a proof-of-concept study for the identification of escapees. Anim Genet. 2010;41(5):515–22.

19. Glover KA. Genetic characterisation of farmed rainbow trout in Norway: intra- and inter-strain variation reveals potential for identification of escapees. BMC Genet. 2008;9:19.

20. Glover KA, Hansen MM, Skaala O. Identifying the source of farmed

escaped Atlantic salmon (*Salmo salar*): Bayesian clustering analysis increases accuracy of assignment. Aquaculture. 2009;290(1–2):37–46.

21. Glover KA, Hansen MM, Lien S, Als TD, Hoyheim B, Skaala O. A comparison of SNP and STR loci for delineating population structure and performing individual genetic assignment. BMC Genet. 2010;11(2).
22. Glover KA. Forensic identification of fish farm escapees: the Norwegian experience. Aquaculture Environ Interact. 2010;1:1–10.
23. Glover KA, Skaala O, Sovik AGE, Helle TA. Genetic differentiation among Atlantic salmon reared in sea-cages reveals a non-random distribution of genetic material from a breeding programme to commercial production. Aquac Res. 2011;42(9):1323–31.
24. Zhang Z, Glover KA, Wennevik V, Svåsand T, Sørvik AGE, Fiske P, et al. Genetic analysis of Atlantic salmon captured in a netting station reveals multiple escapement events from commercial fish farms. Fish Manag Ecol. 2013;20(1):42–51.
25. Madhun AS, Karlsbakk E, Ischsen CH, Omdal LM, Sørvik AGE, Skaala Ø, et al. Potential disease interaction reinforced: double-virus infected escaped farmed Atlantic salmon, Salmo salar L., recaptured in a nearby river. J Fish Dis. 2015;38:209–19.
26. Taylor JF, Bozzolla P, Frenzl B, Matthew C, Hunter D, Migaud H. Triploid Atlantic salmon growth is negatively affected by communal ploidy rearing during seawater grow-out in tanks. Aquaculture. 2014;432:163–74.
27. Frenzl B, Migaud H, Fjelldal PG, Shinn AP, Taylor JF, Richards RH, et al. Triploid and diploid Atlantic salmon show similar susceptibility to infection with salmon lice Lepeophtheirus salmonis. Pest Manag Sci. 2014;70(6):982–8.
28. Leclercq E, Taylor JF, Fison D, Fjelldal PG, Diez-Padrisa M, Hansen T, et al. Comparative seawater performance and deformity prevalence in out-of-season diploid and triploid Atlantic salmon (Salmo salar) post-smolts. Comp Biochem Physiol A-Mol Integr Physiol. 2011;158(1):116–25.
29. Fjelldal PG, Hansen T. Vertebral deformities in triploid Atlantic salmon (Salmo salar L.) underyearling smolts. Aquaculture.

2010;309(1–4):131–6.

30. Leggatt RA, Iwama GK. Occurrence of polyploidy in the fishes. Rev Fish Biol Fish. 2003;13(3):237–46.
31. Aegerter S, Jalabert B. Effects of post-ovulatory oocyte ageing and temperature on egg quality and on the occurrence of triploid fry in rainbow trout, Oncorhynchus mykiss. Aquaculture. 2004;231(1–4):59–71.
32. Flajshans M, Kvasnicka P, Rab P. Genetic studies in tench (*Tinca tinca*) - high incidence spontaneous triploidy. Aquaculture. 1993;110(3–4):243–8.
33. Flajshans M, Kohlmann K, Rab P. Autotriploid tench Tinca tinca (L.) larvae obtained by fertilization of eggs previously subjected to postovulatory ageing in vitro and in vivo. J Fish Biol. 2007;71(3):868–76.
34. Flajshans M, Gela D, Kocour M, Buchtova H, Rodina M, Psenicka M, et al. A review on the potential of triploid tench for aquaculture. Rev Fish Biol Fish. 2010;20(3):317–29.
35. Nomura K, Takeda Y, Unuma T, Morishima K, Tanaka H, Arai K, et al. Post-ovulatory oocyte aging induces spontaneous occurrence of polyploids and mosaics in artificial fertilization of Japanese eel, Anguilla japonica. Aquaculture. 2013;404:15–21.
36. Devlin RH, Sakhrani D, Biagi CA, Eom KW. Occurrence of incomplete paternal-chromosome retention in GH-transgenic coho salmon being assessed for reproductive containment by pressure-shock-induced triploidy. Aquaculture. 2010;304(1–4):66–78.
37. Varkonyi E, Bercsenyi M, Ozouf-Costaz C, Billard R. Chromosomal and morphological abnormalities caused by oocyte aging in Silurus glanis. J Fish Biol. 1998;52(5):899–906.
38. Havelka M, Hulak M, Rodina M, Flajshans M. First evidence of autotriploidization in sterlet (Acipenser ruthenus). J Appl Genet. 2013;54(2):201–7.
39. Havelka M, Hulak M, Rab P, Rabova M, Lieckfeldt D, Ludwig A, et al. Fertility of a spontaneous hexaploid male Siberian sturgeon, Acipenser baerii. BMC Genet. 2014;15:5.
40. Solberg MF, Glover KA, Nilsen F, Skaala Ø. Does Domestication

Cause Changes in Growth Reaction Norms? A Study of Farmed, Wild and Hybrid Atlantic Salmon Families Exposed to Environmental Stress. Plos One. 2013;8(1):e54469.

41. Ozerov MY, Lumme J, Pakk P, Rintamaki P, Zietara MS, Barskaya Y, et al. High Gyrodactylus salaris infection rate in triploid Atlantic salmon Salmo salar. Dis Aquat Org. 2010;91(2):129–36.
42. Haaland ØA, Glover KA, Seliussen BB, Skaug HJ. Genotyping errors in a calibrated DNA -register: implications for identification of individuals. BMC Genet. 2011;12:36.
43. Glover KA, Haag T, Oien N, Walloe L, Lindblom L, Seliussen BB, et al. The Norwegian minke whale DNA register: a database monitoring commercial harvest and trade of whale products. Fish Fish. 2012;13:313–32.
44. Paterson S, Piertney SB, Knox D, Gilbey J, Verspoor E. Characterization and PCR multiplexing of novel highly variable tetranucleotide Atlantic salmon (*Salmo salar* L.) microsatellites. Mol Ecol Notes. 2004;4(2):160–2.
45. O'Reilly PT, Hamilton LC, McConnell SK, Wright JM. Rapid analysis of genetic variation in Atlantic salmon (*Salmo salar*) by PCR multiplexing of dinucleotide and tetranucleotide microsatellites. Can J Fisheries and Aquatic Sci. 1996;53(10):2292–8.
46. King TL, Eackles MS, Letcher BH. Microsatellite DNA markers for the study of Atlantic salmon (*Salmo salar*) kinship, population structure, and mixed-fishery analyses. Mol Ecol Notes. 2005;5(1):130–2.
47. McConnell SK, Oreilly P, Hamilton L, Wright JN, Bentzen P. Polymorphic microsatellite loci from Atlantic salmon (*Salmo salar*) - genetic differentiation of North-American and European populations. Can J Fish Aquat Sci. 1995;52(9):1863–72.
48. Sanchez JA, Clabby C, Ramos D, Blanco G, Flavin F, Vazquez E, et al. Protein and microsatellite single locus variability in *Salmo salar* L (Atlantic salmon). Heredity. 1996;77:423–32.
49. Slettan A, Olsaker I, Lie O. Atlantic salmon, *Salmo salar*, microsatellites at the *SsOSL25, SsOSL85, SsOSL311, SsOSL417* loci. Anim Genet. 1995;26(4):281–2.
50. Grimholt U, Drablos F, Jorgensen SM, Hoyheim B, Stet RJM. The

major histocompatibility class I locus in Atlantic salmon (*Salmo salar* L.): polymorphism, linkage analysis and protein modelling. Immunogenetics. 2002;54(8):570–81.

51. Stet RJM, de Vries B, Mudde K, Hermsen T, van Heerwaarden J, Shum BP, et al. Unique haplotypes of co-segregating major histocompatibility class II A and class II B alleles in Atlantic salmon (*Salmo salar*) give rise to diverse class II genotypes. Immunogenetics. 2002;54(5):320–31.
52. Hernandez-Urcera J, Vera M, Magadan S, Pino-Querido A, Cal RM, Martinez P. Development and validation of a molecular tool for assessing triploidy in turbot (Scophthalmus maximus). Aquaculture. 2012;330:179–84.
53. Darvill B, Lepais O, Woodall LC, Goulson D. Triploid bumblebees indicate a direct cost of inbreeding in fragmented populations. Mol Ecol. 2012;21(16):3988–95.
54. Liebert AE, Johnson RN, Switz GT, Starks PT. Triploid females and diploid males: underreported phenomena in Polistes wasps? Insect Soc. 2004;51(3):205–11.
55. Garner SR, Madison BN, Bernier NJ, Neff BD. Juvenile growth and aggression in diploid and triploid Chinook salmon Oncorhynchus tshawytscha (Walbaum). J Fish Biol. 2008;73(1):169–85.
56. Skaala Ø, Glover Kevin A, Barlaup Bjørn T, Svåsand T, Besnier F, Hansen Michael M, et al. Performance of farmed, hybrid, and wild Atlantic salmon (Salmo salar) families in a natural river environment. Can J Fish Aquat Sci. 2012;69(12):1994–2006.
57. Solberg MF, Zhang ZW, Nilsen F, Glover KA. Growth reaction norms of domesticated, wild and hybrid Atlantic salmon families in response to differing social and physical environments. Bmc Evolutionary Biology. 2013;13:234.
58. Solberg MF, Zhang Z, Glover KA. Are farmed salmon more prone to risk than wild salmon?Susceptibility of juvenile farm, hybrid and wild Atlanticsalmon Salmo salar L. to an artificial predator. Appl Anim Behav Sci. 2015;162:67–80.
59. Jakovlic I, Gui J-F. Recent invasion and low level of divergence between diploid and triploid forms of Carassius auratus complex in Croatia. Genetica. 2011;139(6):789–804.

60. Thorgaard GH, Gall GAE. Adult triploids in a rainbow trout family. Genetics. 1979;93(4):961–73.

61. Cherfas NB, Gomelsky B, Ben-Dom N, Hulata G. Evidence for the heritable nature of spontaneous diploidization in common carp *Cyprinus carpio* L. eggs. Aquac Res. 1995;26:289–92.

62. Schreier AD, May B, Gille DA. Incidence of spontaneous autopolyploidy in cultured populations of white sturgeon, Acipenser transmontanus. Aquaculture. 2013;416:141–5.

63. Yamazaki F. Chromosomal changes in salmonids. Changes in chromosome number and morphology due to over-ripening of eggs and irradiation. Can Trans Fisheries Aquatic Sci. 1983;4962:1–22.

64. Clifford SL, McGinnity P, Ferguson A. Genetic changes in an Atlantic salmon population resulting from escaped juvenile farm salmon. J Fish Biol. 1998;52(1):118–27.

65. Fjelldal PG, Wennevik V, Fleming IA, Hansen T, Glover KA. Triploid (sterile) farmed Atlantic salmon males attempt to spawn with wild females. Aquaculture Environ Interact. 2014;5(2):155–62.

66. Cotter D, O'Donovan V, O'Maoileidigh N, Rogan G, Roche N, Wilkins NP. An evaluation of the use of triploid Atlantic salmon (Salmo salar L.) in minimising the impact of escaped farmed salmon on wild populations. Aquaculture. 2000;186(1–2):61–75.

67. Skilbrei OT, Heino M, Svåsand T. Using simulated escape events to assess the annual numbers and destinies of escaped farmed Atlantic salmon of different life stages, from farms sites in Norway. Ices J Marine Sci. 2014, doi:10.1093/icesjms/fsu133.

CHAPTER

17

ORIGIN OF BROODSTOCK AND EFFECTS ON THE DEFORMITIES OF GILTHEAD SEA BREAM (SPARUS AURATA L. 1758) IN A MEDITERRANEAN COMMERCIAL HATCHERY

John A. Theodorou[1], Costas Perdikaris[2], Barbara Venou[1]

[1]Department of Fisheries and Aquaculture Technology, Technological Educational Institute (T.E.I.) of Western Greece, Mesolonghi, Greece

[2]Department of Fisheries, Regional Unit of Thesprotia, Igoumenitsa, Greece

ABSTRACT

The use of broodstock of different origin as a method to improve fry production performance and consequently to minimize deformities was examined at industrial scale in a commercial gilthead sea bream hatchery. The outcome of fry production from three different broodstock groups (BA: broodfish (Mediterranean) with multiannual hatchery presence, BB: selected offspring originating from the BA group, and BC: broodfish of Atlantic origin) was investigated in the same rearing conditions and feeding protocol. Performance factors assessed were the survival and weaning of the larvae; the mortality rates from the "weaning until the end of the hatchery stage" of the larvae/fry; the percentage of fry without swim bladder; the percentage of fry with skeletal deformities and the feed conversion ratio. In all factors, no statistical differences among the experimental groups were detected. However, due to early rejection of the deformed individuals, benefits are expected from the decrease of the supplied amount of food and the reduced labor cost.

Keywords: Broodstock, Gilthead sea bream, Hatchery, Larval mortality, Skeletal deformities

INTRODUCTION

One of the main stages of marine aquaculture is larval production. Several improvements in hatchery rearing technology and use of broodstock have been occurred in the last decades (Theodorou 2002; Pavlidou 2009). However, problems such as skeletal deformities still represent a major factor of quality degradation and fry losses for the hatcheries. Gilthead sea bream is one of the most important cultured species in the Mediterranean aquaculture. Several studies on fry deformities exist for this species, recording qualitative and quantitative data (Koumoundouros et al. 1995; Divanach et al. 1996; Koumoundouros et al. 1997a, b; Galeotti et al. 1999, 2000; Carrillo et al. 2001; Cahu et al. 2003; Boglione et al. 2013a). There are also studies focusing on the causes of these deformities, which are related to physicochemical, genetic and nutritional factors, as well as on improvement efforts (Johnson and Katavic 1984; Liao et al. 1993; Chatain 1994; Boglione et al. 2001; Cahu et al. 2003; Verhaegen et al. 2007; Boglione et al. 2013b; García-Celdrán et al. 2014, 2015). However, most of the studies include limited data in terms of the economic benefits from the reduction of fry deformity-caused losses. Furthermore, the existing literature on the reduction of fry deformities using broodstock of different origin is contradictory (Liao et al. 1993; Verhaegen et al. 2007).

Consequently, the objective of the present study was to evaluate the possible economic benefits for a large-scale commercial hatchery in Greece from the replacement of the existing broodstock. This broodstock (of Mediterranean origin), routinely operating for a long period, was replaced by other broodstock sources including broodstock of Atlantic origin, taking also into account other improvements during the fry production process. Since the aim of the study was to improve the gilthead sea bream fry production at industrial scale through the reduction of fry losses caused by skeletal deformities, the effort was concentrated on the analysis of the data obtained throughout the entire typical production cycle of the hatchery.

MATERIALS AND METHODS

Production Data and Hatchery Operating Profile

Production data includes the tanks' stocking conditions as well as the larvae feeding regime, the physicochemical conditions in the rearing tanks, as well as data on broodstock characteristics. The experimental broodstock groups used were BA, BB and BC. Eggs from each broodstock group were stocked during 2008 at the same time in identical plastic circular tanks of 16 m^3each, under the same culture conditions (physicochemical, live food quality and quantities). Egg batches were stocked three times for each group (BA1, BA2, BA3; BB1, BB2, BB3 and BC1, BC2, BC3; see also Table 1) over the production period of the hatchery.

Table 1: Gilthead sea bream fry hatchery production data in three different broodstock groups and batches (standard deviations of means in parentheses)

	Experimental groups and batches of gilthead sea bream fry												*p*value
	BA1	BA2	BA3	BB1	BB2	BB3	BC1	BC2	BC3	Mean BA	Mean BB	Mean BC	
Final survival (%)[a]	13	15	12	15	22	15	19	28	13	13.3 (1.5)	17.3 (4.0)	20 (7.5)	0.32
Weaned fry (×10^3)	234.7	260.5	213.0	254.0	388.7	245.0	325.0	434.5	228.0	236.1 (23.8)	295.9 (80.5)	329.2 (103.5)	0.38
Mortality (%)[b]	6	6	6	5	6	8	4	8	15	6 (0.0)	6.3 (1.5)	9 (5.6)	0.52
Without swim bladder (%)	8	11	8	10	6	7	12	2	5	9 (1.7)	7.7 (2.1)	6.3 (5.1)	0.63
Deformed fry (%)	24	19	10	15	15	8	18	12	4	17.7 (7.1)	12.7 (4.0)	11.3 (7.0)	0.46
Total losses (%)[c]	37	36	22	30	27	23	34	23	24	31.7 (8.4)	26.7 (3.5)	27.0 (6.1)	0.58
Fry for sale (%)[d]	63	64	78	70	73	77	66	77	76	68.3 (8.4)	73.3 (3.5)	73.0 (6.1)	0.58
FCR[e]	1.53	1.09	0.97	1.11	1.22	0.99	1.16	1.13	1.04	1.2 (0.3)	1.1 (0.1)	1.1 (0.1)	0.77

[a]As a percentage of the initial stocking of the tank

[b]From the weaning phase to the sale of the fry

[c]These losses refer to the fry after the weaning phase

[d]The remaining fry for sale

[e]Feed conversion ratio was recorded throughout the dry feed supply period

Each experimental tank was stocked with 1 kg of eggs, and the mean number of eggs (1,400,000) was estimated, by measuring the mean egg diameter, with a mean initial rearing density of 88 larvae l^{-1}. The eggs were collected from the broodstock tanks and hatching was completed after two days. The feeding scheme (from hatching until the first 120 days) is presented in Fig. 1. The amounts as well as the type of feed supplied, until D120, were kept the same for all experimental groups (Table 2).

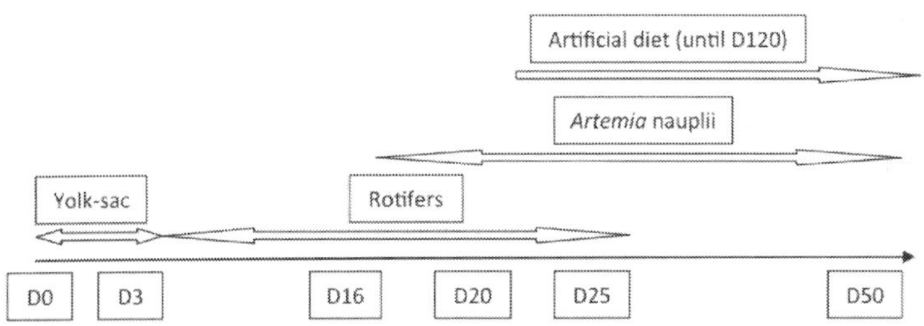

Figure. 1: Feeding scheme of larvae/fry during 120 days post-hatching.

Table 2: Food supply per experimental culture tank of gilthead sea bream fry

Live food type	**Quantity**	**Dry feed (μm)**	**Quantity (kg)**
Algae	11.5 m^3	100–200	2
Rotifers	3200×10^6	200–300	20
Artemia nauplii	5300×10^6	300–500	35
		400–500	65
		500–800	190
		800–1200	350

The physicochemical conditions were the same in all experimental tanks; Temperature until weaning was kept at 19–20 °C; afterwards and until the end of the "hatchery phase" it ranged between 18 and 22 °C. The level of oxygen saturation ranged between 85 and 105 % and pH was kept stable at 7.8. For the purpose of the study, several performance factors were evaluated as follows. The number of the weaned larvae per tank; the final survival expressed as the percentage of the initial stocking of the tank (based on the daily collection of the dead larvae/fry), as well as the percentages of: (1) the mortality (from the beginning of the weaning phase to fry sales); (2) the fry without swim bladder at a mean weight of 0.5–1.0 g which were rejected from the tanks and counted after buoyancy testing (i.e. these individuals sink to the bottom during a salinity increase from 35–36 to 58–60 ppt, after anaesthetized with 2-phenoxy-ethanol); (3) the fry with skeletal deformities which were also rejected by hand, following optical observation and counted during the previous buoyancy test; and (4) the total fry losses after weaning. The type of the skeletal deformities observed in this study were the lack of or the existence of atrophic operculum, snout/mouth dysplasia, lack of dorsal fin and hypoplasia of the caudal fin, with the last two of them being of significant commercial importance (Chatain and Corrao 1992; Morretti et al. 1999). Finally, feed conversion ratio (FCR = the amount of dry feed provided in kg/wet weight gain in kg) was also recorded for the supply period of dry feed.

The mean values of the measurements between broodstock groups were compared by applying one-way analysis of variance (ANOVA) at significance level of $p < 0.05$. All statistical tests were performed using the statistical program Statgraphics 2.1.

The hatchery used the BA broodstock group of Mediterranean origin since the early 2000s' (F_0generation). The typical production cost per fry was estimated based on the financial statements of the hatchery's annual production of 22 million fry (Table 3). For the evaluations of the economic data the "Active Based Costing" analysis of the hatchery costs was used. The results are discussed as "production cost per fry" and "distribution of costs per fry".

Table 3: Typical cost structure of the gilthead sea bream hatchery with annual production of 22 millions fry per year

Type of hatchery costs	Total costs (%)	Values (€)
Staff-labor	40	580,000
Energy	6	85,000

Larvae-weaning feeds	15	215,000
Fry-feeds	7	95,000
Drugs-chemicals	1	18,700
Oxygen	3	41,500
General expenditures	6	102,000
Administrative costs	7	100,000
Depreciations	15	210,000
Total	100	1,447,200
Cost per fry		0.066

A critical part of the decision-making process of the hatchery operating management is the planning for improved outcome of the next annual production cycle through innovations (Theodorou et al. 2015). In this case, the managing staff had to decide, if it is required to replace the given broodstock (BA) of Mediterranean origin with other broodfish groups such as BB broodstock (F_1 generation) from selected offspring originating from the BA broodstock group. Eggs from this group were first used in the production cycle of the hatchery during 2007. The BC group consisted of Atlantic (French) broodfish, originally imported as fry of different geographic origin, and was compared to the previous two groups.

The BA broodstock group consisted of 43 fish of mean weight of 4.9 kg, BB group of 45 fish with mean weight of 2.4 kg and BC group of 40 fish with mean weight of 2.6 kg. In all groups males were less than 10 %. The exact number of males and the effective breeding numbers are not available. Sex ratio in sea bream broodstock tanks is not stable given that the species is a protandrous hermaphrodite. Moreover, hatchery managers avoid disturbing the broodfish with manual handling (i.e., stripping) in spawning tanks in order to reduce the stress levels. The eggs in all groups used in the current experiment were produced in a 7–10 day period (from 4 to 15 of May 2009), under increasing artificial photoperiodic regime, which is suitable for spawning synchronization. The protocol was LD 10:14 during the middle of the spawning period (May), starting from LD 8.5:15.5 at the beginning (15th of March) to LD 12:12 close to the end (30th of June). Hatching took place more or less simultaneously. The typical broodstock selection process was based on growth rate and the desired phenotype (i.e., body shape). The first selection stage was performed during the pre-ongrowing phase, where the 2 % of the fastest growing fry was transferred to floating cages in thc sea. Then, at a body weight >800 g, the best 500 fish were transferred to the hatchery in quarantine and finally the best 50 broodfish were selected at most to become the broodstock group.

RESULTS

Production Results

The production data in the three broodstock groups are presented in Table 1. Fry without swim bladder ranged between 2 and 12 %, the percentage of deform fry varied at 4–24 %, while ready-to-sale fry rates ranged between 63 and 78 %. Although no statistical differences were detected ($p > 0.05$ in all performance factors), the BA group presented the higher percentages of fry without swim-bladder and fry with deformities. Accordingly, total losses were increased and conversely the available fry for sale decreased compared to rest of broodstock groups. The most common deformities detected in all groups were those on the head region, while specimens with opercular deformities represented 2 % of the overall deformities. No significant differences were also evident in the final survival rate, fry rejections due to skeletal deformities and fry without swim bladder as well as in the FCR.

Economic Results

The recording of all economic data for the typical production procedure of the hatchery revealed that production cost was €0.066 per fry when the hatchery was fully operating year-round for a production capacity of 22 million fry, as shown in Table 3.

The distribution of the annual costs in terms of both feeds and personnel costs is presented in Figs. 2 and 3, respectively. *Artemia* cysts together with the enriched food represented 40 % of the total food costs, while starter feeds during the weaning phase represented 22 % of the costs (Fig. 2). The personnel costs represented approximately 50 % of the entire costs of the hatchery (Fig. 3).

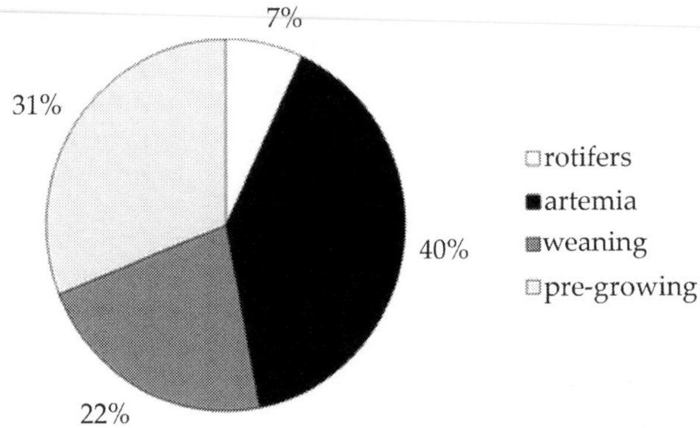

Figure. 2: Hatchery's typical distribution of food costs for gilthead sea bream fry production.

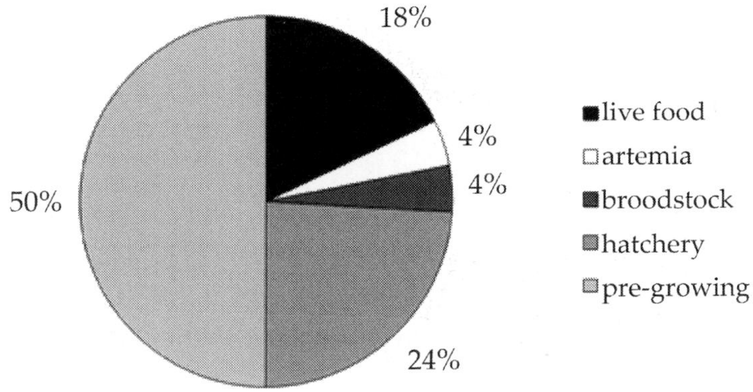

Figure. 3: Hatchery's typical distribution of personnel costs for gilthead sea bream fry production.

Given that there was no statistical differences in the production outcome, the elimination of the deformities between the existing operating broodstock group (BA) and the tested alternative sources (BB and BC groups) is not expected to affect dramatically the economic performance and accordingly the annual returns.

DISCUSSION

Deformities such as the opercular absence affected until recently almost up to 80 % of the farmed gilthead sea bream (Chatain 1994; Andrades et al. 1996; Verhaegen et al. 2007). Increased number of fry without swim bladder was observed in the second batch of BA broodstock group and in the two first batches of BB and BC groups. It was attributed to technical problems in the air supply system that occurred in the hatchery, causing problems to the oil skimmers. This confirmed the importance of the skimmers in inhibiting lipid formation on water surface (Chatain and Ounais-Guschemann 1990).

In the present study, the three different broodstock groups used in the hatchery showed no significant differences on fry deformities and total losses. In addition, all physicochemical conditions and the feeding protocol applied were identical for all groups and no such effects as environmental or nutritional could be detected. Existing literature on the genetic influence on deform fry is contradictory, while interactions between genetic, environmental factors and rearing procedures (e.g., water velocity, temperature and artificial diets are probably the most critical factors) (Sfakianakis et al. 2006; Vatsos and Angelidis 2010) have also been suggested to explain the origin as well as the incidence of fry deformities (Kause et al. 2007; Bardon et al. 2009). Opercular deformities are considered non inheritable in *Tilapia nilotica* (Tave and Handwerker 1994) as well as in gilthead sea bream (Castro et al. 2008). However, inbreeding increased opercular deformities in *Cichlasoma nigrofasciatum* (Winemiller and Taylor 1982). It was also reported that critical performance factors such as fecundity and egg quality vary among different genetic origin of broodstock (Brooks et al. 1997).

The cost analysis revealed qualitative as well as quantitative typical data of the hatchery (Table 3; Figs. 2, 3). Significant economic losses have been attributed to high mortalities during the early developmental stages, which are common in marine aquaculture; thus any effort towards the reduction of the production cost even based only on the reduction of the deformed fry is expected to have significant impact on hatchery economics. The early rejection of deformed fry which took place in the present study reduced both feeding and personnel costs required to prepare live food and to short the fry. The reduction of the personnel occupied either in fry rearing or in live food production could result to a corresponding labour reduction in the pre-growing stage. The decrease of feeding costs became also important, since no differences were apparent in FCR in the present study.

It is concluded that, although no statistical differences were evident between the performance of the different broodstock and fry sources (BA, BB and BC), there could be economic benefits even by the slightly better values of BB and BC groups in critical factors such as the percentage of total losses and eventually the percentage of fry for sale. At any case, the scope for fry improvement through improved broodstock is open-wide; therefore, new trials should be carried out to introduce in the hatchery new possible broodstock candidates of different origins and by genetic selection programmes aiming to pedigree certification.

REFERENCES

1. Andrades JA, Becerra J, Fernandez-Llebrez P (1996) Skeletal deformities in larval, juvenile and adult stages of cultured gilthead sea bream *Sparus aurata* L. Aquaculture 141(1):1–11
2. Bardon A, Vandeputte M, Dupont-Nivet M, Chavanne H, Haffray P, Vergnet A, Chatain B (2009) What is the heritable component of spinal deformities in the European sea bass (*Dicentrarchus labrax*). Aquaculture 294(3):194–201
3. Boglione C, Gagliardi F, Scardi M, Cautaudella S (2001) Skeletal descriptors and quality assessment in larvae and post larvae of wild-caught and hatchery-reared gilthead sea bream (*Sparus aurata* L., 1758). Aquaculture 192(1):1–22
4. Boglione C, Gavaia P, Koumoundouros G, Gisbert E, Moren M, Fontagné S, Witten PE (2013a) Skeletal anomalies in reared European fish larvae and juveniles. Part 1: normal and anomalous skeletogenic processes. Rev Aquacult 5(1):99–120
5. Boglione C, Gisbert E, Gavaia P, Witten PE, Moren M, Fontagné S, Koumoundouros G (2013b) Skeletal anomalies in reared European fish larvae and juveniles Part 2: main typologies, occurrences and causative factors. Rev Aquacult 5(1):121–167
6. Brooks S, Tyler CR, Sumpter JP (1997) Egg quality in fish: what makes a good egg? Rev Fish Biol Fisher 7(4):387–416
7. Cahu CL, Zambonino Infante JL, Takeuchi T (2003) Nutritional components affecting skeletal development in fish larvae. Aquaculture 227(1):1–4

8. Carrillo J, Koumoundouros G, Divanach P, Martinez J (2001) Morphological malformations of the lateral line in reared gilthead sea bream (*Sparus aurata* L. 1758). Aquaculture 192(2–4):281–290
9. Castro J, Pino-Querido A, Hermida M, Chavarrías D, Romero R, Garcia-Cortes LA, Toro MA, Martinez P (2008) Heritability of skeleton abnormalities (lordosis, lack of operculum) in gilthead seabream (*Sparus aurata*) supported by microsatellite family data. Aquaculture 279(1):18–22
10. Chatain B (1994) Abnormal swim bladder development and lordosis in sea bass (*Dicentrarchus labrax*) and sea bream (*Sparus aurata*). Aquaculture 119(4):371–379
11. Chatain B, Corrao D (1992) A sorting method for eliminating fish larvae without functional swimbladders. Aquaculture 107(1):81–88
12. Chatain B, Ounais-Guschemann N (1990) Improved rate of initial swimbladder inflation in intensively reared *Sparus aurata*. Aquaculture 84(3):345–353
13. Divanach P, Boglione C, Menu M, Koumoundouros G, Kentouri M, Cautaudella S (1996) Abnormalities in finfish mariculture: an overview of the problem, causes and solutions. In: Chatain B, Sargalia M, Sweetman J, Lavens P (eds) Sea bass and sea bream culture: problems and prospects. European Aquaculture Society, Verona, pp 45–66 **(Special Publication)**
14. Galeotti M, Beraldo P, De Dominis S, D'Angelo L, Ballestrazzi R, Musetti R, Pizzolito S, Pinosa M (1999) Histological and ultrastructural study of opercular anomalies in gilthead sea bream larvae (*Sparus aurata*). Bull Eur Fish Pathol 19(3):123–126
15. Galeotti M, Beraldo P, de Dominis S, D'Angelo L, Ballestrazzi R, Musetti R, Pizzolito S, Pinosa M (2000) A preliminary histological and ultrastructural study of opercular anomalies in gilthead sea bream larvae (*Sparus aurata*). Fish Physiol Biochem 22:151–157
16. García-Celdrán M, Cutákova Z, Ramis G, Armero E (2014) Effect of the geographical origin on the presence of skeletal deformities and swimbladder anomaly in a cultured gilthead sea bream (*Sparus aurata* L) population. Estimation of genetic parameters. III Workshop en Investigación Agroalimentaria—WiA3.14. Cartagena, 12–13 de mayo de 2014 pp 85–88

17. García-Celdrán M, Ramis G, Manchado M, Estévez A, Afonso JM, María-Dolores E, Peñalver J, Armero E (2015) Estimates of heritabilities and genetic correlations of growth and external skeletal deformities at different ages in a reared gilthead sea bream (*Sparus aurata* L.) population sourced from three broodstocks along the Spanish coasts. Aquaculture 445:33–41
18. Johnson DW, Katavic I (1984) Mortality, growth and swim bladder stress syndrome of sea bass (*Dicentrarchus labrax*) larvae under varied environmental conditioned. Aquaculture 38(1):67–78
19. Kause A, Ritola O, Paananen T (2007) Changes in the expression of genetic characteristics across cohorts in skeletal deformations of farmed salmonids. Genet Sel Evol 39(5):529–543
20. Koumoundouros G, Kiriakos Z, Divanach P, Stefanakis P, Kentouri M (1995) Morphometric relationships as criteria for the evaluation of larval quality of gilthead sea bream. Aquacult Int 3(2):143–149
21. Koumoundouros G, Oran G, Divanach P, Stefanakis P, Kentouri M (1997a) The opercular complex deformity in intensive gilthead sea bream (*Sparus aurata* L.) larviculture. Moment of apparition and description. Aquaculture 156(1–2):165–177
22. Koumoundouros G, Gagliardi F, Divanach P, Boglione C, Cataudella S, Kentouri M (1997b) Normal and abnormal osteological development of caudal fin in *Sparus aurata* L. fry. Aquaculture 149(3–4):215–226
23. Liao I-C, Su M-S, Chang S-L (1993) A review of fish genetic research and conservation issues in Taiwan. In: Main K, Reynolds E (eds) Selecting breeding of fishes in the United States and Asia. The Oceanic Institute Makapuu point Hawaii, Honolulu, pp 162–180
24. Morretti A, Pedini-Fernandez-Criado M, Cittolin G, Guidastri R (1999) Manual on hatchery production of seabass and gilthead seabream, vol 1. FAO, Rome, p 194
25. Pavlidou P (2009) Present and future status of Mediterranean fry. In: Proceedings of Finfish Training Course on Prevention of Malformations in Sea Bass and Sea Bream. Athens, Greece
26. Sfakianakis DG, Georgakopoulou E, Papadakis IE, Divanach P, Kentouri M, Koumoundouros G (2006) Environmental determinants of haemal lordosis in European sea bass, *Dicentrarchus*

labrax (Linnaeus, 1758). Aquaculture 254:54–64

27. Tave D, Handwerker TS (1994) Semi-operculum: a non-heritable birth defect in *Tilapia nilotica*. J World Aquac Soc 25(2):333–336
28. Theodorou JA (2002) Current and future technological trends of European seabass-seabream culture. Rev Fish Sci 10(3–4):529–543
29. Theodorou JA, Perdikaris C, Filippopoulos NG (2015) Evolution through innovation in aquaculture: a critical review of the Greek mariculture industry. J Appl Aquacult 27(2):160–181
30. Vatsos IN, Angelidis P (2010) Water quality and fish diseases JHVMS 61(1):40–48
31. Verhaegen Y, Adrianes D, de Wolf T, Dhert P, Sorgeloos P (2007) Deformities in larval gilthead sea bream (*Sparus aurata*): a qualitative and quantitative analysis using geometrics morphometrics. Aquaculture 268(1):156–168
32. Winemiller KO, Taylor DH (1982) Inbreeding depression in the convict cichlid, *Cichlasoma nigrofasciatum* (Baird and Girard). J Fish Biol 21(4):399–402

SECTION VI

FUTURE: SUSTAINABILITY AND ENVIRONMENTAL IMPACTS

CHAPTER

18

GLOBAL AQUACULTURE PERFORMANCE INDEX (GAPI): THE FIRST GLOBAL ENVIRONMENTAL ASSESSMENT OF MARINE FISH FARMING

John P. Volpe, Jennifer L.M. Gee, Valerie A. Ethier, Martina Beck, Amanda J. Wilson and Jenna M.S. Stoner

Seafood Ecology Research Group, School of Environmental Studies, University of Victoria, Victoria, BC V8W 2Y2, Canada

ABSTRACT

"Sustainable" is among the most sought after of all seafood product adjectives. Ironically it is also one of the most poorly defined and understood. The Global Aquaculture Performance Index (GAPI) is the first tool to assess environmental performance of global marine aquaculture production, permitting direct comparison of disparate species, production methods and jurisdictions. Clear patterns emerge from this analysis; significant variation of environmental performance is driven by the species being farmed, significant room for improvement exists across the entire sector, the worst performing players are also the fastest growing, particularly within Asia, and perhaps most importantly, this work highlights the potential trap awaiting policy makers who focus too narrowly on farm production efficiency alone as a solution to diminishing seafood availability.

Keywords: Aquaculture, efficiency, marine ecology, performance metrics, seafood, sustainable production

INTRODUCTION

With over 87% of global capture fisheries currently fully- or over-exploited [1] aquaculture is looked upon with increasing urgency to fill the growing global demand for seafood. Over the past three decades the growth of aquaculture production has exceeded all other agricultural sectors worldwide (8.8% annual compounded growth since 1980) [1]. Such rapid growth does not come without challenges and production related environmental impacts are diverse and well documented [2,3,4]. To further complicate matters, while overall production grows rapidly so too does the diversity of species being brought into culture [1] thus amplifying the breadth of potential aquaculture × environment interactions.

Seafood is among the most global of commodities with international imports dominating seafood consumption in most developed nations [1]. Environmentally conscious buyers of aquaculture products face a complex calculus in determining how differing species, production regions and production systems all affect "sustainability" of the product. In response, a *de facto* sustainable seafood industry has arisen, aiding buyers, both wholesale and retail, in making informed conservation choices. A review of 63 market-based initiatives concluded the lack of coherence across the plethora of guides, standards and certifications is confusing buyers and the lack of demonstrably improved performance on the water undermines the potential efficacy of the entire approach [5].

In order to bring clarity to the myriad species, locales and production methods characterizing modern aquaculture we adopted the analytical foundation of the Environmental Performance Index (EPI) [6]. The EPI is a globally recognized statistical framework which scores environmental performance of all recognized countries against 10 core environmental issue areas. Each issue area yields a performance score that is the weighted aggregate of multiple independent metrics. EPI country rankings are presented biennially at the World Economic Forum meeting in Davos, Switzerland, and have had a transformative effect on the way global environmental performance is measured and compared. Our tool, retrofitted specifically for assessment of marine finfish aquaculture products is the Global Aquaculture Performance Index (GAPI). Like the EPI, GAPI indicates which products perform best across an array of environmental criteria allowing users to drill down to assess performance within each species, producing country, or individual issue areas. In so doing, clear environmental leaders and laggards (both species and countries) are made apparent. Perhaps most importantly, best performing combinations of species, countries and production systems are identified, and provide clear templates of improvement for those lagging behind.

Quantifying environmental performance of aquaculture production has historically proven difficult, reflecting scarcity of data, inconsistent reporting, incomplete science, and a wide range of potential environmental impacts across a global distribution of production. As significant as these challenges may be, they are insufficient to excuse inaction. However, prerequisite to addressing the challenge is creation of a baseline "state of the industry" performance snapshot designed for clear policy relevance.

EXPERIMENTAL SECTION

Which Metrics Should Be Included and How to Measure Each?

Literally hundreds of metrics could conceivably be employed in the assessment of aquaculture environmental performance, but which should be and how should inclusion/omission be decided? A pilot survey of current seafood sustainability initiatives was undertaken to determine what suite of performance criteria should be included in the GAPI assessment tool. Issues addressed repeatedly across initiatives are, by consensus, considered to be significant and strong candidates for inclusion here. We assessed 30 aquaculture sustainability schemes active in North America and Western Europe. Ten issue areas were consistently addressed across initiatives. These 10 markers of environmental performance (Table 1) were deemed the minimum necessary suite of indicators required for comprehensive assessment of global marine aquaculture and are in fact the product of a *de facto* peer review by the sustainable seafood community. To determine performance on a 0 to 100 scale as we have done, absolute best (100) and worst (0) performance must first be defined. A perfect score of 100 equates to absolutely no measurable environmental impact. Perfection (absolutely no impact in each of the 10 criteria) is clearly unattainable but the closer a player comes, the higher the score. Determining the worst performance (a score of 0) is more complicated. Theoretically, a given product could perform infinitely poorly within one or all indicators. For instance, what would be the worst possible performance for antibiotic or parasiticide use? The scope is, at least theoretically, infinite. To solve this dilemma, GAPI reviews the pool of performances for that indictor and sets "0" as the worst observed actual performance. Thus, like a classroom grading system, GAPI grades on a curve, where a performer's GAPI score is partially dependent on the performance of the pool of players among which it is being assessed. This is consistent with the objective of the tool, which is to generate performance profiles that are informative for making comparisons of two or more products. A further implication is that a GAPI score is only informative relative to another score. GAPI is not a standard with a threshold above which a product is considered "sustainable". Rather GAPI scores reveal strengths and weaknesses in environmental performances relative to other players, and thus avoids the false simplicity of a single absolute sustainability threshold.

Table 1: The 10 Global Aquaculture Performance Index (GAPI) indicators with a brief description and formula for each.

Criteria	Criteria Description	Criteria Equation
Antibiotic	Amount of antibiotics used, weighted by a measure of human and animal health risk.	Σ (Amount Active Ingredient (kg) × WHO-OIE Score) / mT Fish Produced
Antifoulants (Copper)	Estimated proportion of production using copper-based antifoulants.	(BOD (mg O_2) * Area of Impact (km^2)) / mT Fish Produced
Biochemical Oxygen Demand (BOD)	Relative oxygen-depletion effect of waste contaminants (uneaten feed and feces).	∑(Amount (kg) × [(1/LC50) + 1] * Persistence (Days)) / mT Fish Produced
Capture-Based Aquaculture	The extent to which a system relies on the capture of wild fish for production, taking into account the sustainability of these wild fish inputs.	(∑Proportion Fish/Livestock/Plant * Knife Coefficient (Megajoules/mT) * Total Feed Consumed (mT)) / mT Fish Produced
Ecological Energy	Amount of energy, or net primary productivity (NPP), that farmed fish divert from the marine ecosystem through consumption of wild fish ingredients of feed	∑ Net Primary Production of Feed Inputs (g carbon/kg) / mT Fish Produced
Escapes	Number of escaped fish, weighted by an estimate of the per capita risk associated with escapes.	∑Pathogen-specific Wild Losses (mT) / mT Fish Produced
Industrial Energy	Energy consumed in the acquisition and processing of feed ingredients, which serves as a proxy for the total industrial energy used in production.	∑(Amount from Wild Capture (kg) * Sustainability Score) / mT Fish Produced
Parasiticide	Amount of parasiticides used, weighted by measures of environmental toxicity and persistence.	(Production (mT) * % Production Using Copper-based Antifoulants) / mT Fish Produced
Pathogens	Number of on-farm mortalities, weighted by an estimate of wild species in the ecosystem that are susceptible to farm-derived pathogens.	∑(Amount (kg) × [1/LC50 + 1 × Persistence (days)] / mT Fish Produced
Sustainability of Feed	Amount, efficiency, and sustainability of wild fish ingredients of feed.	∑((Proportion of Feed by Species * Sustainability Score of Each Species) * Fish In: Fish Out Ratio * Production (mT)) / mT Fish Produced

* GAPI takes the square root of each indicator formula to make the range of performance values more manageable and disperse the final scores so that differences are more apparent.

Following the identification of the 10 issues which were to become the focus of GAPI, the next step was developing the analytical framework to measure on-the-water performance. Deriving metrics capable of capturing all marine finfish aquaculture production globally is a significant undertaking. Numerous expert workshops were convened, each focused on developing and refining specific ecological indicators. This ensued a multiyear process cumulatively involving substantial input from more than 30 experts, including biologists, producers, statisticians, seafood buyers, and individuals engaged in sustainability assessment. The consensus products of these workshops are presented in Table 1.

Since final GAPI scores are informed by the pool of country-species assessed, it is critical that the pool is representative of the entire peer group. In 2007, 82 marine finfish species or species groups were farmed in 62 countries (FAO 2008a). However, production was dominated by a relatively small number of species. The present assessment was restricted to the top 20 species by production (Table 2), which cumulatively constituted 98.5% of all marine finfish aquaculture production. The remaining 1.5% of global production is spread across an additional 39 species. Just as a small number of species comprise the majority of production, the same is true for producing countries. A species may be farmed in numerous countries but typically the great majority of production occurs in only a few. Thus only those countries that together comprised the top 90% of production of each of the 20 selected species were included. These two decision rules resulted in an assessment of 20 marine finfish species being produced in 22 producing countries which together comprise 94% of marine finfish production by weight (mT) and 91% by value (USD) [7].

Table 2. Species assessed by GAPI in descending order of 2007 production volume.

Species Rank (mT)	Marine Finfish Species (in production in 2007)	% of Global Marine Finfish Production by Weight	% of Global Marine Finfish Production by Value	Producing Countries Included in GAPI Analysis (Representing 90% of Species Production)
1	Atlantic salmon	40.8	50	Canada, Chile, Norway, UK
2	Milkfish	17.1	4.6	Indonesia, Philippines
3	Flathead grey mullet	6.7	3.8	Egypt
4	Japanese amberjack	4.9	9	Japan
5	Red seabream	3.8	3.7	China, Japan
6	Gilthead seabream	3.5	4.7	Greece, Israel, Italy, Spain, Turkey
7	Coho salmon	3.3	3	Chile, Japan
8	Bastard halibut	3.2	4.1	China, Korea
9	European seabass	3	4.3	Greece, Italy, Spain, Turkey
10	Japanese seabass	2.9	0.9	China
11	Groupers	2	1.4	China, Indonesia, Taiwan
12	Large yellow croaker	1.8	0.5	China
13	Red drum	1.5	0.5	China
14	Korean rockfish	1	1.4	Korea, Australia, Indonesia, Malaysia
15	Barramundi	0.9	0.6	Thailand
16	Cobia	0.8	0.4	China, Taiwan
17	Tiger pufferfish	0.5	0.7	China, Japan
18	Atlantic cod	0.3	0.3	Iceland, Norway
19	Chinook salmon	0.3	0.5	Chile, New Zealand
20	Turbot	0.2	0.4	France, Spain
	Total % of Global Marine Finfish production	98.5	95	
	% of Global Marine Finfish production assessed by GAPI	93.7	91	

Deriving the GAPI Score

The derivation of the final GAPI score for each species-country pair assessed (e.g., Atlantic salmon–Norway) consists of eight steps (full methodology detailed in Appendices A and B as well at Global Aquaculture Performance Index).

Select Key Indicators of Environmental Performance

Emphasis has been placed on identifying a suite of indicators that sufficiently describes the major ecological impacts of marine finfish aquaculture while using the fewest indicators possible.

Construct Indicator Metrics

In order to determine how to best measure actual performance we developed specific criteria to ensure that each is:

- relevant and measures direct environmental impact;
- performance oriented and tracks actual, on-the-water performance; (as opposed to aspirational or "best practice")
- transparent (both formulae and data); and
- utilizes the highest quality data available

Details of the process are presented in Table 1 .

Set Targets for Each Indicator

By setting a zero-impact target for each indicator, GAPI permanently sets the environmental performance at the ecological ideal rather than continually recalibrating the goal as the performance of the industry improves or as viewpoints of what is an "acceptable" level of impact shift.

Collect Data

A wide range of data drawn from international organizations, regulatory bodies, conservation organizations, academia, seafood industry groups, and seafood industry trade press were used. All data used are publicly available and traceable. This step represents significant effort not only in acquisition but quality assessment and standardization of data and explains the time lag between production date and scoring.

Winsorization

Winsorization is a common statistical approach [8] to dealing with extreme outliers so those values do not distort the distribution of the entire data set. This is important to maintain the legitimacy of the dataset.

Proximity-to-Target Calculation

In order to directly compare performance among two or more disjunct indicators (*i.e.*, escaping fish and the sustainability of feed sources) in a statistically meaningful way, it is necessary to standardize performance for each on the same 0-to-100 scale. Proximity-to-target calculations quantify how close a performer is to meeting zero-impact for each of the 10 indicators.

Weighting Indicators

The 10 indicators included within GAPI have already been deemed by the conservation community to be important drivers of environmental performance. Principal component analysis (PCA) measures how much of the total variation in the data is explained by each indicator, thus providing a measure of each indicator's relative importance or weight (Table 3). To be clear, PCA does not ascribe "importance". Any user-defined combination of weights would ultimately be subjective. Likewise, assigning equal weights to each of the 10 indicators is itself a subjective weighting. Indeed, indicators used in sustainability assessments typically reflect subjective judgments without mentioning or systematically assessing critical assumptions [9]. The PCA approach ascribes weight according to that criteria's capacity to separate leaders from laggards; an objective statistical technique absent any investigator influence.

Table 3. Example calculation of the final normalized GAPI score for tiger puffer fish produced in China.

A	B	C	D	E
Indicator	**Indicator Performance**	**PCA Derived Weight (%)**	**Weighted Performance (PCA Weight (%) × Indicator Performance)**	**GAPI Score (Sum of Weighted Performances)**
ANTI	37	15	5.6	42
BOD	72	8.3	3.7	
CAP	100	5	5.1	
COP	28	5	2.3	
ECOE	66	15	9.8	
ESC	47	8.3	3.9	
FEED	11	8.3	1.6	
INDE	60	15	5	
PARA	84	5	4.3	
PATH	3	15	0.5	

Note: First, the performer's environmental performance within each indicator (Columns A and B) is determined by calculating the proximity-to-target for each normalized indicator and standardized on a scale of 0 (worst) to 100 (no impact) (Column C). The weight that each of these indicators contributes to the final score is then calculated using Principal Component Analysis (PCA) (Column D). The product of indicator performance (Column C) and PCA-derived weight, expressed as a percentage (Column D) yield the weighted indicator performance (Column E). The final GAPI score (Column F), which describes China's normalized tiger puffer fish aquaculture performance, is the sum of the 10 weighted performance scores in Column E.

Calculating the Final Country Score

The final score is the sum of the 10 criteria scores post-PCA weighted. GAPI reports two scores for each product: a normalized score and a cumulative score.

Normalized scores are standardized to reflect environmental performance per mT of production (Table 3). Normalized scores reflect the inherent production profile of that species and are not influenced by scale of production (as are cumulative scores). Because impacts are standardized per mT of production, policy relevant questions may be asked directly; How does a species produced in one country compare to other countries producing the same species? Which species score consistently poorly and which consistently well *and why*? What specific dimensions of performance can most immediately and cost effectively be improved and which jurisdictions may provide a template for doing so (based on their superior score)?

For instance, China scores particularly poorly in biological oxygen demand (BOD), ecological energy consumption (ECOE) and feed use (FEED) relative to other countries (Table 4). Resources are more likely to yield substantive performance improvement if targeted at one or more of these areas rather than escapes (ESC) or parasiticides (PARA), where China scores relatively well and therefore the scope for improvement is more modest. Normalized scores encourage policymakers to think about regulations that can improve the relative performance of the industry. They provide an "apples to apples" comparison against other industries or countries, regardless of their size.

Table 4: Criteria scores contributing to the overall normalized scores. Cell shading indicates criteria group; Inputs (none), Outputs (light) and Biological (dark) criteria

Species	Country	Overall Score	CAP	ECOE	FEED	INDE	ANTI	BOD	COP	PARA	ESC	PATH
Atlantic Cod	Norway	62	100	64	68	58	98	59	28	45	0	63
Atlantic Cod	Iceland	69	100	64	68	58	98	92	28	88	0	87
Atlantic Salmon	Chile	66	100	72	75	62	68	81	28	77	38	67
Atlantic Salmon	Canada	70	100	74	75	57	90	66	28	99	39	72
Atlantic Salmon	UK	72	100	66	75	60	96	61	49	53	40	86
Atlantic Salmon	Norway	72	100	75	75	64	99	50	28	84	40	78
Barramundi	Thailand	41	100	39	44	40	38	0	28	40	35	46
Barramundi	Indonesia	46	100	39	44	40	42	94	28	5	35	57
Barramundi	Australia	47	100	70	65	0	58	62	28	30	35	24
Barramundi	Malaysia	50	100	39	44	40	77	75	28	0	35	57
Bastard Halibut	China	41	100	59	34	61	37	39	28	69	49	0
Bastard Halibut	Korea,	61	100	51	42	23	82	44	100	4	100	57
Chinook Salmon	Chile	64	100	75	75	55	68	98	28	64	42	53
Chinook Salmon	New Zealand	73	100	70	52	43	100	100	28	100	47	100
Cobia	Taiwan	35	100	0	51	5	24	74	28	6	41	60
Cobia	China	37	100	0	51	5	37	42	28	35	41	60
Coho Salmon	Japan	56	100	68	62	61	13	68	28	79	34	77
Coho Salmon	Chile	63	100	71	75	59	68	90	28	64	37	52
European Seabass	Turkey	61	100	59	74	43	50	66	28	81	53	75
European Seabass	Spain	63	100	60	69	47	57	93	31	78	49	75
European Seabass	Greece	66	100	61	75	50	81	58	28	81	47	75
European Seabass	Italy	69	100	61	72	31	64	97	64	81	74	75
Flathead Grey Mullet	Egypt	71	68	67	99	57	32	79	100	81	41	91
Gilthead Seabream	Turkey	55	100	75	69	55	50	66	28	81	5	44
Gilthead Seabream	Greece	56	100	69	79	60	50	49	28	57	5	56
Gilthead Seabream	Spain	57	100	69	72	57	50	90	31	56	8	56
Gilthead Seabream	Israel	60	100	67	80	41	50	69	49	77	32	56
Gilthead Seabream	Italy	65	100	67	81	37	50	98	64	81	53	56

Groupers	Indonesia	10	0	0	0	0	0	85	28	6	32	3
Groupers	China	15	0	0	0	0	37	21	28	47	35	3
Groupers	Taiwan	28	0	41	45	39	24	58	28	0	35	3
Japanese Amberjack	Japan	45	99	60	39	56	36	0	28	84	43	32
Japanese Seabass	China	32	54	24	18	25	31	4	28	44	35	57
Korean Rockfish	Korea	54	100	55	52	40	82	45	28	0	49	57
Lg Yellow Croaker	China	41	100	23	33	27	31	37	28	89	48	57
Milkfish	Phillippines	59	98	78	90	57	69	37	28	0	46	39
Milkfish	Indonesia	68	98	80	90	61	97	67	28	45	46	39
Red Drum	China	26	100	0	4	5	37	21	28	49	37	37
Red Seabream	China	25	100	0	0	0	37	11	28	65	44	32
Red Seabream	Japan	47	100	59	60	52	36	9	28	72	44	32
Tiger puffer	Japan	39	100	69	11	60	36	80	28	21	47	3
Tiger puffer	China	42	100	66	11	60	37	72	28	84	47	3
Turbot	Spain	63	100	68	83	44	23	82	51	65	61	73
Turbot	France	68	100	68	81	44	52	100	51	67	61	73

In contrast, cumulative scores are not expressed per unit production but instead are weighted by total production to reflect the total aggregate effect of that production along that country's coastline. Cumulative scores look at the aggregate effect of an industry: *i.e.*, what is the overall impact of a country's aquaculture industry? Cumulative performance is rarely if ever reported however it is this perspective that confronts policymakers with important questions of industry scale and carrying capacity.

Both normalized and cumulative measures are important. To use an analogy from climate change, CO_2 emissions have a minor impact on a normalized basis compared to methane (a mT of methane is orders of magnitude more damaging than a mT of CO_2). However, the magnitude of CO_2 emissions have an earth-changing cumulative impact. As a consequence, governments are attempting to address emissions of both gases. In simplest terms, normalized scores are most relevant for industry while cumulative scores are the only relevant metrics from an ecological perspective (ecological processes do not recognize performance per mT, only cumulative performance).

RESULTS AND DISCUSSION

Sustainability must be demonstrated not assumed. Data availability and quality remain preeminent challenges to any assessment of seafood sustainability. However, verification of the sustainability of any production system requires that abundant, high-quality data are available for analysis. We found data deficiencies to be particularly challenging in the traceability of feedstocks, feed formulation, and the cumulative ecosystem effects of both chemical use and escapes. Currently, determinants of sustainability are typically informed by spotty qualitative data leading to questionable conclusions that may reflect vested interests more than actual performance. The long-term ecological and economic viability of the aquaculture industry depends on shifting policy and production decisions toward quantitatively rigorous, performance-based regulatory frameworks such as GAPI.

Not all marine finfish aquaculture is the same. While it might be reasonable to assume significant performance differences across drastically different types of aquaculture such as shellfish farming and marine finfish farming, GAPI scores reveal tremendous variation in environmental performance just within the marine finfish sector (Figure 1 and Figure 2). These variations are highlighted in species-country pair scores, country scores, and species scores. For instance, normalized species-country scores range from a low of 10 (groupers – Indonesia) to a high of 73 (Chinook salmon - New Zealand).

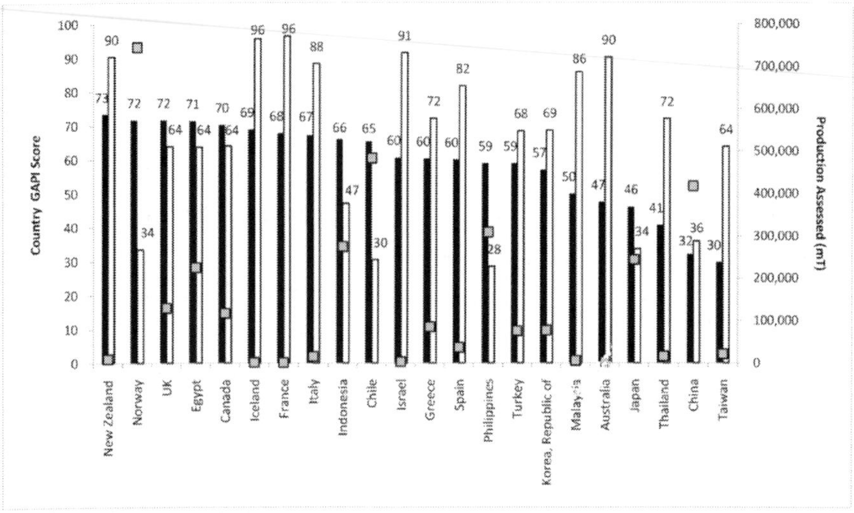

Figure 1: Weighted mean country performance expressed as normalized GAPI scores.

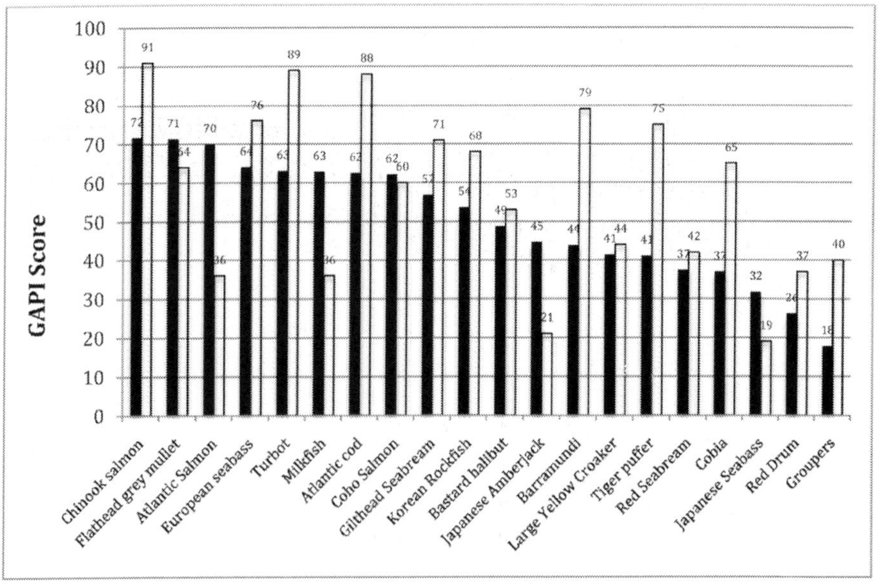

Figure 2: Weighted mean species performance expressed as normalized GAPI scores (performance per mT production; black bars) and cumulative GAPI scores (performance of total global production; open bars). Species arranged by decreasing normalized score.

There remains substantial room for improvement. While there is strong variation in GAPI scores across countries and species, and while GAPI does not define pass or failure scores, the findings strongly suggest there is room for improvement within the entire marine finfish sector. Even the best performers are approximately 30 points away from the aspirational target performance of 100. As aquaculture expands, attention should be paid to ensure that, at a minimum, the industry does not shift further towards the poor performers, at least until performance improve significantly. *The worst performing sectors of the industry are also the fastest growing.* Marine finfish farmed in tropical and subtropical water, such as groupers (normalized score, 18) red drum (normalized score, 26), and cobia (normalized score, 37), have some of the worst scores on both and normalized and cumulative scales (Figure 3), yet production of these species is currently growing at a rate that outpaces all other species worldwide. Low final scores reflect poor performance across indicators rather than isolated to a few problem areas. In particular, warm water species consume large quantities of feed and receive large amounts of antibiotics, often used prophylactically to mitigate questionable production conditions. We estimate that as much as 5.5 million Kg of antibiotic materials (bioactive ingredient only, excludes non-active components) may be used annually. This material, much of which is discharged directly to the marine environment, is comprised almost exclusively of compounds classified as "critical" for human and/or veterinary treatment. Indiscriminant and wide-spread use of antibiotics as described here has long been known to critically threaten efficacy of human theraputants. How such threats may be affected by release of active ingredients into the marine food web is unknown [10].

Performance scores only tell some of the story. While the overall GAPI scores are informative, of even greater interest are the differential performances within each indicator that comprise the final score (Table 4). While the overall GAPI scores reflect aggregated performance trends, individual indicators highlight areas where a particular production system may excel and where improvement may be possible. For instance, turbot from Spain and coho salmon from Chile have divergent performance profiles but both have a normalized score of 63. The turbot scores poorly for antibiotic use (23), only one third the score of coho (68). In contrast, the coho performs poorly with respect to escapes (37) compared to the turbot (61). Similarly, barramundi from Australia and red seabream from

Japan both have a normalized species score of 47. Barramundi scores poorly for parasiticides use (30) compared to the red seabream (72). Conversely, the red seabream performs poorly for biological oxygen demand (9), while the barramundi scores reasonably well (62). A key contribution of GAPI (and similarly designed indices) is the capacity to expose root causes inferior performance. In so doing it makes clear that prescriptive solutions such as reduction of parasiticide use will yield far greater benefit for barramundi than red sea bream - where biological oxygen demand is the low hanging fruit in terms of return on investment.

Asia faces significant sustainability hurdles. Asian countries account for the 15 lowest species-country scores (Figure 3). The trend towards lower normalized scores in Asian countries largely results from the prevalence of poor Inputs performance such as ecological and industrial energies, feed sustainability, and biological oxygen demand. Asian countries also tend to score poorly in the antibiotics and parasiticide indicators as GAPI assumes that performers use the maximum allowable dose or quantity in the absence of actual performance data. Given that Asia is and will continue to be the epicenter of aquaculture growth, the observed performance trends should serve as clear warning.

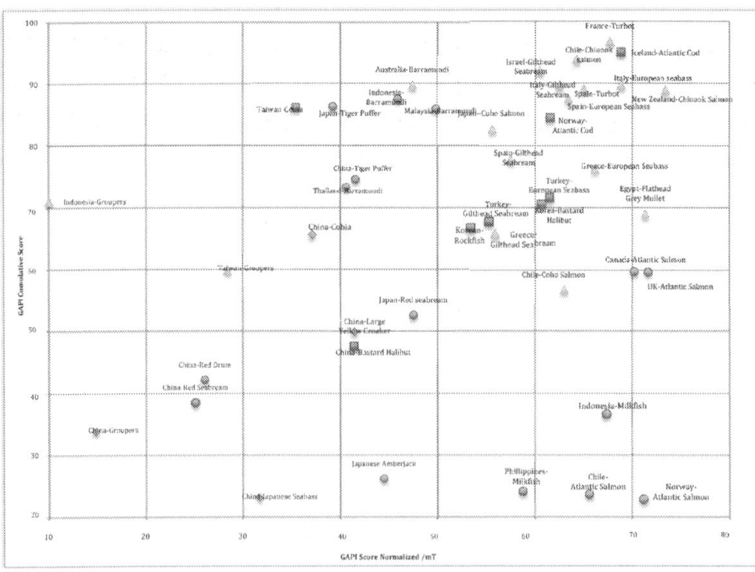

Figure 3. Final normalized scores plotted against final cumulative scores where products are discriminated by age.

Note: Products in the lower left quadrant are the worst performers on both scales. Products here tend to youngest (0–5 years: blue triangles; 6–20 years: red squares) After 20 years a product is of intermediate age and shows improvement in both normalized and cumulative performance (green triangle). With greater age, gained efficiencies of production drive more production such that the eldest industries (

Atlantic salmon performance illustrates that scale is everything. Comparison of cumulative and normalized scores demonstrates sheer scale of production can have drastic effects on environmental performance. Some of the best-performing species on a normalized basis are among the worst on a cumulative basis (Figure 3 and Table 5). For example, Atlantic salmon is the third-highest ranking species on a per mT basis (normalized score, 70), but when production volume is taken into account, Atlantic salmon score drops almost 50% to third worst of the 20 assessed species. In contrast, cobia has one of the biggest environmental footprints of any marine finfish (normalized score, 37) and among the worst performers on a per mT basis. However, because cobia farming is currently a modest sized industry it has a relatively small cumulative impact (cumulative score, 65) compared to production leaders Atlantic salmon and milkfish. In other words, there is a tipping point when production efficiencies no longer yield superior environmental performance and can indeed drive performance decline. This raises a question at the heart of seafood sustainability: How do we expand aquaculture to support the food and protein needs of 9 billion humans without overwhelming the carrying capacity of the marine environment? Clearly, part of the answer lies in selecting the right species, choosing the right environments in which to grow them, and utilizing responsible farming practices. At the same time, regulators need to consider the double-edged nature of production efficiency and how such efficiency, regarded as an industry objective, can lead to unanticipated problems. When normalized performance is plotted as a function of cumulative performance an inverted horseshoe pattern emerges (Figure 3). The left terminus of the horseshoe (poor normalized and cumulative scores) are the youngest products still in the "trial and error" phase of development. The right terminus is comprised of the oldest players, those products that enjoy high normalized performance and as a result have proliferated greatly resulting in large scale production and decline cumulative scores. The belly of the horseshoe (intermediate normalized and cumulative scores) are intermediate aged players that have found some production success but have yet to leverage that into large scale production. Policy decisions informed solely by normalized performances are likely to perpetuate this trend. However it is important to reiterate that normalized performance is irrelevant to the environment. The only environmentally relevant metric is cumulative performance.

Table 5: The ranking of country-species pairs by normalized and cumulative scores

Country-Species	Normalized		Cumulative		% Global Marine	Difference
	Rank	GAPI score	Rank	GAPI score	Finfish Production	Cum-Norm
New Zealand-Chinook salmon	1	73	5	90	0.27	17
Norway-Atlantic Salmon	2	72	38	33	21.00	−39
UK-Atlantic Salmon	3	72	28	64	3.70	−8
Egypt-Flathead grey mullet	4	71	29	64	6.41	−8
Canada-Atlantic Salmon	5	70	27	64	3.30	−6
Iceland -Atlantic cod	6	69	2	96	0.04	27
Italy-European seabass	7	69	7	89	0.24	20
France-Turbot	8	68	1	96	0.02	29
Indonesia-Milkfish	9	68	34	46	7.49	−22
Greece-European seabass	10	66	17	77	1.00	11
Chile-Atlantic Salmon	11	66	42	23	10.80	−43
Italy-Gilthead Seabream	12	65	9	88	0.20	23
Chile-Chinook salmon	13	64	3	94	0.06	30
Spain-European seabass	14	63	10	87	0.26	24
Chile-Coho Salmon	15	63	32	57	2.90	−6
Spain-Turbot	16	63	8	88	0.19	26

Norway-Atlantic cod	17	62	11	87	0.27	25
Turkey-European seabass	18	61	21	70	1.19	9
Korea-Bastard halibut	19	61	22	69	1.17	8
Israel-Gilthead Seabream	20	60	4	91	0.08	31
Phillippines-Milk-fish	21	59	41	28	8.78	−30
Spain-Gilthead Seabream	22	57	18	77	0.58	20
Greece-Gilthead Seabream	23	56	23	68	1.42	12
Japan-Coho Salmon	24	56	16	80	0.38	25
Turkey-Gilthead Seabream	25	55	26	66	0.95	10
Korea-Korean Rockfish	26	54	24	68	1.01	15
Malaysia-Barra-mundi	27	50	13	86	0.16	36
Japan-Red seabream	28	47	33	50	1.91	2
Australia-Barra-mundi	29	47	6	90	0.07	43
Indonesia-Barra-mundi	30	46	12	86	0.13	41
Japan-Japanese Amberjack	31	45	43	21	4.49	−23
China-Tiger puffer	32	42	19	72	0.43	31
China-Large Yellow Croaker	33	41	35	44	1.76	2
China-Bastard halibut	34	41	36	43	1.89	1
Thailand-Barra-mundi	35	41	20	72	0.45	31
Japan-Tiger puffer	36	39	14	85	0.12	46
China-Cobia	37	37	30	62	0.74	24
Taiwan-Cobia	38	35	15	84	0.11	49
China-Japanese Seabass	39	32	44	19	2.86	−13
Taiwan-Groupers	40	28	31	59	0.49	31
China-Red Drum	41	26	37	37	1.40	10

China-Red Seabream	42	25	39	33	1.56	8
China-Groupers	43	15	40	29	1.22	14
Indonesia-Groupers	44	10	25	68	0.18	58

CONCLUSIONS

The GAPI analysis allows us to identify why products score poorly, suggest how their peers have addressed the same challenges and provide insight in policy-relevant contexts ready for decision makers to take action. These initial results beg the question: how does aquaculture grow in a way that both supports the global food industry and mitigates local environmental damage? Part of the answer lies in carefully selecting which species we will farm and in choosing the right environments in which to grow them (*i.e.*, reducing normalized scores). At the same time, regulators need to consider the carrying capacity of local waters, and begin to design and reward operations that demonstrably minimize environmental footprint. Time is of the essence, however. The most current production data available (2011) reveal global marine aquaculture grew 22% [11] during the brief life of this project which began in 2009. The qualitative profile of production has remained constant over this period; both relative ranking of species and production countries have remained stable however the overall scale of production continues to rise rapidly. Therefore, while many of the revelations revealed by the GAPI project are sobering, this current analysis should be interpreted as a conservative snapshot of current industry performance.

GAPI is a work in progress and is intended to both inform and stimulate discussion of the appropriate metrics for evaluating performance and to drive the gathering and sharing of data. We are hopeful that GAPI will transform the way environmental performance is assessed and will aid decision makers—policymakers, producers, buyers, or standard setters—as they continue to address the promise and challenges of marine aquaculture.

Acknowledgments

This work was financially supported by the Lenfest Oceans Program and Pew Environmental Trusts and came to fruition through the dedication and leadership of Chris Mann and Rachel Hopkins. We recognize the EPI research group for initial inspiration of this work. We thank Katherine Bostick, Peter Bridson, Carrie Brownstein, Alice Chiu, Matthew Elliot, Rebecca Goldburg, George Leonard, Corey Peet, and Helene York for their many contributions to the development of the GAPI approach and resultant tool. We thank Vicki Blazer, Claude Boyd, Alejandro Bushmann, Alex de Sherbinin, Daniel Esty, Jennifer Ford, Christine Kim, Kevin Lafferty, Astrid Sholtz, Tanja Srebotnjak, Peter Tyedmers, and Judith Weis for their thoughtful contributions and advice in building criteria-specific assessment modules.

Conflicts of Interest

The authors declare no conflict of interest.

REFERENCES

1. Food and Agriculture Organization (FAO). *The State of World Fisheries and Aquaculture 2002*; FAO Fisheries Department: Rome, Italy, 2003.
2. Hall, S.J.; Delaporte, A.; Phillips, M.J.; Beveridge, M.; O'Keefe, M. *Blue Frontiers: Managing the Environmental Costs of Aquaculture*; The WorldFish Center: Penang, Malaysia, 2011.
3. Scottish Executive Research Unit. *Review and Synthesis of the Environmental Impacts of Aquaculture*; Government of Scotland: Edinburgh, Scotland, 2002.
4. Phillips, S. *Environmental Impacts of Marine Aquaculture Issue Paper*; Pacific States Marine Fisheries Commission: Portland, OR, USA, 2005.
5. Jacquet, J.; Hocevar, J.; Lai, S.; Majluf, P.; Pelletier, N.; Pitcher, T.; Sala, E.; Sumaila, R.; Pauly, D. Conserving wild fish in a sea of market-based efforts. *Orynx* **2009**, *44*, 45–56.
6. Emerson, J.W.; Hsu, A.; Levy, M.A.; de Sherbinin, A.; Mara, V.; Esty, D.C.; Jaiteh, M. *Environmental Performance Index and Pilot Trend Environmental Performance Index*; Yale Center for Environmental Law and Policy: New Haven, CT, USA, 2012.
7. Food and Agriculture Organization (FAO). *FishStat Plus: Universal Software for Fishery Statistical Time Series*; FAO: Rome, Italy.

Available online: http://www.fao.org/fishery/statistics/software/fishstat/en (accessed on 15 November 2010).

8. Emerson, J.; Esty, D.; Levy, M.; Kim, C.; Mara, V.; de Sherbinin, A.; Srebotnjak, T. *Environmental Performance Index*; Yale Center for Environmental Law and Policy: New Haven, CT, USA, 2010.
9. Singh, R.; Murty, H.; Gupta, S.; Dikshit, A. An overview of sustainability assessment methodologies. *Ecol. Indicat.***2009**, *9*, 189–212.
10. Cabello, F. Heavy use of prophylactic antibiotics in aquaculture: A growing problem for human and animal health and for the environment. *Environ. Microbiol.* **2006**, *8*, 1137–1144.
11. Food and Agriculture Organization (FAO)-FIGIS. Available online: http://www.fao.org/fishery/figis/en (accessed on 2 May 2010).

CHAPTER

19

ATLANTIC SALMON POPULATIONS INVADED BY FARMED ESCAPEES: QUANTIFYING GENETIC INTROGRESSION WITH A BAYESIAN APPROACH AND SNPS

Kevin Alan Glover[1], Cino Pertoldi[2,3], Francois Besnier[1] , Vidar Wennevik[1] , Matthew Kent[4] and Øystein Skaala[1]

[1]Section of Population Genetics and Ecology, Institute of Marine Research

[2]Department 18/Section of Environmental Engineering, Aalborg University

[3]Department of Biosciences, Aarhus University

[4]Center for Integrative Genetics, Department of Animal and Aquacultural Sciences, Norwegian University of Life Sciences

ABSTRACT

Background

Many native Atlantic salmon populations have been invaded by domesticated escapees for three decades or longer. However, thus far, the cumulative level of gene-flow that has occurred from farmed to wild salmon has not been reported for any native Atlantic salmon population. The aim of the present study was to investigate temporal genetic stability in native populations, and, quantify gene-flow from farmed salmon that caused genetic changes where they were observed. This was achieved by genotyping historical and contemporary samples from 20 populations covering all of Norway with recently identified single nucleotide polymorphism markers that are collectively diagnostic for farmed and wild salmon. These analyses were combined with analysis of farmed salmon and implementation of Approximate Bayesian computation based simulations.

Results

Five of the populations displayed statistically significant temporal genetic changes. All five of these populations became more similar to a pool of farmed fish with time, strongly suggesting introgression of farmed fish as the primary cause. The remaining 15 populations displayed weak or non-significant temporal genetic changes. Estimated introgression of farmed fish ranged from 2-47% per population using approximate Bayesian computation. Thus, some populations exhibited high degrees of farmed salmon introgression while others were more or less unaffected. The observed frequency of escapees in each population was moderately correlated with estimated introgression per population $R^2=0.47$ $P<0.001$. Genetic isolation by distance existed within the historical and contemporary data sets, however, the among-population level of divergence decreased with time.

Conclusions

This is the first study to quantify cumulative introgression of farmed salmon in any native Atlantic salmon population. The estimations demonstrate that the level of introgression has been population-specific, and that the level of introgression is not solely predicted by the frequency of escapees observed in the population. However, some populations have been strongly admixed with farmed salmon, and these data provide policy makers with unique information to address this situation.

Keywords: Admixture, Aquaculture, Environmental impact, Escapees, Simulation, ABC, Fish farming, Migration

BACKGROUND

Aquaculture production of Atlantic salmon (*Salmo salar* L.) was started in Norway in the early 1970's, and now represents a globally significant industry. Each year, hundreds of thousands of farmed salmon escape into the wild [1]. Some of these escapees enter rivers inhabited by native populations [2, 3, 4], outnumbering wild conspecifics on the spawning grounds of some rivers in some years [5]. The Atlantic salmon is characterized by highly significant population genetic structuring [6, 7]. This reflects evolutionary relationships among populations [8, 9, 10], including the potential for adaptive differences [11, 12]. Consequently, the large-scale invasion of Atlantic salmon populations by domesticated farmed escapees represents one of the most striking examples of human-mediated increased straying rates for any organism. This has raised global concerns for the fitness of native populations [13, 14, 15].

Genetic changes in native Atlantic salmon populations as a result of introgression of farmed escapees have been observed in populations in Ireland [16, 17, 18, 19], North America [20], and Norway [21, 22]. In the most extensive of these studies, six of 21 Norwegian populations investigated displayed significant temporal genetic changes. Based upon several genetic parameters, the authors concluded that the observed changes were primarily driven by introgression of escapees. However, in none of the above-mentioned studies has the accumulated level of introgression, i.e., "admixture", been quantified in a native population. From a management perspective, this is important, if not essential, in order to understand the extent of the problems, and ultimately implement guidelines via the process of risk assessment [23].

Where gene flow arises from a single and definable population or hatchery strain, statistical parameters such as individual-based admixture can be computed to estimate the level of introgression and degree of remaining wild population e.g., [24, 25]. Even in cases of low numbers of populations, it is also possible to infer admixture using a combination of molecular genetic data on real samples in addition to simulations [26, 27]. However, the quantification of genetic introgression of farmed Atlantic salmon into native Norwegian populations represents a more complicated situation than one in which a single or a low number of populations are exchanging genes among themselves [28]. This is because of several factors which are addressed briefly below.

The commercial production of Atlantic salmon in Norway is based upon rearing fish from multiple domesticated strains that were initially founded on wild salmon from more than 40 Norwegian rivers in the 1970's [29]. These domesticated strains have remained genetically isolated from wild salmon since. As a result of founder effects and genetic drift, there are highly significant differences in microsatellite allele frequencies among these farmed strains [30]. Thus, microsatellites provide enough information to distinguish some farmed strains and wild populations in a pair-wise manner [30]. However, the allele frequencies of microsatellites [30] and SNPs [31] display overlap between the farmed strains and wild populations when looking across multiple strains and populations simultaneously. Over time, escapees originate from multiple farms. As a result, the accumulated genetic change in the native population, due to introgression of this pool of farmed salmon, becomes very complicated to quantify, and is potentially underestimated [28]. Adding further to the complexity of this situation is the fact that the domesticated strains have changed greatly over time. Some strains have been terminated, while others have been mixed. Thus, when these points are taken together with the fact that there is non-random distribution of genetic material from the breeding companies to the production farms [32] where the majority of the farmed salmon are held and thus the majority of escapees originate from, it is impossible to accurately reconstruct the allele frequencies of the farmed strains used in Norway in the three to four decade period in which escapees have been observed on the spawning grounds.

A recent genome scan using a 7 k chip identified a set of SNPs that are collectively diagnostic in identifying Norwegian wild and farmed salmon, regardless of their population of origin [31]. These collectively diagnostic markers have the potential to circumvent some of the challenges described above to quantify introgression of farmed salmon in native Norwegian populations. At the same time, statistical approaches such as Approximate Bayesian Computation (ABC) [33] have been used to quantify complex models in population and evolutionary biology [34, 35, 36]. The present study aimed to take advantage of these recently discovered genetic markers and genotype a set of historical and contemporary samples from 20 Norwegian salmon populations that have displayed varying level of farmed escapees on the spawning grounds over the past 2–3 decades. In addition, a pool of farmed salmon was genotyped in order to investigate the direction of any observed temporal genetic changes in the wild populations. Finally, ABC and fixed migration simulation methods were implemented in order to attempt to quantify the level of cumulative introgression of farmed salmon in native populations for the first time.

METHODS

Wild Salmon Samples

Samples of wild salmon were collected from a total of 20 rivers spanning the entire coastline of Norway (Figure 1). Each river was represented by a historical sample that was collected prior to or in the early to moderate stages of the development of the commercial aquaculture industry in Norway, in addition to a contemporary sample that was collected in the period 2000–2009. Most of these samples were based upon fish scales taken from adult fish captured within each river (Table 1). The historical adult samples were primarily collected by the Norwegian Atlantic salmon gene bank. These samples were taken from multiple locations and years within each river to ensure a representative sample of each population. Of the contemporary adult samples, the majority were taken in association with recreational angling. These samples form the basis of the national monitoring program for estimating the frequencies of escapees within Norwegian rivers [1, 4, 5]. Where angling was the primary sampling method, samples were collected from multiple years and locations within each river in order to ensure the samples were as representative for the populations as possible. A few of the river samples were represented by juveniles (parr) collected by electrofishing (Table 1). These samples were collected in multiple locations within each river, and consisted of fish of varying age in order to ensure representative sampling. Due to the fact that the samples upon which this project were based were captured by recreational angling or in association with other previous research projects and monitoring programs (and subsequently donated to this study), no specific licenses were required for this specific study.

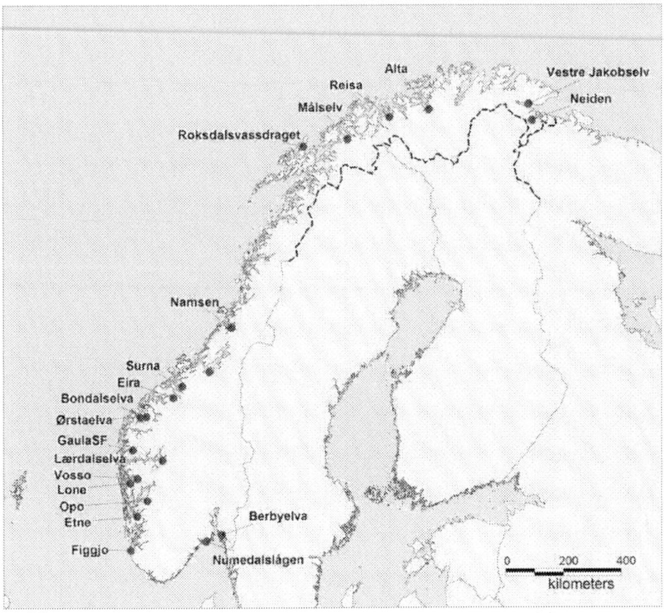

Figure 1: Location of the 20 rivers upon which the study is based.

Table 1: Numbers and types of samples, including some population genetics summary statistics for the 20 Atlantic salmon rivers

Population	**N**	**uH_E**	**HWE**	**LD**	**Sample type**	**Population**	**N**	**uH_E**	**HWE**	**LD**	**Sample type**
Neiden H (1979–82)	70	0.35	2	130	AD	Ørsta H (1986–89)	38	0.38	2	98	AD
Neiden C (2009)	77	0.36	5	112	AD	Ørsta C (2006–08)	31	0.36	4	130	AD
V. Jakobselva H (1989–91)	92	0.35	5	243	AD	GaulaSF H (1987–93)	35	0.36	0	120	AD
V. Jakobselva C (2007–08)	96	0.37	2	183	AD	GaulaSF C (2006–08)	82	0.36	2	131	AD
Alta H (1988–90)	39	0.34	1	85	AD	Lærdalselva H (1973)	90	0.36	1	125	AD
Alta C (2005–2007)	63	0.34	2	102	P	Lærdalselva C (2005–08)	45	0.36	1	120	AD
Reisa H (1986–91)	44	0.35	4	101	AD	Vosso H (1980)	45	0.34	0	98	AD
Reisa C (2006)	55	0.35	1	136	P	Vosso C (2007–08)	43	0.36	0	138	SM
Målselva H (1986–88)	39	0.35	3	102	AD	Loneelva H (1986–93)	59	0.34	0	136	AD

Målselva C (2008)	30	0.36	1	111	P	Loneelva C (2001–07)	50	0.36	3	134	AD
Roksdalsvassdraget H (1987–93)	31	0.37	0	110	AD	Opo H (1971–73)	60	0.35	3	116	AD
Roksdalsvassdraget C (2008)	89	0.37	3	128	AD	Opo C (2010)	61	0.36	3	180	P
Namsen H (1977)	74	0.36	0	129	AD	Etne H (1983)	72	0.35	1	121	AD
Namsen C (2008)	89	0.37	1	140	AD	Etne C (2006–2008)	83	0.36	1	122	AD
Surna H (1986–89)	23	0.36	2	90	AD	Figgjo H (1972–75)	51	0.35	1	118	AD
Surna C (2005–08)	45	0.37	4	122	AD	Figgjo C (2006)	71	0.36	1	119	AD
Eira H (1986–94)	31	0.36	2	108	AD	Numed-alslågen H (1989–93)	42	0.35	1	89	AD
Eira C (2005–2008)	40	0.35	1	123	AD	Numed-alslågen C (2007–08)	68	0.36	3	132	AD
Bondalselva H (1986–88)	39	0.37	3	103	AD	Berbyelva H (1988–93)	44	0.33	1	132	AD
Bondalselva C (2007)	13	0.36	1	70	P	Berbyelva C (2007–08)	87	0.33	5	139	AD
						Farmed pool (2005–2010)	375	0.37	17	491	FA

Rivers are sorted north to south.

Population=name of river with postscript letter H=historical sample, C=contemporary sample. N=number of fish included in the genetic analyses, uHE=unbiased heterozygosity, HWE=number of deviations from Hardy Weinberg Equilibrium $P<0.05$ (over the total of 72 loci), LD=number of times that linkage disequilibrium was observed between pairs of loci within any single sample at P<0.05 (from a total of 72 loci=2556 pair-wise combinations), Sample type=AD adults captured within river by angling, P parr, SM smolt, FA=taken from farm.

Prior to isolation of DNA, all scale samples were first examined for growth patterns in order to exclude any potential farmed salmon that had escaped from a commercial fish farm, using established methods [37]. The habitat and demographic data for the populations included in the present study, including numbers of escapees observed, are presented online.

Farmed Salmon Samples

A total of 375 farmed salmon were analysed in this study. These fish were collected in the period 2005–2010 from 49 separate sources. These included 48 marine cages (approx. 8 salmon per cage) located on 35 commercial farms spanning from the south to the far north of Norway, in addition to eight escapees captured in the sea. These samples were picked from approximately 6000 farmed fish that had been previously genotyped with microsatellites in association with a forensic service conducted by the Institute of Marine Research to identify the farm of origin for escaped salmon for the legal authorities [32, 38, 39, 40, 41]. The aim of this sampling strategy was to generate a pool of farmed fish representing the genetic diversity of farmed fish in Norway.

Genotyping

DNA was extracted in 96-well format using Qiagen DNeasy blood and tissue kit. Each plate contained two or more negative control wells. DNA aliquots of these samples were sent to the Centre for Integrative Genetics (CIGENE) in Norway for SNP (n=99) analysis using a Sequenom platform. A list of the markers, their NCBI assay details and linkage map positions are available.

Seventy of the SNP markers genotyped here were selected from the panel of SNPs that have been suggested to be collectively diagnostic for farmed and wild Norwegian salmon [31]. These 70 SNPs include the ten top ranking loci, 54 of the 60 top ranking loci, and a further 16 selected from the top 200 ranking loci. As the collectively diagnostic loci identified by Karlsson et al. (2011) were ranked based upon their F_{ST} between a pool of farmed and pool of wild Atlantic salmon, and the sequential difference in F_{ST} between each locus was very small, it is not expected there is any specific combination of loci required to create the genetic signal permitting identification of farmed and wild salmon. However, the ability of the collectively diagnostic markers used in the present study to differentiate farmed and wild salmon has been empirically evaluated here (see Results). It is furthermore important to note that seven of the wild populations used in the marker identification study conducted by Karlsson et al. (2011) overlap with populations in the present study (Alta, Namsen, Surna, Lærdalselva, Vosso, Figgjo, Numedalslågen). While this can potentially cause ascertainment bias, this has been considered when interpreting results. In addition to the 70 diagnostic SNPs, a further 29 SNPs were also genotyped. These were selected as putatively neutral SNPs that are known to be polymorphic in Norwegian salmon.

Statistical Data Analysis

In order to investigate temporal genetic stability in the 20 populations, and the direction of any potential changes, the data set was organized into the historical and contemporary samples. The pool of farmed salmon was only used for specific tests and the simulations to quantify gene flow (see below). For all computations, the data sets were divided into the loci that are collectively diagnostic between farmed and wild Atlantic salmon and the randomly selected loci.

Genotypic data was first organized in the program MSA, coding the nucleotides A, C, G and T as alleles 1–4 [42]. MSA was used to compute F_{ST} values (global and pair-wise) and compute significance levels associated with these tests using the Fisher's exact method as implemented in the program. F_{ST} values were all computed using the Weir and Cockerham estimator [43]. Confidence intervals associated with the global F_{ST} values for the historical and contemporary sets of samples were computed from the distribution of 1000 F_{ST} values calculated from 1000 bootstraps where 35% of the individuals from each population were randomly re-sampled. This latter test was computed in the program R (R development team).

Population-specific summary statistics, i.e., numbers of alleles, heterozygosities, numbers of deviations from Hardy Weinberg Equilibrium (HWE), and the numbers of times that linkage disequilibrium (LD) was observed between pairs of loci within each sample were computed in the program GENALEX v6. [44] using program default parameters for these tests. Where appropriate, statistical significance levels were tested against $P<0.05$, $P<0.001$ and Bonferroni corrected threshold levels.

GENALEX v6. was also used to create principal component analysis plots (PCA) based upon a matrix of F_{ST} values. This was conducted using the program's default values in order to investigate spatio-temporal population genetic structure in addition to the direction of any observed temporal changes in relation to the pool of farmed fish. Isolation by

Distance (IBD) analyses on the historical and contemporary set of samples was conducted using the Mantel test as implemented in the R package "Vegan". The test was computed with input data from a matrix of pair-wise F_{ST} values, and, the pair-wise distances between river mouths in kilometers.

Genetic assignment tests, using direct assignment and exclusion and different combinations of samples and loci sets, were conducted in the program GENECLASS2 [45], using a specific algorithm for the computations [46] and probabilities of $P<0.05$ and $P<0.001$ for exclusion. Bayesian clustering analysis was computed in the program STRUCTURE [47, 48]. This was used to look at temporal genetic changes within each wild population one at a time by including the historical and contemporary sample. Analysis parameters included an admixture model, correlated allele frequencies, and assuming no population prior. Each analysis with this program consisted of 5 replicate runs for K=1-5, each with a burn-in of 250 000 replications, and a run length of 500 000 Markov chain Monte Carlo (MCMC) iterations.

In order to investigate the statistical power of the different sets of genetic markers implemented here, several tests were computed. First, the distribution of pair-wise F_{ST} between two groups of 100 randomly selected fish from within each of the groups (wild and farmed) was plotted following 1000 bootstraps. This was conducted for farmed vs. wild, wild vs. wild and farmed vs. farmed using the diagnostic markers and the randomly selected markers. Next, the assignment power of these sets of loci was examined in STRUCTURE. Due to the fact that there were more diagnostic loci than randomly selected loci, and that assignment power is influenced by the number of loci, a sub-set of the diagnostic loci (equal to the number of randomly selected loci) was also used to compute these assignment tests. The following routine was repeated 1000 times for each set of SNPs: 100 farmed fish were sampled as the learning farmed sample, 100 wild fish among all populations were sampled as the learning wild samples, then one fish was sampled randomly to constitute the unknown individual to identify. STRUCTURE was then run with 50 000 burn-in and 500 000 iterations, at K=2, using population information and population flag, so that population information is only used for the learning samples, but not for the unknown individual. The accuracy of assignment of this individual to either the farmed or the wild salmon groups was compared for the different sets of SNPs using the threshold probability of 0.6. The re-sampling simulations were performed in R, and the STRUCTURE runs with ParallelStructure R package (https://r-forge.r-project.org/projects/parallstructure/) [49].

Simulations to Quantify Gene Flow

Two alternative methods were employed to quantify the amount of gene flow that would be required from an alien population to cause the observed temporal genetic change within each wild population, computed as F_{ST} between each wild population's historical and contemporary sample. The simulations were coded and executed in R using previously published scripts to simulate realistic genetic introgression using gametes sampled randomly from the donating and recipient populations [28]. Average generation time for each wild population was set to five years, thus, the number of generations used to simulate gene flow was set as the number of years between the historical and contemporary sample for each population, divided by five. An effective population size (N_E) of 200 for all populations except those displaying an N_E less than this in which case the observed N_E, as reported previously [21], using a one sample linkage disequilibrium based method implemented in LDNE 1.31 software [50], was used.

First, the posterior point estimate of migration rate (M) was inferred by an ABC algorithm [33]. The routine to quantify this gene flow consisted of the following steps: 1. Determine a prior distribution for migration rate M(e.g., M~N(0.1,0.1)). 2. Simulate n scenario of introgression where the value of M is sampled from the prior distribution, and compute the F_{ST} between historical and each simulated population. 3. Calculate the vector $s$of the n differences between simulated and observed F_{ST}: s: (s_1, s_2,... s_n) where $s_i = (F_{STO} - F_{STi})$. 4. Solve the linear regression: $M = \alpha + \beta s + \epsilon$ (Where ϵ is a vector of residuals, s is the vector estimated in step 3, and α is a constant). The estimate of α gives the expected value of M when $F_{STO} - F_{STi} = 0$. 5. Update the prior distribution of M with estimated distribution of α. 6. Repeat step 2–5 until α converges to a stable estimate.

To account for the standard deviation of the observed F_{ST} between historical and contemporary sample in the estimation of M, the posterior distribution of M was also estimated with an alternative "fixed migration rate" approach that consisted of the following steps: 1. Simulating genetic introgression from alien population with a fixed migration rate *M* per generation. Tested values of M were from 1 to 80%, every 1% between 1 and 20, and every 5% between 20 and 80%. 2. From the set of simulated populations (each corresponding to a value of M=1, 2, 3, 4...80% migration), selecting 1000 among those that gave an F_{ST} that fit within the 95% confidence interval of the observed F_{ST}. 3. Mean and standard deviation of migration rate from the 1000 selected scenario provide a posterior distribution of (*M*).

For both ABC and fixed migration rate approach, two alternative scenarios were tested. First, the alien population considered for the simulations was a pool of farmed salmon, and second, to account for possible straying, the alien population was the historical population of the geographically nearest neighbour river. The R package ADEgenet [51] was used to generate PCA plots of historical, contemporary and simulated populations from each scenario based on individual genotypes.

RESULTS

Data Quality

Loci that displayed technically unreliable genotype clustering, or a genotyping coverage <90% in the entire data set, were excluded prior to all statistical analyses. This stringent quality control reduced the total number of loci from 99 to 72. Among the remaining loci were 47 SNPs selected from the collectively diagnostic panel for farmed and wild salmon (including the top ten ranked loci, and 35 of the 60 highest ranking loci) [31], and a set of 25 random SNPs. These panels are hereafter referred to as 47d (diagnostic) and 25r (random). Individual fish displaying genotyping coverage <75% over the remaining 72 loci were also removed from the data set (301 fish removed from 2912 samples). Thus, the final data set for analysis consisted of 2611 individuals genotyped for 72 loci. Within this, a total of 364 178 alleles were successfully scored, giving an overall genotyping coverage of 97%.

Within Population Summary Statistics

A range of population genetics parameters are summarized for all wild populations and the pool of farmed salmon. The unbiased expected heterozygosity (UH_E) over all 72 loci was very even among wild populations, ranging from a low of 0.33 for both the historical and contemporary samples for Berbyelva, to a high of 0.38 for the historical sample representing the river Ørstadelva. None of the populations displayed clear increases or decreases in this parameter between their historical and contemporary samples.

The parameters HWE and LD have the ability to indicate disturbances in populations due to introgression of genetically distinct fish. At the significance level $P < 0.05$, no wild salmon sample displayed more than five deviations from HWE across the 72 loci. This is the number of observations that are more or less expected by chance at this significance level. From a total of 2556 comparisons within each sample (72 loci pair-wise) at $P < 0.05$, LD was observed from a low of 70 times (2.6%) in Bondalselva contemporary sample, to a high of 243 (9.2%) in the historical Vestre Jakobselva sample. The pool of farmed salmon displayed similar summary statistic values to the wild populations, although increased frequency of HWE and LD was observed in this sample due to it being a mixture of fish from multiple sources

The effective population size for each wild population is presented. Most of the samples and populations displayed N_E 100-1000+. Notable examples of low N_E were Vestre Jakobselva (98), Bondalselva (25), Ørstaelva (94), and Opo (57) for the historical samples, and Vestre Jakobselva (75) and Berbyelva (67) for the contemporary samples. The upper and lower 95% confidence intervals surrounding all estimates were large however, with the upper boundary often reaching infinity.

Comparison between 47d And 25r

In order to test the statistical characteristics of the putatively diagnostic loci 47d vs. the randomly selected loci 25r, several comparisons were conducted. The number of loci displaying non-overlapping allele frequencies between the pool of farmed salmon and all 20 historical wild salmon samples separately was 15 for 47d (32%) and 3 for 25r (12%). Thus, while 32% of the loci from 47d displayed non-overlapping allele frequencies, the majority did not. Of these non-overlapping loci, some displayed moderately strong differentiating frequencies but not all. Based upon the distribution of the F_{ST} values between 100 farmed and 100 wild salmon (historical samples) randomly re-sampled from the data set 1000 times, 47d displayed approximately double F_{ST} than 25r.

Genetic assignment values as computed in STRUCTURE revealed that assignment to correct source (farmed pool or wild pool) was higher for 47d than 25 randomly re-sampled loci from 47d ("25resam"), and from 25r (Correctly assigned wild: 69, 63, and 23%; correctly assigned farmed: 82, 58, and 30% when using 47d, "25resam" and 25r respectively). Together, these tests demonstrate that the diagnostic loci contain far greater statistical power to differentiate farmed and wild salmon than the randomly selected loci.

Temporal F_{ST} changes

Over the full set of 72 loci, the number of populations displaying significant temporal genetic changes at $P<0.05$ and $P<0.001$ was 11 and five respectively. For either 47d or 25r at $P<0.05$ (Table 2), the number of populations displaying temporal changes was 12. At the more stringent threshold $P<0.001$, this number dropped to five. Looking specifically at 47d, the three rivers displaying the largest temporal genetic changes were all located on the west of Norway; Opo, Vosso and Loneelva respectively. In all three of these rivers, temporal genetic changes, and measured by F_{ST}, were greater for 47d than 25r, a trend observed in a total of 13 of the 20 populations in the entire data set.

Table 2: Temporal genetic stability in the 20 rivers ordered north to south

Population	Pair-wise F_{ST} historical vs Contemporary				Exclusion of contemporary sample from historical $P<0.001$			
	72 SNPs	47d SNPs	25r SNPs	22 Micros	72 SNPs	47d SNPs	25r SNPs	22 Micros
Neiden	−0.0002	−0.0017	0.0025	0.0009	0%	0%	0%	6%
V. Jakobselv	0.0054**	0.0067**	0.0029	0.0064**	7%	5%	1%	16%
Alta	0.0040*	0.0059*	0.0003	−0.0002	2%	3%	0%	2%
Reisa	0.0023	0.0036	−0.0004	0.0041*	6%	4%	0%	15%
Målselv	0.0038	0.0082*	−0.0043	−0.0026	7%	10%	0%	13%
Roksdalsvass.	0.0037*	0.0066*	−0.0016	0.0014	1%	1%	0%	20%
Namsen	0.0042**	0.0027*	0.0068*	0.0013*	1%	2%	0%	9%
Surna	0.0093	0.0010	−0.0055	0.0025	2%	2%	0%	12%
Eira	−0.0039	−0.0051	−0.0016	0.0005	5%	0%	0%	34%

Bondalselva	0.0034	0.0041	0.0022	0.0043	8%	8%	8%	14%
Ørstaelva	0.0006	0.0005	0.0008	0.0003	6%	3%	0%	6%
GaulaSF	0.0011	0.0025	−0.0015	0.0001	0%	2%	1%	0%
Lærdalselva	0.0005	0.0001	0.0011	0.0015	2%	4%	0%	17%
Vosso	0.0125**	0.0168**	0.0049	0.0070**	12%	16%	2%	58%
Loneelva	0.0071**	0.0105**	0.0003	0.012**	4%	2%	0%	52%
Opo	0.0200**	0.0216**	0.0172**	0.0258**	8%	2%	8%	100%
Etne	0.0038*	0.0062**	−0.0007	0.0006	1%	0%	1%	5%
Figgjo	0.0050*	0.0040*	0.0071*	0.0048**	3%	0%	0%	38%
Numedalslågen	0.0036*	0.0023	0.0058*	0.0032*	4%	0%	0%	29%
Berbyelva	0.0032*	0.0042*	0.0015	0.0053**	1%	3%	0%	16%

Stability is quantified by pair-wise F_{ST} and percent exclusion of the contemporary sample from the historical sample. Both parameters measured using sub-sets of SNPs and previously published data using microsatellites.

Micros=microsatellite data taken from [21]. Note, the contemporary Vosso sample is not the same as the contemporary Vosso sample used in Glover et al., 2012b. *=significant $P<0.05$, **=significant $P<0.001$.

A statistically significant correlation was observed between the degree of within-river temporal genetic change (as revealed by F_{ST}) using 47d and 25r ($R^2=0.36$ $P=0.0049$). However, the degree of within-river change was more strongly related between 47d and previously published results for these populations based upon 22 microsatellites [21] ($R^2=0.63$ $P<0.0001$).

Genetic assignment tests were used to investigate the probability of excluding the composite genotype for each individual fish taken from the contemporary sample from the historical genetic profile of the population. This compliments temporal F_{ST} analyses as it also reflects distribution of genotypes among individual fish in the contemporary sample. At $P<0.001$, the percentage of fish from the contemporary sample that could be excluded from the historical sample ranged from a low of 0% for the river Neiden, to a high of 12% for the river Vosso. Using 47d or 25r only, the percentage of fish from the contemporary sample that could be excluded from the historical profile were generally lower than with the 72 loci (Table 2). In comparison with previous data on these populations using 22 microsatellites [21], the exclusion levels achieved with even the full set of 72 SNPs were strikingly lower (Table 2).

Bayesian clustering analysis for each population separately revealed clear temporal genetic changes in the rivers Opo and Vosso when computed using data from all 72 loci. The results of admixture analysis for the remaining rivers, using either all 72 loci or 47d was either very subtle, or non-existent.

Comparisons of Wild Salmon to the Farmed Salmon

Computed pair-wise, all historical and contemporary samples representing the wild populations displayed statistically significant differences to the pool of farmed salmon, using both sets of SNPs separately, and following adjustment for multiple testing (all $P<0.001$). Several trends can be extrapolated from these comparisons (Figure 2). First, all F_{ST} values computed pair-wise between each historical wild sample and the farmed sample was higher for 47d compared to 25r (Anova $DF_{1:38}$, $F=12.1$, $P<0.001$). This is consistent with the results of the section above dedicated to comparing the statistical properties of 47d and 25r. Second, a geographic trend to the pair-wise difference between each wild population's historical sample and the pooled farmed sample was present. Populations from the far north and south of Norway displayed much greater pair-wise F_{ST} estimates to the farmed sample than populations from mid- and western Norway displayed to the farmed sample. This was detected in 47d and 25r, but more pronounced in the former. These observations suggest that a population in northern Norway is likely to display a greater genetic change compared to a population from mid Norway for the same level of farmed salmon introgression.

Looking specifically at 47d, the pair-wise difference between the wild sample and the pool of farmed salmon decreased with time in 14 of the 20 populations. This was most noticeable for Vosso, Opo and Vestre Jakobselva (Figure 2). For example, the pair-wise F_{ST} between Vosso and the farmed sample dropped from 0.085 to 0.038.

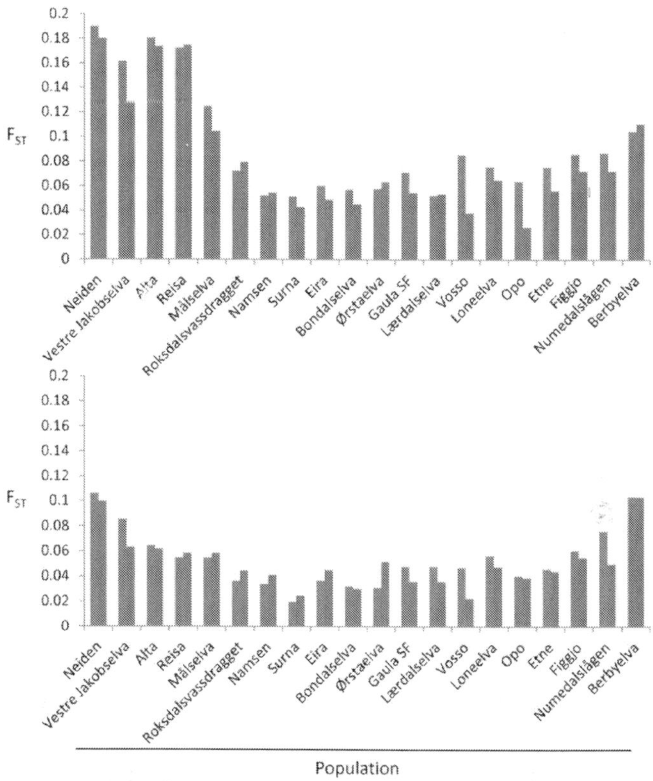

Figure 2: Pair-wise F_{ST} between the historic sample for each wild population and the farm sample (blue–left bar), and the contemporary sample for each wild population and the farm sample (red–right bar), computed using 47 diagnostic SNPs (top), and 25 randomly selected SNPs (bottom). Populations are ordered north to south.

Furthermore, a significant positive relationship between temporal genetic change within each river, and the change in pair-wise F_{ST} between that specific rivers historical sample and the farmed salmon sample, and that rivers contemporary sample and the farmed sample, was observed for 47d (R^2=0.41, P=0.0022), but not for 25r (R^2=0.07, P=0.25) (Figure 3). Thus, analyses based upon 47d demonstrate that where rivers display temporal genetic change, it is largely in the direction of the farm sample.

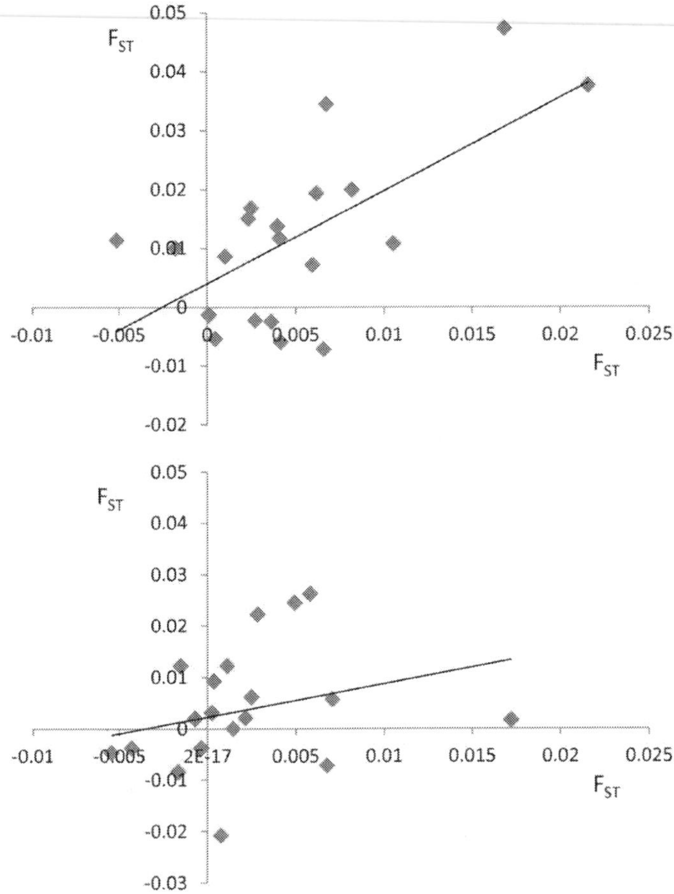

Figure 3: Relationship between the temporal genetic change observed within any given population (i.e., pair-wise F_{ST} between the historical and ceontemporay sample) on the X axis, and the difference in pair-wise F_{ST} between each populations historical sample and the farm sample, and that same populations contemporary sample and the farm sample, placed on the Y axis. These relationships are computed with 47 diagnostic SNPs (top) ($R^2=0.41$ $P=0.0022$) and 25 randomly selected SNPs (bottom) ($R^2=0.07$ $P=0.25$). Each point represents the above mentioned relationship for each of the 20 populations.

Assignment tests were also used to exclude the 375 genotyped farmed fish from each wild population. For both the historical and contemporary samples, the exclusion percentages followed a similar geographic pattern as observed for the pair-wise F_{ST} between each population and the farmed sample (Figure 2). The percentage of farmed fish that could be excluded from each wild fish sample decreased between the full set of 72 loci to 47d and 25r. The relative change for any given population, i.e., the drop in exclusion of the farmed fish between the historical sample and contemporary sample (which is more interesting in this specific context than the absolute level of exclusion) was still very noticeable for 47d.

An alternative way of investigating the direction of the observed temporal genetic changes in the wild populations is to conduct PCA analysis. A PCA plot for the seven populations located in the west of Norway, which includes the three rivers displaying the highest temporal genetic changes in the complete data set for 47d, revealed that six of the seven populations displayed genetic changes in the direction of the farm sample (Figure 4). Notably, none of them displayed temporal genetic changes away from the farm sample. While some other populations located in the other regions of Norway displayed temporal genetic changes in the approximate direction of the farm sample, this was not observed for all samples displaying temporal genetic changes.

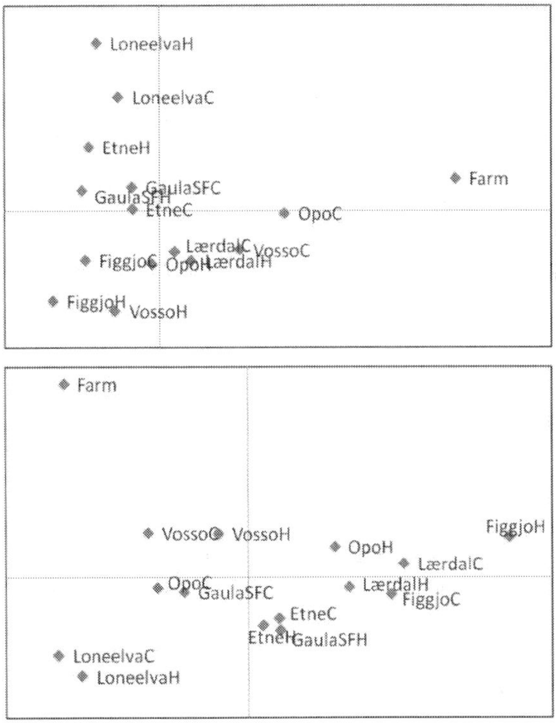

Figure 4: Principal compont analysis for the historical (H) and contemporary (C) samples for the seven rivers grouped into the west of Norway. This region

includes the three rivers displaying the greatest temporal genetic changes in the entire study. Plots are based upon 47 diagnostic SNPs (top) and 25 randomly selected SNPs (bottom). X-and Y axes explain 35, 23%, and 29, 23% for the top and bottom figures respectively. PCA plots for all 20 rivers arranged into four geograpahic regions are also presented online.

Quantification of Gene Flow Required to Cause Observed Changes

The estimated level of farmed salmon introgression required to cause the observed temporal genetic changes in the 20 populations ranged from 2-47% and 7-41% using the ABC and fixed migration methods respectively (Table 3). For the two populations displaying the greatest temporal genetic change (Opo and Vosso), the estimated level of farmed salmon introgression ranged from 36-41% and 33-47%. In contrast, the simulated level of gene flow required from the nearest neighbor population, i.e., straying, to cause the observed temporal changes in these two populations was much greater, approximately 80-82% and 46-61% for the ABC and fixed migration methods respectively. This is far greater than straying rates typical for this species. For both computation methods, the introgression rate required by the straying scenario was always far greater than introgression required by the farmed escapees scenario. This is likely to reflect the genetic similarity of the nearest neighbor population to each recipient wild population, and how distinct each wild population's historical sample was to the farmed sample, the latter of which follows a geographic trend as demonstrated earlier (Figure 2).

Table 3: Estimated percentage introgression of farmed salmon, or the nearest wild population required to cause the observed temporal genetic changes

Population	Admixture from farm salmon				Admixture from nearest neighbor population			
	ABC		Fixed migr.		ABC		Fixed migr.	
	Pr. Gen.	Total	Pr. Gen.	Total	Pr. Gen.	Total	Pr. Gen.	Total
Neiden	0.004	0.022	0.014	0.073	0.029	0.139	0.028	0.142
	±0.009	±0.051	±0.006	±0.031	±0.035	±0.152	±0.023	±0.110
V. Jakobs.	0.031	0.116	0.032	0.116	0.130	0.385	0.175	0.414
	±0.005	±0.030	±0.020	±0.063	±0.011	±0.045	±0.133	±0.223
Alta	0.031	0.116	0.043	0.117	0.114	0.350	0.136	0.359
	±0.004	±0.024	±0.031	±0.062	± 0.013	±0.042	±0.092	±0.198
Reisa	0.017	0.066	0.035	0.126	0.040	0.143	0.116	0.319
	±0.006	±0.028	±0.023	±0.070	±0.017	±0.061	±0.084	±0.190
Målselv	0.054	0.190	0.051	0.167	0.083	0.273	0.097	0.280
	±0.006	±0.032	±0.041	±0.114	±0.004	±0.030	±0.071	±0.175
Roksdalsvass	0.055	0.192 ±	0.063	0.199	0.071	0.239	0.077	0.220
	±0.004	0027	±0.054	±0.146	±0.004	±0.029	±0.052	±0.143
Namsen	0.01	0.062	0.033	0.170	0.056	0.252	0.027	0.126
	±0.013	±0.074	±0.020	±0.090	±0.047	±0.210	±0.037	±0.162
Surna	0.010	0.038	0.083	0.243 ±	0.039	0.137	0.255	0.507
	±0.014	±0.057	±0.073	0179	±0.046	±0.138	±0.188	±0.253
Eira	0.016	0.053	0.042	0.145	0.033	0.116	0.092	0.268
	±0.031	±0.100	±0.032	±0.097	±0.035	±0.141	±0.069	±0.168
Bondalselva	0.026	0.098	0.092	0.263 ±	0.122	0.363	0.301	0.547
	±0.015	±0.055	±0.081	0193	±0.029	±0.074	±0.223	±0.270
Ørstaelva	0.014	0.050	0.070	0.217	0.050	0.165	0.304	0.554
	±0.019	±0.068	±0.060	±0.154	±0.042	±0.129	±0.219	±0.273

GaulaSF	0.022	0.085	0.050	0.165 ±	0.028	0.105	0.097	0.280
	±0.009	±0.038	±0.041	0115	±0.008	±0.034	±0.070	±0.172
Lærdalselva	0.015	0.088	0.027	0.169	0.019	0.115	0.026	0.154
	±0.027	±0.142	±0.013	±0.077	±0.028	±0.144	±0.021	±0.112
Vosso	0.077	0.360	0.102	0.410	0.107	0.459	0.283	0.605
	±0.003	±0.032	±0.015	±0.200	±0.005	±0.033	±0.140	±0.145
Loneelva	0.094	0.307	0.075	0.226	0.124	0.375	0.109	0.311
	±0.010	±0.029	±0.069	±0.166	±0.006	±0.031	±0.071	±0.172
Opo	0.084	0.474	0.061	0.331	0.238	0.817	0.338	0.804
	±0.004	±0.044	±0.053	±0.209	±0.020	±0.029	±0.205	±0.197
Etne	0.044	0.197	0.040	0.170	0.069	0.274	0.098	0.337
	±0.005	±0.033	±0.030	±0.107	±0.033	±0.117	±0.067	±0.189
Figgjo	0.009	0.060	0.029	0.178	0.018	0.120	0.044	0.236
	±0.010	±0.069	±0.013	±0.077	±0.023	±0.125	±0.035	±0.162
Numedals.	0.007	0.030	0.040	0.143	0.021	0.078	0.060	0.191
	±0.006	±0.026	±0.030	±0.089	±0.010	±0.041	±0.046	±0.130
Berbyelva	0.025	0.093	0.040	0.138	0.045	0.163	0.070	0.219
	±0.012	±0.049	±0.029	±0.089	±0.005	±0.030	±0.050	±0.138

Estimations of introgression are based upon approximate Bayesian computation (ABC), or a fixed migration rate based simulation (Fixed migr.). Populations ordered north to south.

ABC=estimation of introgression using approximate Bayesian computation, Fixed migr. = estimation of introgression using a fixed migration rate estimator (see Methods).

The observed temporal genetic changes for the rivers Opo and Vosso, in relation to the simulated genetic change by introgression of farmed fish, or the nearest neighboring population, confirmed results presented earlier that the genetic changes observed in these two populations were clearly directional to the farmed fish, and not likely to be cause by natural straying (Figure 5). Put simply, the observed temporal genetic change in both of these populations overlapped almost perfectly with the simulated genetic change caused by the pool of farmed salmon, but not by the nearest neighbor (Figure 5). For other populations that displayed more modest temporal genetic changes however, this pattern was more difficult to elucidate from the simulation-based PCA plots due to the fact that the changes were small and distributions overlapped.

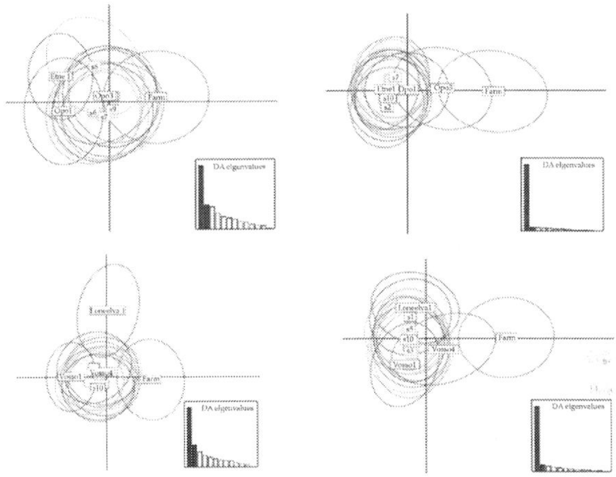

Figure 5: Principal component analysis depicting observed historical and contemporary samples for the river Opo (top panels), and Vosso (bottom panels), including the farmed sample and nearest wild population, following simulated introgression from the farmed sample (left panels), and simulated introgression from the nearest neighbor (right panels). In each case, the results of ten independent simulations, S1-S10 are presented. The text box represents the center point of the observations with the 95% confidence interval represented by the ellipse.

Observed Frequency of Escapees and Genetic Changes

A set of correlations between the frequency of farmed escapees observed on the spawning grounds in the time-period in which the present study is conducted, and various genetic parameters produced with the present study for these populations are summarized (Table 4), and plotted graphically (Figure 6). All genetic parameters gave stronger correlations with the weighted mean frequency of escaped salmon than the un-weighted mean frequency of farmed escaped salmon. This suggests that the weighted estimate more accurately reflects the true numbers of escapees in the populations as it corrects for noise in the estimations (i.e., small sample sizes in some years etc.).

Table 4: Relationships between the frequency of farmed escaped salmon in the spawning population, and observed changes in various genetic parameters

Statistic	Un-weighted mean	Weighted mean
Pair-wise F_{ST} 72	0.19 (0.052)	0.52 (0.0003)
Pair-wise F_{ST} 47d	0.17 (0.067)	0.44 (0.0014)
Pair-wise F_{ST} 25r	0.06 (0.30)	0.41 (0.0022)
F_{ST} to farm 72	0.13 (0.11)	0.22 (0.039)
F_{ST} to farm 47d	0.25 (0.024)	0.37 (0.0045)
F_{ST} to farm 25r	<0.01 (0.72)	<0.01 (0.99)
ABC introgression	0.16 (0.08)	0.47 (0.0007)
Fixed migration rate introgression	0.25 (0.025	0.36 (0.005)

Values computed are R^2 (*P*-value). Number behind statistics refers to number of SNP loci used, introgession computations only compared for 47d.

Pair-wise F_{ST}=observed value between historical and contemporary sample for each river; F_{ST} to farm=absolute difference in pair-wise F_{ST} between the populations historical sample and the farm sample, and between the populations contemporary sample and the farm sample, ABC introgression=estimated introgression of farmed salmon (Table 3) using ABC method, Fixed migration rate introgression=estimated introgression of farmed salmon (Table 3) using the fixed migration rate method. Un-weighted mean=mean% of farmed salmon observed on the spawning grounds for these rivers in the period 1989–2009 [5, 52], weighted mean=weighted average percentage of farmed salmon in the population combining data from both sports-fishing and spawning population samples [53].

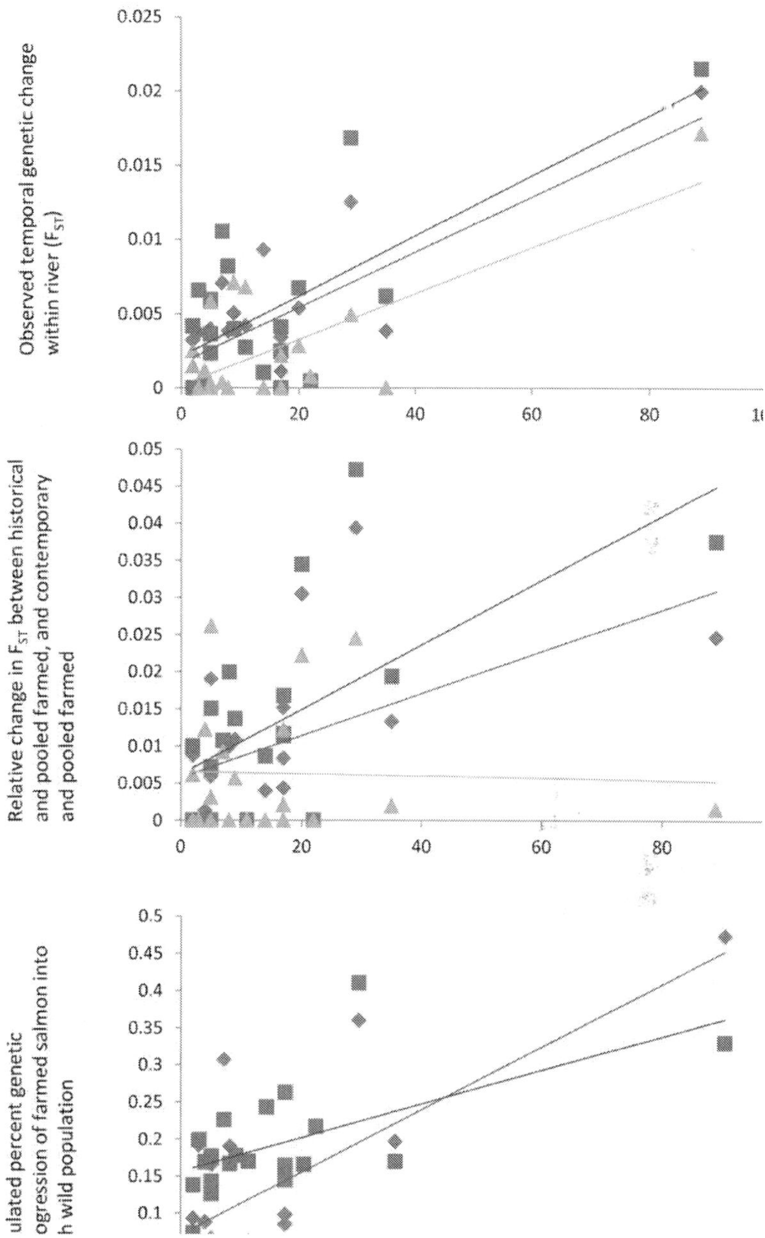

Figure 6: Relationship between the weighted mean number of escapees observed in each of the rivers in the period 1989–2009 (X axis), and the following three sets of estimators of genetic change in the 20 populations (Y axes). Top graph Y axis = the observed within-river temporal genetic change as

measured by pair-wise F_{ST} using all 72 loci (blue diamonds), 47d (red squares) and 25r (green triangles). Middle graph Y axis = the relative change (%) in pair-wise F_{ST} between a population's historical sample and the pooled sample of farmed salmon, and the same populations contemporary sample and the pooled sample of farmed salmon using all 72 loci (blue diamonds), 47d (red squares) and 25r (green triangles). Bottom graph Y axis = the simulated level of farmed salmon genetic introgression required in each population to cause the observed temporal genetic changes as based upon the ABC method (blue triangles) and the fixed migration method (red squares), both of these simulations computed using 47d. Significance levels and R^2 values for all relationships are provided in Table 4.

Spatio-Temporal Variation

The pair-wise F_{ST} values among all samples included in this study, including associated *P* values, are presented online. Using all sets of markers, global F_{ST} estimates among the 20 wild populations decreased significantly with time (Table 5). Nevertheless, a geographic pattern to the genetic structure was still evident with both 47d and 25r, and there was no detectable change in IBD (Table 5). This means that while overall variation among the wild samples decreased with time, this happened "evenly" which did not influence the relationship between genetic and physical distance.

Table 5: Spatio-temporal analysis of population genetic structure (global F_{ST}) including isolation by distance (IBD)

Loci	Historical samples				Contemporary samples				Global F_{ST} change?
	IBD	***P***	**Global F_{ST}(SD)**	***P***	**IBD**	***P***	**Global F_{ST}(SD)**	***P***	***P***
72	0.79	<0.001	0.055 (0.004)	<0.001	0.77	<0.001	0.046 (0.003)	<0.001	≤0.003
47d	0.79	<0.001	0.061 (0.004)	<0.001	0.76	<0.001	0.051 (0.004)	<0.001	≤0.001
25r	0.60	<0.001	0.043 (0.006)	<0.001	0.66	<0.001	0.036 (0.004)	<0.001	0.046

IBD = isolation by distance with associated *P* value, Global F_{ST} = global F_{ST} among the wild populations with standard error in brackets and associated *P* value. Hist. Vs. Cont. = statistical significance whether or not the change in global F_{ST} with time was significant or not.

DISCUSSION

Norway is the world's largest producer of farmed Atlantic salmon, and has over 200 rivers supporting native Atlantic salmon populations. Many Norwegian populations have displayed moderate to high frequencies of domesticated farmed escapees on the spawning grounds for two decades or more [1, 4, 54]. At the same time, Norwegian farmed salmon originated from native Norwegian populations approximately ten generations ago. Thus, it follows that Norway is not only the country where the potential genetic interaction between farmed escaped salmon and wild conspecifics is the most extensive, it represents the country in which the statistical challenges to quantify genetic introgression of farmed escapees are the most demanding. The present study addressed this situation by genotyping a sub-set of SNP markers (47d) that have been reported to be collectively diagnostic for farmed and wild salmon [31], and by implementing ABC simulations to quantify introgression of farmed salmon.

The most important results of this study can be summarized as follows: 1. All populations displaying significant temporal genetic changes with 47d became more similar to the pooled of farmed salmon. Furthermore, the stronger the temporal genetic change with 47d, the more similar it became to the pool. This strongly suggests that where populations displayed clear temporal genetic changes, introgression of farmed fish has been the primary cause 2. This is the first study to estimate cumulative introgression of farmed salmon in any native Atlantic salmon population. Estimations ranged between 2–47% and 7–41% per population using the ABC and fixed migration simulation methods respectively. It is concluded that while the level of introgression has been population specific, farmed salmon have heavily introgressed in some wild Norwegian populations.

Are the Diagnostic SNPs Universally Informative

The panel of diagnostic SNPs used here (47d) represents a sub-set of the markers recently identified as collectively diagnostic for farmed and wild Norwegian salmon [31]. The panel 47d included 35 of the top 60 loci ranked by Karlsson et al. (2011), including the top 10 ranking loci, and a further set of 12 loci taken from the ranks 60–200. While this study has not used the exact combination of markers reported to be collectively diagnostic, which is in large part due to poor genotyping quality for many of those markers, 47d still provides similar characteristics of the panel reported to be collectively diagnostic. This is based upon the comparisons reported here showing the greater level of signal for 47d vs. 25r for a variety of statistic parameters (e.g., Figures 2, 3,).

Seven of the populations investigated here overlap with some of the populations used to identify the diagnostic SNPs in Karlsson et al. [31]. However, there were no sign that ascertainment bias influenced the results of the present study. This is based upon the following observations: 1. The population displaying the greatest temporal change with 47d was not included in the ascertainment panel, 2. Pair-wise F_{ST} between each wild population and the pool of farmed salmon showed a clear geographic trend (Figure 2), with populations in the north of Norway displaying the greatest difference to the farmed pool. This is likely to reflect a combination of the fact that there is a distinct evolutionary divide between Atlantic salmon populations in the north and rest of Norway [21], and that Norwegian farmed strains were largely sourced from wild populations south of the observed evolutionary divide [29, 55]. Thus, it follows that detection of introgression in native populations in Northern Norway should be easier to detect with this set of markers than in populations for example from mid- and western Norway. Consequently, the present study serves to validate the usefulness of the collectively diagnostic markers in populations other than those in which the marker identification was conducted. Given that Atlantic salmon populations display highly significant population genetic structuring throughout the Atlantic [6, 7], it is likely that these markers will also serve useful to identify introgression of Norwegian farmed salmon in native populations outside Norway.

Assignment tests using 47d provided less statistical power to reject individual salmon from the historical baseline than with microsatellites for all 20 populations (Table 2). This is probably due to the fact that assignment power is strongly influenced by total number of alleles in the data set [39, 56], and the fact that a microsatellite data set based upon 22 markers [21] has approximately 3–4 times more alleles than 47 SNPs. Inclusion of more of the collectively diagnostic SNPs identified by Karlsson et al. (2011) would increase these assignment statistics. However, in order to accurately identify the ancestry of hybrids beyond the second admixed generation, it has been suggested that 50 or more ancestry diagnostic markers (i.e., fixed allele differences) are required [57, 58]. Only 15 markers from the panel 47d displayed non-overlapping allele frequencies between the pool of farm salmon and all historical samples for each wild population. Furthermore, the allele frequency differences for these non-overlapping markers were not close to fixation between groups. Thus, while the markers identified by Karlsson et al. (2011) provide more information to differentiate farmed and wild salmon than randomly selected markers, the identified markers are more correctly regarded as "collectively informative" than diagnostic. Clearly, there is a need to identify more informative genetic markers on the domesticated/ wild interface. As farmed salmon outgrow wild salmon approximately 2–3 times under hatchery conditions [59, 60], it is suggested that there is significant potential to identify markers tightly linked with this trait that has been selected for in all farmed strains [29, 61].

ABC and Fixed Migration Estimations of Introgression

Admixture between hatchery fish released deliberately into the wild and native populations has been computed in other species and systems, for example brown trout (*Salmo trutta*) in Danish rivers [25]. Often, admixture has been estimated using Bayesian clustering implemented for example in the program STRUCTURE [47, 48]. Here, clustering analysis was able to reveal temporal changes in the populations Opo and Vosso (the two populations displaying the largest temporal changes), however, in other rivers, also those displaying statistically temporal genetic changes, this analysis did not reveal changes. This is consistent with previous results using microsatellites [21] and is likely to be caused in part by the fact that the observed genetic changes for most of the rivers were low to modest, and therefore under the detection potential for STRUCTURE.

Introgression of farmed salmon was estimated here using the simulation approach. As it is implemented, the ABC routine finds the point estimate of migration rate M that best explains the observed F_{ST} between historical and contemporary population. However this approach does not take into account the possibly large confidence intervals around the observed F_{ST}. In many of the populations studied here, the genetic distance between historical and contemporary samples was not significantly different from zero. Therefore an alternative approach was also employed to better account for the uncertainty around the observed F_{ST}. By choosing a set of scenarios where the genetic distance between historical and simulated population fitted within the 95% confidence interval of the observed F_{ST}, the second approach reflects the range of variation around the posterior mean of *M*, accounting for the uncertainty around the observed F_{ST}. The standard deviation of *M* given by ABC estimation only accounts for the variation induced by random gamete sampling of our simulations and is thus lower than the standard deviation of *M* given by fix migration rate approach. This last approach explores the possible values of *M* stepwise with predefined steps (1,2...20%) and is deemed to be less accurate to estimate the posterior mean of *M* than the ABC routine that converges gradually to an optimum value. We therefore present the results from the two methods as complementary. Other alternative approaches could also have provided reliable estimates of posterior distribution of *M*, such as rejection based Bayesian Inference [33, 62], that would have estimated both posterior mean and standard deviation of *M*, taking into account the uncertainty around the observed F_{ST}. However, algorithms that estimate the value of a parameter based on its posterior likelihood are sensitive to the shape of the likelihood curve or "likelihood landscape"; a leptocurtic curve would result the algorithm to converging rapidly to the optimum value of the estimated parameter, whereas a platicurtic curve would result in the algorithm converging slowly. In the present data, F_{ST} between historical and contemporary samples were close to zero with large standard deviations for some populations. Such combination of small mean value and large standard deviation represents a flat likelihood landscape where the algorithm is slow to converge, and often lacks precision.

Challenges and Alternative Approaches to Quantify Introgression

Some of the challenges to quantify introgression of farmed salmon in native populations have been described in detail in the introduction. Briefly, these include the complicated logistics associated with the distribution of genetic material within the aquaculture industry [38], the fact that there are multiple farmed strains which have and continue to change significantly over time due to splitting and mixing, the fact that over time, gene-flow arises from multiple farmed sources which partially conceals the degree of genetic change in wild populations when studying non-diagnostic markers [28], and finally, the fact that most loci display overlapping allele frequencies between groups of farmed and wild salmon. It is likely that many of these challenges will exist in other countries where farmed and wild salmon coexist, and, in other aquaculture species such as marine fish where interactions between escapees and wild conspecifics have already been registered [63, 64, 65].

In order to investigate the direction of genetic change in wild populations and quantify introgression, a pool of 375 farmed salmon, sampled over a five year period from 48 cages located on 35 farms spanning the Norwegian coastline were included in the present study. While this pool of farmed salmon does not necessarily accurately represent the allele frequencies of escapees entering all of these rivers for the entire study period, they clearly permitted elucidation of the direction of the genetic changes in wild populations (Figures 3 and 4). Furthermore, in the populations Opo and Vosso, which showed the highest estimated introgression rates, the observed direction of genetic change in relation to this pool of farmed salmon, and the simulated direction in relation to gene-flow from the pool of farmed salmon displayed almost perfect overlap (Figure 5). Thus, these results strongly indicate that this approach is valid. Nevertheless, it is suggested that a more accurate estimation of introgression of farmed salmon could be achieved if samples of farmed fish entering each specific river were collected together with samples of adult wild fish. If this was conducted yearly, together with sampling offspring from the subsequent generation at different life history stages (Figure 7), this would provide the most robust estimations of the allele frequencies for the native and intruding farmed fish. This cohort-based sampling could then be combined with the ABC and fixed migration simulations presented here to quantify introgression per year and thereafter per generation.

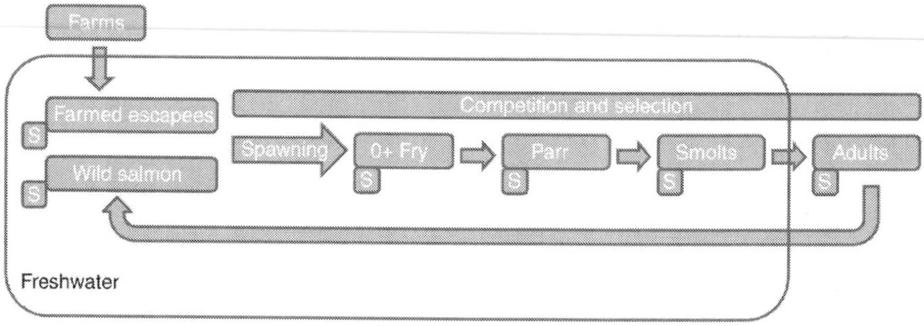

Figure 7: Illustration of a single-cohort design to quantify introgression of farmed escaped salmon in a wild population, and estimate selection at the different life-history stages. S = sampling for genetic analysis.

The offspring of farmed salmon display genetically-based lower survival in the natural environment compared to wild salmon [66, 67, 68, 69]. Thus, it is likely that the frequency of farmed and hybrid salmon in a cohort will decrease with time. While this was not tested here, it is noteworthy that the two populations displaying the greatest introgression levels, i.e., Opo and Vosso, were represented by juvenile as opposed to adult samples. It is therefore possible that the level of farmed salmon introgression in these two populations would have been less if the contemporary samples for each of these rivers were adults. The cohort-based design (Figure 7) would permit addressing this issue, and potentially provide estimations in strength of selection against farmed fish in the different rivers, for the different life-history stages.

Implications for Management

The direct translation of molecular genetic data into aquaculture or fisheries management and regulation has not been without its challenges [70, 71]. Good examples include identification of the farm of origin for escapees [38, 41], poaching from protected populations [72], fishing competition fraud [73], and regulation of harvest at the individual [74] and population level [75]. Nevertheless, there is still a need to improve the translation of molecular genetic data into a "currency" that governing bodies can implement if these techniques are to find a routine place in the management of aquaculture and fisheries resources. While it is acknowledged that the methods for quantification of introgression presented here are still in need of further refinement, they provide management authorities with the first multiple-population estimations of introgression of farmed salmon, an essential early step in risk assessment [23]. Nevertheless, other important management issues, such as what early warning indicators for introgression exist, and what are the biological consequences of introgression, remain.

One of the limitations in using genetics techniques to quantify introgression of farmed salmon in native populations is that fact that it can only be used to validate introgression. While this is important, the obvious management target for mitigation prior to introgression is the frequency of farmed salmon in the native population. However, the correlation between some of the genetic parameters investigated here and the frequency of farmed salmon observed in these populations was at best modest. This is consistent with observations from a previous study [21], and is likely to be associated with the fact that there are large gaps in the data reporting the numbers of escapees in rivers, and that the density of the native population appears linked with resilience of the population to introgression. The implications of this for managers is that monitoring the frequency of farmed salmon in wild populations, or modeling genetic changes in the populations based upon the observed frequency of escapees [53, 76], will not provide an accurate estimation of introgression for all populations.

There are widespread concerns regarding the genetic integrity and long-term fitness of native populations where large numbers of escapees have been observed [13, 14, 15]. However, while introgression of farmed escaped salmon has been documented in rivers in several countries, biological consequences of introgression has thus far not been reported for any wild Atlantic salmon population. This provides several challenges for managers. At what level should acceptable thresholds be set in the absence of fitness-consequence data in wild populations? And if a very low management threshold is set (for example <5% introgression), will the analytical methods and genetic markers available today, or in the near future, be able to accurately quantify introgression at such low levels? Given that there have been major advances in the use of sterile triploid Atlantic salmon for the commercial aquaculture, and 100% containment of aquaculture fish is ultimately unrealistic, management authorities should consider increasing efforts to convert the industry over to the use of sterile fish.

CONCLUSION

This study is the first to quantify cumulative introgression of farmed salmon in any native Atlantic salmon population. Based upon ABC and fixed migration estimations, it has been demonstrated that introgression of farmed salmon in wild Norwegian Atlantic salmon populations has been population-specific, ranging from no detectable impact in some populations to strong introgression in others. Furthermore, where populations displayed clear temporal genetic changes, they all became more similar to a pool of farmed salmon. While the level of farmed salmon introgression was partially correlated with the frequency of escapees observed in the population, it is concluded that other mechanisms, such as the density of the recipient native population, is likely to influence the relative success of farmed fish. These data provide policy makers with unique information to address the influence of farmed escaped salmon on native populations.

DECLARATIONS

Acknowledgements

A large number of people and organizations have been involved in collecting and donating samples for this study. These include the Norwegian Gene bank run by the Norwegian Directorate for Nature management (DN), the Norwegian Institute for Nature Research (NINA), Erlend Waatevik, Eero Niemelä, Reidar Borgstrøm and Svein Jakob Saltveit. In addition, fishermen, river owners and local volunteers are acknowledged for assisting the collection of contemporary samples. We would like to thank Michael M. Hansen, Terje Svåsand, Ove T. Skilbrei and Geir Lasse Taranger and anonymous referees for constructive advice on earlier drafts of this paper and Anne G. E. Sørvik for isolating DNA. This study was primarily financed by the Norwegian Ministry of Fisheries. We thank the Danish Natural Science Research Council for financial support to CP (grant numbers: 11–103926, 09–065999, 95095995) and the Carlsberg Foundation (grant number 2011-01-0059).

REFERENCES

1. Anon: Oppdaterte rømmingstall. 2012, http://www.fiskeridir.no/statistikk/akvakultur/oppdaterte-roemmingstall In Norwegian
2. Milner NJ, Evans R: The incidence of escaped Irish farmed salmon in English and Welsh rivers. Fish Manag Ecol. 2003, 10 (6): 403-406. 10.1111/j.1365-2400.2003.00348.x.
3. Walker AM, Beveridge MCM, Crozier WO, Maoileidigh N, Milner N: Monitoring the incidence of escaped farmed Atlantic salmon, *Salmo salar* L., in rivers and fisheries of the United Kingdom and Ireland: current progress and recommendations for future programmes. ICES J Mar Sci. 2006, 63 (7): 1201-1210. 10.1016/j.icesjms.2006.04.018.
4. Fiske P, Lund RA, Hansen LP: Relationships between the frequency of farmed Atlantic salmon, *Salmo salar* L., in wild salmon populations and fish farming activity in Norway, 1989–2004. ICES J Mar Sci. 2006, 63 (7): 1182-1189. 10.1016/j.icesjms.2006.04.006.
5. Fiske P, Lund RA, Østborg GM, Fløystad L: Rømt oppdrettslaks i sjø- og elvefisket i årene 1989–2000. NINA oppdresgsmelding. 2001, 704: 1-26. In Norwegian
6. Ståhl G: Genetic population strcuture of Atlantic salmon. Population genetics and fishery management. Edited by: Ryman N, Utter F. 1987, Seattle: University of Washington Press, 121-140.
7. Verspoor E, Beardmore JA, Consuegra S, De Leaniz CG, Hindar K, Jordan WC, Koljonen ML, Mahkrov AA, Paaver T, Sanchez JA, et al: Population structure in the Atlantic salmon: insights from 40 years of research into genetic protein variation. J Fish Biol. 2005, 67: 3-54.
8. Dionne M, Caron F, Dodson JJ, Bernatchez L: Landscape genetics and hierarchical genetic structure in Atlantic salmon: the interaction of gene flow and local adaptation. Mol Ecol. 2008, 17 (10): 2382-2396. 10.1111/j.1365-294X.2008.03771.x.
9. Perrier C, Guyomard R, Bagliniere JL, Evanno G: Determinants of hierarchical genetic structure in Atlantic salmon populations: environmental factors vs. anthropogenic influences. Mol Ecol. 2011, 20 (20): 4231-4245. 10.1111/j.1365-294X.2011.05266.x.

10. Dillane E, McGinnity P, Coughlan JP, Cross MC, De Eyto E, Kenchington E, Prodohl P, Cross TF: Demographics and landscape features determine intrariver population structure in Atlantic salmon (*Salmo salar* L.): the case of the River Moy in Ireland. Mol Ecol. 2008, 17 (22): 4786-4800. 10.1111/j.1365-294X.2008.03939.x.
11. Garcia De Leaniz C, Fleming IA, Einum S, Verspoor E, Jordan WC, Consuegra S, Aubin-Horth N, Lajus D, Letcher BH, Youngson AF, et al: A critical review of adaptive genetic variation in Atlantic salmon: implications for conservation. Biol Rev. 2007, 82 (2): 173-211.
12. Taylor EB: A review of local adaptation in salmonidae, with particular reference to Pacific and Atlantic salmon. Aquaculture. 1991, 98 (1–3): 185-207.
13. Naylor R, Hindar K, Fleming IA, Goldburg R, Williams S, Volpe J, Whoriskey F, Eagle J, Kelso D, Mangel M: Fugitive salmon: assessing the risks of escaped fish from net-pen aquaculture. Bioscience. 2005, 55 (5): 427-437. 10.1641/0006-3568(2005)055[0427:FSATRO]2.0.CO;2.
14. Ferguson A, Fleming IA, Hindar K, Skaala Ø, McGinnity P, Cross T, Prodohl P: Farm escapees. The Atlantic salmon genetics, conservation and management. Edited by: Verspoor E, Stradmeyer L, Nielsen JL. 2007, Oxford, UK: Blackwell, 357-398.
15. Hindar K, Ryman N, Utter F: Genetic effects of cultured fish on natural fish populations. Can J Fish Aquat Sci. 1991, 48 (5): 945-957. 10.1139/f91-111.
16. Clifford SL, McGinnity P, Ferguson A: Genetic changes in Atlantic salmon (*Salmo salar*) populations of northwest Irish rivers resulting from escapes of adult farm salmon. Can J Fish Aquat Sci. 1998, 55 (2): 358-363. 10.1139/f97-229.
17. Clifford SL, McGinnity P, Ferguson A: Genetic changes in an Atlantic salmon population resulting from escaped juvenile farm salmon. J Fish Biol. 1998, 52 (1): 118-127. 10.1111/j.1095-8649.1998.tb01557.x.
18. Crozier WW: Escaped farmed salmon, *Salmo salar* L., in the Glenarm River, Northern Ireland: genetic status of the wild population 7 years on. Fish Manag Ecol. 2000, 7 (5): 437-446. 10.1046/j.1365-

2400.2000.00219.x.

19. Crozier WW: Evidence of genetic interaction between escaped farmed salmon and wild Atlantic salmon (*Salmo salar* L) in a Northern Irish river. Aquaculture. 1993, 113 (1–2): 19-29.
20. Bourret V, O'Reilly PT, Carr JW, Berg PR, Bernatchez L: Temporal change in genetic integrity suggests loss of local adaptation in a wild Atlantic salmon (*Salmo salar*) population following introgression by farmed escapees. Heredity. 2011, 106 (3): 500-510. 10.1038/hdy.2010.165.
21. Glover KA, Quintela M, Wennevik V, Besnier F, Sørvik AGE, Skaala O: Three decades of farmed escapees in the wild: a spatio-temporal analysis of population genetic structure throughout Norway. PLoS One. 2012, 7 (8): e43129-10.1371/journal.pone.0043129.
22. Skaala O, Wennevik V, Glover KA: Evidence of temporal genetic change in wild Atlantic salmon, *Salmo salar* L., populations affected by farm escapees. ICES J Mar Sci. 2006, 63 (7): 1224-1233. 10.1016/j.icesjms.2006.04.005.
23. Extracts from "Fisken og havet, 3-2010. Risk assessment–environmental impacts of Norwegian aquaculture. Edited by: Taranger GL, Boxaspen KK, Madhun AS, Svåsand T. 2011, Norway: Institute of Marine Research, Printed August 2011
24. Hansen MM, Bekkevold D, Jensen LF, Mensberg KLD, Nielsen EE: Genetic restoration of a stocked brown trout salmo trutta population using microsatellite DNA analysis of historical and contemporary samples. J Appl Ecol. 2006, 43 (4): 669-679. 10.1111/j.1365-2664.2006.01185.x.
25. Hansen MM, Mensberg KLD: Admixture analysis of stocked brown trout populations using mapped microsatellite DNA markers: indigenous trout persist in introgressed populations. Biol Lett. 2009, 5 (5): 656-659. 10.1098/rsbl.2009.0214.
26. Perrier C, Baglinière JL, Evanno G: Understanding admixture patterns in supplemented populations: a case study combining molecular analyses and temporally explicit simulations in Atlantic salmon. Evol Applications. 2012, 6: 218-230. 10.1111/j.1752-4571.2012.00280.x.
27. Hansen MM: Estimating the long-term effects of stocking

domesticated trout into wild brown trout (Salmo trutta) populations: an approach using microsatellite DNA analysis of historical and contemporary samples. Mol Ecol. 2002, 11 (6): 1003-1015. 10.1046/j.1365-294X.2002.01495.x.

28. Besnier F, Glover KA, Skaala O: Investigating genetic change in wild populations: modelling gene flow from farm escapees. Aqua Envi Interactions. 2011, 2: 75-86. 10.3354/aei00032.
29. Gjedrem T, Gjoen HM, Gjerde B: Genetic-origin of Norwegian farmed Atlantic Salmon. Aquaculture. 1991, 98 (1–3): 41-50.
30. Skaala O, Hoyheim B, Glover K, Dahle G: Microsatellite analysis in domesticated and wild Atlantic salmon (*Salmo salar*L.): allelic diversity and identification of individuals. Aquaculture. 2004, 240 (1–4): 131-143.
31. Karlsson S, Moen T, Lien S, Glover KA, Hindar K: Generic genetic differences between farmed and wild Atlantic salmon identified from a 7K SNP-chip. Mol Ecol Resour. 2011, 11: 247-253.
32. Glover KA, Skaala O, Sovik AGE, Helle TA: Genetic differentiation among Atlantic salmon reared in sea-cages reveals a non-random distribution of genetic material from a breeding programme to commercial production. Aquacult Res. 2011, 42 (9): 1323-1331. 10.1111/j.1365-2109.2010.02720.x.
33. Beaumont MA, Zhang WY, Balding DJ: Approximate Bayesian computation in population genetics. Genetics. 2002, 162 (4): 2025-2035.
34. Excoffier L, Estoup A, Cornuet JM: Bayesian analysis of an admixture model with mutations and arbitrarily linked markers. Genetics. 2005, 169 (3): 1727-1738.
35. Sousa VC, Fritz M, Beaumont MA, Chikhi L: Approximate bayesian computation without summary statistics: the case of admixture. Genetics. 2009, 181 (4): 1507-1519. 10.1534/genetics.108.098129.
36. Estoup A, Beaumont M, Sennedot F, Moritz C, Cornuet JM: Genetic analysis of complex demographic scenarios: Spatially expanding populations of the cane toad, *Bufo marinus*. Evolution. 2004, 58 (9): 2021-2036.
37. Lund RA, Hansen LP: Identification of wild and reared Atlantic salmon, *Salmo salaar* L., using scale characters. Aquac Fish Manag.

1991, 22: 499-508.

38. Glover KA: Forensic identification of fish farm escapees: the Norwegian experience. Aquacult Envi Interactions. 2010, 1: 1-10.
39. Glover KA, Hansen MM, Lien S, Als TD, Hoyheim B, Skaala O: A comparison of SNP and STR loci for delineating population structure and performing individual genetic assignment. BMC Genet. 2010, 11 (2):
40. Glover KA, Hansen MM, Skaala O: Identifying the source of farmed escaped Atlantic salmon (*Salmo salar*): Bayesian clustering analysis increases accuracy of assignment. Aquaculture. 2009, 290 (1–2): 37-46.
41. Glover KA, Skilbrei OT, Skaala O: Genetic assignment identifies farm of origin for Atlantic salmon *Salmo salar* escapees in a Norwegian fjord. ICES J Mar Sci. 2008, 65 (6): 912-920. 10.1093/icesjms/fsn056.
42. Dieringer D, Schlotterer C: Microsatellite analyser (MSA): a platform independent analysis tool for large microsatellite data sets. Mol Ecol Notes. 2003, 3 (1): 167-169. 10.1046/j.1471-8286.2003.00351.x.
43. Weir BS, Cockerham CC: Estimating -statistics for the analysis of population structure. Evolution. 1984, 38 (6): 1358-1370. 10.2307/2408641.
44. Peakall R, Smouse PE: GENALEX 6: genetic analysis in excel. Population genetic software for teaching and research. Mol Ecol Notes. 2006, 6 (1): 288-295. 10.1111/j.1471-8286.2005.01155.x.
45. Piry S, Alapetite A, Cornuet JM, Paetkau D, Baudouin L, Estoup A: GENECLASS2: a software for genetic assignment and first-generation migrant detection. J Hered. 2004, 95 (6): 536-539. 10.1093/jhered/esh074.
46. Rannala B, Mountain JL: Detecting immigration by using multilocus genotypes. Proc Natl Acad Sci USA. 1997, 94 (17): 9197-9201. 10.1073/pnas.94.17.9197.
47. Pritchard JK, Stephens M, Donnelly P: Inference of population structure using multilocus genotype data. Genetics. 2000, 155 (2): 945-959.

48. Falush D, Stephens M, Pritchard JK: Inference of population structure using multilocus genotype data: linked loci and correlated allele frequencies. Genetics. 2003, 164 (4): 1567-1587.

49. Besnier F, Glover KA: ParallelStructure: A R package to distribute parallel runs of the population genetics program structure on multi-core computers. PLoS One. 2013, 8 (7): e70651-10.1371/journal.pone.0070651.

50. Waples RS, Do C: LDNE: a program for estimating effective population size from data on linkage disequilibrium. Mol Ecol Resour. 2008, 8 (4): 753-756. 10.1111/j.1755-0998.2007.02061.x.

51. Jombart T, Devillard S, Balloux F: Discriminant analysis of principal components: a new method for the analysis of genetically structured populations. BMC Genet. 2010, 11:

52. Anon: Vedleggsrapport merd vurdering av måloppnåelse for de enkelte bestandene. Rapport fra Vitenskapelig råd for Lakseforvaltning nr 2b (In Norwegian). 2010

53. Diserud OH, Fiske P, Hindar K: Forslag til kategorisering av laksebestander som er påvirket av rømt oppdrettslaks. NINA Rapport 782 32s (In Norwegian). 2012

54. Gausen D, Moen V: Large-scale escapes of farmed Atlantic salmon (*Salmo salar*) into Norwegian rivers threaten natural populations. Can J Fish Aquat Sci. 1991, 48 (3): 426-428. 10.1139/f91-055.

55. Gjøen HM: Genetic composition of the broodstock at AKVAFORSK. Estimates of heterozygosity, effects of strain differences and heterosis. Thesis. 1989, Ås: Agricultural University of Norway, 55- (In Norwegian)

56. Kalinowski ST: Genetic polymorphism and mixed-stock fisheries analysis. Can J Fish Aquat Sci. 2004, 61 (7): 1075-1082. 10.1139/f04-060.

57. Fitzpatrick BM: Estimating ancestry and heterozygosity of hybrids using molecular markers. Bmc Evol Biol. 2012, 12 (131):

58. Fitzpatrick BM, Johnson JR, Kump KD, Shaffer BH, Smith JJ, Voss RS: Rapid fixation of non-native alleles revealed by genome-wide SNP analysis of hybrid tiger salamanders. Bmc Evol Biol. 2009, 9 (176):

59. Glover KA, Ottera H, Olsen RE, Slinde E, Taranger GL, Skaala O: A comparison of farmed, wild and hybrid Atlantic salmon (*Salmo salar* L.) reared under farming conditions. Aquaculture. 2009, 286 (3–4): 203-210.

60. Solberg MF, Glover KA, Nilsen F, Skaala O: Does domestication cause changes in growth reaction norms? a study of farmed, wild and hybrid Atlantic salmon families exposed to environmental stress. PLoS One. 2013, 8 (1): e54469-10.1371/journal.pone.0054469.

61. Gjedrem T: The first family-based breeding program in aquaculture. Rev Aquac. 2010, 2 (1): 2-15. 10.1111/j.1753-5131.2010.01011.x.

62. Tavare S, Balding DJ, Griffiths RC, Donnelly P: Inferring coalescence times from DNA sequence data. Genetics. 1997, 145 (2): 505-518.

63. Bekkevold D, Hansen MM, Nielsen EE: Genetic impact of gadoid culture on wild fish populations: predictions, lessons from salmonids, and possibilities for minimizing adverse effects. ICES J Mar Sci. 2006, 63 (2): 198-208. 10.1016/j.icesjms.2005.11.003.

64. Glover KA, Dahle G, Jorstad KE: Genetic identification of farmed and wild Atlantic cod, *Gadus morhua*, in coastal Norway. ICES J Mar Sci. 2011, 68 (5): 901-910. 10.1093/icesjms/fsr048.

65. Moe H, Dempster T, Sunde LM, Winther U, Fredheim A: Technological solutions and operational measures to prevent escapes of Atlantic cod (*Gadus morhua*) from sea cages. Aquacult Res. 2007, 38 (1): 91-99. 10.1111/j.1365-2109.2006.01638.x.

66. Fleming IA, Hindar K, Mjolnerod IB, Jonsson B, Balstad T, Lamberg A: Lifetime success and interactions of farm salmon invading a native population. Proc Roy Soc Lond B Biol Sci. 2000, 267 (1452): 1517-1523. 10.1098/rspb.2000.1173.

67. McGinnity P, Prodohl P, Ferguson K, Hynes R, O'Maoileidigh N, Baker N, Cotter D, O'Hea B, Cooke D, Rogan G, et al: Fitness reduction and potential extinction of wild populations of Atlantic salmon, *Salmo salar*, as a result of interactions with escaped farm salmon. Proc Roy Soc Lond B Biol Sci. 2003, 270 (1532): 2443-2450. 10.1098/rspb.2003.2520.

68. McGinnity P, Stone C, Taggart JB, Cooke D, Cotter D, Hynes R, McCamley C, Cross T, Ferguson A: Genetic impact of escaped

farmed Atlantic salmon (*Salmo salar* L.) on native populations: use of DNA profiling to assess freshwater performance of wild, farmed, and hybrid progeny in a natural river environment. ICES J Mar Sci. 1997, 54 (6): 998-1008.

69. Skaala Ø, Glover Kevin A, Barlaup Bjørn T, Svåsand T, Besnier F, Hansen Michael M, Borgstrøm R: Performance of farmed, hybrid, and wild Atlantic salmon (*Salmo salar*) families in a natural river environment. Can J Fish Aquat Sci. 2012, 69 (12): 1994-2006. 10.1139/f2012-118.
70. Waples RS, Punt AE, Cope JM: Integrating genetic data into management of marine resources: how can we do it better?. Fish Fish. 2008, 9 (4): 423-449. 10.1111/j.1467-2979.2008.00303.x.
71. Reiss H, Hoarau G, Dickey-Collas M, Wolff WJ: Genetic population structure of marine fish: mismatch between biological and fisheries management units. Fish Fish. 2009, 10 (4): 361-395. 10.1111/j.1467-2979.2008.00324.x.
72. Withler RE, Candy JR, Beacham TD, Miller KM: Forensic DNA analysis of Pacific salmonid samples for species and stock identification. Environ Biol Fishes. 2004, 69 (1–4): 275-285.
73. Primmer CR, Koskinen MT, Piironen J: The one that did not get away: individual assignment using microsatellite data detects a case of fishing competition fraud. Proc Roy Soc Lond B Biol Sci. 2000, 267 (1453): 1699-1704. 10.1098/rspb.2000.1197.
74. Glover KA, Haag T, Oien N, Walloe L, Lindblom L, Seliussen BB, Skaug HJ: The Norwegian minke whale DNA register: a database monitoring commercial harvest and trade of whale products. Fish Fish. 2012, 13: 313-332. 10.1111/j.1467-2979.2011.00447.x.
75. Shaklee JB, Beacham TD, Seeb L, White BA: Managing fisheries using genetic data: case studies from four species of Pacific salmon. Fish Res. 1999, 43 (1–3): 45-78.
76. Hindar K, Fleming IA, McGinnity P, Diserud A: Genetic and ecological effects of salmon farming on wild salmon: modelling from experimental results. ICES J Mar Sci. 2006, 63 (7): 1234-1247. 10.1016/j.icesjms.2006.04.025.

CITATION

CHAPTER 2

Christophe Béné, Manuel Barange, Rohana Subasinghe, Per Pinstrup-Andersen, Gorka Merino, Gro-Ingunn Hemre. Meryl Williams, Feeding 9 billion by 2050 – Putting fish back on the menu, DOI 10.1007/s12571-015-0427-z.

CHAPTER 3

Håstein, T., Hellstrøm, A., Jonsson, G. et al. Acta Vet Scand (2001) 42(Suppl 1): S43. https://doi.org/10.1186/1751-0147-42-S1-S43.

CHAPTER 4

Marek S. Ziętara, Dar'ya Lebedeva, Gabriela Muñoz, Jaakko Lumme, A monogenean fish parasite, Gyrodactylus chileani n. sp., belonging to a novel marine species lineage found in the South-Eastern Pacific and the Mediterranean and North Seas, DOI 10.1007/s11230-012-9379-2.

CHAPTER 5

Sadok Khouadja, Faouzi Lamari, Amina Bakhrouf, Characterization of Vibrio parahaemolyticus isolated from farmed sea bass (Dicentrarchus labrax) during disease outbreaks, https://doi.org/10.1186/2008-6970-5-13.

CHAPTER 6

MA Freeman, M Eydal, M Yoshimizu, K Watanabe, AP Shinn, K Miura, K Ogawa, Molecular identification and transmission studies of X-cell parasites from Atlantic cod Gadus morhua (Gadiformes: Gadidae) and the northern black flounder Pseudopleuronectes obscurus (Pleuronectiformes: Pleuronectidae), doi:10.1186/1756-3305-4-15.

CHAPTER 7

Haakon Hansen, Trygve T Poppe, Turhan Markussen, Egil Karlsbakk, Seatrout (Salmo trutta) is a natural host for Parvicapsula pseudobranchicola (Myxozoa, Myxosporea), an important pathogen of farmed Atlantic salmon (Salmo salar), DOI 10.1186/s13071-015-0828-z.

CHAPTER 8

Saraya Tavornpanich, Mathilde Paul, Hildegunn Viljugrein, David Abrial, Daniel Jimenez, Edgar Brun, Risk map and spatial determinants of pancreas disease in the marine phase of Norwegian Atlantic salmon farming sites, https://doi.org/10.1186/1746-6148-8-172.

CHAPTER 9

Toshio Takeuchi, Progress on larval and juvenile nutrition to improve the quality and health of seawater fish: a review, DOI 10.1007/s12562-014-0744-8.

CHAPTER 10

Moldal et al.: Substitution of dietary fish oil with plant oils is associated with shortened mid intestinal folds in Atlantic salmon (Salmo salar), doi:10.1186/1746-6148-10-60.

CHAPTER 11

Tacchi et al.: Transcriptomic and physiological responses to fishmeal substitution with plant proteins in formulated feed in farmed Atlantic salmon (Salmo salar), doi:10.1186/1471-2164-13-363.

CHAPTER 12

Sharen Bowman, Sophie Hubert, Brent Higgins, Cynthia Stone, Jennifer Kimball, Tudor Borza, Jillian Tarrant Bussey, Gary Simpson, Catherine Kozera, Bruce A. Curtis, Jennifer R. Hall, Tiago S. Hori, Charles Y. Feng, Marlies Rise, Marije Booman, A. Kurt Gamperl, Edward Trippel, Jane Symonds, Stewart

C. Johnson, Matthew L. Rise, An Integrated Approach to Gene Discovery and Marker Development in Atlantic Cod (Gadus morhua), DOI 10.1007/s10126-010-9285-z.

CHAPTER 13

Morais et al.: Effects of genotype and dietary fish oil replacement with vegetable oil on the intestinal transcriptome and proteome of Atlantic salmon (Salmo salar), doi:10.1186/1471-2164-13-448.

CHAPTER 14

Gardner et al.: Microarray gene expression profiles from mature gonad tissues of Atlantic bluefin tuna, Thunnus thynnus in the Gulf of Mexico, doi:10.1186/1471-2164-13-530.

CHAPTER 15

Cantas et al.: The culturable intestinal microbiota of triploid and diploid juvenile Atlantic salmon (Salmo salar) - a comparison of composition and drug resistance. BMC Veterinary Research 2011 7:71.

CHAPTER 16

Kevin A Glover, Abdullah S Madhun, Geir Dahle, Anne G E Sørvik, Vidar Wennevik, Øystein Skaala, H Craig Morton, Tom J Hansen and Per G Fjelldal, The frequency of spontaneous triploidy in farmed Atlantic salmon produced in Norway during the period 2007–2014, DOI 10.1186/s12863-015-0193-0.

CHAPTER 17

John A. Theodorou, Costas Perdikaris, Barbara Venou, Origin of broodstock and effects on the deformities of gilthead sea bream (Sparus aurata L. 1758) in a Mediterranean commercial hatchery, DOI 10.1007/s40071-016-0140-3.

CHAPTER 18

John P. Volpe, Jennifer L.M. Gee, Valerie A. Ethier, Martina Beck, Amanda J. Wilson and Jenna M.S. Stoner, Global Aquaculture Performance Index (GAPI): The First Global Environmental Assessment of Marine Fish Farming, doi:10.3390/su5093976.

CHAPTER 19

Glover et al.: Atlantic salmon populations invaded by farmed escapees: quantifying genetic introgression with a Bayesian approach and SNPs, doi:10.1186/1471-2156-14-74.

INDEX

A

B

C

D

E

F

Q

R

S

T

V

W

X